Industrial Entomology

Omkar
Editor

Industrial Entomology

Springer

Editor
Omkar
Department of Zoology
University of Lucknow
Lucknow, Uttar Pradesh, India

ISBN 978-981-10-3303-2 ISBN 978-981-10-3304-9 (eBook)
DOI 10.1007/978-981-10-3304-9

Library of Congress Control Number: 2017932072

Printed on acid-free paper

This Springer imprint is published by Springer Nature
The registered company is Springer Nature Singapore Pte Ltd.
The registered company address is: 152 Beach Road, #21-01/04 Gateway East, Singapore 189721, Singapore

Preface

While dealing with certain topics in entomology to undergraduate and postgraduate students of zoology over the last three decades, I realized a dearth of information on certain topics indicating a strong need of a specialized book dealing with those topics. In this way, the title Industrial Entomology came into existence in my mind. *The term industrial entomology refers to the use of insects, their activities, or products on industrial scale for the human welfare.* In the same context, I contacted with certain real experts for the selected topics covered under the umbrella of industrial entomology. The book starts with a brief write up on industrial entomology, followed by honey bee diversity, bee keeping, diseases, and natural enemies of honey bees; lac insects, lac cultivation, and lac culture; mulberry sericulture; diseases and natural enemies of *Bombyx mori*; tropical tasar culture; temperate tasar culture; eri-culture and muga sericulture; insect pollinators, insects as food, and insect pharmaceuticals; and mass production of natural enemies for biocontrol of insect pests. To my knowledge, there is no such book dealing with these specialized topics by the real experts. I am sure that the book will be quite useful in studies not only to undergraduate and postgraduate students but to all those interested in practicing these areas besides the policy planners.

I am thankful to all the contributors and subject experts including Professors M.R. Srinivasan, D.D. Barsagade, and M.S. Khan and Drs. Pradyumn Kumar, K.K. Sharma, P. Kumar, T.P.S. Chauhan, R.K. Goel, Mukesk Tayal, Babulal, A.A. Siddiqui, Bijoy K. Singh, Geetanjali Mishra, and Sunita Yadav, among others, from the inner core of my heart for instantly agreeing to my proposal and sparing time from their routine and hectic schedule to finalize and submit the chapters assigned to them within the time frame.

I am especially thankful to my research team consisting of Dr. Geetanjali Mishra (associate professor), M/S. Shashwat Singh, Desh Deepak Chaudhary, Garima Pandey, Swati Saxena, Apoorva Shandilya, Priya Singh, and Chandani Verma for continued support during this venture. I greatly appreciate the last few weeks' support from Dr. Geetanjali Mishra in finalizing this book.

I am also thankful to my wife, Mrs. Kusum Upadhyay, for her encouragements and silent sacrifice during the entire course of this project.

Thanks are also due to Dr. Mamta Kapila, Ms. Manjusha Nalamolu, Mr. Raman Shukla, and Mr. Joseph Daniel from Springer Group for motivating me and also for keeping me alert from time to time, thus helping me in completing this book in time.

Lucknow, Uttar Pradesh, India Omkar

Contents

Contributors

S.A. Ahmed Central Muga Eri Research & Training Institute, Central Silk Board, Lahdoigarh, Assam, India

R. Aruna Department of Agricultural Entomology, Tamil Nadu Agricultural University, Coimbatore, Tamil Nadu, India

Babulal Central Sericultural Research and Training Institute, Central Silk Board, Pampore, Jammu and Kashmir, India

D.D. Barsagade Department of Zoology, MJF Educational Campus, Rashtrasant Tukadoji Maharaj Nagpur University, Nagpur, Maharashtra, India

T.P.S. Chauhan Ex-Scientist, Central Silk Board, Niranjanpur, Dehradun, Uttarakhand, India

S. Chouhan Central Sericultural Research and Training Institute, Central Silk Board, Pampore, Jammu and Kashmir, India

Rakesh K. Goel Central Silk Board, Haridwar, Uttarakhand, India

Babul Lal Jat Department of Entomology, CCS Haryana Agricultural University, Hisar, Haryana, India

Jaswinder Kaur Indian Institute of Maize Research, Indian Agricultural Research Institute Campus, New Delhi, India

H.D. Kaushik Department of Entomology, CCS Haryana Agricultural University, Hisar, Haryana, India

M.S. Khan Department of Entomology, College of Agriculture, G.B. Pant University of Agriculture & Technology, Pantnagar, Uttarakhand, India

M. Kishan Tej Department of Agricultural Entomology, Tamil Nadu Agricultural University, Coimbatore, India

P. Kumar Indian Institute of Natural Resins and Gums, Namkum, Ranchi, Jharkhand, India

Pradyumn Kumar Indian Institute of Maize Research, Indian Agricultural Research Institute Campus, New Delhi, India

Yogesh Kumar Department of Entomology, CCS Haryana Agricultural University, Hisar, Haryana, India

Soujanya P. Lakshmi Indian Institute of Maize Research, Indian Agricultural Research Institute Campus, New Delhi, India

Geetanjali Mishra Department of Zoology, University of Lucknow, Lucknow, Uttar Pradesh, India

Y.D. Mishra (deceased)

Omkar Department of Zoology, University of Lucknow, Lucknow, Uttar Pradesh, India

J.C. Sekhar Indian Institute of Maize Research, Indian Agricultural Research Institute Campus, New Delhi, India

K.K. Sharma Lac Production Division, Indian Institute of Natural Resins and Gums, Namkum, Ranchi, Jharkhand, India

A.A. Siddiqui Former Scientist, Central Silk Board, Kalyanpur, Lucknow, Uttar Pradesh, India

B.K. Singh Central Muga Eri Research & Training Institute, Central Silk Board, Lahdoigarh, Assam, India

M.R. Srinivasan Department of Agricultural Entomology, Tamil Nadu Agricultural University, Coimbatore, Tamil Nadu, India

S.B. Suby Indian Institute of Maize Research, Indian Agricultural Research Institute Campus, New Delhi, India

Mukesh K. Tayal Regional Sericultural Research Station, Central Silk Board, Miransahib, Jammu, Jammu and Kashmir, India

Sunita Yadav Department of Entomology, CCS Haryana Agricultural University, Hisar, Haryana, India

Manish Kumar Yogi Department of Entomology, College of Agriculture, G.B. Pant University of Agriculture & Technology, Pantnagar, Uttarakhand, India

About the Editor

Professor Omkar has been actively engaged for over three decades in teaching and for nearly four decades in research. He is currently working as professor and head at the Department of Zoology, University of Lucknow, Lucknow, India. He is a fellow of the National Academy of Sciences, India, and seven more professional bodies. He is a recipient of the Young Indian Zoologist of 20th Century Gold Medal by the Zoological Society of India (2000); Prof. T. N. Ananthakrishnan Foundation Award by the Prof. T. N. Ananthakrishnan Foundation, Chennai (2012); Rescholar Award of Excellence in Agricultural Entomology by the Association of Entomologists, Patiala (2014); and Prof. G. S. Shukla Gold Medal by the Academy of Environmental Biology, India (2014). He has authored books including *Concepts of Toxicology*, *Experimental Animal Physiology and Biochemistry*, *Kin Recognition and Oviposition Strategies in Aphidophagous Ladybirds*, and *Monitoring and Management of Mango Fruit Fly* and edited books such as *Pesticides, Man and Biosphere*, *Prof. S. B. Singh Commemoration Volume of Zoological Society of India*, *Modern Approaches to Insect Pest Management*, *Predaceous Ladybirds of Uttar Pradesh*, and *Ecofriendly Pest Management for Food Security*. Professor Omkar has published about 150 research papers in international journals and 62 in Indian journals, 21 book chapters, and 14 popular science articles. Most of his research revolves around identifying and harnessing the potential of beneficial insects, in particular ladybird beetles, for which he is globally recognized. He has supervised 20 Ph. D. theses and completed about ten research projects funded by state and central government agencies. He is a member of the editorial board of the *International Journal of Tropical Insect Science*, and chief editor of the *Journal of Applied Bioscience* besides being president of the International Society of Applied Biology. He works as a referee for about 40 international and 20 Indian journals of repute in addition to working as a subject expert in grant applications for the Swiss National Science Foundation, Switzerland, and a few central and state government funding agencies. He is the recipient of prestigious travel awards from the Indian National Science Academy, through which he has traveled, collaborated, and worked with global experts on ladybird biology at the University of East Anglia, UK, and the University of South Bohemia and Czech Academy of Sciences, Czech Republic.

1 An Introduction to Industrial Entomology

Omkar

As we trace the footsteps of man, from the time of his evolution till the present day, we are able to pinpoint many milestones that grab attention. These milestones are noteworthy because they stand out as discoveries that have been astoundingly successful in lifting the life of human from that of a hunter and gatherer to the human of today, a highly sophisticated and civilized version. Each of these milestones, such as discovery of fire, development of instruments, wheel, agriculture, etc., have been aimed at easing the life of humans in utilizing the natural resources. Some of the most recent milestones have been related to the mechanization of processes that were erstwhile labour intensive, resulting in increased output as well as enhanced and sustained quality of products. Mechanization of agriculture, hand in hand with scientific input has revolutionized and led to a multifold increase in agricultural yield globally. Large-scale/industrial rearing and culture of insects for obtaining their products or for their direct use in food, pharmaceuticals, pest management, etc. can be referred to as industrial entomology. Industrial entomology can also be described in other way as use of insects, their activities and products on industrial scale for human welfare.

The relationship of insects and humans has been largely a harmful or uneasy one. Due to their diverse and at times weird structures, seemingly indestructible armours, huge numbers, an almost ubiquitous presence, through folklore and mythology, negative spin of the media as well as silence of scientists, insects have obtained a largely villainous identity in the human minds at large. This is despite the fact that harmful insect species are less than 1 % of the total known insect species. It is also because of this fear that insects remain one of the most under exploited natural resources. This is despite the fact that their diverse structures as well as habitat adaptations make their study and utilization a necessity for further civilizational advancements.

Omkar (✉)
Department of Zoology, University of Lucknow, Lucknow 226007, Uttar Pradesh, India
e-mail: omkaar55@hotmail.com

Omkar (ed.), *Industrial Entomology*, DOI 10.1007/978-981-10-3304-9_1

Of the insects, honeybees, silkworms and lac insects have been identified as friendly and utilized on a large scale. The potential benefits of these insects and the utilization of their products have been known across centuries. The process of obtaining these insect products started from their collection in the wild to small-scale cottage industries and now larger concerns. While the processing of the produce of these insects has increasingly become more refined and utilized, there are still miles to go in increasing the efficacy of rearing these insects so as to increase the output multifold. Several insects work industriously to increase our crop productivity by facilitating and enhancing the crop pollination. Likewise, there are many that are quite helpful to farmers at global levels in suppressing the population of various crop pests, which affect crop production, being their natural enemies, i.e. parasitoids and predators.

Other than these, there are other insects that have the potential of providing beneficial produce for the humans, such as in food and medicines. While insects have over the centuries been seen as a food source in multiple communities across the globe, it is only in recent years that the potential of insects as a sustained food source, especially a protein one, is being taken seriously. There is an increasing realization that our capacities to increase the production of the food that we currently eat are highly constrained by the limited agricultural land available to us, the limit on the yield capability of plants, the sustaining capacity of the soil, the rapidly changing climates as well as the burgeoning human population. There is not much scope of increase and we are slowly reaching a plateau. It is in this purview that the potential of insects as food of the future is increasingly being approved. Governmental as well as private think tanks and scientific organizations are increasingly supporting research into areas exploring the worth of insects as food. Owing to their small size, high reproductive outputs and faster development, insects can be reared much faster and more efficiently than other livestock. This efficacy is in addition to their known higher nutritional contents, which are much higher than most of our current common food items. In view of this need, there is an increased need to refine and industrialize the process of utilizing insects as food.

Similarly, insects are also a source of medicines, though the hunt of compounds of medicinal value has till recently largely concentrated on botanicals. This is despite the amazing defensive chemical armoury that insects have at their disposal. These days, when antibiotic resistance has reached scary proportions, our hopes for future sustenance depend on the identification of novel chemicals from novel unexploited resources. Insects are potential resources here, and there is a lot of work to be done on their identification followed by the industrialization of their rearing and utilization.

It is these aspects of entomology, that is, the industrialization of rearing and utilization of beneficial insects and their products, that come under the purview of industrial entomology. While there are many books of entomology that describe apiculture, lac culture, sericulture, etc. to the best of my knowledge, there is no book that talks about industrial entomology per se, exclusively and in depth. Also topics such as entomophagy, entomotherapy, mass rearing of biocontrol agents, etc. are almost completely out of the purview of most books. In this book, an attempt has been made to approach the three common cultures, i.e. apiculture, sericulture and lac culture in detail, with expert views, and also new potentials and exciting aspects of insects have also been given due to attention under the umbrella of industrial entomology.

Insects will be the sustaining force of the future. They will not only be our sources, of food, medicines, clothes, etc., but will also be inspirations for further modernization and improvement of many of our current industrial processes, agricultural production, recreation or amusement. After all what can be more efficient than the production system of honeybees; the storage and traffic management system of ants; the architecture of the bees, termites and other insects; the richness of colours of the butterflies; the water and dust resistance of beetles; and many such other wonders. Insects should and will be the future inspirations.

Honeybee: Diversity, Castes and Life Cycle

2

Sunita Yadav, Yogesh Kumar, and Babul Lal Jat

Contents

2.1 Introduction

Honeybees are insects that come under order Hymenoptera and family Apidae and showed complete metamorphosis. Honeybee species are characterized by particular functional traits that facilitate pollination services to a greater or lesser degree (Bluthgen and Klein 2011). Based on morphometric, behavioural and biogeographical studies, 26 subspecies have been identified (Ruttner 1988; Sheppard et al. 1997; Sheppard and Meixner 2003; Engel 2004; Arias and Sheppard 2005). *A. mellifera* occurs naturally over different geographical areas, extending from Scandinavia in the north to the Cape of Good Hope in the south and from Dakar in the west to Oman in the east. According to varying climatic conditions, the population is adapted to the particular geographical region (Ruttner et al. 1978). The colonies of *A. mellifera* are found from sea level to 1000 m above sea level in temperate zones and from sea level to 3700 m in the tropics, and they also survive in extreme hot and arid zone conditions of Oman (Dutton et al. 1981). *A. dorsata* is distributed in Pakistan, through the Indian subcontinent and Sri Lanka to Indonesia and parts of the Philippines in the east, and its distribution is similar to the dwarf honeybee. The rock bees from Nepal and the Himalayas have recently been reclassified as *Apis laboriosa*. However, it is not clear whether the rock honeybees of Sikkim and

S. Yadav (✉) • Y. Kumar • B.L. Jat
Department of Entomology, CCS Haryana Agricultural University, Hisar, Haryana, India
e-mail: sunitayadav10@rediffmail.com

Omkar (ed.), *Industrial Entomology*, DOI 10.1007/978-981-10-3304-9_2

Assam in northern India, western Yunnan Province in China and northern Burma should be classified as *A. dorsata* or as *A. laboriosa*. *A. florea* distribution is generally confined to warm areas, i.e. Pakistan, Iran, Oman, India and Sri Lanka. Its distribution almost ceases at altitude above 1500 m and is absent in north of the Himalayas. Frequently, it can be seen in tropical forest and cultivated areas. In contrast to all of the above, the distribution of *A. cerana* was found more in tropical, subtropical and temperate areas of Asia, occurring in the Indian subcontinent and Sri Lanka in the west, through South East Asia, to Indonesia and the Philippines in the east. For ages, colonies of the oriental honeybee, *A. cerana*, have provided mankind with honey and beeswax, as well as furnishing invaluable service in the pollination of agricultural crops. Among all the *Apis* species, only *A. cerana* and *A. mellifera* are domesticated by man because of their hidden nesting habit. Behavioural limitations of the dwarf and giant honeybees, particularly their practice of open-air nesting, prevent their being kept in man-made hives for reasonably long periods, while hiving colonies in specially constructed containers is essential in that it enables the colonies to be manipulated.

2.2 Honeybee Diversity

Honeybees comprise the genus *Apis* in the family Apidae, order Hymenoptera. The European honeybee is classified as *A. mellifera*, the Indian honeybee is *A. cerana*, the Koschevnikov's honeybee is *A. koschevnikovi*, the dwarf honeybee is *A. florea*, the andreniform dwarf honeybee is *A. andreniformis*, the giant honeybee is *A. dorsata* and the giant mountain honeybee is *A. laboriosa*. Three of these are native to Asia and one is native to the Euro-African region. All of these are similar in appearance, though there are differences in size and colour. All build vertical combs that are two cells thick. All bees feed on pollen and nectar and make nests with wax secreted from their bodies.

2.2.1 *Apis dorsata* Fabricius (Figs. 2.1 and 2.2)

This wild bee is also known as the giant or rock honeybee and found in Asia and occurs all over India, in wild state, at low altitude in the sub-mountain region (up to 1067 m above sea level) and in dry forests and wet regions as well as in the mangrove forests of Sundarbans (Beeson 1941). These bees build a single-comb, exposed nest measuring about 1.5–1.8 m from side to side and 0.6–1.2 m from top to bottom. However, the highest dimensions of a single comb are recorded to be of 2.1 × 1.2 × 0.3 m (Pant 1985). Nests are often seen hanging from branches of trees, roofs or ceilings and at higher altitudes, e.g. Shimla 2100 m (Singh 1962) and 3000 m (WOI 1988). Nest can occur singly or in groups in a single tree; such tree is called bee tree. The adult bees hang in curtains around the nest to control nest conditions. Due to its defensive behaviour, there is little potential for development in the management of either species, though there is often potential for improving the quality of the honey by using more care in processing. The giant honeybee is the most spectacular and ferocious of the four *Apis* species. The workers are the size of

Fig. 2.1 *Apis dorsata*

Fig. 2.2 *Apis dorsata* hive

a wasp and the queen is slightly bigger. The bees are a common sight on sweet shops in Indian markets and are seen sucking sugar syrup and also juices of grapes and other overripe fruits. In this process, they remain unmindful of their occasional brushing away with a piece of cloth. They rarely sting the shopkeeper and the customers unless accidentally crushed or trampled upon.

Distribution and Nesting Behaviour It is found all over India in the plains and in the hilly areas up to 2100 m above sea level. Geographical distribution extends to Pakistan, Sri Lanka, Malaysia, the Indo-China region, the Philippines, China and Indonesia. During summer, the colonies migrate to mountains up to the elevation of about 2100 m to avoid extreme heat or in search of flora. The colonies arrive in hills in March and April and return to plains in June/July before the monsoons. During migration, the swarms are known to make short halts. During winters, they are abundantly found in plains. The migration improves colony survival and provides several crops for honey in the areas during the year.

The nesting behaviour of this wild honeybee has been studied with much interest, and efforts have been made to hive it also. Under natural conditions, *A. dorsata* generally make their nests higher than 10 m, but some are seen at lesser heights, particularly in areas where there are only a limited number of tall trees. The honeybees build single combs on the projections of rocks or those of tall buildings in cities and more commonly on tall trees on the roadside, in gardens or in forests. Honeybee colonies prefer consistently old comb remnant sites for nest building and utilize them as an index and as visible signal to bee swarms for new colonization (Thakur 1991). More than a hundred colonies can be counted in the Taj Mahal gardens at Agra. Fletcher (1952) reported 156 colonies on a bee tree in southern India. They can make nests on any tree species including the thorny *Acacia catechu*, but they seem to show preference for the tall *Ficus* trees that have thick limbs wide apart, allowing free flight to the bees.

Comb Structure and Castes Giant honeybee colonies are known to exhibit a migratory lifestyle with several colony movements during an annual cycle (Koeniger and Koeniger 1980; Oldroyd et al. 2000; Woyke et al. 2000). Vertical single combs are built by these bees. The comb area can be up to 1 m^2. Individual combs can be very large, with reported dimensions of 1.5 m × 1 m (Oldroyd and Wongsiri 2006). The egg placement in the cells, the concentric arrangement of brood, the raising of queen cells, the issue of swarms, the communication by dances, etc. are broadly similar to *A. mellifera* and *A. cerana indica*. About 7000 bees would weigh 1 kg. Average tongue length is 6.68 mm. The queen is darker in colour than the workers and about one-fifth long as the workers and about 2 mm broader.

The combs made by this honeybee are large and semicircular, quite thick on top, where honey is stored and honey harvesters go for it. Combs are attached to the strong limbs of a tree, and the attachment is quite wide and strong to be able to carry the weight. On an average, there are 23,000 cells on the comb, constructed bilaterally on either side of the comb. The cells are hexagonal in shape and 5.35–5.64 mm in diameter and they slope upwards. The depth of a brood cell is 16 mm.

Brood and honey stores are in the same comb. In a comb, honey is stored on the top with thicker sides, and pollen is stored below, followed by the brood area. The queen cells are made at the lower end. In a colony, 80–90 % of the bees arrange themselves in a layer of 3–6 bees thick, and they are oriented in the same direction with their heads upwards and their wings half spread. This cover of bees is like a bag that touches the comb at the edges, leaving a hollow space of 1–2 cm between the comb surface and the cover of the bees. This space is used by the nurse bees to move around and feed the brood. At the lower end of the bag, there is a mouth open for the worker bees to enter and to come out. The field bees perform dances at the exposed portion of the comb on the vertical plane.

In a colony, the brood is reared in worker cells. Worker and drone cells are equal in size. Both giant honeybee species, *A. dorsata* and *A. laboriosa*, are unique in the genus *Apis* in that the hexagonal cells used for rearing drone and worker brood are the same size, albeit this is a debated point, i.e. Tan (2007) reports a statistically significant size difference between drone and worker cells; however, previous research reports no consistent difference (Oldroyd and Wongsiri 2006). The capping of the drone cells is raised, whereas those of the worker brood are flat. The queen cells are made on the lower rim of the comb. This honeybee like other *Apis* species also produces reproductive swarms. Once swarming starts, it may continue for some time, and a new swarm may be issued every 3–4 days. The new swarms generally settle at some distance in the area. A unique feature of this honeybee is that a nucleus colony may sprout out of the mother colony and settle at a distance of 1 m on the same limb of the tree. A new comb is built, and at first, the workers keep moving between the mother colony and the nucleus, but after a few days, the nucleus becomes independent. It appears this honeybee is capable of nocturnal orientation.

Behaviour The behaviour of the species is unpredictable and they will not live inside a hive. Generally, these bees show ferocious temperament and are provoked by slight disturbance. Worker bees attack in mass and follow the victim over long distances. Worker bees cover the comb like a curtain and orient upwards for insulation and protection. They are sensitive to smoke which is normally used by honey hunters.

In the normal working state of the colony, it is not at all aggressive. A strong wave of "shimmering" is passed every time they are disturbed by hand. It is reported that most of the small animals that might approach the colony as enemies are repelled by this shimmering.

They ignore the people who may walk or sit under the trees. The bees attack only when they are disturbed by a stone, stick or any other thing. The suspected enemy (quite often an innocent traveller) is then attacked in scores, within a radius of hundreds of yards of even up to 1 km. After stinging men, cattle and animals for hours, it takes them a long time, in fact an overnight, to calm down moderately. Even the next morning, some excited bees are seen scouting around. It appears that colony defence in the open is an evolutionary behaviour. During this state of excitement of the colony, many passers-by are stung by multiples of bees or by a cloud of bees, and the victim writhes on the ground in agony and even loses the presence of mind to call for help. In many cases near the colony, 80–100 stings per victim are a common occurrence. The very first step to be taken is to save the victim from shock and subsequent death, by administering antidotes, through regulated injections. In case a single alarmed bee makes a run, as many as 5000 bees may be alerted out of the curtain of bees. Only one bee has to sting, and the smell of isopentyl acetate released because of stinging will convey to the other bees where to locate the "enemy". The smell lasts for 10–15 min and within that time, the victim is at a great peril. The bees chase him menacingly, and the only escape for him is to decoy and hide behind a hedge, to lie on the ground with face downwards, to cover himself with a sheet of cloth or a blanket and to protect his naked body parts.

Honey-Gathering Behaviour The giant honeybee has a great economic value in India, as this species is a good honey gatherer. Honey is stored in the top half portion of the comb. It can, therefore, be harvested without destroying the colony by cutting the honey-storing portion of the comb. Honey yield is as high as 50–80 kg honey per colony, which can be obtained from a single comb. A sizable portion of honey produced in India comes from *A. dorsata* species.

Before the introduction of *A. mellifera*, it contributed 60–70 % to the honey crop and practically all the commercial wax (ghedda) in the country. In quality, honey is like the hive honey provided extraction is done properly. In Nepal and eastern and central India, it plays an important role in the economy and tribal life of the people. Those who have learnt the art of honey harvesting climb a tree or a rock with a bucket or a basket. A small fire and smoke is started underneath, and they reach the comb, cut out the thick top corner full of honey, pour it in a basket or a bucket and come down. When the bees migrate in the hot summer before the monsoon, the entire comb is brought down for wax. Among the tribes, that art is developed to varying degrees, and there is a wastage and unnecessary killing of the bees. From the cave paintings found in central India, it is evident that honey harvesting has been practised for thousands of years.

Foraging Behaviour Honeybees were found as predominant pollinators (85.23 %) in the seed production plot. Earlier, Sharma et al. (1974) reported that honeybees were the predominant pollinators (42.1 %) of cauliflower. Sinha and Chakrabarti (1980) reported that honeybees constituted 79, 82.4 and 83.3 %, respectively, in 3 consecutive years. Kakkar and Sharma (1991) observed that honeybees constituted 38.7 % on cauliflower bloom. The foraging efficiency of honeybees directly affects the crop production and productivity, and it depends on the availability of bee forage, conditions of the colony and foraging range of worker bees (Pudasaini and Bahadur 2014). Observations on the foraging behaviour of the species showed that their behaviour is similar to domesticated *Apis* species. But in contrast to domesticated bees, *A. dorsata* bees have been reported to forage during night also. Selvakumar et al. (2015) reported that among bees, *A. dorsata* (10.16) visited more flowers than *A. mellifera* (8.68), *A. c. indica* (7.16) and *A. florea* (4.86). Depending upon the flower structure, they act both as top and side workers. This species is very hard working and very efficient pollinator of crops in India. This honeybee is also common on *Brassica* crops, *Eruca sativa*, sunflower and cucurbits in Punjab, Haryana and Himachal Pradesh. Pudasaini and Bahadur (2014) reported that *A. dorsata* visited 9.33 and 15.83 flowers at 10:00 am and 2:00 pm of the day, respectively, with the peak foraging hours of around 1200–1400 h.

Seasonal Migration The migratory open-air nesting *A. dorsata* honeybee migrates at least twice a year (Deodikar et al. 1977; Koeniger and Koeniger 1980; Reddy 1983; Venkatesh and Reddy 1989; Underwood 1990; Dyer and Seely 1994; Kahono et al. 1999; Thapa et al. 2000; Lipiński 2001; Woyke et al. 2001; Liu et al. 2007). The same nesting sites are occupied year after year. Neumann et al. (2000) and Paar et al. (2000) genotyped the migrating *A. dorsata* bees and showed that the same swarms return to their natal nesting sites. The seasonal migration of *A. dorsata* is quite spectacular. As forage decreases towards the end of the season, colonies abandon their combs and migrate to lower elevations, establishing new nests there for the mass flowering of the monsoon season (Ahmad 1989; Dyer and Seeley 1994). They descend from the mountains to the plains of India after the monsoons or from the highlands to the coastal areas in South India and flourish on the winter flora of crops, trees and shrubs. They raise their brood and gather honey (some is harvested by man), and in the summer they are ready to migrate to altitudes less than 1500 m. It is almost certain that they perceive and forecast the oncoming events and prepare to move to safer areas before the crushing monsoons in southern, eastern and northern India.

With the coming of summer, the desert areas and the plains of northern India burn with heat. The atmospheric pressures fall and the dust storms are ready to whirl around in the countryside, and it appears the bees are set to avoid them, and hence they must migrate to safer areas to the mountainsides where there is thick vegetation cover. The way the mighty swarms move from their abodes on tall trees as if at a short notice, by some elaborate system, the scientific phenomenon is worth studying and recording further.

The migration of *A. dorsata* from the plains of Kalka and up into the valley reaching Kasauli is very interesting. Around 15th of May every year, overpowering humming sounds are heard almost all day, and groups of swarms of *A. dorsata* bees are seen coming up the valley almost following the contours of the terrain, from 5 to 30 m above the ground moving upwards and settling into the jungle up to 1500 m high. The swarms when they come are a sight to be seen. Many times a swarm arrives late in the day and stays overnight, settling on a low tree, and the next morning they are gone. This commotion goes on day after day for weeks. It seems that the migrating swarms somehow excite others on the way, and they make quick preparations (using up honey reserves) to move along. In the empty combs, hardly any honey is left behind since it is used up by the bees before leaving.

2.2.2 *Apis florea* Fabricius (Figs. 2.3 and 2.4)

This bee is also known as the dwarf or little honeybee and is found in Asia. Like *A. dorsata*, these bees build a single-comb, exposed nest. Nests are built around a twig of a shrub or a branch of a tree. They are the smallest honeybee species both in size of the body and the nest.

Distribution These are generally confined to warm climates. This species is found in the plains of India up to about 300 m above sea level. This species is distributed in India, Sri Lanka, Pakistan, the Indo-China region, Malaysia, the Philippines, Indonesia and up to Iran and Oman in the west. This is a wild bee, but attempts to keep it in specially designed hives have met with partial success in India. *A. florea* is highly migratory, but long-distance migrations are unknown. This honeybee also has a habit of nest shifting within short distances. The bees may shift within 2 months in dry season and within 5 months in the rainy season. When they shift, they carry away all the provisions, including wax. When disturbances cause them to

Fig. 2.3 *Apis florea*

Fig. 2.4 *Apis florea hive*

desert the comb, then they leave behind honey, brood and pollen stores. But the bees continue to return to old comb to take away comb wax, honey and pollen.

Comb Structure and Castes The little bee builds a single vertical comb nest, and the comb is constructed around the stem of a bush, branches of bushes, hedges, trees or a dried thick stick in the shaded places. The nesting location of *A. florea* is unique, not easily accessible to animals including mankind that could help avoid animals including human interferences and vehicular traffic. Accordingly, *A. florea* builds its colony at interior side away from the road. *A. florea* avails various plant species including human-built structures for nesting under shady places on the twigs/branches. Shady places help protect the colony members from bright light, strong winds and inclement weather conditions. Even though *A. florea* nests ranged between ground level and up to 50 ft, it preferred ground level to up to 15 ft height more (Vaudo et al. 2012; Woyke et al. 2001; Manunath 2008). Sometimes, the combs are constructed in a protected place in the hollow of a hedge, a stack of sticks or even a hollow in a building structure. At the top the nest encircles the strong stem so as to give it a good strength. The portion of the comb that encircles around the stem (or on the flat of ceiling in a building) is thick, but as the comb is built further down, it becomes thin in depth, although as broad as the upper portion. In the autumn, they move short distances to unshaded nesting sites. The adult bees hang in curtains around the nest to control nest conditions. In the same comb, brood is present in the lower section and the honey is found in the upper section. The comb is broad at the top and it serves as a landing place for the foragers.

The comb itself is less than 1 ft in length and contains about 14,000 cells. The size of all the three types of cells varies with geography and becomes smaller as we go from north to south. In a colony, there are about 6000 bees, and the queen lays 350 eggs per day in a brood area of 600 sq. cm. The worker brood cells are made on the flat surface in the middle of the comb. The comb is always covered by a curtain of bees. Drone cells are 1.5 times larger in diameter than worker cells and are found on the lower part of the comb. Queen cells are raised in spring and autumn at the bottom edge of the comb and are quite large. Compared to the other species of honeybees, its workers have a long life of 61.2 days, that is, 2.5 times that of an *A. mellifera* worker.

Since this honeybee shifts a lot, it would appear its biology is adapted to multiplication, swarming and dissolution of the colony with the purpose of reassembling under a new situation. During a short span of 2–5 months, the nest is built, brood is reared, honey is stored and 12–16 queen cells are built for further propagation. When the first virgin queen emerges, half the number of workers leaves along with the mother queen. Subsequently, the virgin queens mate and smaller swarms leave every few days. In the end, a handful of bees remain on the comb, and eventually they also swarm away or simply disperse, leaving the comb bare and deserted.

Behaviour These bees are very prone to swarming. They are gentle in temperament; however, they do sting when irritated. Colonies can be shifted to crops at blooming time for pollination. In areas of its distribution, it lives along with *A. cerana* and *A. dorsata*; the ratio of their size is 1.0:1.27:2.09, respectively. The ability to survive in a very hot and dry climate is its special trait, and it can live in deserts (50 °C) without any harm. The heat tolerance ability is further demonstrated by the fact that worker bee's daily activity starts at 18 °C and continues up to 40 °C. It is no wonder that this bee is a relentless visitor of flowers of crops, trees, shrubs and the annuals. Her small size restricts her to a shorter flight and hence she exploits the flowers intensively. In one study, on *Brassica* crops this bee constituted 73–74 % of the insect visitors observed. The swarms forming the new colonies generally settle at a distance of less than 100 m, after shifting their site a few times. Some swarms do go far but not more than a few hundred metres. It has been observed that if a queen is removed or it dies naturally, the workers build a queen cell by modifying a worker cell somewhere in the middle of the comb.

Honey-Gathering Behaviour Due to small size of its comb, *A. florea* is a poor honey yielder, and a comb yields 200–900 g of honey. The honey is thin in consistency. Honey hunters take away the whole comb and thus destroy large number of colonies. The honey produced by this species is believed to have special medicinal qualities, but there are no scientific studies to support this belief. The medicinal value, if any, must be attributed to the nectar of the plants in the locality that are available to other bee species also. *A. florea* honey fetches higher price in countries like Oman.

A. florea even though quite small in body size can be quite aggressive and resort to robbing the nests of *A. mellifera* which is much bigger in body size and greater in colony strength. Lately, some researches on this honeybee have also been conducted in Oman and in Iran showing interest in its preservation as a species.

Foraging Behaviour Foraging behaviour is similar to other bees and shows consistency in foraging on a single crop during visit. They are top workers. Abrol (2010) reported that the dwarf honeybee *A. florea* was the most abundant flower visitor and comprised more than 94 % of the total visitors. Commencement of flight activity occurred when a minimum threshold of environmental variables was exceeded, while the cessation was governed mainly by decline in light intensity and radiation. They do not forage at night as seen in *A. dorsata.* They can forage better on crops with small flowers or in umbelliferous crops. On average, *A. florea* visited 1.33 ± 0.26 and 6.17 ± 0.58 umbels and flowers/min, during different hours of the day.

2.2.3 *Apis mellifera* Linnaeus (Figs. 2.5 and 2.6)

This is a domesticated bee species, also known as European or Euro-African honeybee. These bees can be kept in hives, and methods have been devised to allow for a more rational utilization of their potential. The bee is similar in habits to the Indian honeybee in that it builds multiple combs parallel to each other with uniform bee space. Combs are built in hollows of trees, in walls or in shady places. It is with this species that a potential for beekeeping development exists.

Distribution It is native to Western Asia, Europe and Africa. This species is found all over Europe and spread to other continents during the last five centuries.

Fig. 2.5 *Apis mellifera*

Fig. 2.6 *Apis mellifera* hive

European races of the western hive bee have been introduced into most parts of the world, including America, Australia and Asia. *A. mellifera* is the most widespread of these species, occurring throughout Europe, Africa, Northern and Western Asia (e.g. Ponto-Caspian and as far east as the Tien Shan), the Levant, Caucasia and the Iranian Plateau (Ruttner 1988, 1992; Ruttner and Kauhausen 1985; Sheppard et al. 2003), as well as adventive in the Americas and Australia (Kerr 1957; Sheppard 1989; Engel 1999; Moritz et al. 2005). Now, it is found almost in every country. This bee has been studied intensely from both a strict biological and a beekeeping viewpoint. There are many well-recognized races and strains of *mellifera* which greatly differ in appearance.

Comb Structure and Castes This species of *Apis* normally build multi-comb nests in enclosed cavities. Tongue length varies from 5.5 to 7.2 mm. It has many desirable traits. It maintains a prolific queen, swarms less, has gentle temperament and is good honey gatherer. This race has achieved a great success in some states of India and has proved to be superior performer than *A. cerana*.

Races There is tremendous variation in this bee across its range, and at least 20 different subspecies or "races" are recognized, broadly divided into European and African groups. Several races of this bee are considered especially desirable for beekeeping. The western honeybee or European honeybee (*A. mellifera*) is a species of honeybee comprised of several subspecies or races. At least 29 subspecies of *A. mellifera* have been delineated on the basis of morphometry (Ruttner 1988; Engel 1999; Sheppard et al. 2003). These subspecies are now typically divided into four major groupings, supported by morphometric and genetic studies in addition to analyses of ecological, physiological and behavioural traits: group A, which includes subspecies throughout Africa; group M, which includes subspecies from

Western and Northern Europe; group C, which includes subspecies from Eastern Europe; and group O, which includes species from Turkey and the Middle East (Ruttner et al. 1978; Ruttner 1988; Garnery et al. 1992; Arias and Sheppard 1996; Franck et al. 2001; Miguel et al. 2011). There are many different races of *A. mellifera*, some tropical and others temperate.

African bees are termed as killer bees. They were imported into Brazil in an attempt to establish an industry in some of the tropical regions. Some of these bees accidentally escaped and became established. They have continued to expand their range in the tropical lowlands, and in most cases, they have actually replaced the existing European bees. The establishment of African bees in tropical America has caused a great disruption of the beekeeping industry. The African bee is noted for its defensiveness and unpredictability. These are characteristics considered undesirable from a beekeeper's point of view.

Behaviour They are gentle in temperament, but when irritated they do sting although workers die after stinging.

Honey-Gathering Behaviour Field honeybees collect flower nectar. On entering the hive with a full honey sac, which is an enlargement of the oesophagus, the field bee regurgitates the contents into the mouth of a young worker, called the house or nurse bee. The house bee deposits the nectar in a cell and carries out the tasks necessary to convert the nectar to honey. When the honey is fully ripened, the cell is sealed with an airtight wax capping. Both old and young workers are required to store the winter supplies of honey. In migratory beekeeping, it produces an average of 50–60 kg honey/year/hive. Yields of 100 kg/year or better are possible under optimum conditions.

Foraging Behaviour Foraging is a social enterprise (Seeley 1985) in which bees collect pollen, nectar, water and propolis from plants. The act of collecting all these is called foraging and the bee is a forager. Under normal colony conditions, the forager bees are workers with an age of over 21 days, at which time they shift to perform out-colony tasks including water, nectar, pollen or resin collection (Sharma 2014). Pollen is carried into the nest or hive on the hind legs of the field bees and placed directly in the cells. The pollen of a given load is derived mostly from plants of one species, which accounts for the honeybee's outstanding role as pollinator. The anticipation of the commencement of foraging is associated with an increased titre of juvenile hormone (JH) in foragers which is not affected by foraging experience but by diurnal variations (Elekonich et al. 2001). If it flew from one flower species to another, it would not be effective in the transfer of pollen, but by confining its visits on a given trip to the blossoms of a single species, it provides the cross-pollination required in many varieties of plants. Like *A. dorsata*, foraging behaviour depends upon the flower structure. Generally, the foraging skills and the number of forager workers increase with age (Dukas and Visscher 1994).

2.2.4 *Apis cerana* Fabricius (Figs. 2.7 and 2.8)

The eastern or Indian or Asian hive bee (*A. cerana*, synonym of *A. indica*) is native to Asia. Beekeeping is developed with this bee in different regions of Asia, since it can be easily hived in fabricated containers. There is a lot of variation in the eastern hive bee across its range, and little work has been done towards selecting more desirable strains from a beekeeping point of view. Techniques of beekeeping with this bee are similar to those used with the western hive bee, though the hives used are smaller.

Distribution This species had been the base of Indian beekeeping and found throughout India except the plains of North India. *A. cerana* is the Asiatic honeybee or the oriental honeybee because they are only found in Asia, from Iran in the east to Pakistan in the west and from Japan in the north to the Philippines in the south. Thus, *A. cerana* does not live only in tropical and subtropical areas of Asia but also in colder areas such as Siberia, Northern China and the high mountain area of the

Fig. 2.7 *Apis cerana*

Fig. 2.8 *Apis cerana hive*

Himalayan region (Koeniger 1976). It thrives up to 2500 m above sea level. In China and Japan, the indigenous domesticated *A. cerana* has now mostly been replaced by *A. mellifera* which was introduced some years ago.

In India, there was hardly any beekeeping with *A. cerana* in Punjab, Haryana and the plains of Uttar Pradesh, but *A. mellifera* is doing very well in these areas. The Indian race of the species is *A. c. indica*, and there are many distinct strains present in different geographical regions. It occurs naturally up to 46°N in Asia east of Iran and in the valleys of Hindukush and Himalayas up to 2000 or even 3000 m in the southern valleys of these mountain ranges.

Comb Structure and Castes Feral colonies of *A. cerana* are found in hollows of trees and holes in the rocks or the walls of houses. In the indigenous methods of beekeeping in India, the nesting situations are simulated, and bees are kept in log hives (Figs. 2.9) and in hollowed out house walls with entrance hole opening on the outside and window opening on the inside. Records of this method of beekeeping in Kashmir are available dating back 1470 A.D. In nature, a colony has 6–8 combs with 6000–7000 workers in subtropical areas and 10,000–20,000 in temperate areas like Japan.

The diameter of a worker cell and worker bee varies a great deal in various parts of Asia. The cell size is 4.87 mm in Peshawar, 4.78 in Japan, 4.67 in Peking, 4.25 in the plains of India and Thailand and 4.2 in Sri Lanka. This honeybee has the habit of gnawing wax of old cells down to the middle sheets. Thus, wax debris accumulates at the bottom and provides medium for the development of wax moth. The standard Langstroth hives by itself are not suited for this bee, but if the bee space in between the combs and the cell size is suitably altered, the hive will be accepted. For the tropical South India, a smaller hive containing 6–8 movable frames has been

Fig. 2.9 *Apis cerana* log hive

devised, commonly known as the villager hive; it has many variations in the dimensions of frames and hives.

The drone cells of *A. cerana* have characteristic holes like pores in the capping which is not the case in *A. mellifera.* The caste system, the colony structure, the parallel comb arrangement and the biology of the two species are very similar, but the fanning position at the hive entrance is the opposite; in *A. cerana*, the workers keep their head away from the entrance, whereas in *A. mellifera* the workers keep their head directed towards the entrance. The sequence of functions by the workers is the same in the two, but *A. cerana* workers perform the various functions a little early in life, e.g. cell cleaning starts before the third day; brood care, pressing the pollen in cells and comb construction start on the third day; development of pharyngeal glands starts between the 4th and 16th day; the wax glands start functioning on the third day, but the maximum secretion is between the 12th and 16th day; and the orientation flights are between the 7th and 11th day.

Behaviour *A. cerana* is a bee with gentle temperament and it responds to smoking. It is frugal in habits but lack of flora is quickly reflected into absconding. It also has a strong tendency for swarming; a colony may issue up to 5–6 swarms in a year. Management and manipulation practices to prevent and control swarming also do not always work. Due to incessant swarming, ravages of bee enemies, lack of honey flora and resultant absconding, one comes across a large number of weak colonies. It is poor propolizer, and practically no propolis (bee glue) is brought to the nest for reinforcing the combs or sealing the cracks, etc.; therefore, wax moth does considerable damage.

Although yields are considerably lower than with the western hive bee, this bee has the advantage of being well adapted to the area. It is more resistant to some of the disease and pest problems found in the area. Therefore, it is better able to survive under the minimal management conditions that often characterize beekeeping at the small-farmer level.

In recent years, there has been an effort in Asia to replace the eastern honeybee with European races of the western honeybee. This has been successful only in temperate regions and only for large-scale, capital-intensive operations where the technology is available to control disease and parasite problems of the European races. In any small-scale beekeeping development effort, the existing bee resource of the area should be used. Importing bees for such a project is far riskier than it is worth. Imported bees often are not adapted to the areas into which they are introduced. Importing bees also risks the introduction of exotic bee diseases and parasites.

As compared with the western honeybee, *A. cerana* bees are more excitable when disturbed and are prone to sting more readily, particularly when the weather outside is chilly. The smell of body sweat or a perfume also excites them. When a hive is knocked, the bees inside make a hissing sound produced by quick movements of wings. The same sound is produced when one blows at an exposed frame;

the bees move away exposing the cells which can then be examined. When the bees are attacked by enemies like a wasp or a large black ant, they raise the abdomen and collectively make violent lateral shakings. The hissing sound and the shaking movements frighten the enemies. If a giant wasp (*Vespa mandarinia*) comes to the colony as a predator, he has to face the wall of bees that may ball around him and even kill him without suffering a single casualty.

A. cerana has the habit of absconding during the dearth period, and the attack of enemies like wasps and wax moth further aggravates absconding behaviour. In order to check absconding, the colonies should be shifted to areas where there is good nectar flow and there are comparatively fewer enemies. Competition from other honeybees (*A. mellifera*) may also cause commotion and absconding.

Mated *A. mellifera* queens are accepted by *A. cerana* workers, who raise the brood, and the workers of both the species live together for some time and eventually *A. cerana* die out of old age and the colony becomes pure *A. mellifera*. The reverse queen introduction is not likely to be a success owing to pheromone differences of the queens; chemically, their pheromones are alike, but *A. cerana* lacks a component, whereas that of *A. mellifera* has a broader spectrum. Queen rearing, queen emergence, mating and reproductive swarming are similar in the two species. In *A. cerana*, an average of nine queen cells has been observed in the swarming season.

Mating Behaviour Reproductive swarming is a regular feature of *A. cerana*, and it starts when the colony strength is around 20,000 bees. The drones fly between 1115 and 1515 h, and the maximum number comes out between 1215 and 1415 h. The flight activity of drones and queens of *A. cerana* (Ruttner et al. 1972) was observed in an isolated place; eight queens mated, three of them on two flights. Successful mating flights were longer than those of *A. mellifera* (average of 30.8 min). In body development, *A. cerana* drones are smaller, and they contain around 1 million spermatozoa as compared with 10–11 million in *A. mellifera*. The queen of *A. cerana* on emergence goes out for mating flight after 4 days, between 1330 and 1530 h, and most probably multiple mating (8–10) takes places until the spermatheca of the queen is filled with 3.5 million spermatozoa. The queen starts laying eggs within 2–3 days after mating.

As in other *Apis* species, drones in *A. cerana* develop from unfertilized eggs and the workers and queens from fertilized eggs. On the loss of a queen, the colony becomes a laying worker, and from the unfertilized eggs, drones develop and ultimately the colony perishes. The cell size on the foundation sheets is smaller suiting this species.

Honey-Gathering Behaviour Honey yields of up to 15–20 kg/year are obtained in some areas, but the average is much lower. Because of small size, short foraging distance, absconding, swarming, etc., this species is a low honey yielder. On an average, a colony yields 3–5 kg of honey per year in plains, but the yields are as high as 20–25 kg in Kashmir. The selective breeding does help in improving the desirable traits but of little practical value because of no control on parentage.

Foraging Behaviour These bees have short foraging distance but, efficient pollinators of different crops of coastal and hilly areas. These are top workers. In its natural area of distribution, *A. cerana* is a common visitor of a number of crops like *Brassica*, clovers, sunflower, *Eucalyptus*, pome fruits, coconut and other palms, rubber, etc. It is a gatherer of honey on most of these plants and is an important pollinator of pome fruits. *A. cerana* bees started their foraging activities early in the morning (06.14 ± 0.004) and ceased late in the evening (17.28 ± 0.011) (Singh 2008). The total duration of foraging activity was 1000 h and the average duration of foraging trip was 4.5 ± 0.14 min. Two peaks of foraging activities were observed between 0830 and 1030 (peak I) and 1130 and 1330 (peak II). The peak I period was the main foraging period, and peak II was the second foraging period; both were very useful from the pollination point of view.

As a honey gatherer and, possibly, an exploiter of flora, *A. cerana* seems to be frugal in her behaviour; she can sustain herself at lower colony strength, fewer resources and lower colony activity as compared with *A. mellifera.* It starts its daily activity at temperatures as low as 6–8 °C in winter, whereas *A. mellifera* is active at warmer temperatures with stronger colony strength. A further distinction of behaviour is that *A. cerana* visits a greater variety of cultivated and wild plants as in a natural habitat of a forest area, while *A. mellifera* sticks to fewer but rich resources.

Races In India, based on morphological features, two "races" of *A. cerana* are identified: a black "hill" morph that is often said to live at higher elevation and a yellow "plain" morph found at lower elevations (Kapil 1956; Narayanan et al. 1960a, b). The intra-specific classification of the Asiatic honeybee species, *A. cerana*, is in a state of flux and uncertainty (Hepburn et al. 2001). Studies carried out by the International Centre for Integrated Mountain Development (ICIMOD) reveal that *A. cerana* populations can be divided into three subspecies, namely, *Apis cerana cerana*, *Apis cerana himalaya* and *Apis cerana indica.* Of these, *A. cerana cerana* is distributed over the north-west Himalayas in India, North-West Frontier Province of Pakistan and Jumla region of Nepal. *A. cerana himalaya* is found in hills of Nepal, Uttar Pradesh, the north-east Himalayas and Bhutan; *A. cerana indica* is found in the plain areas and foothills of the region. However, based on morphometric observations, four major races of this honeybee have been described. *A. c. indica* is the smallest and is distributed in South India, Sri Lanka, Bangladesh, Burma, Malaysia, Thailand, Vietnam, Indonesia and the Philippines. The standard size *A. c. himalaya* is distributed in Afghanistan, Jammu and Kashmir, North India, China and North Vietnam. *A. cerana cerana* is distributed in most parts of China. There is a large size race *A. cerana japonica* that is distributed in Korea and Japan.

2.2.5 *Apis koschevnikovi*

Newly identified honeybee species is very similar to *A. cerana*. This species was first described by Buttel-Reepen, who dedicated it to Koschevnikov, a nineteenth-century pioneer of honeybee morphology (Gupta 2014). The species was described again by Maa in 1953, this time with the name *Apis vechti*. It was finally rediscovered by Tingek et al. in 1988. Its other name is the red bee (this species was named for a short period as *Apis vechti*). It also nests in natural cavities and builds multiple parallel combs. It has been reported from only some parts of Indonesia.

2.2.6 *Apis andreniformis* (Fig. 2.10)

Another small honeybee species is recently reported and is similar to *A. florea* in many ways. The dwarf honeybees, *A. andreniformis* and *A. florea*, are sister species with a partially sympatric distribution in southern Asia. It also builds a single comb in the open like the latter and is absent from colder climates where the more widespread multiple-comb, cavity-dwelling honeybee species occur. Although their propensity for and frequency of swarming and migration vary regionally, it is almost always associated with the sequence: rainfall > flowering > swarming or migration (Hepburn and Radloff 2011). It has been reported from Thailand, China and Malaysia. The dwarf honeybee, *A. andreniformis*, extends from the eastern foothills of the Himalayas eastward to Indochina, Sundaland and the Philippines.

2.2.7 *Apis laboriosa* (Fig. 2.11)

This species has also been identified recently. It is similar to giant honeybee in many ways. The giant honeybees of Nepal and the Himalayas have recently been reclassified as *A. laboriosa* (Gupta 2014). Although minor variations in anatomical, physiological and behavioural characteristics exist among the different geographical races of the giant honeybees, they are essentially similar in all their major biological

Fig. 2.10 *Apis andreniformis* (Source: Nicolas Vereecken, https://www.flickr.com/photos/nico_bees_wasps/5592188371)

Fig. 2.11 *Apis laboriosa* (Source: https://www.pinterest.com)

attributes. It also builds a single comb in the open. Its presence has been reported from Western China, parts of the Himalayas, Nepal, Bhutan, Tibet and India.

It is called as the largest honeybee of the world and distributed in Nepal, Bhutan, Sikkim and Yunnan between altitudes of 1200 and 1400 m. It is distinctly larger than the common *A. dorsata* and remains active at comparatively lower temperatures and greater heights. It has exposed combs, and in its area of distribution, the temperatures range between 10° and −5 °C. The dense coat of long dark body hair appears to have good survival value.

2.3 Honeybee Castes, Colony Organization and Life Cycle (Figs. 2.12 and 2.13)

The honeybee is a social insect with three different types of individuals or castes in the colony: queens, drones and workers. Each caste has its special function in the colony. The workers are undeveloped females, the drones are known as males and the queen is the fully developed female. The queen's job is to lay eggs, as many as several hundreds in a day. These larvae develop into drones, workers or new queens, depending on how the workers treat them.

All young larvae of less than 2 days are fed with royal jelly by the massive provisioning scheme. The different feeding schemes determine the caste of the adult bee. Thus, any female egg or larva less than 2 days old has the potential to become either a queen or a worker. Each caste has a different developmental time and thus reared in a distinct type of cell. In this chapter these three castes, their cells and functions are discussed in detail.

Queen Bee The queen, a true mother bee, is the only female that is completely developed sexually. This is a result of a total diet of royal jelly during a developmental period. She is distinguished by her long, slender appearance, due to the full development of the ovaries in her abdomen. In the colony, she is found in the area of the brood nest.

Fig. 2.12 Honeybee castes © 2006 Encyclopædia Britannica, Inc.

Morphological Features Either she is longer than a drone or worker, being over seven-eighths of an inch long and has a tapering abdomen. The queen's tongue is short, her jaws weak, eyes like the workers and wings short, hardly more than half the length of the abdomen. She has no pollen baskets, but possesses a sting without barbs that resembles that of the bumblebee, in being curved, yet, she seldom makes use of it.

Cell Shape The queen is reared in a specially constructed royal or queen cell. Queen cells appear similar to peanut shells that hang from the surface of the comb. They can be located along the edges of the comb or within the comb area. The colony constructs queen cells when there is a need to rear queens, though cells are sometimes started and then abandoned. These are called false queen cups.

Feeding The developing queen larva is always surrounded by royal jelly, a special, highly nutritious food produced by head glands of the workers. This feeding scheme, called massive provisioning, is unique to the queen and continues throughout her entire developmental period.

Developmental Period The developmental time of the queen, 16 days, is the shortest. The queen, like the workers, is developed from an impregnated egg, which comes from a fertile queen. The eggs are placed in queen cells, which are usually built on the edge of or around an opening in the comb and extend either vertically or diagonally downwards. These resemble a peanut in form and size. The eggs are placed in these cells, either by the worker bees, which transfer them from the worker cells, or else by the queen. The queen may be developed from an egg or from a worker larva less than 3 days old, which will then be transferred from a worker cell to a queen cell. The development of the queen is much the same as that of a worker, though she is fed richer and more quantities of food, called royal jelly. The condi-

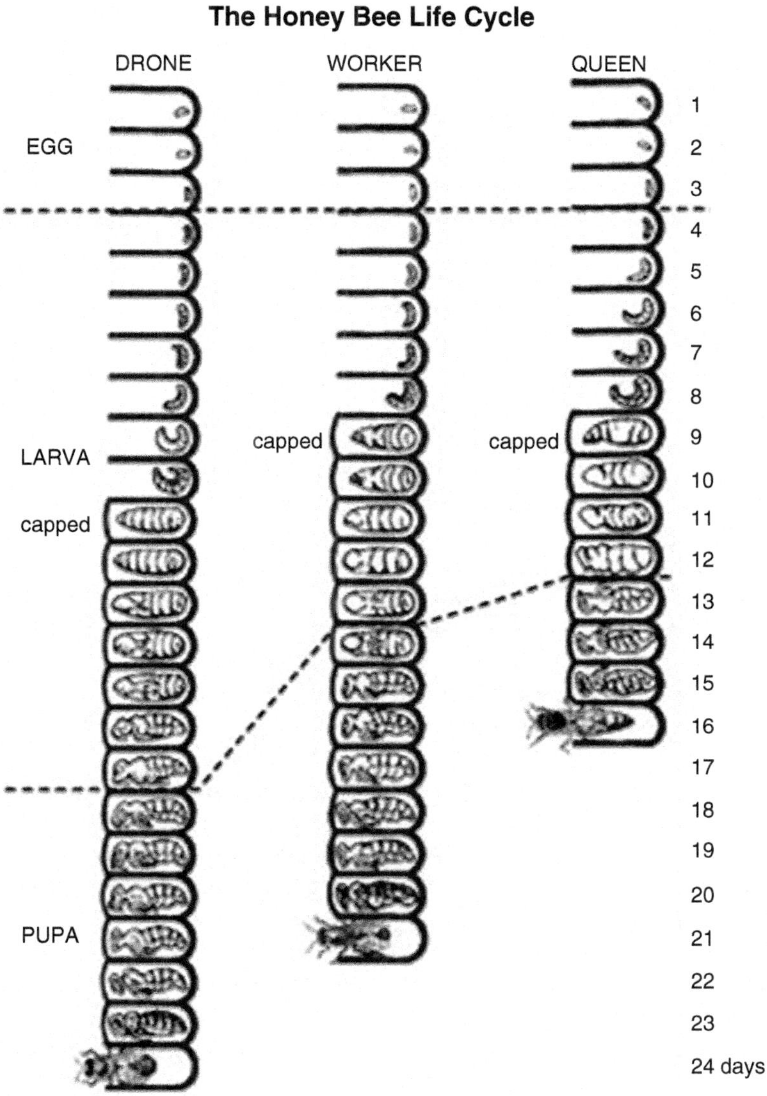

Fig. 2.13 Life cycle of honeybees (Source: https://www.pinterest.com)

tions which usually lead to the building of the queen cells are loss of queen, inability of queen to lay fertile eggs and too great numbers of bees or too little room in the hive, which is likely to be true in times of great honey secretion.

Mating Approximately, 5 days after emerging from her cell, if the day is pleasant, the virgin queen begins to take a series of mating flights, and otherwise she will wait for the first pleasant day for this purpose. She takes a number of such flights over a period of 2–3 days and may mate with ten or more different drones. The sperm is stored in a special organ, the spermatheca, and the queen never mates again after this period.

If the queen is observed upon her return from her wedding tour, it may be easily determined whether she has been successful. If she has been successful, she will bear the organ of the drone suspended to her body. If the queen lays any eggs before meeting the drone, or if for any cause she fails to meet the drone, the eggs will only produce drone bees.

Egg Laying and Behaviour About 5 days after taking her mating flights, the queen begins to lay eggs. During favourable periods, a good queen can lay more than 1500 eggs per day. Factors which affect egg laying are the weather, the nectar and pollen flows, the size of the queen and the condition of the colony. The number of eggs laid varies with the annual cycle as available resources of nectar and pollen vary. Large amounts of incoming resources stimulate workers to give the queen more food, which in turn stimulates her to lay more eggs.

Several of the queen's glands produce a complex of compounds called the queen substance. It is distributed throughout the colony by workers that care for the queen and pass it on to other workers. The queen substance is a combination of pheromones, chemical compounds that serve to control the behaviour of other individuals of the same species. Pheromones produced by the queen and by the other individuals of the colony serve to harmonize colony behaviour.

Number and Function Normally, there is only one queen per colony, though sometimes two queens are present when the old queen is being superseded. The function of the queen is simply to lay eggs and thus keep the colony populous; and this she does with an energy that is startling. A good queen in her best state will lay 1500 eggs a day.

The queen controls the sex of her offspring. When an egg passes from her ovary to her oviduct, the queen determines whether the egg is fertilized with sperm from the spermatheca. A fertilized egg develops into a female honeybee, either worker or queen, and an unfertilized egg becomes a male honeybee or drone.

Longevity The queen, when considered in relation to the other inhabitants of the colony, possesses a surprising longevity. It can live for up to 4 years, but in the tropics, where the yearly laying period is longer, the queen does not live as long. Depending upon their vigour and excellence, older queens either cease to be fertile or else become impotent to lay drone eggs, the spermatheca having become emptied of the seminal fluid. In such cases, the workers usually supersede the queen, that is, they destroy the old queen and start queen cells for the purpose of rearing young, fertile and vigorous queens.

Older queens do not have the laying capacity of younger queens; therefore, young, vigorous queens are preferred by beekeepers in intensive beekeeping; queens are replaced after every year.

Worker Bee Workers are females that are not fully developed sexually. They do the work of the colony and maintain it in good condition. Workers have special structures and organs which are associated with the duties they perform.

Morphological Features They are the smallest members of the colony, measuring a little more than one-half inch in length. They also possess peculiarities of structure that at once distinguish them from both the queen and drones. Their tongues are almost twice as long as in either the drone or queen, their jaws are much stronger and their wings attain the extremity of the body. The last joints of the posterior legs, known as the tibia and tarsi, are hollowed out to form pollen baskets. The presence of pollen baskets differs them from both the drones and the queen. The eyes are smaller than those of the drone but do not differ with the eyes of the queen.

The workers also possess a natural weapon of defence, the sting, which they use when the occasion requires, and usually die after stinging. The mechanism of this organ is very interesting. At its base is a double gland, which secretes the poison which, when secreted, is poured into an ample poison sack, which is as large as a flax seed. The sting proper is an organ, consisting of three sharp spears, very smooth and of exquisite polish, which lie side by side and make up the sting as seen by the naked eye. The central lancet is hollow, a little shorter than the others are. The central opening connects with the poison sack, so that all the poison passes through this part of the sting. The sidepieces are marvellously sharp, and each barbed at the end with teeth, of which seven are prominent and which extend out and back like the barb of a fishhook, so that the sting cannot be withdrawn when once fairly used and with its loss the bee's life is sacrificed. These sidepieces are worked alternately by small muscles at the base of the sting, and when fairly inserted the poison is intruded through the central piece. The workers also possess a honey stomach or crop, in which honey is carried to the hive.

Cell Shape Workers are reared in the same type of cell that is used to store honey and pollen. This type of cell makes up the majority of the comb in the colony. The capping on sealed worker cells is opaque and flat.

The size of the cells of naturally built (i.e. without embossed foundation) worker comb is useful for distinguishing between species and some races of *Apis* commonly kept in hives. The distance across ten cells of comb built by the eastern hive bee (*A. cerana*) in the Philippines averages 4.1 cm, and in southern India, the distance is 4.3–4.4 cm. The African races of the western hive bee build comb with measurements of 4.7–4.9 cm across ten cells, while the distance in comb constructed by common European races is 5.2–5.6 cm.

Feeding Worker bees are raised in the multipurpose, horizontally arranged cells of the comb. Future workers receive royal jelly only during the first 3 days, compared to future queens, who are fed royal jelly throughout their larval life. This difference accounts for the great variation in anatomy and function between adult workers and queens.

After the third day, worker larvae are gradually switched to a progressive feeding scheme where they are fed with a mixture of royal jelly, honey and pollen. With progressive feeding, the larvae are fed periodically; thus, food is not always available to them.

Pollen is used to feed older brood and is eaten in large quantities by nurse bees that are producing royal jelly from the head glands. It is the protein, vitamin and mineral component in the bee diet.

Developmental Period The adult worker emerges from the cell 21 days after the egg is laid. The workers always hatch from an impregnated egg, which can only come from a fertile queen and is always laid in the small horizontal cells. The eggs are in the form of a short, slightly curved cylinder and are fastened by one end to the bottom of the cell. They can be easily seen by holding the comb so that the light will shine into the cells. The eggs hatch in about 4 days.

The larva is white and footless and lies coiled up, floating in a whitish fluid previously placed in the cell. This food is composed of pollen and honey and is all consumed by the larva. In about 5 days, the cell is capped over by the bees. The cap is composed of pollen and wax, so that it is darker, more porous and more easily broken than the honey caps. It is also more convex. The larva, now full grown, commences to surround itself with a thin cocoon made of fine silk and in 3 days assumes the pupa state, when it is called a nymph. It now looks like the imago of fully developed bee, except that the legs, wings and tongue are folded on the breast, and the insect is now colourless. Upon the 21st day, the bee emerges from the cell.

Number and Function These are the most numerous individuals of the hive that make up about 98 % of the colony. Their number varies from 20,000 to 40,000 in every good colony. In a strong colony, maximum number of bees present on one side of the hive is 2000.

The tasks that the adult workers perform change with their age (Table 2.1). This is correlated with the physiological development of various glands. The function of the worker bees is to do all the manual labour of the hive. They secrete the wax, which forms in small pellets beneath the abdomen, build the comb, feed the young larvae and cap the cells, whether they may be brood or honey cells. This work is done by the younger bees. The older bees gather the honey; collect the pollen (beebread); bring in the propolis (bee glue), which is used to close up openings as a cement; supply the hive with water; defend the hive from all improper intrusion; destroy drones when their day of grace is past; kill and arrange for replacing

Table 2.1 Schedule of a worker bee in the hive

Days after emergence	Task
1–2	Clean cells and warm the brood nest
3–5	Feed older larvae with honey and pollen
6–10	Feed younger larvae with products of the head glands
11–18	Ripen nectar, produce wax and construct comb
19–21	Guard and ventilate the hive, take exercise and orientation flights to learn to fly and locate the hive
22+	Forage for nectar, pollen, water or propolis

worthless queens; and lead forth a portion of the bees when the conditions impel them to swarm. Pollen is collected just above the brood by all the honeybee species followed by honey on top of the frame. However, this scheme of work division is not absolutely fixed; workers can change tasks to meet the needs of the colony.

Longevity The life span of worker adults varies greatly with the time of year, but they never attain a great age. During periods when the colony is relatively inactive (dearth periods), workers may live 3 months or more, but when the colony is active, few workers live for as long as 6 weeks. During these active periods, about 3 weeks are spent as a hive bee and the remainder as a forager. In some cases, those reared in autumn may live for 9 months, while those bred in spring will wear out in 3 months. The life span of workers of tropical races of the western hive bee and the eastern hive bee is shorter. Their longevity depends upon their activity and hence upon the time of year in which they live.

Drone Bee Drones, the males of the colony, are produced from unfertilized eggs. The queen can control whether or not the egg is fertilized as she lays it. The body of the drone is larger than that of the worker or queen. The eyes are large and cover practically the whole head. The end of the abdomen is blunt and is covered with a tuft of small hairs. Drones cannot sting. As the sting is a modified structure of the female genitalia, drones do not have stings. They also do not have any of the structures necessary to collect nectar and pollen.

Morphological Features The male bees are only found in the hive from October until March, when there will be a few hundred, though the number may be controlled by the apiarist and should be greatly reduced. These are longer than the workers, being nearly of an inch in length and more bulky than either the queen or workers. Their flight is heavy, and they may be known by their deep, low hum. Their tongue is short, jaws weak and their posterior portion destitute of pollen baskets. The eyes meet above and are very prominent. The drones, too, have no defence organ, the sting being absent.

Cell Shape Drones are reared in cells of the same shape as worker cells only larger. Drone cells are sealed with dome-shaped capping. The capping of the drone cells is very convex and protrudes beyond the worker and darker in colour. Both drone and worker brood cells are very readily distinguished from honey cells.

Feeding Worker bees mix the honey with pollen and feed drone larvae. Future drones receive royal jelly for the first 3 days. After that, they are shifted to progressive feeding as discussed in worker feeding.

Developmental Period The developmental period of drones is 23 days. The male bees come from unfertilized eggs. These eggs may come from an unfertilized queen, from a fertile worker or from a fertilized queen, which may voluntarily prevent fertilization. Such eggs are placed in the larger horizontal cells in the same manner as the worker eggs are placed in the smaller cells. The development of the drones from egg to larva, to pupa and to imago is essentially like that of the workers, though they do not come forth till the 24th day from the laying of the egg. The difference of temperature and other conditions may slightly advance or retard the development of any brood in the different stages. The drones, like other bee castes, appear as grey, soft and unsophisticated when they first emerge from the cells.

Number and Function A strong colony can have about 300 drones. The longevity of the male bee is about 6 months. However, during periods when resources are scarce, the workers run the drones out of the colony. They die as they cannot fend for themselves. The sole function of the drones is to fertilize the queen. The mating of honeybees takes place in the air, away from the colony. When the weather is good (warm shiny day), mature drones leave the colony during the afternoon and congregate in certain areas where they wait for virgin queens to fly. After mating, the drone organs adhere to the queen, and their abstraction is fatal to the life of the drone. They die after mating. They have no sting, do not carry pollen and are unable to produce wax.

Other drones sometimes return to colonies that have a virgin queen. Such colonies will accept drones from other colonies and will tolerate a large drone population while the queen is a virgin. However, after a queen mates, the workers run many of the drones out of the colony.

The following table (Table 2.2) summarizes the developmental periods, starting at the time the egg is laid. Depending on the type (species and/or race) of honeybee, the weather conditions or the time of year, the figures may vary a day or so.

Table 2.2 Developmental periods of honeybee castes

Developmental period (days)	Worker	Queen	Drone
Egg hatches after	3	3	3
Cell is sealed after	9	8	10
Adult emerges after	21	16	24

Table 2.3 Organs, their location and function in worker bee

Structure of organ	Location	Function
Head glands	Front of the head	Produce brood food and royal jelly
Wax glands	Under the abdomen	Produce wax
Odour glands	Near the upper tip of the abdomen	Produce scent to orientate bees when the colony is disturbed
Sting and associated glands	Tip of the abdomen	Defend the colony
Long tongue	Head	Gathers nectar
Honey stomach	Enlarged area of the oesophagus	Carries nectar and water
Pollen comb, press and basket	Hind legs	Comb pollen from the body, press it into pellets and carry it to hive. Also used to carry propolis

Worker bee possesses different organs/glands to perform various functions. These are summarized in the following (Table 2.3).

Difference Between Eggs of Normal and Laying Workers When a colony has become queenless and there are no young larvae or female eggs from which to rear a new queen, laying workers can develop. The ovaries of some workers in the colony develop because of the absence of queen substance, and they start to lay eggs. Since workers do not have the body structure or behaviour necessary to be fertilized, all of the eggs are unfertilized and thus produce drones. Laying workers can be suspected in a colony if there are an excessive number of drones present. Close examination of brood comb can verify this. Worker cells that contain drone brood (i.e. worker cells capped with a domed cap) and cells that contain a number of eggs of varying sizes laid in a haphazard fashion confirm the presence of laying workers. A good queen lays only one egg per cell which is placed in the centre of the base of the cell.

2.4 Conclusions

Honeybees are one of the most fascinating organisms across phyla. Their usefulness to the human race can in no way be overestimated. Other than contributing directly to human well-being by producing a wide variety of products, honeybees are also responsible for enhancing the agricultural production through their pollination services. Honeybees are also a model system for understanding social behaviour, communication as well as extreme industriousness.

References

Abrol DP (2010) Foraging behaviour of *Apis florea* F., an important pollinator of *Allium cepa* L. J. Apic Resi 49(4):318–325

Ahmad R (1989) A note on the migration of *Apis dorsata* in the Andaman and Nicobar Islands. Bee Wrld 70:62–65

Arias MC, Sheppard WS (1996) Molecular phylogenetics of Honeybee subspecies (*Apis mellifera* L.) inferred from mitochondrial DNA sequence. Mol Phylogenet Evol 5:557–566

Arias MC, Sheppard WS (2005) Phylogenetic relationships of honey bees Hymenoptera: *Apinae: Apini* inferred from nuclear and mitochondrial DNA sequence data. Mol Phylogenet Evol 37:25–35

Beeson CFC (1941) Book: The ecology and control of the forest insects of India and the neighbouring countries

Bluthgen N, Klein AM (2011) Functional complementarity and specialization: the role of biodiversity in plant–pollinator interactions. Basic Appl Ecol 12:282–291

Deodikar GB, Ghatge AL, Phadke RP, Mahindre DB, Kshirsagar KK, Muvel KS, Thakar SS (1977) Nesting behaviour of Indian honeybees III. Nesting behaviour of *Apis dorsata* Fab. Indian Bee J 39:1–12

Dukas R, Visscher PK (1994) Lifetime learning by foraging honey bees. Animal Behaviour 48:1007–1012

Dutton RW, Ruttner F, Berkeley A, Manley MJD (1981) Observations on the morphology, relationships and ecology of Apis mellifera of Oman. J Apic Res 20:201–214

Dyer FC, Seeley TD (1994) Colony migration in the tropical honeybee, *Apis dorsata* F. (Hymenoptera: Apidae). Insect Soc 41:129–140

Elekonich MM, Schulz DJ, Bloch G, Robinson GE (2001) Juvenile hormone levels in honey bee (*Apis mellifera* L.) foragers: foraging experience and diurnal variation. J Insect Physiol 47:1119–1125

Engel MS (1999) The taxonomy of recent and fossil honeybees (Hymenoptera: Apidae; Apis). J Hymen Res 8(2):165–196

Engel MS (2004) Geological history of the bees (Hymenoptera: Apoidea). Revista de Tecnologia e Ambiente 10(2):9–33

Fletcher L (1952) Bee Craft 34(9):139–140

Franck P, Garnery L, Loiseau A, Oldroyd BP, Hepburn HR, Solignac M et al (2001) Genetic diversity of the honeybee in Africa: microsatellite and mitochondrial data. Heredity 86:420–430

Garnery L, Cornuet JM, Solignac M (1992) Evolutionary history of the Honeybee *Apis mellifera* inferred from mitochondrial DNA analysis. Mol Ecol 1:145–154

Gupta RK (2014) Taxonomy and distribution of different honeybee species (Eds) Beekeeping for Poverty Alleviation and Livelihood Security, pp 63–101

Hepburn HR, Radloff SE (2011) Honeybees of Asia. Springer, Berlin, xii + 669 pp

Hepburn HR, Smith DR, Radloff SE, Otis GW (2001) Infraspecific categories of *Apis cerana*: morphometric, allozymal and mtDNA diversity. Apidologie 32:3–23

Kahono S, Nakamura K, Amir M (1999) Seasonal migration and colony behavior of the tropical honeybee *Apis dorsata* F. (Hymenoptera: Apidae). Treubia 31:283–297

Kakkar KL, Sharma PL (1991) Studies on the role of honeybee, *Apis cerana indica* F. in the pollination of cauliflower, *Brassica oleracea var. botrytis*. Indian J Ent 53:66–69

Kapil RP (1956) Variation in the biometric characters of the Indian honeybee *Apis cerana indica*. Indian J Ent 28:440–457

Kerr WE (1957) Introdução de abelhas africanas no Brasil. Brasil Apicola 3:211–213

Koeniger N (1976) Interspecific competition between *Apis florea* and *Apis mellifera* in the tropics. Bee Wrld 57:110–112

Koeniger N, Koeniger G (1980) Observations and experiments on migration and dance communication of *Apis dorsata* in Sri Lanka. J Apic Res 19:21–34

Lipinski Z (2001) Essence and mechanism of nest abandonment by honeybees swarms. Blenam, Olsztyn, p 291

Liu F, Roubik DW, He D, Li J (2007) Old comb for nesting site recognition by *Apis dorsata* Field experiments in China. J Insect Sci 54:424–426

Maa T (1953) An inquiry into the systematics of the tribus Apidini or honeybees (Hyn.). Treub 21(3):525–640

Manjunath MG (2008) Studies on nesting behaviour of giant honeybee, *Apis dorsata* F. in Mysore, Karnataka. M. Phil, Dissertation, pp 1–103

Miguel I, Baylac M, Iriondo M, Manzano C, Garnery L, Estonba A (2011) Both geometric morphometric and microsatellite data consistently support the differentiation of the *Apis mellifera* evolutionary branch. Apidol 42:150–161

Moritz RFA, Hartel S, Neumann P (2005) Global invasions of the western honeybee (*Apis mellifera*) and the consequences for biodiversity. Eco Sci 12(3):289–301

Narayanan ES, Sharma PL, Phadke KG (1960a) Studies on biometry of the Indian bees III. Tongue length and number of hooks on the hind wings of *Apis cerana indica* F. collected from Madras State. Indian Bee J. 23:3–9

Narayanan ES, Sharma PL, Phadke KG (1960b) Studies on biometry of the Indian bees. 1. Tongue length and number of hooks on the hind wings of *Apis cerana* indica F. Indian Bee J 22:81–88

Neumann P, Koeniger N, Koeniger G, Tingek S, Kryger P, Moritz RFA (2000) Home-site fidelity in migratory bees. Nature 406:474–475

Oldroyd BP, Wongsiri S (2006) Asian honey bees: biology, conservation, and human interactions. Harvard University Press, Cambridge, MA

Oldroyd BP, Osborne KE, Mardan M (2000) Colony relatedness in aggregations of *Apis dorsata* Fabricius (*Hymenoptera, Apidae*). Insects Soc. 47:94–95

Paar J, Oldroyd BP, Kastberger G (2000) Giant honeybees return to their nest sites. Nature 406:475

Pant NC (1985) Bee keeping. In: Handbook in animal husbandry. ICAR, New Delhi, pp 692–711

Pudasaini R, Bahadur TR (2014) Foraging behavior of different honeybee species under natural condition in Chitwan, Nepal. Eur J Acad. Essays 1(9):39–41

Reddy CC (1983) Studies on the nesting behaviour of *Apis dorsata* F. In: Proceedings of the 2nd international conference on apiculture in tropical climates, New Delhi, pp 391–397

Ruttner F (1988) Biogeography and Taxonomy of Honeybees. Springer, Berlin/Heidelberg, p. 284

Ruttner F (1992) Naturgeschichte der Honigbienen. Ehrenwirth, Munich, 357

Ruttner F, Kauhausen D (1985) Honeybees of tropical Africa: ecological diversification and isolation. In: Proceedings of the third international conference on apiculture in tropical climates, Nairobi, Kenya, 5–9 November, 1984, pp 45–51

Ruttner F, Tassencourt L, Louveaux J (1978) Biometrical-statistical analysis of the geographic variability of *Apis mellifera* L. Apidol 9:363–381

Ruttner F, Woyke J, Koeniger N (1972) Reproduction in *Apis cerana*: Mating Behaviour. J Apic Res 11(3):141–146

Seeley TD (1985) Honeybee ecology. Princeton University Press, Princeton, p. 201

Selvakumar P, Sinha SN, Pandita VK, Srivastava RM (2015) Foraging behaviour of honeybee on parental lines of hybrid cauliflower pusa hybrid- 2. Standing Commission of Pollination and Bee Flora. http://apimondiafoundation.org/foundation/files/070.pdf

Sharma HFA (2014) The foraging behaviour of honey bees, *Apis mellifera*: a review. Veterinarni Medicina 59(1):1–10

Sharma AK, Dhaliwal HS, Kakar KL (1974) Insect visitors and pollinators of cauliflower (*Brassica oleracea var. botrytis*) seed crop bloom Himachal. J Agril Res 2:74–78

Sheppard WS (1989) A history of the introduction of Honeybee races into the United States. Am Bee J 129:617–619

Sheppard WS, Meixner MD (2003) *Apis mellifera pomonella*, a new honey bee subspecies from Central Asia. Apidologie 34:367–375

Sheppard WS, Arias MC, Meixner MD, Grech A (1997) *Apis mellifera ruttneri*, a new honey bee subspecies from Malta. Apidologie 28:287–293

Sheppard WS, Meixner MD, Hepparda WSS, Eixnera MDM (2003) *Apis mellifera pomonella*, a new Honeybee subspecies from Central Asia. Apidol 34:376–375

Singh S (1962) Beekeeping in India: 1–214. Indian Council of Agriculture Research, New Delhi

Singh MM (2008) Foraging behaviour of the Himalayan honeybee (*Apis cerana* F.) on flowers of *Fagopyrum esculentum* M. and its impact on grain quality and yield. Ecoprint 15:37–46

Sinha SN, Chakrabarti AK (1980) Bee pollination and its impact on cauliflower seed production. In: Proceedings of the second international conference on apiculture in tropical climates, New Delhi, Feb 29th–Mar 4th, 1980, pp 513–527

Tan NQ (2007) Biology of *Apis dorsata* in Vietnam. Apidol 38:221–229

Thakur ML (1991) Honey and the Honeybees. Indian Council of Forestry Research and Education, Dehradun

Thapa R, Wongsiri S, Oldroyd BP, Prawan S (2000). Migration of *Apis dorsata* in northern Thailand. In: Proceedings of the 4th Asian Apiculture Association conference, Kathmandu, pp 39–43

Tingek A, Mardan MB, Rinderer TE, Koeniger N, Koeniger G (1988) Rediscovery of *Apis vechti* (Maa, 1953): the Saban honeybee. Apidologie 19:97–102

Underwood BA (1990) Seasonal nesting cycle and migration patterns of the Himalayan honeybee *Apis laboriosa*. Natl Geogr Res 6:276–290

Vaudo AD, Ellis JD, Cambray GA, Hills M (2012) Honey bee (*Apis mellifera capensis/A. M. scutellata* hybrid) nesting behavior in the Eastern Cape, South Africa. J South Africa 59(3):323–331

Venkatesh G, Reddy CC (1989) Rates of swarming and absconding in the giant honey bee, *Apis dorsata* F. Proc Indian Acad Sci 98:425–430

Wealth of India (1988) Honeybees, 75–101pp

Woyke J, Wilde J, Wilde M (2000) Swarming, migrating and absconding of *Apis dorsata* colonies. In: Proceeding of the 7th international conference on tropical bees, management and diversity and 5th Asian Apicultural Association Conference, 19–25 March 2000, Chiang Mai, Thailand

Woyke J, Wilde J, Wilde M (2001). Swarming, migration and absconding of *Apis dorsata* colonies. In: Proceedings of the 7th international conference on apiculture in tropical climates, Chiang Mai, pp 183–188

Beekeeping in India

3

M. Kishan Tej, R. Aruna, Geetanjali Mishra, and M.R. Srinivasan

Contents

3.1 Introduction

Beekeeping is an art and a mesmerizing science. In India beekeeping is mostly practised as a full-time occupation and an engrossing hobby to produce handsome income and table honey. Honeybees are special gift to mankind because beekeeping can be done for both their pollination services and their cherished products such as honey, beeswax, propolis, bee venom, etc. These products have their widespread use in different small and large scale industries in India. The only bitter part of beekeeping is the bee sting. Honeybees sting to defend their colony, but this bitterness

M. Kishan Tej • R. Aruna • M.R. Srinivasan (✉)
Department of Agricultural Entomology, Tamil Nadu Agricultural University,
Coimbatore 641003, Tamil Nadu, India
e-mail: mrsrini@tnau.ac.in; mrsrini@gmail.com

G. Mishra
Department of Zoology, University of Lucknow, Lucknow 226 007, Uttar Pradesh, India

Omkar (ed.), *Industrial Entomology*, DOI 10.1007/978-981-10-3304-9_3

will be only in the initial stage of beekeeping, and after one gets habituated to keep bees, he will only taste the sweetness of honey. Most beekeepers develop a tolerance for bee venom over time and have reduced sensitivity to pain and swelling. So understanding honey bee science is to know and unravel nature's most industrious as well as most fascinating insects.

As of now seven species of *Apis* have been described; India is an exclusive country which habitats four of these; two domesticated species, viz. *Apis cerana* (oriental honeybee) and *A. mellifera* (occidental or European honeybee) and two wild species, viz. *Apis dorsata* (giant/rock honeybee or dumna) and *A. florea* (dwarf honeybee). Among the four species, *A. mellifera* is an introduced species to India because it is resistant to Thai sacbrood virus (TSBV) and also highly suitable for commercial beekeeping.

Because of the different climatic zones in India, there is a massive multiplicity of flora which helps in potential beekeeping. People of India have a long connection with beekeeping and honey since ancient times. Ancient Indians gifted some records about beekeeping as paintings or carvings on rocks. Honey and its medicinal uses were mentioned in the old Ayurveda books of India. After independence, the government of India took policy decision to revive various traditional village industries and an All India Khadi and Village Industries Board (KVIB) was formed in 1954. Through harmonized efforts of well-joined organizations like KVIC (Khadi Village Industries Commission) and State KVIBs, Beekeepers' Cooperatives and Public Institutions, the beekeeping industry came into limelight of village industries in India within two decades. In view of the budding importance of beekeeping, in 1981, an All India Coordinated Research Project (AICRP) on Honey bee Research and Training was launched by ICAR involving Agricultural Universities (Ramchandra et al. 2012; Sivaram 2012). Later a Central Sector Scheme entitled "Development of beekeeping for improving crop productivity" was launched by the Ministry of Agriculture in 1994–1995 during the eighth 5-year plan. The scheme targets production and distribution of honeybee colonies, organizing trainings and awareness programmes. A Beekeeping Development Board also worked to organize the beekeeping activities. The scheme was approved for continuation during the ninth 5-year plan. However, the scheme got incorporated under the Macro Management Scheme. Right now approximately there are about 1.5 million bee colonies in India, which produce 55,000 tonnes of honey annually. India is one of the honey-exporting countries. The major markets for Indian honey are Germany, the USA, the UK, Japan, France, Italy and Spain.

3.2 Honeybee Species in India

3.2.1 Rock Bee *(Apis dorsata)*

They are huge and ferocious bees that construct a single comb in the open usually about 3-4 feet tall. They can be seen all over the subcontinent mainly in the forests and also in concrete jungles. In hilly regions they construct their nest up to an altitude of 2700 m. Rock bees habitually shift their places. Nearly 50–80 kg of honey can be squeezed from a single colony of rock bee per year (Mishra 1995). They occur from Pakistan (and, perhaps, parts of southern Afghanistan) in the west (Crane 2004), through the Indian subcontinent and Sri Lanka to Indonesia and parts of the Philippines in the east. Its north-south distribution ranges from the southern part of China to Indonesia; it is found neither in New Guinea nor in Australia. The giant honeybees of Nepal and the Himalayas have recently been reclassified as belonging to another species of *Apis*, as *A. dorsata laboriosa* (Akratanakul 1990). *Apis dorsata binghami* is another subspecies of *dorsata* distributed in restricted areas of the north-east, namely, in Khasi Hills, Sikkim and Meghalaya (Allen 1995; Otis 1996).

Mostly these bees construct their combs at a height of more than 20 ft from the ground, but in some cases, we can also see the colonies hanging from branches just above 2 ft from the ground. Colonies of *A. dorsata* may occur singly or in groups. The lower part of the comb is the energetic area in which the foraging and scout bees will take off and land. As these bees are aggressive, they will attack the intruders (Ramchandra et al. 2012), and every so often they will chase even up to 100 m. Sometimes these bee stings can turn fatal to the humans. Because of danger involved in harvesting rock bee honey, it is generally priced high locally. Some trained bee hunters prefer to work at night. Smoke is used to pacify the bees, and in some places professionals add chicken feathers to the smoke produced by burning charcoal which irritates the rock bees and causes them to move out owing to odour produced due to addition of chicken feathers, allowing easy honey extraction. There is general concern that the total number of *A. dorsata* nests all over Asia is declining, partly due to shrinking forest areas, the use of toxic pesticides in foraging farm lands and bee hunting.

3.2.2 Little Bee (*Apis florea*)

Apis florea or dwarf honeybee is also a wild honeybee spp., but these bees are small and less ferocious when compared to the rock bees. These bees build single vertical combs (Hepburn and Radloff 2011; Wongsiri et al. 1996). They also construct palm-sized combs in the bushes, hedges, buildings, caves, empty cases, etc. The major difference between the rock bee and little bee comb is that the little bees construct combs encircling the twigs while the rock bees construct the comb on the undersurface of the branch. The honey produced by these bees is dramatically less when compared to the rock bee as these bees produce only about half a kilo of honey per year per hive. However, in the Kutch area of Gujarat, large quantities of honey from *A. florea* are harvested (Soman and Chawda 1996). As these bees also have a habit of shifting their colonies frequently, they are also non-rearable, but attempts in India have brought partial success (Mishra 1995). These bees are found only in plains and not in hills above 450 MSL. Compared to other honeybees, these bees are attractively coloured with red to brown colouration having white bands. They are excellent pollinators, which give them an important ecological role in the places they inhabit. These bees are well known for their distinctive defensive behaviours and camouflage in dense forests. A more touching example is the specific behavioural response they exhibit against their chief predator *Oecophylla smaragdina* (weaver ant); when these ants are in close proximity, the bees produce adhesive barriers to obstruct the ant's path. *Apis florea* is also identified for their hissing sounds when they see a predator. This hissing sound is audible to human ear.

3.2.3 Indian Bee (*Apis cerana*)

Indian honeybee or Eastern honeybee is a well-known bee species in India. Prior to the introduction of Italian bee, this was the only rearable *Apis* bee spp. in India. It is also found and has been domesticated in Pakistan, Nepal, Burma, Bangladesh, Sri Lanka and Thailand. These are comparatively non-aggressive and rarely shift locations. These bees construct multiple parallel combs in dark places such as clay pots, logs, wall, tree openings, etc. and produce 7–9 kg of honey per colony per year. Ruttner (1988) classified *Apis cerana* into subspecies based on the living habitats and genetic diversity; of these *Apis cerana indica* and *Apis cerana cerana* occur in India. In India, the subspecies *Apis cerana indica* is recognized into two morphotypes like "hills bee" (black coloured) and plains bee (yellow coloured) (Ramchandra et al. 2012). Presently beekeeping with Indian bees is mostly done in south India and particularly in Kanyakumari district of Tamil Nadu, with more than 50,000 beekeepers involved.

Since these bees have built their colonies in dark cavities, it enables man to keep them in specially constructed movable frame hives. The combs of *A. cerana* colony are built parallel to each other and at uniform distance known as the "bee space", which is respected between them. Compared to rock bees and Italian bees, these are small in size but bigger than the dwarf bees. Brood comb consists of cells of two sizes: smaller for the worker brood and larger for the drone brood. The queen cells are built on the lower edge of the comb. Like other bee species, these bees also store honey in the upper part of their hive. Because of this behaviour, the bee boxes are designed in such a way that the super chamber or the honey chamber is in the upper part of the hive where these bees store honey which helps in easy honey extraction.

3.2.4 European Bee/Italian Bee (*Apis mellifera ligustica*)

Italian bee (*Apis mellifera ligustica*) is one of the sub species of *A.mellifera* and is not native to India and was introduced from Europe during the second half of 20th century. The introduction was primarily because the native Indian bee colonies were vanishing because of the Thai sacbrood virus. Presently they are well established in India and mostly present in northern India because of the rich flora such as mustard, safflower, sun flower, etc. As rice is the major crop in south India, these bees don't get enough amount of food they need. In south India, beekeeping with Italian bees is hardly practised; for commercial beekeeping, these Italian bees have to be migrated by floral mapping. They are also similar in habits to Indian bees, which build parallel combs in dark places and store honey at the upper portion of their colony (Akratanakul 1976; Maa 1953; Otis 1990; Tirgari 1971). They are bigger than all other honeybees except *Apis dorsata*. They produce 25–40 kg of honey per colony per year. Probably these bees are the one of the most studied animals. The introduction of *A. mellifera* to India created problems such as the interspecies transmission of bee pests and diseases. But the introduction of these bees to India can be recorded as success story because it created employment for many people in India with profitable income and also by the pollination service these bees done to Indian flora.

3.2.5 Stingless Bee

Stingless or dammar bees are of smallest size compared to other honey-yielding bees (less than 5 mm). They belong to the family Apidae and subfamily Meliponinae. It consists of two genera *Melipona* and *Trigona*. Meliponinae includes eight genera, having 15 subgenera and more than 500 species (Wille 1983). These bees are widely known as dammar bees in India (Rasmussen 2013) (dammar is a resin from among dipterocarp trees) with additional local names commonly applied, e.g. "putka" in Sikkim and Nepal (Gurung et al. 2003; Singh et al. 2011; Lepcha et al. 2012); "ngap siwor", "ngap hamang" and "ngap khyndew" in Khasi language (Pugh 1947); and "cherutheneecha" and "arakki" in Kerala (Nair 2003). As the name implies, these bees can't sting as their stingers are highly reduced, but they try to defend their colony from intruders using their mandibles (Michener 2000). The stingless bees are important pollinators of various food crops and can be domesticated. But the honey yield per hive per year is very low approximately 100 g. As in other regions where stingless bees occur, colonies can be kept in tree logs, wooden boxes and clay pots for harvesting small quantities of highly prized medicinal honey, wax and propolis, used for its household and curative properties. The materials used for nest building are mainly pure wax or cerumen, a mixture of wax and propolis, resins, plant fibres and clay (Rasmussen and Camargo 2008).

3.3 Biology and Society

Honeybees are one of the most brilliant products of nature. One of the most superior characters which honeybees demonstrate is eusociality in which they take care of their young ones with cooperative brood care and have other advanced ways of communications and defensive mechanisms. Honeybees have three developmental stages (egg, larvae and pupa) and an adult stage. In adult stage there are three castes (single queen, hundreds of drones and thousands of workers). The queen is a fertile, functional female that can produce males and females, the worker is an unfertilized female capable of only producing males (due to the haplodiploid sex determination system found in honey bees) and the drone is male (Tribe and Fletcher 1977; Winston 1979). The food they are fed during larval stage decides their caste; queen larvae is fed with royal jelly by nurse bees throughout its larval period. Recently, Kamakura (2011) found that a 57 kDa protein royalactin, present in the royal jelly, is a reason for the larvae to become queen.

3.3.1 The Queen

Queen bee is the mother of all other bees in the colony. It can be identified with its long abdomen and short wings. The duty of the queen is to lay eggs. The queen maintains the colony by its pheromones. Her productivity depends on the amount of food the workers bring in and the amount of brood space in the colony. She can lay more than 1500 eggs a day. If it is a honey flow season and if there are enough cells available, she will lay up to 2500 a day (Winston 1992). Queen emerges from queen cell which is situated at the bottom portion of the comb and looks like a small cup, and in India it is famously known as cow's teat because of its structural resemblance. After emergence the newly emerged queen destroys the remaining queen cells in the colony and fights any other queens she finds. The virgin queen will typically stay in the colony for a few days in order to feed and gain strength and allow her reproductive organs to mature a little further (Mackensen 1943; Winston 1992; Woyke 1963, 1969, 1973). After 6–8 days, the queen will leave the colony for her nuptial flight, which occurs 30 m above the ground where she mates with many strong drones who can fly with the queen as she flies better than drones. Postmating, the queen returns to the hive to lay the eggs (Mackensen 1943; Winston 1987, 1992). Average life span of queen is about 5 years, but the egg laying capacity will be only up to 2 years.

3.3.2 The Worker

There are thousands of workers in a colony, and they perform all the duties in the colony including foraging, defending, brood rearing and cleaning activities. They are smaller than the queen and drones. There are about 8000–25,000 workers in *A. florea* colony, 40,000–50,000 workers in *A. mellifera* colony, 20,000–40,000 workers in *A. cerana* and 50,000–80,000 in *A. dorsata* colony (Winston 1992; Wongsiri et al. Wongsiri et al. 1991; Wongsiri et al. 1996). For defending the colonies, worker bees possess sting which is a modified ovipositor, and venom is pumped out at the time of stinging. Workers may lay eggs, under certain conditions, which develop into drones since workers never mate and they have no sperm to fertilize their eggs (Anderson 1963; Mackensen 1943). However, in a normal queen right colony, worker regulation occurs and workers consume eggs produced by other workers (Ratnieks 1993). In *A. cerana*, unlike *A. mellifera*, there can be a relatively large number of egg laying workers (Partap and Verma 1998). Workers at their young stages perform indoor duties, and they will get license to go out for foraging only when they are old enough (normally after 21 days) (Winston and Fergusson 1985).

3.3.3 The Drone

Drones can be easily identified by their dark colour and eyes touching at the top of their head. Their only function is to fertilize the queen and enjoy the food inside the hive. They do not sting as they lack stingers. Drones are "haploid", and they only possess one-half of the pairs of genes found in the "diploid" workers and queen (Anderson 1963; Mackensen 1943; Winston 1992). In a colony there are about hundreds of drones, drone cells differ from worker cells with enlarged cappings, and in India commercial beekeepers decap these drone cells as drones consume the stored honey. They have excellent navigation abilities when compared to the other two castes because they have around 75–80 % more facets in their compound eyes than the workers or queen (Gary 1963; Koeniger 1969, 1970; Ruttner 1966; Winston 1992).

3.4 Honeybee Foraging

India is a vast country with different climatic zones providing rich flora for honeybees. By foraging, honeybees collect pollen and nectar where pollen is a protein source and nectar is carbohydrate source which together fulfils their nutrient requirements (Seeley, 1985 and Winston 1987). As the honeybees have division of labour, foraging will be only performed by the forager bees (Von Frisch 1967; Suwannapong 2000). Among the foraging bees, there are two types: nectar collectors and pollen collectors. For collecting the full load of nectar or pollen, they have to visit hundreds of flowers (Akratanakul 1976). In addition to these, the foragers also collect water (Farnesi et al. 2009; Marcucci 1995; Bankova et al. 1983, 2000) and propolis (plant resins) in case of Italian and stingless bees. Bees are the most effective pollinators of crops and natural flora and are reported to pollinate over 70 % of the world's cultivated crops. It has also been reported that about 15 % of the 100 principal crops are pollinated by domestic bees (Kenmore and Krell 1998; Abrol 2012). Table 3.1 lists some commercial crops benefitted by honeybee pollination in India.

As the honeybees in India have vast floral diversity, shortlisting of the flora is difficult, and these are some of the crops with rich source of pollen and nectar from which bees were benefitted.

Anacardium occidentale, *Nephelium litchi*, *Azadirachta indica*, *Callistemon citrinus*, *Glycine max*, *Cajanus cajan*, *Hevea brasiliensis*, *Acacia catechu*, *Dalbergia sissoo*, *Eucalyptus* sp., *Syzygium cumini* and *Nephelium litchi* are some of the rich nectar sources. *Zea mays*, *Psidium guajava*, *Sesamum indicum*, *Sorghum bicolor* and *Helianthus annuus* are some of the rich sources of pollen. *Citrus sinensis*, *Coriandrum sativum*, *Cucumis melo*, *Eucalyptus* spp., *Musa* spp. and *Pongamia pinnata* are some of the plants that provide high levels of pollen and nectar to bees. *Carica papaya*, *Cocos nucifera* and *Musa* spp. provide food source to bees throughout the year (Kishan et al. 2014).

3.4.1 Foraging Distances

Foraging distance of *A. cerana* is around 200–300 m from the hive (Pratap, 2011 and Koetz 2013). Some studies showed that the Indian bee can forage up to 900 m (Hyatt 2011). Maximum foraging range of *A. cerana* is 1500–2500 m (Dhaliwal and

Table 3.1 Some of the commercial crops benefitted by honeybee pollination in India

Fruits and nuts	Almond *Prunus dulcis*, apple *Malus* spp., apricot *Prunus armeniaca*, peach *Prunus persica*, strawberry *Fragaria* spp., citrus *Citrus* spp. and litchi *Litchi chinensis*
Vegetable and vegetable seed crops	Cabbage *Brassica oleracea* var. *capitata*, cauliflower *Brassica oleracea* var. *botrytis*, carrot *Daucus carota*, coriander *Coriandrum sativum*, cucumber *Cucumis sativus*, melon *Cucumis melo*, onion *Allium cepa*, pumpkin *Cucurbita* spp., radish *Raphanus sativus* and turnip *Brassica rapa* subsp. *Rapa*
Oilseed crops	Sunflower *Helianthus annuus*, niger *Guizotia abyssinica*, rapeseed *Brassica napus*, mustard *Brassica juncea*, safflower *Carthamus tinctorius*, gingelly *Sesamum indicum*
Forage seed crops	Lucerne *Medicago sativa* and clover *Trifolium* spp. (Ragumoorthi et al. 2007)

Sharma 1974). *Apis mellifera* have a great foraging range and can go even up to 10 km (Abrol 2011). But most of the foraging range of *A. mellifera* is below 6 km (Visscher and Seeley 1982).

3.5 Beekeeping Equipment

As beekeeping has changed over the centuries, the related equipment has also changed. Traditionally beekeepers in India used to practice beekeeping in baskets, wooden logs, underground beehive, clay pots for keeping stingless bees, etc., but the Langstroth bee space (1851), Johannes comb foundation (1857) and honey extraction techniques by Frang von Hruschka concepts had a great impact on beekeeping in India which made a dramatic change and urged the beekeepers of the subcontinent to switch over to movable frames, as beekeeping with movable frames is user-friendly and also the modern beekeeping equipments make work easy for commercial handlers. (Mishra 1995; Ramchandra et al. 2012; Singh 2014).

3.5.1 Honeybee Hive

A beehive is a place in which a single colony of honeybee exists containing and performing various functions for their livelihood; it contains various parts like hive stand, bottom board, brood chamber, super chamber, inner and outer cover. The hive stand consists of a wooden pole or iron stand fixed to the ground; it may be of single legged made up of PVC (polyvinyl chloride) or iron or four-legged stand made up of iron. Each had their advantages and limitations as it is easy to attach an ant pan for single stand, while the four-legged stand is easy to carry from one place to another because it is not completely fixed to the ground like single legged stand. The bottom board rests on the stand and is separable from the hive stand. Above the Bottom board, one can find the brood chamber that consists of brood frames which is a home to honeybees where they rear their larvae in the comb constructed in the brood frames; it also contains pollen and some honey for their daily consumption; these frames are also made of wood and are arranged vertically and parallel to one another. A brood box normally contains one queen bee. The queen lays eggs, placing one each inside a cell of the comb. The eggs hatch to larvae and the larvae mature into the adult bees. When the brood chamber is well populated with bees, the beekeeper fixes a super chamber on the top of the brood chamber. Like the brood chamber, the super chamber also consists of frames arranged parallel and vertical, and it is the place of our interest as it is a honey-storing place for bees. Queen excluder is placed between the brood and super chambers for prevention of queen to enter into the super chamber. However, if the queen is a prolific egg layer, the beekeeper can use the option of fixing a second brood chamber to the first before fixing the super chamber. In such cases there will be great yield of honey because of more number of worker bees gathering nectar. The honey and pollen stored in the brood chamber are meant only for the developing larvae and not for extraction by the beekeeper. The standard height of the super chamber is three-fourth of the brood chamber. The number of super chambers will increase in the honey flow season as the bees will collect more nectar in that season. If the beekeeper extracts honey from the hive at short intervals, there is no need for fixing a second or third super chamber to the first. Over the super chamber, there will be top cover which acts as roof to the hive (Fig. 3.1).

Types of Beehive
There are different types of hives used in India such as Langstroth hive for *A. mellifera*, BIS hive (Bureau of Indian Standards) for *A. mellifera* and *A. indica* and Newton hive and Marthandam hive for *A. cerana.*

Langstroth Ten-Frame Hive
Stand: Any four-legged stand 15–25 cm high will do. Its upper dimensions should be such as to support the bottom board properly (Fig. 3.2).

Bottom board: It can be made either by taking a piece of wood 550 mm long, 406 mm broad and 22 mm thick or by joining two wooden boards together, nailed in position with wooden rods. Along each end of the longer side is nailed a wooden rod 550 mm long, 22 mm broad and 22 mm thick, and another wooden rod 363 mm x 22 mm is nailed at the back. The front is provided with entrance rod which is 363 mm x 22 mm x 22 mm, and this has an entrance 75 mm long and 22 mm deep in its middle. Two wooden blocks, to be used for shortening the entrance, when necessary, should also be prepared, each block being 75 mm x 38 mm X 22 mm.

Brood chamber: It is a rectangular box without top and bottom and is made of 22 mm thick wood. Its length on the outside is 500 mm and on the inside 456 mm, its breadth on the outside is 406 mm and on the inside 363 mm and its height is 238 mm. A rabbet 16 mm deep and 13 mm wide is cut along the entire length of its width planks.

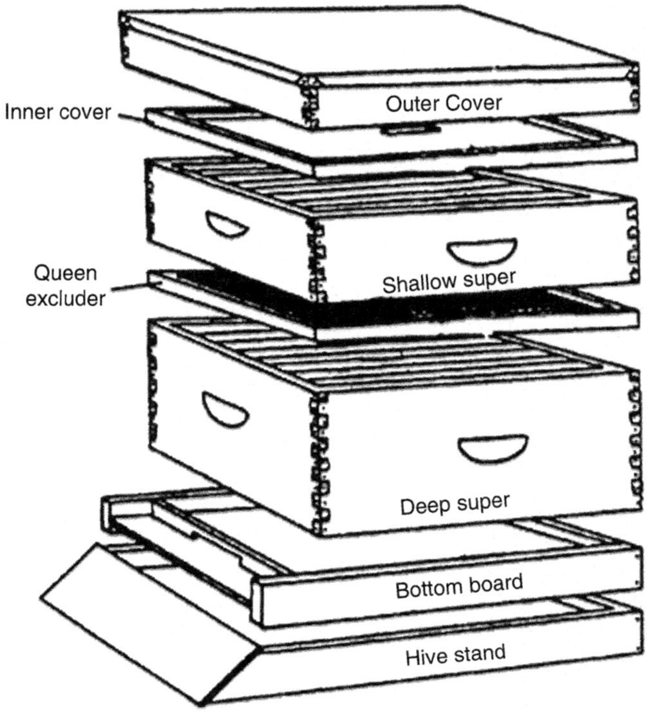

Fig. 3.1 Components of a standard beehive

Frame: Consists of top bar, two side bars and a bottom bar.

(i) Top bar*:* 475 mm long, 25 mm wide and 22 mm thick. It is cut to 9 mm thickness on both sides for a length of 25 mm. It has a groove in the middle of its lower side for fixing the comb foundation sheet.
(ii) Side bar*:* Each is made of 9 mm thick wood and is 226 mm long. The upper part of each is 34 mm wide and lower part 25 mm wide. Each is cut out from the middle portion at either end to accommodate the top and the bottom bars, respectively. There are four holes in each side bar for wiring the frame.
(iii) Bottom bar*:* 440 mm long, 19 mm wide and 9 mm thick. The outside measurements of the frame are 440 mm x 228 mm.
(iv) Two 15 mm staples should be driven into the top bar on its opposite side so that the frames stand 34 mm apart. One should make all frames either Hoffman or staple-spaced type. Tinned wire of 28 gauge should be used in wiring the frame.

Super: The dimensions of the super and the super frames should be the same as those of the brood chamber and the brood chamber frames, respectively.

Inner cover: This is wooden board to cover the brood chamber or the super as the case may be. It is 500 mm long, 406 mm broad and 9 mm thick wood. It has 9 mm thick and 22 mm wide wooden bar nailed onto each of its four sides.

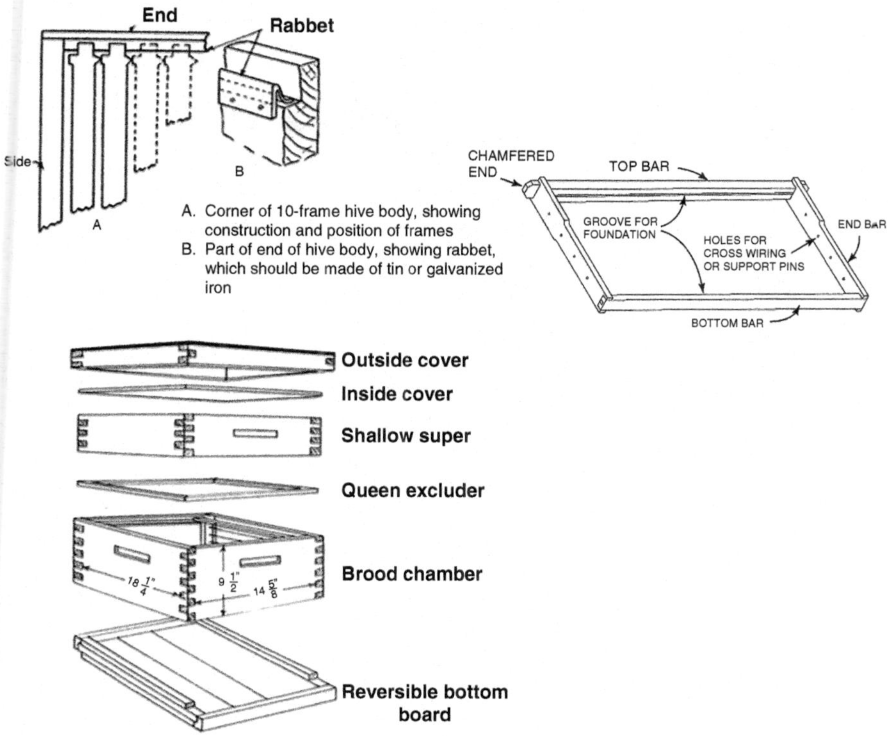

Fig. 3.2 Langstroth ten-frame hive

Top cover: It is made up of 9 mm thick wooden board nailed to a rectangular frame 50 mm high, all covered over with a metallic sheet so as to make it impervious to rainwater. Its inside measurements are 525 mm x 425 mm. It rests loosely over the hive.

Newton Hive (Fig. 3.3)

Stand: A log of wood of about 10 cm in diameter and well soaked in wood preservative is buried deep into ground. A length of about 20–30 cm is left above the ground, and a board 40 x 30 cm is fixed on its top with long nails and screws. The hive is placed on this platform on the log.

Bottom board: It is a plank slightly wider and 25 mm longer than the brood chamber with beadings on three sides into which the hive body fits in tightly. The extension of the front serves as the alighting board.

Brood chamber: It is a box without top and bottom and is made of 22 mm thick planks with outer dimensions 278 mm x 256 mm x 160 mm and inner 234 mm x 225 mm x 160 mm. Along the top of the front and rear planks, a groove of 6 mm depth and 9 mm width is made for resting the frames, and a clearance of about 6 mm is provided between the lower extremity of the frames and the bottom board. The front plank has an opening 88 mm x 9 mm at its lower side to serve as an entrance.

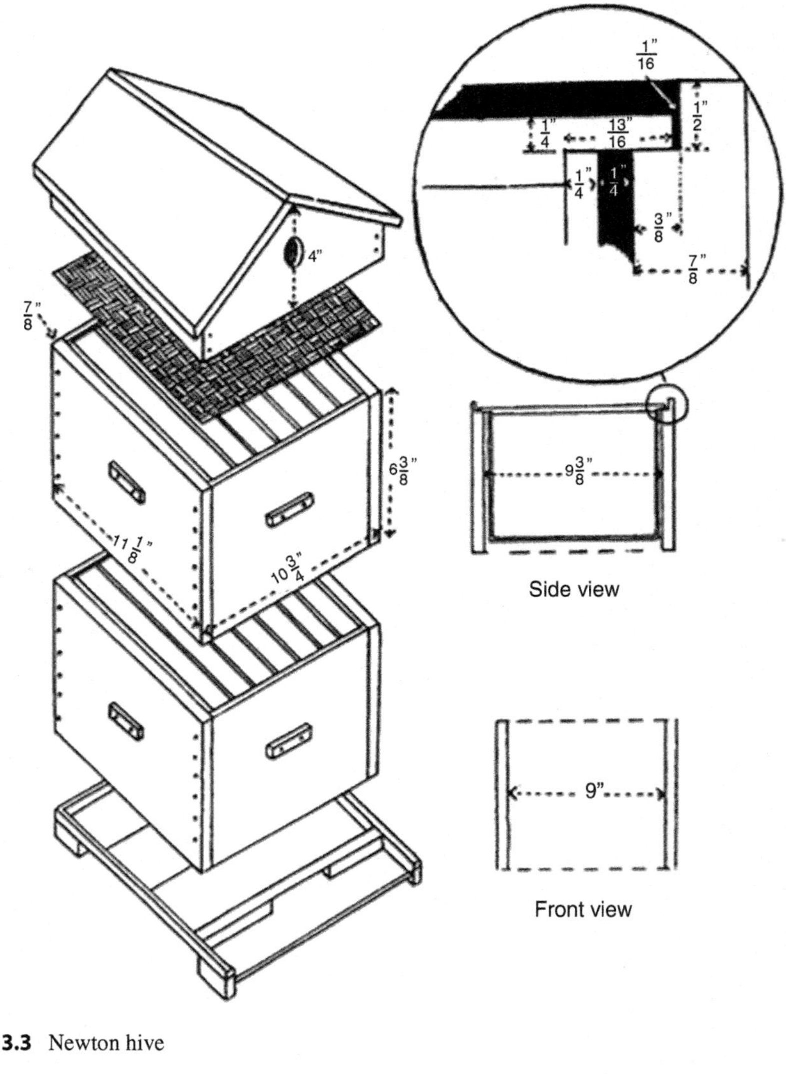

Fig. 3.3 Newton hive

Brood frame: Self-spacing (i) top bar breadth 22 mm, length 250 mm and thickness 3 mm; (ii) side bar height 144 mm, width at the top 28 mm and width at the bottom 12 mm; and (iii) inner length of frame 206 mm and inner height of frame 144 mm.

An extension of 3 mm is given on either side of the side bar, and a clearance of 6 mm is ensured when two frames are kept side by side. There are seven frames in a brood chamber.

Super and super frame: It has the same length and breadth as the brood chamber, but its height is 78 mm. The dimensions of the super frame are those of the brood frame, but the internal height is 62 mm.

Fig. 3.4 Beekeeping tools

Top cover: It has sloping planks on either side. An opening of 87 mm square, fitted with wire gauge, is made on the low ceiling plank to provide ventilation. Two holes in the front and rear planks of the top provide the necessary draught. Care should be taken to provide a clearance of about 6 mm between the ceiling plank and the frames below.

For the manufacture of hives, light, well-seasoned, good quality timber should be used. The wood used should not have a strong smell. *Kail* (*Pinus excelsa*), *teak* (*Tectona grandis*) and *toon* (*Toona ciliata*) are some of the woods suitable for the purpose. The hives should be preferably painted white or aluminium on the outside to protect the timber from weathering agencies. The hive parts should be accurately cut so that they may be interchangeable throughout the apiary and the particular part of the country.

3.5.2 Other Beekeeping Equipments (Fig. 3.4)

Smoker

The smoker is a metal cylinder in which smoke is produced by igniting fire. The smoker is attached with a bellow to blow air into the fire. The regulated smoke that comes out of the nozzle is directed into the hive to make the bees docile and less prone to stinging.

Honey Extractor

It consists of a metal drum with a centrifugally rotating device, for the extraction of honey from the frames. Four frames filled with honey from the super chamber can be placed in the extractor at one time to extract honey by rotating it with the help of a handle, and the honey gets dislodged through a pipe attached at the bottom of the extractor, and the honey can be collected in jar, cylinder, etc. The use of the extractor does not cause any damage to the combs, and they can be placed back in the super chamber after honey extraction. Two-frame extractor is also available in which only two frames can be placed at a time; it will be helpful for small-scale and household beekeeping.

Comb Foundation Sheet

In nature, bees build new combs from beeswax secreted by them and make parallel combs which are attached to the ceiling of the cavity or box. The combs may be built in the direction of the entrance, at right angles to it or in an oblique fashion. In the movable frame hive, it is imperative that straight combs be built in the frames so that when shifted from hive to hive, they may maintain the correct bee space between them. These comb foundation sheets are prepared by using comb foundation mill which uses wax sheets. These comb foundation sheets will save the energy of honeybees for building their combs as we are providing the basement.

Decapping Knife

Honeybees seal the cells in honeycomb once the honey is stored in it, so for extracting the honey by using extractor, we have to decap the sealed portion by means of decapping knife. The decapping knife may be a normal plain steel knife or an electric heated knife.

Hive Tool

It is a piece of flattened iron with hammered down edges and is used for prying apart the frames in the hive and for scraping bee glue and superfluous pieces of comb from the various parts of the hive.

Bee Veil

It is worn over the face for protection against stings. It should be made of black light material such as nylon nettings so that we can get a better picture. Veils should be made to fit snugly around the hat and to fit tightly to the shoulder leaving enough space between the veil and face.

Gloves

They give much protection as the honeybees mostly sting on the fingers and hands while handling them. They may be made of heavy canvas or rubberized cloth and are useful for beginners to develop confidence.

Bee Brush

A bee brush or a whisk broom is often employed to brush off bees from honeycombs before it is used for honey extraction.

Feeders
Various kinds of feeders for feeding sugar syrup to bees are used by beekeepers. The division board feeder, a wooden trough of the regular frame dimensions with shoulders so made that it may hang in the hive just like any other frame and with a wooden strip to serve as a float, is a useful appliance. A sugar syrup filled tin with holes in the lid is also a good type of feeder.

Queen Cell Protector
It is a cone-shaped structure made of a piece of wire wound spirally. It fits around a queen cell. A queen cell which may have to be introduced from a queen right to a queenless colony is often protected in a queen cell protector until its acceptance by the bees.

Dummy or Division Board
It is a wooden partition which serves as a movable wall and helps to reduce the size of the brood chamber so that bees can keep the hive air conditioned and well protected from bee enemies or inclement weather.

Embedder
It is a small tool with a spur or round wheel on the top. It is used to fix the comb foundation sheet on the wires of the frame. Electric wire is also used for this purpose which is useful to reinforce the comb and give extra strength to the comb.

Drone Trap
It is a rectangular box with one side open. The other side is fitted with queen excluder sheet. At the bottom of the box, there is a space for movement of worker bees. There are two hollow cones at the bottom wall of the box. Drones entering through the cones into the box get trapped. The narrow end of the cone is wide enough to let the bees pass out but not large enough to attract their attention or re-entry. This device is used at the entrance to reduce the drone population inside the hive.

Pollen Trap
Pollen-trapping screen inside this trap scrapes pellets from the legs of the returning foragers. It is set at the hive entrance. The collected pollen pellets fall into a drawer type of receiving tray.

3.6 Bee Products

Besides providing the treasured pollination services, honeybees gift some valuable merchandises to the mankind. As the name indicates, the first and foremost gift by honey bee is honey; other products which the honey bee provides include beeswax, pollen, royal jelly, propolis and bee venom.

Table 3.2 Average composition of honey (White 1962)

Water	17.2
Fructose	38.2
Glucose	31.3
Sucrose	1.3
Maltose	7.3
Higher sugars	1.5
Free acids	0.43
Lactone	0.14
Total acid	0.57
Ash	0.169
Nitrogen	0.041
pH	3.91
Diastase value	20.8

3.6.1 Honey

Honey is the substance made when the nectar and sweet deposits from plants are gathered, modified and stored in the honeycomb by honeybees (Singh et al. 2012). Quality of honey varies depending upon the types of floral and extrafloral nectar. The honey gathered can be classified as uni- and multifloral. Though mono-floral honeys are not common, yet honeys can be categorized on the basis of floral source such as litchi honey, berseem honey, eucalyptus honey, *Brassica* honey, etc. (Mishra 1995). Honey is a rich carbohydrate source which mainly contains fructose and glucose. Water is the other main constituent of honey, and it also contains numerous other types of sugars, acids, vitamins, proteins and minerals (White 1980). Honeybees seal the honey in comb cells after evaporating the excess moisture to reduce it to less than 20 % (Mishra 1995). The average composition of honey is shown in Table 3.2.

Honey is harvested in two ways: in case of *Apis dorsata* by squeezing the combs which contains some impurities like pollen and larva, but in case of domesticated bees (*A. mellifera* and *A. cerana*) honey is extracted with honey extractor without impurities using centrifugal force. Honey may remain in liquid form or may crystallize and hence can be presented to consumers as liquid honey or granulated honey (Mishra 1995). Honey is processed by a two step process. First the honey is indirectly heated in a water bath and kept at 60 °C for 30 min to kill the yeast cells responsible for fermentation of honey. Later it is filtered while still warm through a two layer cloth filter (when the viscosity of honey is lesser), cooled and bottled in glass bottles.

3.6.2 Beeswax

Wax is the other product produced by honeybee. The wax is produced by the wax glands when the worker bee is about 14–18 days old. The wax is used for building their nest by bees, and the normal colour of the wax is white, but the colour may change because of the influence of pollen source. Specific gravity of beeswax is 0.95 and melting point is 65 °C. Beeswax contains complex esters of monoatomic alcohols, 70.4–74.7 % of fatty acids and 13.5–15.0 % and 12.5–15.5 % of saturated hydrocarbons (Phadke and Phadke 1975). Normally wax is obtained from the damaged combs. Wax extraction can be done by two types: extracting wax using hot water bath is the most common method in India, and solar extractor is also in use which uses sun energy for melting of wax. Wax is used in preparing comb foundation sheet, candles, polishes, furniture, pharmaceutical industry and perfume industries, and it is a vital constituent of cosmetics like cold creams, lipsticks and rouges because it adheres better to the skin (Mishra 1995).

3.6.3 Bee Venom

The main components of bee venom are proteins and peptides. Urtubey (2005) mentioned the use of bee venom in apitherapy in China, India, Egypt, Babylon and Greece. Bee venom is present in the venom sac and will be injected using sting. The bee venom can be collected using venom extractor which possesses mild electric current, and the bees get irritated with this current so they try to sting, and the venom will be collected in the bottom glass plate. The USA is the leading producer of venom and had produced only 3 kg of venom in the past 30 years (Mraz 1982; Abrol 2012). Apitherapy is an age-old practice followed in India and some other countries for curing of joint pains in which bees are made to sting the patient by holding the bee from its wings with thumb and index. The venom collected by the above method can be made for subcutaneous injections. Ointment made by mixing apitoxin, Vaseline and salicylic acid (1: 10: 1) can be applied on the affected areas. The salicylic acid makes the skin soft and increases penetration (Mishra 1995).

3.6.4 Propolis

Propolis is used in construction and adaptation of honey bee nest and also to cover the cracks and crevices of the hive; it is a sticky dark-coloured material (Burdock 1998). The colour of the propolis may vary in temperate climates; it ranges from a light yellow or brown to a dark-brown colour, often with a reddish hue. Propolis tends to become darker the longer it is in the hive. The colour of propolis also varies according to the trees and plants from which it is harvested (Fearnley 2005). It can be used to treat wounds, infections, dermatitis and cancer. It is a strong fungicide and disinfectant (Ghisalberti 1979). It has an inhibitory activity against bacteria, fungi and yeast (Aspay 1977and Olivieri et al. 1981). Among the *Apis* spp., only *A. mellifera* is known to forage for propolis. The tropical stingless bees do collect a resinous substance similar to propolis, which they use to seal up the hive and to create honey and pollen storage vessels (Fearnley 2005).

3.6.5 Royal Jelly

Royal jelly is a milky white cream. It is strongly acidic and rich in protein, sugars, vitamins, RNA, DNA and fatty acids and is secreted by the nurse bees at the age of

6–12 days (Abrol 2013). It is also a very nutritious food for human beings as it increases vigour and vitality. Royal jelly is rich in amino acids such as alanine, arginine, aspartic acid, glutamic acid, glycine, isoleucine, lysine, methionine, phenylalanine, tryptophan, tyrosine and serine. Eight of the essential amino acids required for human beings are present in royal jelly. Besides this, it also contains vitamins A, B and C, iron, copper, phosphorus, silicon and sulphur. It can be harvested by Doolittle or grafting method in which artificial queen cell cups made from pure wax are attached to a brood frame which consists of bars holding small wax blocks. Then one- or two-day-old larvae will be placed in the queen cell cups and kept inside the hive. The nurse bees feed the larva with royal jelly which can be harvested (Mishra 1995). As the royal jelly is more nutritious, it can be helpful to mankind in many ways.

3.7 Rearing Methods and General Management of Honey Bees

The apiary or the place where bees are kept must be dry without dampness. Natural or artificial water source in the vicinity for honey bees, shade such as trees or artificial structures under which the bee hives will be placed and sufficient bee forage or the plants that provide pollen and nectar to honey bees are essential prerequisites.

Hive inspection must be taken up at least twice a week to look for the presence of queen, eggs and brood, honey and pollen storage and bee enemies like wax moth, mites and, diseases. Brood net expansion must be done by providing comb foundation sheet in empty frame during honey flow period. Sugar syrup feeding must be provided inside the hive during dearth period by dissolving sugar in water at 1:1 dilution. Supering or addition of frames in super chamber is done when brood chamber is covered with bees on all frames. Comb foundation sheet or constructed comb is provided in super chamber. At the time of honey extraction, the bees bushed away using brush, cells are uncapped using uncapping knife and honey is extracted using honey extractor and the combs are replaced in hive for reuse.

Swarming is a natural method of colony multiplication in which a part of the colony migrates to a new site to make a new colony. Swarming occurs when a colony builts up a considerable strength or when the queen's substance secreted by queen falls below a certain level. When bees swarm, it is possible that a beekeeper may lose a part of his colony and hence swarm management must be done by removing brood frames from strong colonies and providing to weak ones, pinching off the queen cells during inspection, dividing strong colonies and trapping and hiving the primary swarm.

The honey bees have to be managed during the honey flow season that normally coincides with spring season. Providing more space for honey storage by giving CFS or built combs, confining queen to brood chamber using queen excluder, prevent swarming as explained earlier, building sufficient population prior to honey flow by provide sugar syrup, dividing strong colonies into two or three new ones. – if colony multiplication need and taking up queen rearing technique to produce new queens for new colonies.

During severe summer the management methods include, providing sufficient shade, sprinkling water on gunny bag or rice straw put on hive to increase RH and reduce heat, increasing ventilation by introducing a splinter between brood and super chamber and providing sugar syrup, pollen supplement or substitute and water. To overcome winter the management methods comprise maintaining strong and disease free colonies, provide new queen to the hives and providing winter packing in cooler areas (Hilly areas). At the time of rainy season the management methods are to confine bees to the hive, providing sugar syrup, avoiding dampness in apiary site and providing drainage.

Honeybees are attacked by many pest, diseases and viruses, and they are strong enough to defend their colonies from various pests, but when they are weak, beekeeper should assist the bees for defending the colony.

3.7.1 Pest and Predators of Honeybee

These include minute mites to gigantic bears, in fact bears are very much fond of honey. Major opponents in this category include wax moths, birds, mites, ants, hive beetles and bears (Morhe 1999).

3.7.1.1 Wax Moth

Wax moths are major problem to beekeeping in Asia and predominantly in India. Wax moth occurs because of the poor management practices by beekeeper. *Galleria mellonella* (greater wax moth) and *Achroia grisella* (lesser wax moth) are the major damaging wax moth species. *Vitula* spp. (dried fruit moth), *Plodia interpunctella*, *Ephestia kuehniella* and *E. cautella* are the other moths associated with colonies of honeybees (*Apis cerana*, *A. mellifera*, *A. dorsata* and *A. florea*) (Kumar 1996).

Greater Wax Moth (*Galleria mellonella*)

In India it is the major pest of *A. cerana* causing the colonies to abscond. It causes considerable damage if the beekeeper doesn't follow the proper storage of empty combs, rendered wax, comb foundation sheets and bee-collected pollen. Larval stage is the damage-causing stage, and the larva is about 3–30 mm long, lives in silken tunnels and feeds on pollen, nectar and newly emerged honeybees. Larval period is between 22 and 60 days. In India they are active during March to October (Garg and Kashyap 1998), but from June to October which is considered as a dearth period in India, their activity is high.

The Lesser Wax Moth (*Achroia grisella*)

As the name suggests, it is smaller than greater wax moth, but it is widely distributed and is also seen in higher altitudes. The length of the larvae is about 15–20 mm and the larva feeds on the same food as the greater wax moth do. Their highest activity will be during June to October in India and completes three to four generations during that period (Singh 1962).

Control Measures

Maintaining hygienic beekeeping practices is the best way to prevent or control the honeybee colonies from wax moth attack. In south India, the Indian bee boxes are mainly of Marthandam type which readily produces cracks, and this provides a suitable habitat for the wax moth. Closing the cracks and crevices and reduction of hive entrance could stop the wax moth arrival (Ramachandran and Mahadevan 1951). However, in north India as the beekeeping is mainly with Italian bees, they have less problem with the wax moth as the Italian bee itself closes the cracks and crevices by using propolis. By keeping the infested combs in hot water (60 °C) for 4–5 h, the larvae can be killed. Fumigation of the affected wax combs with paradichlorobenzene (PDB) will be effective (Casanova 1992). The use of biocontrol agents like *Bacillus thuringiensis*, *Galleria* nuclear polyhedrosis virus (GNPV), oviposition attractants and genetic manipulation are some of the measures for keeping the wax moth population in check. *Apanteles galleriae* is a larval parasite of wax moth. Major work was carried on *Bt* formulations (Ali et al. 1973). Initial larval stages of *Galleria* were more susceptible to *Bt* treatment than the later stages.

3.7.1.2 Other Minor Pests

Hive Beetle *(Aethina tumida)*

This is small and black beetle which is present in and around the bee colonies; it will eat and destroy the cells constructed by the bees and also feed on pollen, eggs and small honey bee grubs (Lundie 1940 and Caron 1990). When the population of these beetles is high, they will cause considerable damage. For controlling, fume boards can be placed over the beehive, and the hive may be kept on a concrete floor as the beetle is a soil pupator.

Ants and wasps

Ants are enemies to both honeybee and beekeeper as they cause pain to the beekeeper by their influential bites and they will feed on everything they get from the honeybee colony which includes the dead and the live bees, honey and the brood. (Akratanakul 1986; Buys 1990; Abrol 1997; Abrol and Kakroo 1994). When the ant population is high in number, they even cause the *A. mellifera* and *A. cerana* colonies to abscond. In India, engine oil or grease is applied to the ant stand for obstructing their movement. Chemicals like ethyl or methyl alcohol, sodium fluoride, borax powder, salt or powdered sulphur can be used for ant control (Nikiel 1972).

Large social wasps *Vespa sp* prey honey bees with ease. *V. orientalis*, *V. magnifica* and *V. cincta* are some of the species that devor honey bees and weaken the colonies. Peak activity of the wasp was reported from August to November in Himachal Pradesh (Rana et al. 2000). The bees kill wasps through shimmering behaviour forming balls around wasps. The intruders is killed either being stung or due to high temperature at centre of ball (43–46 °C) and suffocation (Abrol and Kakroo 1994).

Destruction of wasp nest, use of protective screens and bait lures have been suggested for managing wasps.

Birds

As the honeybees have aerial movement, they are prone to hunting by birds. *Merops apiaster* (European bee-eater), *M. orientalis* (small green bee-eater), *M. leschenaulti* (chestnut-headed bee-eater), *M. superciliosus persicus* (blue-cheeked bee-eater), *M. philippinus philipinnus* (blue-tailed bee-eater), *Indicator indicator* (honey guides), *Dicrurus macrocercus*, *D. aster* (drongo/King crow), *Cypselus* spp., *Apus spp.* (swifts) and *Lanius* spp. (shrikes) are some of the major bird enemies attacking bees (Gulati and Kaushik 2004; Ramchandran et al. 2012). Among them, bee-eater poses major threat to honeybees (Dyer and Fry 1980). For control of these birds making high pitch noises, producing scaring sounds by beating the empty tins can be done so that the birds get scare (Gulati and Kaushik 2004).

Indian Bear (*Melursus ursinus*)

The apiaries located near to the hilly regions are more susceptible to bear attack. Hanging of beehives to tree branches can be done to control. Besides hanging normal hives, top bar hives can also be used.

3.7.2 Viruses

Honeybees in India are affected by *Apis* iridescent virus, Thai sacbrood virus (TSBV) and Kashmir bee virus. Among them TSBV was an introduced virus and a major one. TSBV was first detected in Meghalaya in 1978 (Kshirsagar et al. 1982). This virus caused a disastrous outbreak and devastated more than 90 % of *A. cerana* colonies in India (Mishra 1995; Devanesan and Jacob 2001). Both the larval and pupal stages are susceptible to this disease, but the adult is an immune stage (Ramchandran et al. 2012). Because of viral attack, brood will die in prepupal unsealed stage, dead larvae can be seen with tip of the head capsule turned upwards, dead prepupae turn into saclike structure and the colour of the affected larvae also changes from white to yellow or grey (Mishra 1995). There are no packed materials for the virus control. The disease can be avoided to certain extent by avoiding, replacing or mixing bee colonies and hive equipments from TSBV-affected apiaries. Recently RT-PCR based method of diagnosis of TSBV has been developed (Aruna et al. 2016).

3.7.3 Bacterial Diseases

American foulbrood (AFB) disease caused by *Paenibacillus larvae* and European foulbrood (EFB) caused by *Melissococcus pluton* are the dangerous bacterial disease infecting honeybee colonies (Nakamura 1996; Oldroyd and Wongsiri 2006; Bailey and Collins 1982).

3.7.3.1 American Foulbrood Disease

AFB is the one of most widely spread bacterial disease (Gochnauer 1981).AFB-infected larvae normally die after their cell is sealed. Caps of these dead brood cells are usually darker than the caps of healthy cells. The entire population of the hive gets infected. For control of the disease, sterilization of equipments can be done using formalin at 6 ml per litre, and Terramycin at 250–400 mg in 5 l of sugar syrup can be fed to diseased colony twice at weekly intervals for effective control (Mishra 1995).

3.7.3.2 European Foulbrood Disease

European foulbrood is less harmful compared with American foulbrood, and it infects the midgut of infected bee larvae (Suwannapong et al. 2011). In India it was first observed in Maharashtra in 1971 by Diwan and his coworkers on *A. cerana indica.* The honeybee colonies which are attacked by the *Varroa* are highly susceptible to EFB disease as this bacterial disease is a stress-related disorder (Bailey and Collins 1982). Young larvae of 4–5 days old are highly susceptible to EFB, and the colour of the larvae also changes from shiny white to yellow or brown in colour. For control of the EFB disease, Terramycin and formalin can be used as mentioned in AFB control. Other mechanical methods namely shookswarm (where the adult bees are shaken into new hives discarding the infected brood combs) can be adopted to avoid use of antibiotics.

3.7.4 Mite Enemies of Honeybee

Mites are important adversary of honey bee in India; they spread from one place to another as the beekeeper moves the colonies to floral-rich source and also because of migratory beekeeping (Anderson 1999, Boecking et al. 2000; Oldroyd and Wongsiri 2006). Tewarson et al. (1992) was the first person to study about life cycle of *Varroa destructor* on *A. cerana* in India.

3.7.4.1 *Varroa jacobsoni*

The mite was first reported in India by Phadke et al. (1966) from Delhi. It is a native pest to *A. cerana* in India, but after the introduction of *A. mellifera* to India, it started affecting the Italian bee colonies also (Mishra 1995). Mite is reddish brown in colour and female mite is about 1.1 mm in length and 21.6 mm (Sammataro et al. 1994; Sammataro 1997). It can pierce and tear open the host's integument and feed on the haemolymph of the honeybee (Suwannapong et al. 2011; Delfinado and Baker 1987). Symptom is called as varroasis and larval stage of honeybee is the most susceptible stage. For controlling, sugar powder can be dusted over the honeybees and in the frames; as the bees tend to groom and by the process, the *Varroa* mite can be dislodged. Other than sugar dust, sulphur and *Acorus calamus* (sweet flag) powder can also be dusted. Ritter (1981) and De Jong et al. (1982) suggested synecar, a mixture of sugar powder + chloropropylate or bromopropylate at the rate of 50–100 mg per colony, depending upon the strength, can be dusted in passages between the frames. Presently some commercial products are available such as Coumaphos®, Bayer Bee Strips® or CheckMite® (Suwannapong et al. 2011; De Jong 1997; Gerson et al. 1988; Le Conte et al. 1989) which are hardly practised in India.

3.7.4.2 *Acarapis woodi*

Acarapis woodi was first reported in India (Singh 1957) from *A. cerana* colonies. This mite was first named as *Tarsonemus woodi* (Rennie 1921; Rinderer et al. 1999), but later it was renamed as *Acarapis, Acar* from *Acarus* (mite) and *Apis* from bee (Suwannapong et al. 2011). This mite attacks the tracheal system of honeybee; it attacks all the three castes of honeybee. The typical symptom is “K”-winged condition, where the bees cannot fly and the wings are disjoined in condition. This mite also feeds on haemolymph and the life span of the honeybee is reduced (Hirschfelder and Sachs 1952; Mishra 1995). For controlling of this mite, formic acid, menthol or thymol can be applied; fumigation using Folbex strips can be done (Atwal 1971).

3.7.4.3 *Tropilaelaps clareae*

It was first reported in India from *Apis dorsata* (Bhardwaj 1968). This mite species attacks all five species of honeybee but is primarily found on *A. dorsata* and *A. mellifera* (Atwal and Dhaliwal 1969; Laigo and Morse 1969). It was first discovered in the rat (Delfinado and Baker 1961). The mite attacks the pupae and prepupae stages of bees. Mature female mites attach on and suck the haemolymph from the larvae and adults. Infected honey bees have poor wings; irregular brood pattern is a typical symptom of this mite (Mishra 1995; Suwannapong et al. 2011). Sulphur dusting is an effective control method (Atwal and Goyal 1971). The use of organic products like formic acid, oxalic acid and other essential oils at the right time can be effective for all mite species (Cramp 2008). *T. clareae* is difficult to control compared to other mite species, as this mite is readily available with *A. dorsata* colonies, but the professional beekeepers remove the brood frames from their hives so that the female mite will starve to death as this mite can live only 7 days without food.

3.8 Migratory Beekeeping in India

Migratory beekeeping provides good returns to the beekeeper as the returning bees to the hive are maximum, because of abundant flora in that region. For doing commercial migratory beekeeping, the beekeeper has to map the floral resources available and do planned migration accordingly.

In northern India, commercial beekeepers shift the colonies between plains and hills for migratory beekeeping. During October–November, colonies are migrated to the plains of Uttaranchal, Uttar Pradesh, Haryana, Punjab and Rajasthan to exploit rapeseed and mustard. During December–January, colonies are migrated to eucalyptus plantation of Uttar Pradesh and Haryana. Bee colonies will also be migrated to litchi orchards at Ramnagar and Dehradun from February to March. Some beekeepers will also migrate to sunflower fields of

Punjab and Haryana. Beekeepers will also migrate to forest plantations of Uttar Pradesh for shisham till May.

In southern India, migration of bee colonies from southern Tamil Nadu (mainly Marthandam of Kanyakumari District) to Kerala during January–March is a renowned practice. The commercial beekeepers migrate the colonies to rubber plantations which are spread over about 0.40 million hectares. During that period, beekeepers will harvest tons of honey and store to sell when they get better price. Rubber is considered as the third major source of honey next to rapeseed/mustard and sunflower in India. Beekeepers from Kerala and Tamil Nadu migrate their colonies mainly to Quilon, Kottayam, Changanacherry, Trichur, Palghat, Kozhikode and Cannanore districts for rubber–honey flow. In Tamil Nadu, during May–June, beekeepers migrate the colonies for harvesting nectar from tamarind flowers. Colonies are also migrated to high ranges of Devikulam, Peermedu, Idukki and other districts to cardamom estates.

3.9 Pesticide Usage and Honeybees

Pesticide use has become inevitable in modern agriculture. With pesticide consumption increasing several folds during the last four decades, the side effects are also increasing, one of which is the toxicity to honeybees. Pesticides, alone and in combination with other factors, have had a devastating effect on honeybees and wild pollinators. Pesticides commonly found in lawn and garden products and used in agriculture are known to be hazardous to bees, some killing bees outright and others with subtle effects that reduce a bee's ability to thrive. Besides increasing agricultural production, they cause undesirable environmental effects including the effect on nontarget species, such as honeybees and other pollinators. Hence, the safety of poisonous agrochemicals must be ensured.

The use of pesticides for pest control on the one hand and the use of honeybees for cross pollination are not always compatible, as honeybees are susceptible to many of commonly used pesticides for the control of insect pests (Johansen 1977; Mac Kenzie and Winston 1989; Poehling 1989; Stark et al. 1995; Russell et al. 1998; Cuningham et al. 2002; Sundararaju 2003). The major constraint confronting pollinator–plant interaction is the indiscriminate and excessive use of pesticides for controlling insect pests (Bisht et al. 1983; Rana and Goyal 1991; Zhong et al. 2004). The loss of honeybees directly affects beekeeping through loss of honey production and indirectly the crop production due to inadequate pollination. Reduction of population of these beneficial insects due to insecticides, therefore, incurs significant environmental, ecological and economic costs (Pimental 1980; Crane and Walker 1983).

3.9.1 Impact of Pesticides on Bees

The use of pesticides affect the bees in several ways:

- The use of herbicides eliminates the weed flora which serves as very good food source for bees especially during dearth period.
- Direct exposure to insecticidal sprays result in the death of bees and sometimes lead to the total destruction of bee colonies.
- Contamination of water resources affect water carriers.
- Contamination of nectar and pollen causes brood mortality.
- Widespread use of chemicals also contaminates the hive products.
- Indiscriminate use of pesticides threatens the integrity of bee–flower mutualistic system.

3.9.2 Minimizing Pesticide Hazards to Bees/Management Practices

Proper understanding of above-mentioned principles can go a long way in reducing pesticide hazards to honeybees. The basic principle, of course, is that honeybees should not get exposed to the toxic effects of insecticides as far as possible. Reducing pesticide injury to honeybees requires communication and cooperation between beekeepers and farmers, since both mutually benefit from honeybees, the beekeeper in terms of its products and the farmer in terms of increased production of crops. While it is unlikely that all poisoning can be avoided, a balance must be struck between the effective use of insecticides, the preservation of pollinators and the rights of all – the beekeeper, farmers and the community.

3.9.3 Guidelines to Beekeepers

- It is most desirable that bee colonies should be maintained where use of pesticides or drift from pesticides is minimum. For this, the beekeeper should be fully conversant with the type of pesticides used in their locality, which in turn depends upon the cropping pattern and the pest complex. He/she should also be aware of normal wind currents prevalent in that area to protect against the harmful effects from drift.
- If ever disinfestation of beehives becomes necessary, he/she should use only the recommended chemicals, safe to the bees, for the purpose.
- During bloom if the crops in the surrounding areas are being sprayed with the insecticides, it is always advisable to confine the bees within the hives. If it is apprehended that the spray programme will continue for a longer period, it is better to move the hives away to the safe location free from the drift in advance.
- Apiarists and farmers should have close cooperation so that beneficial activity of bee is not jeopardized by the irrational use of pesticides by the latter.
- Feeding of colonies with sugar syrup following pesticide application to reduce bee foraging may help substantially in reducing the exposure of bees to pesticides.

3.9.4 Guidelines for Farmers

The golden principle for the farmers is to use insecticides only when necessary. For this purpose, integrated pest management approaches are available on most crops, which should be strictly practised. It is in the mutual interests of both that the farmer should intimate the spray programme in advance to the beekeeper:

- If there is a choice for insecticides, the use should be restricted to the chemicals in the less hazardous groups.
- The spray operation in the evening is always preferable as it not only gives better deposit and distribution but also bee activity subsides.
- Apply granules or sprays in preference to dusts. Baits used for fruit fly control, should be discouraged as far as possible during the crop in bloom.
- Examine fields and field margins before spraying to determine if bees are foraging on flowering weeds. Where feasible eliminate weeds by mowing or tillage.
- Give careful consideration to position of bee colonies relative to wind speed and direction. Changing spray nozzles or reducing pressure can increase droplet size and reduce spray drift.

3.10 Constraints in Beekeeping in India

As discussed above, indiscriminate use of pesticides poses a major threat to honeybees (Shinde and Phadke 1995; Kaur 1998; Kumar 2000). Lack of honeybee professionals and trained bee labours allows poor management of colonies. Many commercial beekeepers face problems including interference of police and octroi people during the migration of their colonies. A survey conducted in Punjab resulted 37.5 % of beekeepers are facing these problems. The transport costs are also high in migratory beekeeping (Kaur 1998; Sharma et al. 2014). Depletion of floral resource because of growing concrete jungles is one of the major concerns in beekeeping, as the bee boxes are placed in field where 24-h care is not possible, theft of boxes occur (Bansal et al. 2013). Many beekeepers report that the cost of the equipments is too high and this will also discourage the entrepreneurs in this field. There is no separate market for honey and beekeepers sell their honey to local markets. For exporting honey, most of the commercial beekeepers are troubled by the international standards as the beekeepers have poor knowledge of the standards (Bansal et al. 2013; Sharma et al. 2014). Honey from the rubber plantations is the major source in southern India, and the beekeepers from this part cannot export their honey as it has high moisture content. Producer price for honey and other products from beekeeping is very low compared to retailer price, and this always irritates the beekeepers (Singh 2000).

3.11 Overcoming the Constraints

As per FAO (2016), the world total honey production during 2013 was 2.13 million tonnes. China ranks first in honey production with 466,300 tonnes while India ranks seventh with 61,000 tonnes. India has vast potential for beekeeping. The diversity in flora provides more opportunities for the development of beekeeping industry. It is said that based on the area under cultivation in India and bee forage crops, India has a potential to have about 100 million bee colonies while the current figure is less than one million colonies.

Beekeeping should be recognized as an important agricultural activity for increasing the productivity of agricultural/horticultural crops, and a section of beekeeping should be developed within the line departments of the states. Forest department should take an initiative for planting bee flora and should allow the beekeepers to use it. Free trainings on beekeeping with latest improvements can help the beekeepers for updating their knowledge. Effective cooperation among beekeepers, traders, exporters, extension agencies and government should be established. Intensive efforts should be made to improve the domestic consumption of honey through developing honey-based food/consumer products and intensive generic promotion of honey through media. There is a need to conduct effective promotional and awareness campaigns to remove the myths about honey and bees. Government must take steps for selling the honey at best price which helps in beekeeper's economy. The concept effect of pesticides on honeybee should be undewrstood by the farmer so the beekeeper will be forewarned by the farmer before spraying. Presently government is encouraging organic farming (recently Sikkim was declared as a first organic state in India) which is a good initiative for saving bee health and consuming organic honey (Bansal et al. 2013; Sharma et al. 2014).

References

Abrol DP (1997) *Bees and Beekeeping* in *India*. Kalyani Publishers, Ludhiana, 450pp

Abrol DP (2011) Foraging. In: Hepburn HR, Radloff SE (eds) Honeybees of Asia. Springer, Heidelberg/Berlin, pp 257–292

Abrol DP (2012) Pollination biology biodiversity conservation and agricultural production. Springer, Dordrecht/Heidelberg/London/New York

Abrol DP (2013) Beekeeping a comprehensive guide to bees and beekeeping. Scientific publishers, Jodhpur, p 499

Abrol DP, Kakroo SR (1994) In: Veeresh, GK, Virktamath CA (eds) Proceedings of the 1st National Seminar Cum workshop on honeybee diseases, diagnosis and management, Oxford and IBH Publication, Ludhiana, 450

Akratanakul P (1976) Honeybees in Thailand. Am Bee J 116:120–121

Akratanakul P (1986) FAO Bull. Common wealth Institution, IBRA, London, p. 56

Akratanakul P (1990) Beekeeping in Asia. FAO Agricultural Services. Bulletin 68/4 Food and Agriculture Organisation of the United Nations, Rome

Ali AD et al (1973b) J Apic. Restor Dent 12:125–130

Allen MF (1995) Bees and beekeeping in Nepal. Bee World 76:185–194

Anderson RH (1963) The laying worker in the Cape bee, Apis mellifera capensis. J Apic Res 2:85–92

Anderson DL (1999) Genetic and reproductive variation in *Varroa jacobsoni*. Proc. XIIIInt. Congr. IUSSI, Adelaide, p. 33

Aruna R, Srinivasan MR, Selvarajan R (2016) Comparison of genomic RNA sequences of different isolates of Thai Sac Brood virus attacking Indian honey bee, *Apis cerana indica* Fab. Tamil Nadu by reverse transcriptase-polymerase chain reaction method. J Entomol Res 40(2):117–122

Aspay E (1977) Selective effect of propolis in the isolation of *Listeria monocytogenes*.in Nordisk. Vet Dermatol 29(10):440–445

Atwal AS (1971) Acarine disease problem of the Indian honeybee. Apisindica F Am Bee J 111(134–136):196–187

Atwal AS, Dhaliwal GS (1969) Some behavioural characteristics of *Apis indica* F. and *A. mellifera* L. Indian Bee J 31(1):1–8

Atwal AS, Goyal NP (1971) Infestation of honeybee colonies with *Tropilaelaps* and its control. J Apic Res 10(3):137–142

Bailey L, Collins MD (1982) Reclassification of *Streptococcus pluton* (White) in a new genus *Melissococcus pluton*. J Appl Bacteriol 53:215–217

Bankova VS, Popov SS, Marekov NL (1983) A study on flavonoids of propolis. J Nat Prod 46:471–474

Bankova V, De Castro SL, Marcucci MC (2000) Propolis: recent advances in chemistry and plant origin. Apidologie 31:3–15

Bansal K, Singhand Y, Singh P (2013) Constraints of apiculture in India. Int J Life Sci Res 1(1):1–4

Bhardwaj. RK (1968) A new record of mite, *Tropilaelaps clareae* from *Apis dorsata* colonies. BeeWorld 49(3):115

Bisht DD, Naim M, Mehrotra KN (1983) Studies on the role of honeybees in rapeseed production. In: Proceedings of 2nd international conference of apiculture in tropical climate. Indian Agricultural Research Institute, New Delhi, pp 491–496

Boecking O, Bienefeldand K, Drescher W (2000) Heritability of the Varroa specific hygienic behaviour in honey bees (Hymenoptera: Apidae). J Anim Breed Genet 117(6):417–424

Burdock GA (1998) Review of the biological properties and toxicity of bee propolis. Food Chem Toxicol 36:347–363

Buys B (1990) Applied myrmecology. In: RK V et al (eds) A world perspective. Westview Press Inc, Boulder, pp 519–524

Caron DE (1990) Honeybee pests, predators and disease. Cornell University press, Ithaca/New York, pp 171–172

Casanova Ostos RA (1992) Three methods for control of the wax moth, *Galleria mellonella* (L.), in stored honeycombs. Revista Cientifica UNET 6(1):5–16

Cramp D (2008) The practical manual of beekeeping. Spring-hill House, Oxford, pp. 202–201

Crane E (2004) A short history of knowledge about honey bees (Apis) up to 1800. Bee World 85(1):6–11

Crane E, Walker P (1983) The impact of pest management on bees and pollination. International Bee Research Association, London, 107pp

Cuningham SA, Fitzgibben F, Heard TA (2002) The future of pollinators for Australian agriculture. Agriculture 53(8):893–900

De Jong D (1997) Mites: varroa and other parasites of brood. In: Morse RA, Flottum K (eds) Honey bee pests, predators and diseases, 3rd edn. A.I. Root Company, Medina, pp 278–327

De long D, Roger AM, Eiclcwort, George C (1982) Mite pests of honeybees. Ann Rev Enl 27:229–252

Delfinado MD, Baker EW (1961) Tropilaelaps, a new genus of mite from the Philippines (Laelapidae: Acarina). Fieldiana-Zool 44(7):53–56

Delfinado MD, Baker EW (1987) A new *Varroa* (Acari: Varroidae) from the nest of *Apis* cerana (Apidae). Int J Acarol 13:233–237

Devanesan, S., Jacob A (2001). Thai sacbrood virus disease of Asian honeybee *Apis cerana indica* Fab, in Kerala, India. Proceedings 37th international apiculture congress, 28th October–1st Nov 2001, Durban, South Africa

Dhaliwai HS, Sharma PL (1974) Foraging range of the Indian honeybee. J Apic Res 13(2):137–141
Dyer M, Fry CH (1980) Acta 17 International Organisation Congress. Velrag, D.O-G., Berlin, pp. 44–46
FAO (2016) FAOSTAT. Accessed on 1 Nov 2016
Farnesi AP, Aquino-Ferreira R, De Jong D, Bastos JK, Soares AEE (2009) Effects of stingless bee and honey bee propolis on four species of bacteria. Genet Mol Restor Dent 8(2):635–640
Fearnley J (2005) Bee propolis natural healing from the hive. Souvenir Press
Garg R, Kashyap NP (1998) In: Mishra RC (ed) Perspectives in Indian apiculture, Agro Botanica, Bikaner, pp 290–303
Gary NE (1963) Observation of mating behavior in the honeybee. J Apic Res 2:3–13
Gerson U, Lensky Y, Lubinevski Y, Slabezki Y, Stern Y (1988) Varroa jacobsoni in Israel, 1984–1986. In: Neeedham GR, Page RE, Delfinado-Baker M, Bowman CE (eds) Africanized honey bees and bee mites. Ellis Horwood Limited, Chichester, pp 420–424
Ghisalberti EL (1979) Propolis: a review. Bee World 60:59–84
Gochnauer TA (1981) The distribution of Bacillus larvae spores in the environs of colonies infected with American foulbrood disease. Am Bee J 121:332–335
Gulati R, Kaushik HD (2004) Enemies of honeybees and their management – a review, Agric. Revue 25(3):189–200
Gurung MB, Ahmad F, Joshi SR, Bhatta CR (2003) The value of *Apis cerana* beekeeping for mountain farmers in Nepal. Bee Dev J 69:13
Hepburn HR, Radloff SE (2011) Honeybees of Asia. Springer, Berlin
Hirschfelder H, Sachs H (1952) Recent research on the acarine mite. Bee World 33:201–209
Hyatt S (2011) Asian honey bee (*Apis cerana javana*) in Cairns, Far North Queensland. Department of Employment, Economic Development and Innovation (DEEDI)
Johansen CA (1977) Pesticides and pollinators. Annu Rev Entomol 22:177–192
Kamakura M (2011) Royalactin induces queen differentiation in honeybees. Nature 473:478–483. doi:10.1038/nature 10093
Kaur B (1998) Evaluation of selected training course organised by Punjab Agricultural University, Ludhiana. Ph.D. dissertation, Punjab Agricultural University, Ludhiana
Kenmore P, Krell R (1998) Global perspectives on pollination in agriculture and agroecosystem management. In: International Workshop on the Conservation and Sustainable Useof Pollinators in Agriculture, with Emphasis on Bees. Brazil, Sao Paulo
Kishan Tej M, Srinivasan MR, Rajadurai G (2014) Proceedings of the national symposium on emerging trends in eco- friendly insect pest management. AE publications, Coimbatore, p 283
Koeniger N (1969) Experiments concerning the ability of the queen (*Apis mellifera* L.) to distinguish between drone and worker cells. XXII International Beekeeping Congress Summit, p. 138
Koeniger N (1970) Factors determining the laying of drone and worker eggs by the queen honeybee. Bee World 51:166–169
Koetz AH, Hyatt S (2013) Asian honey bee (Apis cerana) remote nest treatment. Asian Honey bee transition to management program, Department of Agriculture, Fisheries and Forestry (DAFF); Cairns, Queensland, Australia
Kshirsagar KK, Saxena UC, Chauhan RM (1982) Occurrence of sacbrood disease in *Apis cerana indica* F. in Bihar, India. Indian Bee J 44:8–9
Kumar Y (1996) Advanced training course in apiculture, CAS, Department of Entomology. CCS Haryana Agricultural University, Hisar, pp. 201–207
Kumar P (2000) Present status, problems and prospects of beekeeping in Ludhiana district of Punjab, M.Sc. thesis, Punjab Agricultural University, Ludhiana
Laigo FM, Morse RA (1969) Control of the bee mite, *Varroa jacobsoni* Oudemans and *Tropilaelaps clareae*. Delfinado and Baker with Chlorobenzilate. Philippines Entomol 1:144–148
Le Conte Y, Arnold G, Trouiller J, Masson C, Chappeand B, Ourisson G (1989) Attraction of the parasitic mite Varroa to the drone larvae of honey bees by simple aliphatic esters. Science 245:638–639

Lepcha SR, Gurung R, Arrawatia ML (2012) Traditional Lepcha craft Sumok-thyaktuk (Lepcha hat) and its conservation in Dzongu Tribal Reserved Area (DTRA), Sikkim, India. Indian J Tradit Knowl 11:537–541

Lundie AE (1940) The small hive beetle *Aethina tumida*, Science Bulletin 220. Department of Agriculture Forestry, Government Printer, Pretoria

Maa TC (1953) An inquiry into the systematics of the Tribus Apidini or honeybees (hymenoptera). Treubia 21:525–640

Mackensen O (1943) The occurrence of parthenogenetic females in some strains of honeybees. J Econ Entomol 36:465–467

Mackenzie KE, Winston ML (1989) Effect of sublethal exposure to diazinon on longevity and temporal division of labor in the honeybee (Hymenoptera: Apidae). J Econ Entomol 82(1):75–82

Marcucci MC (1995) Propolis: chemical composition, biological properties and therapeutic activity. Apidologie 26:83–99

Michener CD (2000) The bees of the world. Johns Hopkins University Press, Baltimore

Mishra RC (1995) Honeybees and their management in India, publications and information division Indian council of agricultural research. Krishi Anusandhan Bhavan, Pusa, New delhi

Morhe RA (ed) (1999) Honeybee pests, predators and diseases. Cornell University Press, Ithaca/New York

Mraz (1982) Bee venom for arthritis – an update. Am Bee J 22:121–123

Nair MC (2003) Apiculture resource biodiversity and management in southern Kerala. Ph. D. thesis. Mahatma Ghandhi University, Kottayam, 277pp

Nakamura LK (1996) *Paenibacillus apiarius* sp. nov./nt. J Syst Bacteriol 46:688–693

Nikiel AH (1972) Glean Bee Cult 100:330

Oldroyd BP, Wongsiri S (2006) Asian honey bees. In: Biology, conservation and human interactions. Harvard University Press, Cambridge, Massachusetts

Olivieri M, Ginocchi G, Melzi R (1981) Antimicrobial effect of propolis. Cronache Farmacentiche 24(2):94–96

Otis GW (1990) Diversity of Apis in Southeast Asia. In: Vearesh GK, Malik B, Viraktanathan H (eds) Social insects and Environment. IBA, Oxford, pp 725–726

Otis GW (1996) Distribution of recently recognized species of honey bees (Hymenoptera: Apidae: *Apis*) in Asia. J Kansas Entomol Soc 69:311–333

Partap U (2011) The pollination role of honeybees. In: Hepburn HR, Radloff SE (eds) Honeybees of Asia. Springer, Heidelberg/Berlin, pp 227–255

Partap U, Verma LR (1998) Asian bees and bee keeping: issues and initiatives, Asian bees and bee keeping progress of research and development. Proceeding of fourth Asian apicultural association international conference, Kathmandu, 23–28 March 1998, pp 3–14

Phadke RP, Phadke RS (1975) Physico-chemical characteristics of waxes from Indian honeybees II Viscosity and refractive index. Indian Bee J 37(1/4):15–18

Phadke KGDS, RB B, Sinha P (1966) Occurrence of mite, *Varroa jacobsoni* Oudemans in the brood cells of honeybee *Apisindica* F. Indian J Ent 28:411–412

Pimental D (1980) Environmental and social cost of pesticides: a preliminary assessment. Oikos 34(2):126–140

Poehling HM (1989) Selective application strategies for insecticides in agricultural crops. In: Jepson PC (ed) Pesticides and nontarget invertebrates. Intercept, Wimborne, Dorset, pp 151–175

Pugh BM (1947) Varieties of bees found in Assam (Extract from the report of the Assistant Agriculture Officer, North East Frontier Agency, Assam, on "Agriculture in the Mynner (Jirang) state, Khasi Hills, Assam"). Indian Bee J 9:62

Ramachandran S, Mahadevan V (1951) The wax moths and their control. Indian Bee J 14:758–785

Ramchandra TV, Subashchandran MD, Joshi NV, Balachandran C (2012) ENVIS technical. report: 49, Centre for ecological sciences. Indian Institute of Science, Bangalore

Ragumoorthi KNMR, Srinivsan V, Balasubramani, Natrajan N (2007) Applied entomology. A.E. Publications, Coimbatore, pp. 26–28
Rana BS, Goyal NP (1991) Field toxicity of methyl demeton and dimethoate to the foragers of honeybee (*Apis cerana indica* F.) visitors to *Brassica chinensis* L. Indian Bee J 53(1–4):73–77
Rana BS, Kumar A, Gupta JK (2000) Influence of environmental factors on the population of predatory wasps of honey bees and evaluation of methods for controlling the wasps. Indian Bee J 62(1–2):47–54
Rasmussen C (2013) Stingless bees (Hymenoptera: Apidae: Meliponini) of the Indian subcontinent: diversity, taxonomy and current status of knowledge. Zootaxa 3647(3):401–402pp
Rasmussen C, Camargo JMF (2008) A molecular phylogeny and the evolution of nest architecture and behavior in *Trigona Spp.*(Hymenoptera: Apidae: Meliponini). Apidologie 39:102–118
Ratnieks FLW (1993) Egg-laying, egg-removal, and ovary development by workers in queenright honey bee colonies. Behav Ecol Sociobiol 32:191–198
Rennie J (1921) Acarine disease in hive bees: its cause, nature and control. N Scotland Coll Agric Bull 33:3–34
Rinderer TE, de Guzman LI, Lancaster VA, Delatteand GT, Stelzer JA (1999) Varroa in the mating yard: I. The effects of *Varroa jacobsoni* and Apistan on drone honey bees. Am Bee J 139:134–139
Ritter W (1981) Varroa disease of the honeybee *Varroa mellifera*. Bee World 62:141–153
Russell D, Meyer R, Bukowski J (1998) Potential impact of micro encapsulated pesticides on New Jersey apiaries. Am Bee J 138(3):207–210
Ruttner F (1966) The life and the flight activity of drone. Bee World 47:93–100
Ruttner F (1988) Biogeography and taxonomy of honeybees. Springer, Berlin
Sammataro D (1997) Report on parasitic honey bee mites and disease associations. American Bee Journal 137:301–302
Sammataro D, Cobey S, Smith BH, Needham GR (1994) Controlling tracheal mites (Acari: Tarsonemidae) in honey bees (Hymenoptera: Apidae) with vegetable oil. J Econ Entomol 87:910–916
Seeley TD (1985) Honeybee ecology. Princeton University Press, Princeton
Sharma SK, Kumar Y, Rana MK (2014) Status and prospects of beekeeping in Haryana. Workshop on promotion of honeybee keeping in Haryana held on June 24 2014
Shinde SG, Phadke RP (1995) Beekeeping in India: history, present status and future. Indian Bee J 57:–37
Singh S (1957) Acarine disease in the Indian honey bee (Apis indica F.). Indian Bee J 19:27–28
Singh S (1962) Beekeeping in India. ICAR, New Delhi, pp. 1–214
Singh D (2000) A focus on honeybees in the tropics. Curr Sci 79:1155–1157
Singh AK (2014) Traditional beekeeping shows great promises for endangered indigenous bee *Apis cerana*. Indian J Tradit Knowl 13(3):582–588
Singh KK, Gaira KS, Rai LK (2011) Agricultural scenario vis-a-vis the pollinator elements of the Sikkim Himalayan region. In: Arrawatia ML, Tambe S (eds) Biodiversity of Sikkim-Exploring and Conserving a Global Hotspot. Information and Public Relations Department, Government of Sikkim, Sikkim, pp 427–441
Singh MP, Chourasia HR, Agarwal M, Malhotra A, Sharma M, Sharma D, Khan S (2012) Honey as complementary medicine. Int J Pharm Bio Sci 3(2):2012
Sivaram V (2012) Status, prospects and strategy for development of organic beekeeping in the South Asian Countries. www.apiservices.com/…/23_02_12_concept_organic_south_asia.pdf
Soman AG, Chawda SS (1996) A contribution to the biology and behaviour of dwarf bee, *A. florea* F. and its economic importance in Kutch, Gujarat, India. Indian Bee J 58(2):81–88
Stark DJD, Jepson PC, Mayer DF (1995) Limitation to use of topical toxicity data for production of pesticide side effects in the field. J Econ Entomol 88(5):1081–1088
Sundararaju D (2003) Occurrence of bee fauna and extent of pollination in insecticide sprayed ecosystem of cashew. J Palynol 39:121–125

Suwannaong G (2000) Ultrastructure and pheromones of the mandibular glands of honeybee foragers in Thailand. Ph. D Thesis, Chulalongkorn University. pp 177

Suwannapong G, Seanbualuang P, Gowda SV, Benbow EM (2010) Detection of odor perception in Asiatic honeybee, *Apis cerana* Fabricius, 1793 workers by changing in membrane potential of the antennal sensilla. J Asia Pac Entomol 13(3):197–200

Suwannapong G, Benbow ME, James CN (2011) Biology of Thai honeybees: natural history and threats. Nova science publishers, New York, pp 1–98

Tewarson NC, Singh A, Engels W (1992) Reproduction of *Varroa jacobsoni* in colonies of *Apis cerana indica* under natural and experimental conditions. Apidologie 23:161–171

Thomas D, Pal N, Chawda SS, Rao KS (2001) Bee flora and migratory routes in India Proceedings 37th International Apiculture Congrress, 28 October–1 November 2001, Durban, South Africa

Thomas D, Pal N, Subba Rao K (2002) Bee management and productivity of Indian honeybees. Apidologie 3:1–5

Tirgari S (1971) On the biology and manipulation of *Apis florea* F. in Iran. 23rd International Beekeeping Congress, pp 330–332

Tribe GD, Fletcher DJC (1977) Rate of development of the workers of Apis mellifera adansonii L. In: Fletcher DJC (ed) African bees: their taxonomy, biology, and economic use. Apimondia, Pretoria, pp 115–119

Urtubey N (2005) Apitoxin: from bee venom to apitoxin for medical use. Termas de Rio Grande Santiago del Estero, Santiago del Estero

Visscher PK, Seeley TD (1982) Foraging strategy of honeybee colonies in a temperate deciduous forest. Ecology 63(6):1790–1801

Von Frisch K (1967) The dance language and orientation of bees. Harvard University Press, Cambridge

Von Frisch K (1971) Bees, their vision, chemical senses and language, Ithaca. Cornell University Press, NewYork

White JW (1962) Composition of American Honeys. Technical Bulletin 1261. Agricultural Research Service, U. S. Department of Agriculture, Washington, DC

White JW Jr (1980) Detection of honey adulteration by carbohydrate analysis. JAOAC 63(1):11–18

Wille A (1983) Biology of the stingless bees. Annu Rev Entomol 28:41–64

Winston ML (1979) Intra-colony demography and reproductive rate of the Africanized honeybee in South America. Behav Ecol Sociobiol 4:279–292

Winston ML (1987) The biology of honeybee. Harvard University Press, Cambridge

Winston ML (1992) The honeybee colony: life history; The hive and the honeybee. Dadant& Sons, Michigan, pp. 73–101

Winston ML, Fergusson LA (1985) The effect of worker loss on temporal caste structure in colonies of the honeybee (*Apis mellifera* L.). Can J Zool 63:777–780

Wolfgang R (1981) Varroa disease of the honeybee, Apis mellifera. Bee World 62:141–153

Wongsiri S, Rindererand TE, Sylvester HA (1991) Biodiversity of Honeybees in Thailand. Prachachon, Bangkok, pp. 50–63

Wongsiri S, Lekprayoon C, Thapa R, Thirakupt K, Rinderer T, Sylvesterand H, Oldroyd B (1996) Comparative biology of Apis andreniformis and *Apis florea* in Thailand. Bee World 77(4):25–35

Woyke J (1963) What happens to diploid drone larvae in a honeybee colony? J Apic Res 2:73–75

Woyke J (1969) A method of rearing diploid droned in a honeybee colony. J Apic Res 8:65–74

Woyke J (1973) Reproductive organs of haploid and diploid drone honeybee. J Apic Res 12:35–51

Zhong HE, Mark L, Sheve P, Cate B (2004) Minimizing the impact of the mosquito adulticide naled on honeybees, *Apis mellifera* (Hymenoptera: Apidae): Aerial Ultra volume application using a high-pressure nozzle system. J Econ Entomol 97(1):1–7

Diseases and Enemies of Honeybees

4

Sunita Yadav and H.D. Kaushik

Contents

4.1 Introduction

Like all other living creatures, honeybees also suffer from several diseases and are attacked by different enemies at`` all the stages of their life cycle. Bees have two distinct life forms (brood and adult), and most diseases are specific to either one stage or the other, but the most virulent diseases are those of the brood. The diseases in brood and adult honeybees are caused by bacteria, fungi, viruses, rickettsiae and protozoa. In India, brood diseases such as American foulbrood, European foulbrood, Thai sac brood and adult bee diseases, viz. acarine, *Nosema* and clustering disease, have been reported in Asiatic hive bee, *Apis cerana* Fab. European honeybees, *Apis mellifera* L., have been reported to suffer from European foulbrood, sac brood and chalkbrood diseases. Honeybees' enemies are those organisms which cause disturbances and nuisance in functioning of honeybee colony. Major enemies of honeybees are wax moths, mites, ants, wasps, birds, bee lice, hive beetles, mice, skunks and bears (Morhe 1999). Cockroaches, leafcutter bees, death's head moth, robber flies, dragonflies, praying mantis, spiders, etc. are some of the minor pests which causes nuisance in bee colony (Thakur and Sharma 1984). These pests cause great loss to honeybee colonies by destroying the raised combs, hives and hive parts.

S. Yadav (✉) • H.D. Kaushik
Department of Entomology, CCS Haryana Agricultural University,
Hisar 125004, Haryana, India
e-mail: sunitayadav10@rediffmail.com

Omkar (ed.), *Industrial Entomology*, DOI 10.1007/978-981-10-3304-9_4

Killing bees and brood, adversely affecting colony development, eating away the food reserves and causing nuisance to the bees and beekeeper and thus, reducing the colony productivity and returns per colony to the beekeepers. Although honeybees have a strong defence mechanism involving 'the sting' against most of the enemies, sometimes they need assistance from the beekeepers against the enemies. Hence, to minimize the losses due to diseases and enemies, proper management practices should be applied in time. The various enemies and diseases causing loss to honeybee colonies are discussed in detail in this chapter.

4.2 Diseases of Honeybees

Bees have two distinct life forms (brood and adult), and most diseases are specific to either one stage or the other, but the most virulent diseases are those of the brood. A brief account of various diseases and their management is given below:

4.2.1 Brood Diseases

They are generally easier to recognize as a group than adult diseases but are more difficult to control. These are either caused by bacteria or fungi or virus.

4.2.1.1 Bacterial Diseases

1. *American Foulbrood (AFB)*

This is the most dreaded and highly infectious bacterial disease of *A. mellifera* brood in many Western countries (Brodsgaard et al. 1998). In India, except an isolated report on *A. cerana* in 1961, it has not been reported in the hive bees from any part of the country. American foulbrood is caused by a spore-forming, rod-shaped, gram-positive bacterium called *Paenibacillus larvae* (Genersch et al. 2006). Young honeybee larvae become infected when they consume *P. larvae* spores in their food. The spores germinate in the gut; bacteria then move into the tissues, where they multiply enormously in number. Infected larvae normally die after their cell is sealed, and millions of infective spores are formed in their remains. In temperate and sub-tropical regions, it is the most virulent brood disease; heat and drought-resistant spores are capable of surviving for more than 35 years and germinate as they get the favourable environment (Chantawannakul and Dancer 2001). In tropical Asia, where sunlight is abundant and temperatures are relatively high throughout the year, the disease seldom causes severe damage. Adult honeybee workers are not affected by AFB spores but spread the spores to larvae while feeding them food contaminated with the infectious spores. The younger the larvae, the more susceptible they are to infection (Riessberger-Galle et al. 2001; Brodsgaard et al. 1998). Death of an infected larva takes place after the cell has been sealed and the cocoon has been spun. AFB kills the infected honeybees and proves lethal to the colonies if present in large quantities (Ratnieks 1992; Genersch 2008).

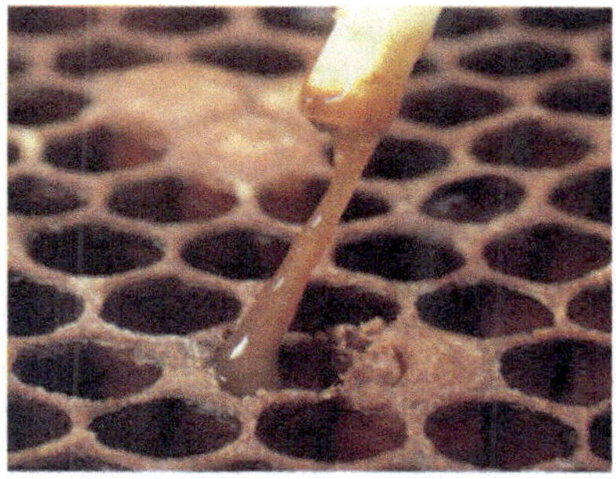

Fig. 4.1 Ropy thread due to AFB (Source:http://txbeeinspection.tamu.edu/american-foulbrood)

Symptoms

- At death, the diseased larva changes from a normal pearly white colour to a creamy brown and then gradually darkens.
- When a matchstick is thrust into the cell of the decomposed pupa, it draws out a ropy thread of several centimetres in length (Fig. 4.1).
- As the larva dries up, it becomes dark brown or black, rather rough scale that lies uniformly on the lower side of the cell. These scales stick very tightly to the cell wall and can be removed only with great difficulty (Genersch 2008).
- The decomposed brood has an unpleasant smell.
- The normal convex cell cap becomes moist, dark and sunken and later perforates. The perforation of the capped cells is the result of the attempt by the workers to uncap it to remove the decomposing remains.
- The brood combs of an affected colony become patchy in appearance (Fig. 4.2), owing to the presence of the intermixed diseased and healthy ones.

2. *European Foulbrood (EFB)*

It is less serious than AFB and is distributed in almost all areas where *A. mellifera* is present. In India, it has also been reported from *A. cerana* colonies of Maharashtra in 1971. Recently, there are reports of incidence of such type of disease in *A. mellifera* colonies in different parts of India. European foulbrood is caused by non-spore-forming bacterium called *Melissococcus plutonius* (Bailey 1983). The bacteria invade the midgut of 4–5-day-old larvae and multiply there, competing with the larva for its food. They remain in the gut and do not invade the larval tissue; larvae that die from the disease do so because they have been starved of food. This is a disease of unsealed (open) brood as the worker bees may leave the cell containing the dead larva uncapped. Sometimes the infected larva does not die until it is sealed, and this may result in sunken and perforated cappings. Subsequently other species of bacteria may multiply in the remains of dead larvae as 'secondary invaders', such as *Paenibacillus alvei*, *Enterococcus faecalis*, *Brevibacillus laterosporus* and *Lactobacillus eurydice*.

Fig. 4.2 Irregular brood pattern of AFB-infected comb

Symptoms

- Infected larvae move inside the cell instead of staying in the normal curled position; as a result, the dead larva is found in an unnatural coiled position across the mouth of its cell.
- Dead larvae become soft, watery and dull yellow. Their breathing tubes are prominent at this watery stage. The affected larvae are discoloured, first creamy yellow and then turn to light brown and then dark brown and occasionally black (Bailey 1961).
- The infected larvae lay upright attached with sidewalls of the cells and sometimes appear melted down at the base of the cells.
- Dead larvae finally dry and become brown removable rubbery scales at the bottom of the cell. Brood pattern becomes irregular (Figs. 4.3 and 4.4).
- The bacteria may not cause any odour in infected colonies. However, secondary invasion by other bacteria could cause sour or foul smell.

Spread of Bacterial Diseases

Within the colonies, these diseases are spread by nurse bees through their feeding activities. Besides the inoculum is also spread in the colonies by in-house/house-cleaning bees which pick up this inoculum during cleaning of the infected larvae from the colonies. However, due to the following reasons, the diseases are spread:

1. Intercolony substitution of bees, brood, queen bee and pollen or food reserves from infected colony to healthy colony
2. Robbing among the colonies
3. Drifting of bees of adjacent apiaries or through the infected foragers of the adjacent apiary by leaving inoculum on the crop flowers which are also visited by bees from healthier colonies of other apiary
4. Drifting of the bees among the colonies in the apiary
5. Use of contaminated equipment in the healthy colonies
6. Hiving of swarms of unknown origin in one's apiary may be a source of inoculum of diseases for their spread to the other healthy colonies

Fig. 4.3 European foulbrood in *A. cerana*

Fig. 4.4 European foulbrood in *A. mellifera*

Management

1. Weak colonies and colonies with a high proportion of diseased brood are destroyed, as with American foulbrood, but lightly diseased colonies may be treated with an antibiotic (0.5–1.0 g oxytetracycline in 500 ml concentrated sugar syrup/colony).
2. Sterilize the combs and other hive parts with formalin at 150 ml/l water, for 48 h at 43 °C in fumigation chambers.
3. Sterilize the combs with ethylene oxide at 1 g/l for 48 h at 43 °C in fumigation chambers.
4. Breeding disease-resistant strains of bees is one of the best measures for the disease management.
5. Burning of colonies including swarm shook (before destroying infested combs, bees are shaken into new comb foundation), coupled with provisioning of either brood alone or brood + pollen combs from the healthy colony, is effective in controlling the disease (Wilkins et al. 2007). This method is commonly followed in European countries.

4.2.1.2 Fungal Diseases

Two fungal diseases are important, viz. chalkbrood and stonebrood.

Fig. 4.5 Chalkbrood-infected comb

1. *Chalkbrood*

This is an extremely common brood disease caused by spore-forming fungus, *Ascosphaera apis* (Maassen ex Claussen) Olive and Spiltoir (Spiltoir and Olive 1955). The threadlike, vegetative growths ('hyphae') of the fungus invade the body tissues of infected larvae, killing them after they have been capped over in their cells. The spores of fungus remain viable for years. The disease is most prevalent in the spring when the brood area is expanding, and the weather is still cool and there are not enough nurse bees to maintain the brood nest temperature (Flores et al. 1996; Borum and Ulgen 2008). Its endemic infection is damaging; otherwise it is a less serious disease. It affects only the brood. Brood cells can be sealed or unsealed. The disease causes significant reduction in bee strength and honey production (5–37 %) but is not usually fatal for the entire bee colony (Wood 1998; Zaghloul et al. 2005; Aronstein and Murray 2010). Workers, drones and queens are all susceptible to the disease. Three–four-day-old larvae and those on periphery of brood area are more susceptible.

Symptoms

- Diseased larvae are stretched out in their cells in an upright position. Dead larvae from chalkbrood disease are chalk white and are often covered with cottony filaments, hence the name 'chalkbrood' (Fig. 4.5).
- Sometimes the diseased larvae can be mottled with brown or black spots, especially on the ventral sides. The white coloration may eventually turn into grey or black, depending on the life stage of the fungus (Aronstein and Murray 2010).
- Chalkbrood mummies, once dry, are loose in the cell and can be removed easily. Often, a few of these mummies are visible on the ground at the entrance to the hive.

2. *Stonebrood*

Stonebrood disease is caused by several fungi from the genus *Aspergillus* which causes mummification of the brood of a honeybee colony. *Aspergillus* infects the bee through the cuticle as well as through the gut if the spores are ingested. *Aspergillus* also produce aflatoxins which are the major reason of death in stonebrood-infected honeybees.

Symptoms

- Its spores are ingested with food and germinate in the gut, growing rapidly to form a collar-like ring near the head.
- After death the larvae turn black and become difficult to crush, hence the name 'stonebrood'.
- Eventually the fungus erupts from the integument of the larvae and forms a false skin.
- The affected adult bees show restlessness, feebleness and paralysis, and the abdomen gets dilated and then mummified.
- Younger bees die earlier.

Management of Fungal Diseases

- There are no specific treatments available for chalkbrood. The most effective control results from avoiding the conditions favourable to *A. apis* increase by maintaining strong and vigorous colonies and bees that show marked hygienic behaviour. Use clean equipment and maintain hives clean and well ventilated (Gochnauer et al. 1975).
- In severe cases, re-queening with a queen from a chalkbrood-free colony is recommended.
- Avoid using antimicrobial chemicals inside bee colonies as they are detrimental for bee health (Bogdanov et al. 2004; Frazier et al. 2008). Apiguard (a varroacide treatment based on thymol) has been reported effective for the management of this fungal infection.
- Removal of mummies by bees results in natural control of the diseases. Assist the bees by collecting and burning the mummified larvae manually.
- For maintaining good colony strength, supplement the colony with good-quality feed.
- Avoid transferring of combs between colonies and change old, blackened brood combs annually as these may anchorage chalkbrood spores (Malonova and Titera 1995; Flores et al. 2005a, b).
- Use irradiation to sterilize contaminated beekeeping equipment, old frames and honeybee combs (Baggio et al. 2005).

4.2.1.3 Viral Diseases

1. *Sac Brood Virus (SBV) and Thai Sac Brood Virus (TSBV)*

Sac brood is a very common virus disease affecting brood. Thai sac brood virus (TSBV) of *A. cerana* brood was noticed for the first time in Thailand during 1976 and in India from Meghalaya during 1978. This disease has caused great economic

Fig. 4.6 Sac brood-infected comb

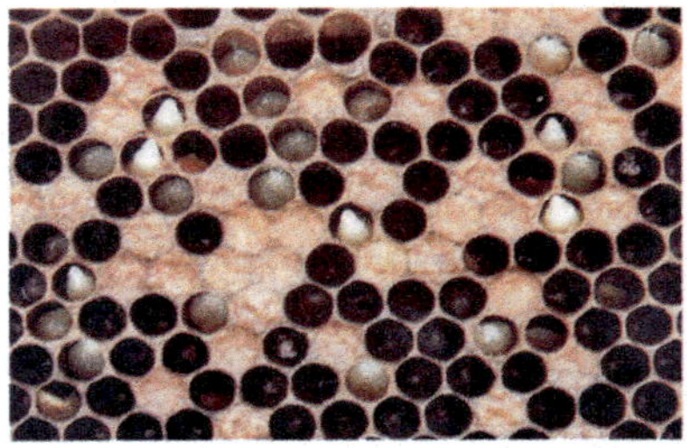

losses to apiculture industry by killing more than 95 % of *A. cerana* colonies in India. It is more destructive than that of very closely related sac brood virus (SBV) disease of *A. mellifera* brood (White 1917; Bailey et al. 1964). However, the signs can sometimes be mistaken for those of AFB. It often goes unnoticed since it usually infects only a small portion of brood, and adult bees will usually detect and remove infected larvae.

Symptoms

- Larva in late stage or near cell sealing or as perusal stage dies.
- Its body stretched on the back.
- Cell cappings are sunken, and brood is patchy. Sometimes the infected larva does not have cell capping (Fig. 4.6).
- Dead larva remains upright in the cell. Its skin becomes tough and colour changes to pale yellow to brown.
- Head and thorax regions are darker.
- Larva, if pulled out with a tweezers, comes out like a sac.
- Dried scale is boat shaped.
- Sometimes bees leave the colony.

Infection, Multiplication and Spread

Virus is ingested with food. Two-day-old larvae are more susceptible. Virus multiplies in body tissues. One dead larva may contain 1 mg virus which is sufficient to infect 1000 colonies. Disease in the colony is spread by nurse bees and among the colonies through swarming, drifting and robbing (Singh et al. 2010).

2. *Chronic Bee Paralysis Virus (CBPV)*

Chronic bee paralysis virus causes chronic bee paralysis disease in honeybees. This disease is also called hairless black syndrome/little black robbers. Infected bees die within a week. The casual viral particles are irregular in shape. Enzyme ribonuclease found in nectar destroys viral RNA. Hence, there can be recovery with flow of nectar.

Symptoms

- Adult bees appear black, hairless and shiny initially and suffer nibbling attacks from healthy bees of their colony.
- Quivering of the wings and bodies of adult bees.
- Adult bees are unable to fly and exhibit K-winged symptoms and crawl along the ground and found clinging to blades of grass in the immediate vicinity of the hive.
- The bees may also have a swollen abdomen, causing dysentery, and will die within a few days (Genersch and Aubert 2010). Severely affected colonies collapse suddenly.

3. *Iridescent Virus*

Apis iridescent virus (AIV) is only the example of an iridovirus from honeybee which was isolated from Indian hive bee, *A. cerana*. This virus was reported from Kashmir and other parts of North India in 1975 and causes 'clustering disease' in adults of *A. cerana* as well as in *A. mellifera* bee colonies. It multiplies in the alimentary tract, hypopharyngeal glands, fat bodies and ovaries of bees.

Symptoms

- During summer, infected bees form 'cluster' initially inside and later outside the hive walls.
- Bees are found crawling on the ground.
- Egg laying, brood rearing and foraging activities are nearly stopped, and contaminated colonies die within 2 months.

4. *Kashmir Bee Virus (KBV)*

KBV was similarly discovered in 1974 as an impurity in preparations of *Apis* iridescent virus from the Asian hive bee (*Apis cerana*) that multiplied to high titres when injected or fed to adult *Apis mellifera* (Bailey et al. 1976, 1979). It is considered the most virulent virus under laboratory conditions (Allen and Ball 1996). When KBV is injected into adult bee haemolymph, death occurs in just 3 days (de Miranda et al. 2012). The Kashmir bee virus usually develops in bee families affected by the Varroa mite as mites have the ability to transmit viruses due to the piercings they make into the cuticle of adult bees. It can kill colonies even at moderate levels of mite infestation (Todd et al. 2007). The Kashmir bee virus is very similar with the virus that causes acute paralysis in honeybees.

Symptoms

- Bees are found crawling in front of the hive and form cluster outside the hive.
- Bees move irregularly in circles, their bodies tremble, and they die in huge numbers, especially during December and January.
- Death of infected bees occurs within a period of 1 week.

Management of Viral Diseases

- For viral pathogens, there is no chemical control.
- Affected colonies should be isolated beyond their flight range.
- Adopt all the management operations to keep colonies strong.
- Provide proper ventilation to reduce humidity.
- Cage the queen for a week and then re-queen.
- Use sterilized equipment/combs.
- Check robbing, drifting and swarming.
- Provide supplement feeding.
- Undertake selective breeding for natural resistance.

4.2.2 Adult Bee Diseases

The diseases of adult bees are caused by protozoa which are single-celled animals and form spores or cysts. They multiply by sexual or asexual methods. Their infection reduces vitality of bees and shortens their life and fecundity. Protozoan is perfect parasites as they do not kill the host immediately. These diseases are difficult to diagnose, though inability to fly, unhooked wings and dysentery can be treated as general symptoms of an unhealthy bee. Microscopic examination is often necessary for a definite diagnosis.

4.2.2.1 *Nosema* Disease

This disease is caused by *Nosema apis* Zander. It is a disease of adult bees. It parasitizes all the castes. Their spores germinate in the ventriculus of the host. Pathogen multiplies in epithelial cells, and it checks RNA synthesis in the host cells. Its spores are shed in the lumen of the digestive tract of the affected bee and are then excreted out. One affected bee may contain 180 million spores. Hypopharyngeal glands of the diseased bee are atrophied. Colony strength dwindles down. Infection spreads through ingestion of food contaminated with faecal matter.

Symptoms

- Bees start foraging at younger age.
- Bees feel fatigued, are less able to fly and fall down during their return journey.
- Bees crawl up the grass blades and fall down on the ground, and such affected fatigued bees gather in depressions/ditches.

Fig. 4.7 Nosema disease in *A. cerana*

- The abdomen is distended with faecal matter.
- Body hairs are lost and bees become shiny.
- The mid-intestine is swollen and, if dissected, shows dull greyish-white contents.
- Bees soil the hive entrance (Fig. 4.7).

Management

- Provide fresh running water. Drain off stagnant water from the apiary.
- While transporting queens, select healthy attendant bees.
- Provide upward ventilation to reduce humidity.
- Feed fumagillin in concentrated syrup. It inhibits DNA replication of the pathogen.
- Disinfect the empty hives with ethylene oxide or acetic acid fumigation at 120 ml/hive.

4.2.2.2 Amoeba Disease

It is caused by *Malpighamoeba mellificae*. The infection is caused by ingesting the cysts along with contaminated food. Cysts germinate, and amoeba migrates to Malpighian tubes and feeds on cell contents. Amoebae multiply by binary fission and form cysts within 18–28 days of ingestion. Cysts accumulate in the midgut/rectum. Peak infestation occurs during April–May. Spring dwindling of colony strength can be experienced in such case.

Management

- Ensure proper hygienic conditions.
- Scarp off the bottom board and disinfect it with 2 % carbolic acid.
- Disinfection of hives and equipment with acetic acid is also helpful.

4.2.3 General Practices for the Management of Diseases

Honeybees could be affected by diseases, and the real cause of abnormality or any disease present in the honeybee broods needs to be ascertained before taking up any control measures. It is best to contact the researchers/scientists/beekeeping experts at the nearest centre or university or government department working on honeybees. After the exact diagnosis of the causative agent of the particular disease, the guidelines/recommendations given by the expert should be followed in true letter and spirit. However, general advisory for the management of common diseases of honeybees is given below:

1. Select good site to locate the apiary preferably in an open, dry place with shade.
2. Adopt general colony hygiene in the apiary like cleanliness in the beehives including cleaning the bottom board frequently.
3. Select and multiply honeybee colonies only from disease-resistant stocks.
4. Keep colonies with good prolific queens.
5. Create broodlessness in colony for at least 15 days by enclosing the queen in a queen cage.
6. Check the colonies periodically for any abnormalities or changes in behaviour of bees.
7. If you observe any colonies with disease, isolate them from healthy ones. Handle diseased and healthy colonies separately.
8. Keep the colonies strong by adding sealed brood comb or worker population only from healthy colonies and also by providing adequate food during dearth periods.
9. Prevent robbing, drifting and absconding and avoid migration of bee colonies when you notice disease symptoms.
10. Follow 'shook swarm' or shaking method to remove contaminated combs completely by transferring entirely new combs in one operation to the colonies with disease symptoms. Destroy the removed combs by burning.
11. Sterilize the combs and equipment by any one of the following methods:
    (a) Disinfect the empty combs and equipment with 80 % acetic acid at 150 ml per hive body in piles for few days at a protected place. Air the treated materials before use.
    (b) Dip the contaminated equipment and combs in soap solution containing 7 % formalin for 24 hours. Then wash the treated material with water, dry and use.
    (c) Disinfect the combs with UV rays in protected chambers/UV chambers, where possible.
12. Use of antibiotics to control honeybee diseases is likely to result in contamination of honey, causing problems in export of honey.

4.3 Enemies of Honeybees

Honeybees are attacked by a number of enemies which take a heavy toll of bee life, and their destructive activities result in desertion of hives by bees. Bee enemies are described under five categories, namely, insects, arachnids, aves, reptiles and mammals.

4.3.1 Insects

The insect enemies of the honeybees belong to phylum Arthropoda and class Insecta.

4.3.1.1 Wax Moth (Lepidoptera: Pyralidae)

The wax moth is regionally called as the bee moth, the wax (or bee) miller or a webworm. Wax moths are the major pest of beehives and can cause substantial losses to combs, damage the beehive materials and spoil beehive products. They are capable of causing a lot of damage in a very short period of time. *Galleria mellonella* L. (greater wax moth) and *Achroia grisella* F. (lesser wax moth), *Vitula* spp. (dried fruit moth), *Plodia interpunctella* (Hbn.) (Indian meal moth), *Ephestia kuehniella* (Zell.) and *E. cautella* (Mediterranean flour moth) are associated with colonies of honeybees (Ritter and Akratanakul 2006). Among these, two species of wax moths, viz. *G. mellonella* and *A. grisella*, are responsible for enormous damage in beekeeping industry, so these are discussed in detail. Out of these two species, *G. mellonella* is more damaging. Damage occurs mainly in the warm and hot months of the year when wax moths are most active. However, considerable damage can still occur during the cool part of late autumn and early spring as greater wax moth can produce a large amount of metabolic heat which can raise the immediate temperature around them by up to 25 °C above the normal environment temperature. Stored and deserted combs, improperly cleaned wax, weak or poorly managed colonies and deserted combs of wild bees are constant sources of wax moth population. Depending upon the availability of food, temperature and habitat of the pest, several overlapping generations can be produced in a year. Wax moths are active from March to October, but its peak activity has been observed from June to November (Ramachandaran and Mahadevan 1951). In South India, maximum infestation of this pest was noted during the dearth period (Viraktamath 1989). It hibernates in larval (about 70 %) and pupal stages (about 30 %) in stored combs.

Identification and Life Cycle of Greater Wax Moth Greater wax moth adult are heavy bodied, brownish grey, 10–18 mm in length (Fig. 4.8). The females are lighter in colour, larger and heavier than males. In the females, the outer margin of forewing is smooth, while semilunar notch is found in males. Labial palp in females is extended forward and head appears beak-like. Its life cycle is completed in four stages, viz. egg, larva, pupa and adult. The female moth starts laying eggs 4–10 days after emergence. Eggs are laid in clusters of 50–150 in small cracks and crevices. Single female

Fig. 4.8 Adult of *Galleria mellonella*

lays on an average of 300–600 eggs (the number may reach up to 1800) in its lifetime of 2 weeks (Milum and Geuther 1935; Khanbash and Oshan 1997). Eggs are smooth, spherical in shape and pinkish to creamish white in colour with size ranging from 0.4 to 0.5 mm. Larva is white to dirty grey in colour, 3–30 mm in size. It lives in long silken tunnels, and after hatching, it feeds on honey, nectar and pollen. The larva makes burrows/tunnels in combs and extends to the midrib of comb. The larva moults 4–6 times in its life. Larval period is between 22 and 60 days (Jyothi and Reddy 1994; Khanbash and Oshan 1997) sometimes extending up to 100 days (Allegret 1975) depending on abiotic factors. The larva moves to the hive's body and makes a small depression in the wood in which it pupates. These cocoons may be found on inner walls of chamber, on inner cover and on frames. Pupation takes place in silken cocoons spun around them by last instar larvae. Pupa is brownish white to dark brown, 14–16 mm long in size. The cocoons are present in clusters and are generally white in colour. Pupal period is of 7–60 days (Kapil and Sihag 1983; Jyothi and Reddy 1993; Brar et al. 1996). Life cycle completes in 6 weeks to 6 months. The effect of temperature, relative humidity and diet on development and metamorphosis of *Galleria* has also been observed by various workers (Burkett 1962; Bogus and Cymborowski 1977; Chauvin and Chauvin 1985; Kumar 2000).

Identification and Life Cycle of Lesser Wax Moth The lesser wax moth is far more widespread and abundant than greater wax moth and is found comparatively at higher altitudes. It is troublesome particularly in stored combs. It is smaller than the greater wax moth with slender body about 13 mm in length and silver grey in colour without markings on wings (Fig. 4.9). Egg stage varies between 2 and 4 days, larval period between 34 and 48 days and pupal period between 5 and 12 days; adult longevity is about 7 days. *A. grisella* larvae are 15–20 mm in size and are usually white with a brown head. They feed on combs, pollen and litter found on the hive floor. They are usually solitary, whereas greater wax moth larvae often congregate in large numbers. Female moth may lay 250–300 eggs. They complete 3–4 generations during active season (Singh 1962).

Nature and Extent of Damage Many people consider greater wax moth as a useful insect because its larvae are used as fish bait in many countries. However, it

Fig. 4.9 Adult of *Achroia grisella*

Fig. 4.10 Damage done by wax moth

causes major losses to commercial beekeepers every year. Almost all colonies of Asian honeybee are prone to the moth infestation (Adlakha and Sharma 1975; Viraktamath 1989). *Apis mellifera* is less prone to the attack of wax moth than other *Apis* species because of the habit of collecting more propolis. Wax moth population starts building from March, reaches its peak in August (99–100 %) and then shows decline till February (Thakur 1991). During dearth and monsoon period, damage is increased to many folds. The moth infests combs with all stages of brood, cells, pollen and honey, converting it into mass of silken webbings (Fig. 4.10). Wax moth larvae spun silken galleries around them near the mid-rid of the brood comb causing galleriasis, a condition in which adult bees are unable to come out of cells as their legs get entangled in the silken galleries underneath. Sometimes wax moth larvae chew through the cappings, usually in a straight line. Worker bees chew the remainder of the capping thereby fully exposing the heads of the pupae, a condition known as bald brood. Wax moth larvae can reduce the combs to a mass of webbings and debris. Severe infestation leads to suspension in brood rearing, foraging activity and ultimately desertion of colony. Weak colonies (53 %) are more susceptible to wax moth infestation as compared to strong colonies (11 %) (Thakur 1991). Other workers (Newton 1917; Jyothi et al. 1990) reported 100 % infestation in deserted combs. Nielson and Brister (1977) found that wax moth attraction was more towards strong and active than weak colonies.

Management Once infestation sets in, it is difficult to control the pest. To prevent and manage wax moth infestation in apiary and stores, the following management practices can be undertaken:

- Best defence is to maintain strong, healthy colonies, closing all cracks and crevices of the hive and reduction of entrance, which gives effective control of wax moth (Ramachandaran and Mahadevan 1951). Unlike *A. mellifera*, the Asian honeybees are very poor propolizers; therefore, closing of cracks and crevices through artificial material is often recommended.
- Apiary hygiene is essential to prevent the infestation of wax moth, and with regular inspection of beehives, it can be managed timely. Good sanitation inside the hive particularly bottom board and control of diseases and other pests are a must to keep the colonies strong.
- Pesticides use should be avoided as they reduce the strength of bees and brood area and kill forager bees (Abrol and Kumar 2000; Menon 1992; Rana and Goyal 1991).
- Excess combs/frames in the hive should be removed especially during dearth period.
- Keep empty infested combs in the sun for a few minutes.
- During initial stages of infestation, destroy the tunnels to kill wax moth larvae.
- During advanced stage of infestation, the whole comb has to be destroyed to check wax moth population.
- Do not leave the empty combs in open for a long period at the apiary site.
- Use newer raised combs or foundation sheet and follow a regular comb renovation programme. The extent of damage caused to combs by *G. mellonella* increases significantly with the age of combs from 6 to 18 months. Thus, the younger the combs, the lesser the damage suffered from *G. mellonella.*
- Artificial cold at –6.7 °C for 4–5 h or –12.2 °C for 3 h or –15 °C for 2 h is effective in killing all stages of wax moth. The least comb area was damaged (40.82 cm^2) in the combs placed in deep freezer followed by sulphur fumigation, NSKE spray and spray of Bt var. kurstaki with comb area damaged being 65.31, 70.04 and 78.31 cm^2, respectively (Yadav et al. 2012).
- All stages of the greater wax moth are killed by exposing empty combs at a temperature of 46 °C for 80 minutes or a temperature of 49 °C for 40 minutes. Be careful not to expose honey combs to temperatures in excess of 49 °C. Naini and Bisht (1972) advised exposure of brood and super chambers to 55 °C temperature with 100 watt bulb for 1 h to kill larvae and pupae of *A. grisella*. Heat-treat only those combs having very little or no honey (combs softened at high temperatures may sag and become distorted). Provide adequate air circulation for the heat to be evenly distributed throughout the comb.
- Wax moth trap can be used for capturing the adult moths. Wax moth trap can be prepared by cutting one 1.25 in. diameter hole just below the slope on the bottle neck of a plastic bottle (2 l). Fill the bottle with the mixtures of vinegar and sugar (1:1). Add one finely chopped banana peel and fill the bottle with water up to 75 % of its capacity. Keep the bottle for 15 days for fermentation as freshly prepared

baits are more attractive to bees and result in high bee mortality. The adult of wax moth enter through the hole of a trap and were not able to escape outside resulting into death after drowning in the liquid (Chhuneja and Yadav 2009).

- Stored combs in airtight rooms/chambers can be fumigated with chemicals to kill the wax moth larvae. Smouldering sulphur, ethylene bromide or acetic acid, *para*-dichlorobenzene (PDB), methyl bromide, calcium cyanide, phosphine, etc. are some of the fumigants which provide effective protection against wax moth infestation (Casanova Ostos 1992). PDB crystals are least hazardous, but they don't kill eggs and also cannot be used on comb honey. But they kill effectively the young larvae emerging from combs placed in storage, repel moths and also prevent egg laying. Supers should be placed in open before using them on colonies.
- Carbon dioxide (CO_2) at concentrations above 95 % can also effectively control wax moth.
- Use of birational methods to check wax moth population particularly the use of *Bacillus thuringiensis* and *Galleria* nuclear polyhedrosis virus (GNPV). Among these major work has been carried out on *Bt* formulations (Ali et al. 1973a, b; Battu and Singh 1977; Cantwell and Shieh 1981). Smaller larvae of *Galleria* were more susceptible to *Bt* treatment than large larvae. (Burges and Bailey 1968). Likewise third instar larvae took more time to get killed than second instar larvae. LC50 values for second and third instar larvae against *Bt* formulations were calculated by Kumar (2000) which were 2.13–2.45 g/l and 2.51–3.46 g/l, respectively. *Bt* formulations were found to be more toxic when injected into haemolymph as compared to its administration in food (Schmid and Berg 1969; Vankova and Leskova 1972). Verma (1995) recorded highest wax moth mortality (98.7 %) in the comb sprayed with *Bt* suspension at 10 g per litre of water which remained effective for 5.5 months. However, combs dipped in *Bt* suspension provided protection against wax moth for 13 months. Similarly *Bt* formulation application at 0.5 g/l/hive gave the highest control (73–85 %) of wax moth followed by sulphur dusting (70.00–75 %) as and when wax moth appeared in the bee boxes (Deka et al. 2010).
- Some hymenopteran parasitoids specifically *Apanteles galleriae* Wilkinson (Braconidae) attack the caterpillars of *G. mellonella* (Ahmad et al. 1983). *Apanteles galleriae* started parasitizing only from second instar larvae of *G. mellonella*. The percent cumulative parasitization during different larval instars revealed that the third instar larvae of *G. mellonella* were preferred the most for parasitization (88.89 %) followed by the fourth instar larvae in which 68.89 % parasitization was recorded (Chhuneja and Yadav 2011).

Other Moths In addition to greater and lesser wax moths, other moths attacking honeybee colonies are dried fruit moth, *Vitula edmandsae*. Its larvae commonly feed on pollen and honey in unprotected stored combs but occasionally found in the combs of strong colonies also (Okumura 1966; Wilson and Brewer 1974). They are mottled grey in colour and 20 mm long. The development from egg to adult requires about 88 days (Okumura 1966). They can complete their development on combs

Fig. 4.11 Fire ants *Monomorium* sp. attacking *Apis mellifera* colonies (Source: Khan and Srivastava 2010)

without destroying the midrib or the entire comb, unlike *G. mellonella.* Pheromone traps (Scott et al. 1984) or sulphur dioxide fumigation (Szabo and Heikel 1987) was found to be effective for preventing its infestation in stored combs.

Indian meal moth, *Plodia interpunctella*, larvae were also found feeding on the pollen, cocoons or dead brood in stored combs (Eckert and Shaw 1960; Wilson and Brewer 1974). The life cycle is completed in 4–6 weeks. In case of severe infestation, loose flimsy webbing across the face of combs is seen. Cold storage at subfreezing temperature in airtight containers is the best method of protecting stored pollen from *Plodia* attack (Whitefoot and Detroy 1968).

Mediterranean flour moth *Ephestia* sp. also sometimes attacks stored combs that contain pollen, but this moth cannot develop on empty brood combs or dead insects (Eckert and Bess 1952).

4.3.1.2 Ants (Hymenoptera: Formicidae)

Ants are not usually serious pests in honeybee colonies (Fig. 4.11 and 4.12, Table 4.1). Occasionally, however, certain species may enter colonies in search of food or establishing nesting sites. Ants are typically found between the inner and outer covers of the hive and in pollen traps. Both the domesticated species of honeybees may face problem due to ants' attack. The weak colonies are more vulnerable and may sometime abscond due to ants' attack. Ants' attack is usually recorded in pre-monsoon period. The ant species damaging honeybee colonies are listed in Table 4.1. Various researchers have observed different ant species, which attack honeybee colonies for honey, pollen and brood (Buys 1990; Woodward and Jones 1991; Abrol 1997). Argentine ants, *Iridomyrmex humilis* Mayr, are capable of destroying strong, populous colonies. In South Africa, it is known as a serious pest of honeybees (Buys 1990). Persistent attacks by ants induce absconding in *A. mellifera* and *A. cerana* colonies. Poneroid ants especially *Eciton* sp., *Anomma* sp. and *Dorylus* sp. kill honeybees. Ants attack in groups of thousands, which can destroy an entire apiary within few hours (Dubois and Collart 1950). *Camponotus compressus* F. (carpenter ant) is occasionally a serious pest of bee colonies in India (Singh 1962; Thakur 1991) and the USA (Walshaw 1967). In India, Singh and Nairn (1994)

Table 4.1 Species of ants associated with honeybees

S. no.	Common name	Scientific name	Subfamily	Nesting site	Hosts
1.	Weaver ants	*Oecophylla smaragdina*	Formicinae	Leaves of trees (arboreal)	Primarily insects including bees
2.	Fire ants	*Solenopsis* spp., *Monomorium* spp.	Myrmicinae	Mounds in open areas or under object such as timber, logs, rocks, brick, etc.	Plants, seeds, crickets, honeybees; bite and sting
3.	Wood ants/ mould ants/ field ants	*Formica* spp.	Formicinae	Large mounds over soil or under stones or logs or in stumps	Bites and secreting formic acid
4.	Black carpenter ant	*Camponotus* spp.	Formicinae	Dead trees and soft dead wood	Omnivorous
5.	Army ants	*Eciton* spp.	Ecitoninae	Temporary nests (nomadic)	Carnivorous all animal material
6.	Driver ants	*Dorylus* spp.	Dorylinae	Temporary nests (nomadic)	Carnivorous all animal material

(Source: Khan and Srivastava 2010)

Fig. 4.12 Fire ants *Monomorium* sp. (Source: Khan and Srivastava 2010)

also reported *Tetraponera rufonigra* as pest of honeybees, whose attack resulted in partial (8–18 %) to complete (8–9 %) destruction of *Apis cerana* colonies. The small red household ant, *Dorylus labiatus*, and small black ants, *Monomorium indicum* and *M. destructor*, are some of the other ant species which visit bee colony for food purposes. Besides causing direct damage to honeybee colonies, ants may be nuisance to beekeepers and may sometimes cause painful irritation through their bites and stings. Even though majority of ant species seldom disturb the bees, these can be nuisance to the beekeeper.

Management

- Keep the colonies strong in the apiary as weak colonies are more vulnerable to ants' attack.
- Colonies capable of defending by fanning should be selected and used as breeder colonies for mass rearing of queen bees.
- Though bees use propolis to fill cracks and crevices, but mud and plaster of paris can also be used to seal cracks and crevices.
- Effective control of ants was recorded when legs of hive stands were smeared with corrosive mercuric chloride sublimate (May 1961) or spent engine oil and grease. Legs of the stand in broad earthen cups containing water also check upward movement of ants.
- Searching ants' nest in the vicinity of the apiaries and disturbing or driving away the ants by using repellents such as ethanol, sodium fluoride, sulphur, borax, kerosene oil, etc. are effective in reducing their attack.
- Keep the apiary clean by removing the dead logs, rotten woods and stones and cut the grass regularly.
- The use of chemicals like ethyl or methyl alcohol, sodium fluoride, borax powder, salt or powdered sulphur for ant control is also available in literature. Carbon disulphide fumigation (2–4 tsp.) or 0.1 % aldrin solution to destroy underground nests of ant (Thakur et al. 1981) is also in practice. Woodward and Jones (1991) recommended the use of pyrethroids and organophosphates for the control of ants. Many workers advised dusting with turmeric powder (Abrol and Kakroo 1998) to keep ants away from the hives.

4.3.1.3 Wasps and Hornets (Hymenoptera: Vespidae)

Several species of wasps and hornets prey on honeybees causing severe damage to bee colonies which leads to loss of entire apiaries (Dave 1943; Subbiah and Mahadevan 1957, Sharma and Deshraj 1985). In commercial apiaries, the wasp and hornet invasion starts after spring and continues during monsoon season causing maximum damage to colonies during July–September. About 20–25 % of bee colonies desert their nest every year due to wasp attack (Adlakha and Sharma 1975). The largest of the social wasps, *Vespa* sp., are physically capable of preying on honeybees with ease. *Vespa orientalis* L. (yellow-banded brown wasp), *V. magnifica* Smith (large black wasp), *V. cincta* F. (yellow-banded wasps), *V. ducalis* Smith and *V. auraria* Smith (golden wasps) are some of the species which destroy the weak and queenless colonies for honey and the brood in apiaries (Kshirsagar and Mahindre 1975) and foragers in field (Abrol 1994; Abrol and Kakroo 1998; Sihag 1992). Shah and Shah (1991) observed *V. velutina* as a serious pest of honeybees in Kashmir. A group of 30 *V. magnifica* was able to kill 25,000–30,000 bees in Japan in just three hours (Akre and Davis (1978). Hirschfelder (1952) estimated that a single female wasp consumes 60–80 bees as a food during her lifetime. Other wasp species associated with bee colonies include *Philanthus ramakrishna* T. and *Palarus orientalis* (Kohl), also known as bee hunter wasps (Thakur 1991). *Vespa tropica* is a fast flier wasp and mostly catches the forager bees (Garg and Kashyap 1998).

Fig. 4.13 *Vespa orientalis*

Fig. 4.14 *Vespa tropica*

Fig. 4.15 *V. mandarinia*

Identification and Life Cycle In North India, four species, viz. oriental hornet, greater banded hornet, Asian giant hornet and yellow-legged hornet, have been identified creating economic problems to beekeepers (Figs. 4.13, 4.14, 4.15 and 4.16). Their nesting site and identification characteristics are given in Table 4.2. A single mated queen (foundress queen) establishes new colony during spring season, which becomes populous during monsoon and attains a peak during autumn. After coming out from hibernating sites (dim and moist places, e.g. cavities in the trees, soil, under the roofs, etc.), they fly to seek tree sap. After taking tree sap, they move to new nesting sites and start making 'embryo nests' (15–60 cells till the first emergence of the workers). After the emergence of the first batch of workers, division of labour takes place, and subsequently the colony develops rapidly. The emergence of new queens and males takes place. The production of males and queens continues for 1–2 months. The foundress queen survives till the emergence of new queens.

Fig. 4.16 *V. velutina*

Table 4.2 Wasps and hornet species associated with honeybees in Asia

Sl. no	Common name	Scientific name	Nesting site	Identification marks
1.	Oriental hornet	*Vespa orientalis*	Arboreal in hollow trees	Medium-sized hornet. Workers generally 18–25 mm. Fully light brown or reddish-brown ground colour. Parts of the head (viewed from the front) yellow. Third and fourth abdominal segments mostly yellow with distinctive markings (Fig. 4.13)
2.	Greater banded hornet	*Vespa tropica* (*Vespa cincta*)	Usually underground, tree hollows	Medium to large sized
				Queen – 30 mm
				Male – 26 mm
				Worker – 24–26 mm, second segment bright yellow, rest abdomen black. Head black. (Fig. 4.14)
3.	Asian Giant hornet	*V. mandarinia*	Arboreal	Largest hornet (worker avg. length 35–39 mm), head yellow. Thorax usually black, abdomen mainly brown with yellow bands and last segment yellow (Fig. 4.15)
4.	Yellow-legged hornet	*V. velutina*	Arboreal (low in bushes and shrubs and underground)	Avg. body length: Worker, 20 mm, thin yellow line at end of abdominal segment, head black, fifth and sixth segment black, pronotum reddish brown, third and fourth segment dark yellow orange (Fig. 4.16)
5.	Lesser-banded hornet	*Vespa affinis*	Arboreal, open looks similar to *V. velutina*	First and second abdominal segment yellow
6.	Yellow-vented hornet	*Vespa analis*	Arboreal, 6–10 ft. height above ground	Usually black with only tip of abdomens (sixth segment) yellow; however variation occurs

(continued)

Table 4.2 (continued)

SL no	Common name	Scientific name	Nesting site	Identification marks
7	European hornet/ old word hornet/ brown hornet	*Vespa crabro*	Arboreal in hollow trees, hollow wall, attics, barns	Adult nocturnal, avg. body length (queen) 25–35 mm; antenna has 13 segment (male) and 12 segment (female). Seven segmented abdomen (male) and six segmented (female)
8.	Bee hunter wasp	*Palarus orientalis* Kohl.	–	Black in colour with transverse yellow lines on the abdomen
9.	–	*Vespa binghami*	–	Mainly brown with lighter yellow head, prominent three ocelli between compound eyes. Nocturnal in habit
10.	Bee hunter wasp	*Phyllanthus ramakrishna* T.	–	–

Source: Khan and Srivastava 2010)

Males and new queens are fed by workers and ingest the secretions from last larval stage. On a sunny day, they leave the nest in the morning. Males fly and wait at the nest entrances waiting for the virgin queens. They capture it and mate on the ground and do not return to the nest. In spite of high male density, only 30–50 % queens mate successfully. Single mating by the queen is observed. Males clean themselves and rejoin the group. The queen then starts locating site for hibernation. Sometimes the queens hibernate in groups.

Nature and Extent of Damage The wasps mostly attack the colonies in a flight radius of 800 m–1.5 km distance. During the attack, they do not use the sting to kill the bees. Attacking behaviour of all the species of *Vespa* is perhaps quite similar. Initially a 'hunting phase' is observed during which a hornet captures and kills bees one at a time. Later a 'slaughtering phase' takes place in which the wasps sit in-group at the entrance of the hive and kill the bees en masse. Finally when this phase has continued long enough and the colony under attack has lost most of its defenders, the wasps invade the hive, which has been called the 'occupation phase'. They consume brood and nectar and take it to their nests. However, they adopt different strategies to catch the prey at entrance or other openings in the hive during 'hunting phase'.

Wasps prefer the thorax portion of adult bees and discard the head and abdomen of the bees. Morse and Nowogrodzki (1990) discussed attack behaviour of different *Vespa* sp. in detail. Some species wait on the back of the hive and capture bees coming out of crevices; others attack regularly at the entrances of hives to deplete colony's field force. Some species enter the hive after killing guard bees and feed on the brood and young bees. *Vespa orientalis* capture bees that approach crevices, alighting board or hive entrance, whereas *V. magnifica* adopt group predation strategy

(Abrol and Kakroo 1998). Rana et al. (2000) reported peak predatory wasp activity from August to November in Himachal Pradesh (avg. 208–252 wasps/day), whereas it was July to September in Jammu (avg. 13.5 wasps/day) (Abrol and Kakroo 1998). *Vespa auraria* was most abundant in Himachal Pradesh, and *V. orientalis* was predominant in Jammu. In Punjab, peak population was observed in the month of September. Two species, viz. *V. basalis* and *V. mandarinia*, start attacking bee colonies in the month of September till November (Garg and Kashyap 1998). Bhalla and Dhaliwal (1980) observed peak wasp attack during dearth periods. The Indian honeybee *A. cerana* is, however, able to resist the attack of wasps to some extent as many of the foragers or soldier bees attempt together to scare away the approaching wasp(s) and are able to defend their colony. This kind of behaviour has not been recorded in *A. mellifera*, and as a result, the wasps and hornets may cause heavy losses in apiaries of *A. mellifera*.

Management

- The best and only effective way is to kill fecund females.
- *A. mellifera* and *A. cerana indica* bees kill wasps through shimmering behaviour forming balls around wasps. Intruder is killed either being stung or due to high temperature at the centre of ball (43–60 °C) and suffocation. This natural defence mechanism has been studied by various researchers (Abrol and Kakroo 1998: Matasuura and Sakagami 1973).
- Destruction of wasp nests is also one of the control measures, studied by various workers (Abrol and Kakroo 1998; Kshirsagar 1971; Kshirsagar and Mahindre 1975; Subbiah and Mahadevan 1957). Burning of nests with kerosene or spraying insecticide at night when all of them are in the nest is another way of controlling wasp damage. Calcium cyanide fumigation (0.5 g) at the nest entrance gave 100 % mortality in hornet nest. The use of protective screens either in the form of wire screen/cage (Rana et al. 2000) or queen gate/queen guard board (Subbiah and Mahadevan 1957) fitted at hive entrance provided physical barrier against wasp attack. However, this method was not found useful against *V. auraria*, which easily adapted to this modification and started lifting bees just outside the cage (Rana et al. 2000).
- Elimination of a lighting board is also recommended to reduce wasp attack (Sharma et al. 1985).
- One of the other common control measures is the use of flap in the apiary. Abrol and Kakroo (1998) observed the reduction in wasp visit from 66–76 wasps/day to 20–37 wasps/day. They noticed that continuous flapping for half an hour keeps hornets away from the apiary for more than 3 hours.
- Different types of baits as lures (Kshirsagar and Mahindre 1975; McGovern et al. 1970; Shah and Shah 1991; Wagner and Reirson 1969) or poison baits (baits mixed with insecticides) (Ennik 1973; Mishra et al. 1989) have been used with limited to high success in different parts of the world. Shah and Shah (1991) could trap 11,483 hornets in Kashmir by making mixture of fermented honey and water, resulting in reduction of wasp attack from 125 to 0–3 wasps/hive. Rana

et al. (2000) tried different baits in the form of fermented honey, honey solution (50 %) and ether extract of *V. auraria* with fermented honey but found fresh fish as the most attractive bait in which 61.5 wasps/day (*V. auraria* and *V. basalis*) were captured followed by mutton bait (30 wasps/day) (Abrol 1994; Abrol and Kakroo 1998).
- Among the poison baits, Mishra et al. (1989) prepared poison bait in jaggery with 1000 mg fenitrothion/kg. Sixteen loads (110 mg/load) of poison bait were sufficient to reduce *V. cincta* visits from 630 (pretreatment count) to 0 on the sixth day of the treatment. Some biotic agents have been known to affect wasp colonies, which have been reviewed by Gupta et al. (1998).

4.3.1.4 Bee Louse (Diptera: Braulidae)

Bee louse (*Braula coeca*) is small, reddish-brown, wingless fly (Bradbear 1988; Smith and Caron 1984). Several adult flies may live on a queen, usually only one will be found on a worker. Bee lice apparently do little harm. Smith and Caron (1984) observed that *B. coeca* prefer young worker bees over old ones, queens over workers or drones and mated queens over virgin queens. *Braula* move rapidly over the body, setting on the dorsal surface at the junction of the bee's thorax and abdomen. They remain there until a hunger response causes them to crawl up to the bee's head near its mouthparts. This movement seems to cause the bee to regurgitate a drop of nectar. The bee louse then inserts its mouthparts into those of its benefactor and takes its food. The louse lays its eggs on the capping of honey storage cells. Upon hatching, the young larvae burrow into the capping. As the larvae grow, their tunnels lengthen and broaden. The larva pupates inside the tunnel. Soon after emergence, the young adult crawls upon a bee. It completes its life cycle from 16–23 days (Hassanein and Abd E-Salaam 1962) to 63–67 days. Some internal parasite of the fly is also observed, but there is not much work on it.

Management

- Blow tobacco smoke in hive, remove bottom board immediately and clear it off with blow lamp (Phillips 1925).
- Atakishiev (1971) recommended cutting the capping of infested combs to control immature lice.

4.3.1.5 Hive Beetles (Coleoptera: Nitidulidae)

The small hive beetle (SHB) *Aethina tumida* is a minor pest of honeybees (Caron 1990; Fig. 4.17) and is a small (about one-third the size of a bee), black beetle covered with fine hairs (Lundie 1940). The beetle lays its eggs on or near combs. The eggs hatch, producing small larvae, similar in appearance to wax moth larvae having three sets of legs just behind the head, but larvae lack the series of paired prolegs that run the length of the wax moth larva's body. Some larvae consume pollen, comb and larval honeybees. After completing the larval stage, they crawl out of the hive and pupate in the soil. Mating and egg laying occur in warmer months. Taber (1999) recorded egg, larval and pupal period as 2–3, 10–20 and 25–60 days,

Fig. 4.17 *Aethina tumida*

respectively. Adult beetles feed on honeybee eggs (Eischen et al. 1999). The beetle is extremely quick moving and can fly, which contributes to its rapid spread among bee colonies and apiaries. Beetle larvae and adults feed voraciously on bee brood primarily drone brood. Beetle larvae do the most damage, as their diet is mostly the same as of the wax moth larvae. Compounding the losses of comb structure, food reserve, repellent nature of beetle faeces and 'slime layer' to adult honeybees force the bees to abscond from the hive to seek a more suitable nesting site (Stanghellini et al. 2000). The beetle spoils the honey by defecating in it and also alters the honey quality by causing fermentation. The most vulnerable are weak hives with stored honey or full honey supers either in storage or above bee escapes. Apiaries adjacent to various fruit trees (mainly citrus) were found to be heavily infested with beetles (Eischen et al. 1999). In absence of their preferred food, they were observed feeding on selected fruits. *Platybolium alvearium* and *Bradymerus* sp. (tenebrionid beetles) also thrive well under unhygienic conditions of the hives (Thakur 1991). *Protaetia aurichalcea, P. impavida* and *Anomala dimidiata* were also found feeding on stored pollen in *A. mellifera* and *A. cerana* hives (Gulati and Kaushik 2004).

Management

- Keep strong and healthy colonies with good hygienic conditions. Clean and tidy hives together with regular examination of empty combs will reduce the beetle incidence.
- Destroy these beetles as soon as they are observed.
- Extract honey from filled supers within 1–2 days as beetles can build up rapidly in stored honey and wax in supers.
- Freeze the frames to –12°C for 24 hours for killing the eggs, grubs and adults.
- Place hives on rock or hard clay-based soil rather than light sandy soil which enable the mature grubs to pupate into soil.
- Maintain boxes in good condition as wrapped, cracked or rotten boxes provide hiding places for adult beetles and make the detection difficult.
- Trap beetles by using non-toxic oil which suffocates them.

Fig. 4.18 *Varroa destructor* (ventral view)

- Expose the beetle pupae to extreme sun heat and natural enemies by racking the soil around the beehive.
- Monitor colonies for hygienic behaviour (ability to get rid of beetles and larvae). These beetle-resistant queen lines can be propagated to produce SHB-resistant colonies.
- Treat soil around infested hives with some insecticide.
- Taber (1999) advocated two effective control measures against beetles. Firstly, placing fume boards over the colony results in flying out of beetles through the hive entrance. Secondly, since beetle larvae pupate in soil, placing the infested colony on a concrete floor or black roofing paper/plastic sheet would prevent the larvae from finding soil ultimately causing a percent of mortality.

4.3.2 Bee Mites

The bee mites given below belong to phylum Arthropoda and class Arachnida. These may be parasitic or nonparasitic (scavengers, pollen feeders, predators on scavengers and phoretic on bees that live in flowers). Three species of mites, namely, two ectoparasitic mites, i.e. *Varroa* mites and *Varroa destructor*; brood mite, *Tropilaelaps clareae*; and a tracheal mite, *Acarapis woodi* (endoparasite), are of economic importance.

4.3.2.1 Varroa Mite, *Varroa destructor* (Parasitiformes: Varroidae)

Identification and Life Cycle Adult female mite is dorsoventrally flattened, brown to dark brown and shining in colour, shaped like a tiny crab, sideways oblong (broader than length), measuring 1.00–1.77 mm in length and 1.50–1.99 mm in width (Figs. 4.18 and 4.19). Gravid female mite enters a brood cell just before it is capped. It may enter either a worker or drone cell, but drone cell is more preferred. After crawling under the larva, the female mite submerges itself in the brood food under the larva where it will remain until the cell is capped by other worker bees. While submerged, the mite erects its peretrimes which serve as breathing tubes, allowing the mite to breathe while it is submerged. Once worker bees have capped the cell, the larva consumes the remaining brood food, thus freeing the mite. The freed mite climbs onto the larva and begins feeding on its haemolymph. Egg laying begins in about 60 h after the cell has been sealed, and mite lays 2–8 eggs. The first egg develops into a male and subsequently eggs develop into females. A six-legged

Fig. 4.19 *Varroa destructor* (dorsal view)

Fig. 4.20 Perforated cell cappings of infested brood

larva develops into an eight-legged protonymph which moults into a deutonymph and finally to the adult form. Females mature in 6.5–6.9 days. The male mite does not eat and its sole purpose is to mate with females, and it afterwards dies and remains in the cell. Old mother and new gravid *Varroa* mites after feeding sometime on an adult worker or drone bee seek another bee larval cell to repeat the life cycle. Intercolony spread takes place mainly through shifting of bee combs, robbing, drifting of worker and drone bees and long-distance migration and also by unscientific beekeeping.

Symptoms of Infestation In the infested colonies, adult mites can be seen on adults, exposed larvae and pupae of honeybees. In severely infested colonies, a large number of dead and live mites can also be seen in the debris on the bottom board. The infested brood has perforations in their cell cappings (Fig. 4.20). Heavily infested colonies usually show symptoms of unsealed (bald) brood cells, and the dead pupa at its different stages of development can be seen. Dead or dying newly emerged bees with malformed wings, legs, abdomens and thoraces may be found on the ground in front of the hive entrance. If left unchecked, the mite can cause dwindling and loss of the affected colonies. Mite-infested weak colonies are prone to robbing by stronger colonies which further results in weakening of the colonies.

Management

Non-chemical measures should be preferred over the use of chemicals to avoid any risk of their residues in honey and other bee products and any adverse effects on brood or adult bees:

- Destroy drone brood by burying deep into the soil as *Varroa* mite has higher multiplication rate on drone brood. It helps in bringing down mite population and its carry over to the next brood cycle.
- Cage the queen bee for 3 weeks to create broodless conditions. It has an adverse effect on the development and multiplication of the mite.
- Shaking bees from infested colony onto frames with comb foundation or broodless combs in another hive for about 2 weeks and destruction of infested brood on the original combs are also helpful to free the bees from the mite.
- *Varroa* adult mites adhering to the body of adult bees often are fallen down by grooming of the bees. These mites fallen on the bottom board will climb up again and move to the bee/brood combs. The placement of a sticky paper on a bottom board covered with eight mesh plastic screen prevents the mite to return to the brood combs as the mites get stuck in these sticky papers.
- Screened floor boards on high-legged hive stands would result the mites to fall through on the ground and starved to death. This is considered to be continuous and effective control. However, the robbing and prevailing temperature conditions must be viewed while following this method.
- Dusting very fine/icing sugar (particle size <5 μm) at 20 g per ten bee frames uniformly between the combs through the bee space in the late evening is an effective measure to check this mite menace. Sugar particles adhere to the ambulacrum of the adult female mite rendering it unable to cling to anything resulting in the fall of the mites. This method of controlling mite can be used throughout the year, even during honey flow, when other chemical means are not allowed.
- Formic acid (85 %) at 5 ml per day continuously for 2 weeks is to be administered to the infested colonies. Five millilitres of formic acid is put in an injection vial, and a thick wick is inserted into the vial to check the drowning of bees and facilitate its slow evaporation. This vial has to be kept inside the infested colony near the entrance towards combs with bees. The formic acid's vapours will dissipate inside the hive and kill the mites even when developing inside the capped brood. Its higher doses result in bees and queen bee losses and also affect egg laying by the queen bee. This chemical is prohibited to be used during honey flow. Formic acid is corrosive and can cause burns. Rubber gloves and safety glasses should be worn, and inhalation of vapours must be avoided. Formic acid 85 % (5 ml) applied by cotton swab method provided 85.3 % control of *V. destructor* over untreated hives (Asha and Sharma 2012).
- The highest average percent recovery of mites (87.5 %) was observed in garlic paste (80 g) placed on the bottom board in a Petri dish for 48 h followed by eucalyptus oil and thymus oil (1 ml/day/colony placed in Petri dishes soaked on cotton swabs), against the infestation of ectoparasitic mite, *Varroa destructor*, in *A. mellifera* L. colonies (Sapna et al. 2010).
- Deosi and Chhuneja (2013) tested non-chemical control measures against *V. destructor* during monsoon season at PAU, Ludhiana. They found that mustard oil (10 and 20 %), canola oil (20 %) and icing sugar (four treatments at 7 days interval) treatments were promising against *V. destructor*.

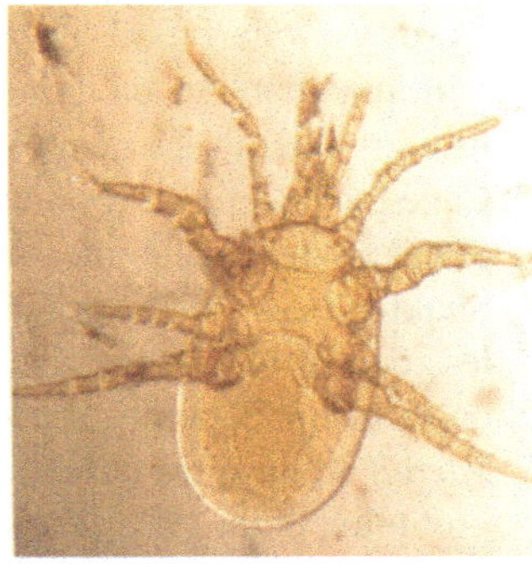

Fig. 4.21 *Tropilaelaps clareae*

4.3.2.2 Brood Mite, *Tropilaelaps clareae* (Parasitiformes: Laelapidae)

Tropilaelaps was first reported on giant honeybee, *Apis dorsata.* In India, it was reported in 1968 on *Apis mellifera*. It is an obligatory brood mite.

Identification and Life Cycle The female mites are about 1 mm long and 0.6 mm wide (Fig. 4.21). The male is slightly smaller. These mites are difficult to detect because of small size and their brownish colour, which blends perfectly with the brood cappings and comb. However, they can be observed under a magnifying lens or a dissecting microscope. They are fast-moving, reddish-brown ectoparasites of bee brood and adults. It is smaller than *Varroa* mite. The life cycle is similar to *Varroa* mite but is relatively shorter. Mite lays 3–4 eggs on a mature bee larva shortly before capping, and progeny feeds only on bee brood. It takes 2 weeks for its total development. The drone brood is most preferred by brood mite, and almost 100 % parasitization has been reported by Burgett et al. 1983. When a bee colony is infested by both *T. clareae* and *V. destructor*, the former may outcompete the *Varroa* mite because of short life cycle (Burgett et al. 1983; Ritter and Schneider-Ritter 1988).

Symptoms of Infestation These tiny mites can also be seen moving on the brims of comb cells in the infested colony. The mite feeds on the haemolymph by its piercing mouthparts. Infested bee pupae get deformed and have dark-coloured spots (Fig. 4.22). The abdomen of the bees attacked gets reduced in size. In the heavily infested colonies, bees with deformed wings and legs or in incompletely developed individuals with stubby wings and smaller or missing eyes can be observed crawling near the vicinity of the hive entrance and on the comb surface (Atwal and Goyal 1971).

Management

- Avoid exchange of infested brood frames.
- Sulphur dusting at 1 g/frame on the top bars of frames two times at an interval of 21 days effectively controls these mites (Atwal and Goyal 1971).
- Formic acid (85 %) at 5 ml per day for 2 weeks (Garg et al. 1984).
- Caging the queen bee for 3 weeks to create broodless condition is also effective to check its infestation (Woyke 1985, 1993).

Fig. 4.22 Deformed adult

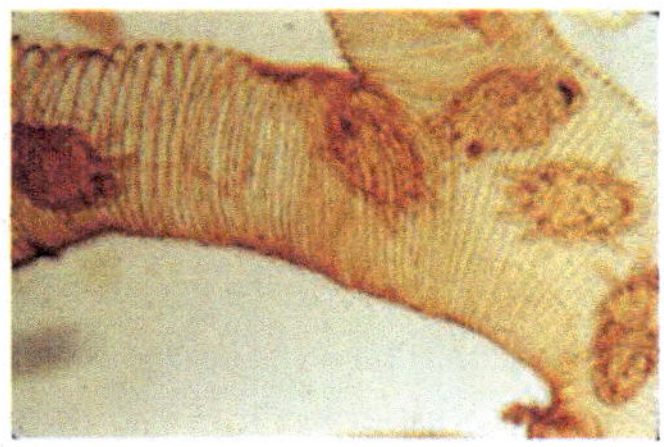

Fig. 4.23 *Acarapis woodi* inside the tracheae of infested bee (Source: Wikipedia)

4.3.2.3 Tracheal Mite, *Acarapis woodi* (Acariformes: Tarsonemidae)

It causes crawling in bees (acarine disease). In India, it was first reported from Himachal Pradesh in 1956.

Identification and Life Cycle The mite is too small to be seen with the naked eye. The female of *A. woodi* is 143–174 μm long, while the male is 125–136 μm long. The body is oval and broadest between the second and third pairs of legs and whitish or pearly white in colour. The cuticle is smooth and shining. A few long hairs/setae are present on the body and legs. This mite has an elongate, beak-like gnathosoma for feeding on the host. It completes its life cycle within the prothoracic tracheae of honeybees. All stages, egg, larvae, nymph and adult, may be found in the tracheae at one time. The infective female lays 5–7 eggs in the tracheae after 3–4 days of mating. Eggs hatch after 3–4 days of laying. The first male is seen on the 11th or 12th day, whereas the first female is seen on the 14th or 15th day. Mating takes place within the tracheae, and one generation is completed in 2 weeks.

Symptoms of Infestation As this mite is mostly found inside the tracheae, so infested bee has to be dissected for examination (Fig. 4.23). Mite feeds on bee haemolymph by piercing the tracheae with their close-ended sharply pointed stylets. It is very difficult to diagnose mite attack on reliable visible symptoms, but there are some unreliable symptoms like dwindling colony strength; a lot of crawling bees with disjointed hindwings called K-wings are seen in front of the hive entrance. Irregular dark stains initially develop on the infested tracheae, which eventually get blackened. Due to mite attack, longevity of bee is reduced.

Management

- Use of menthol crystals at 25 g placed on the top of a colony for up to 2 months is effective in controlling mite infestation.
- Use of formic acid as in case of *Varroa* mite can also control this mite.

4.3.3 Birds

The bird enemies of the honeybees belong to phylum Chordata and class Aves. Major birds associated with honeybees are *Merops apiaster* (European bee eater), *M. orientalis* (small green bee eater), *M. leschenaultia* (chestnut-headed bee eater), *M. superciliosus* (blue-cheecked bee eater), *M. philippinus* (blue-tailed bee eater), *Indicator indicator* (greater honey guides) and *I. variegatus* (honey guides). Minor birds are*Picus viridis* (green woodpecker), *Picoides major* (great spotted or variegated woodpecker), *Parus major* (common great tit), *P. caeruleus* (blue tit), *P. major karelini* (South Caspian tit), *Aegithalos caudatus major* (Caucasian long-tailed tit), *Passer domesticus* (house sparrow) and *Dicrurus macrocercus*, (drongo/king crow). According to their feeding habits, they are broadly classified into two categories: bee eaters and honey guides. Out of 74 families of Indian subcontinent, 24 families largely feed on insects. The major portion of their diet is honeybees or beeswax (Cobb 1979). They visit apiaries occasionally on cloudy days and prey upon bees. The heavy traffic of bees flying in and out of the hives of commercial apiaries provides an exceptional opportunity for insectivorous birds. Among various apivorous birds, bee eaters are most prevalent and devastating.

4.3.3.1 Bee Eaters

Bee eater birds belong to the Meropidae family and include 24 species. Most of the species are migratory in nature and are found in temperate, tropical and sub-tropical regions (Dyer and Fry 1980). They remain in flocks of 15–20 and feed on venomous hymenoptera with the exception of *M. nubicus* and *M. albicollis*, which feed on locusts and flying ants, respectively. A bee eater attacks or catches the flying bees, devouring them by beating against perch. Breeding season of *M. orientalis* and *M. leschenaulti* is from February to June. They form nests in the form of tunnels in earth mound or sandy cuttings (Khan 1996). Bee eaters are gregarious; therefore, large numbers of these holes are often seen at one place. Their droppings get accumulated at the nest entrances forming prominent white streaks. Eggs are 4–9 in number, pure white, oval to round with an average size that varies from 19.3–26.2 × 17.3–21.9 mm. The green bee eater (*Merops orientalis*) and European bee eater (*Merops apiaster*) cause serious losses to beekeepers in sub-tropical areas (Figs. 4.24 and 4.25). The green bee eater is 16–18 cm long, bright green and tinged with blue chin, crown and upper back tinged with golden rufous. European bee eaters are 27–29 cm long, with brown and yellow upper parts, and wings are green and the beak is black.

Fig. 4.24 *Merops orientalis*

Fig. 4.25 *Merops apiaster*

Nature and Extent of Damage Before eating its meal, a bee eater removes the sting by repeatedly hitting and rubbing the bees on a hard surface. During this process, pressure is applied to the bees thereby extracting most of the venom. Notably, the birds only catch prey in flight. The level of damage caused by the apivorous birds varies considerably. An attack by a single bird or by a few together rarely constitutes a serious problem. When a relatively large flock descends upon a few colonies or an apiary, a substantial decline in the worker population may be observed. The degree of damage to the commercial apiaries by predatory birds depends upon the number of predators and intensity of the attack. The mere presence of a few predators in apiaries engaged in queen-rearing can inflict serious losses. The bee eaters sit on tree or telegraph wires near an apiary and pick the bees from the wings and do much harm. Sometimes as many as 40 bees have been found in the stomach of a bird. In one season a pair of birds was reported to consume nearly 30,000 bees (Borchert 1974). Sharma and Khan (1995) studied the predation rate of small green bee eater on *A. mellifera* foragers and found that, on an average, 708 ± 111.2 foragers/day are eaten by the bird. Predation was maximum between 1000 and 1300 h, whereas it was minimum between 1600 and 1900 h.

Management

- Scaring the birds is most effective in checking their visits to beehives which include sound in high pitch with different notes, beating the drums and empty tins or throwing pieces of stones/pebbles through gulel or hand.
- Using sulphur-potash mixture, hanging of 2–3 dead bee eaters at 5 m height and producing distress call/voice of injured bee eater by recording audio cassette and played on the amplifier are other ways of controlling their visits.
- Use reflective tapes to distract birds. Reflective tapes of different colours (1 m × 3.5 cm) fixed on string at a distance of 20–30 cm at a height of 5 m on two poles/stems are generally used.
- Keep beehives under thick canopy of trees and restrict flight activity of honeybees by provision of water near apiary.

4.3.3.2 Honey Guides

This is the second major bird group of predators of honeybees which comprises 11 species under four genera. Among them, 9 species are found in Africa and 2 species in Asia of which one is found in Nagaland and Manipur, while the other is found in Thailand, Burma, Sumatra and Malaya (Khan 1996). They exhibit guiding behaviour and symbiotic relationship with mammals. They prefer beeswax over honey or honeybee larvae:

(a) *Woodpeckers and tits*: They are present throughout the world except in Madagascar and Australia. Green woodpeckers are found in Haryana, Rajasthan, Gujarat and Orissa, whereas great spotted woodpeckers are found in north-eastern hill states. Control: Woodpeckers are generally beneficial so no control is required. They can be frightened or screened away from the apiary as no one advises to destroy tits.
(b) *Drongos/king crows*: These are known as occasionally predators of honeybees. In Hisar region, they are observed as regular predators of bees.
(c) *Tyrants*: *Tyrannus tyrannus* (the eastern king bird or bee marten) is a major problem in queen-rearing operations in the USA where it preys on larger drones and queen but not on workers (Khan 1996).

4.3.4 Reptiles

Among the reptile species that are regularly observed in commercial apiaries are the tokay, which can be as much as 35 cm long, *Calotes* spp., *Acanthosaura* spp. and the skink, *Sphenomorphus* spp. Arboreal reptiles such as many geckos and skinks can attack bees either near the hive entrance or on the branches of flowering trees visited by forager bees. Smaller lizards, *Hemidactylus frenatus*, often hide in the empty space between the outer and inner covers of the hive. The activity of the 'home lizard' may cause great concern to the beekeeper. A serious lizard problem may lead to absconding of bees. Even lizards not living near the hive will feed on the bees once they can locate the apiary. Some snakes are also known to eat bees. But they do not cause much damage to the colony.

Fig. 4.26 Bear

Management

- The beekeepers can do little to prevent the loss of bees from the highly mobile arboreal reptiles, usually well hidden in the trees.
- Beehives should be placed on a platform, with metal cones nailed on the legs about 70 cm above the ground, to prevent lizards from reaching the hives.
- The use of engine oil or grease on the legs of the stands may deter the reptiles from climbing up to the hive entrance.
- Keep bee boxes away from dense bushes, shrubs and tall grasses, which provide safe hiding places to the predators.
- No reliable chemical control of reptiles is available for use in apiaries.

4.3.5 Mammals

4.3.5.1 Bears

The bears belong to family Ursidae (Fig. 4.26). It has been said that once a bear has tasted honey and brood, it is almost impossible to keep it away from apiaries. Protecting colonies from bear attack is usually difficult, particularly when the animals are large and strong. Bear damage is rather easy to see. Hives are shattered to bits by the bears to get the brood comb and honey. They scatter the equipment around the yard. Placing the apiary in a location out of the bear's path reduces its attack. Electrified barbed-wire fences are often used where bears represent a common problem. Moving hives closer to human habitation may be much more effective. Shooting and trapping them are other possible way but very temporary control measures. To prevent damage by bears particularly during autumn and winter in hilly areas, keep the hives in cheap wall enclosures or suspend them from the horizontal branches of trees (Thakur et al. 1981).

4.3.5.2 Skunks

The skunks belong to family Mustelidae (Fig. 4.27). They scratch the bottom board or the front of the hive body to get the bees coming out of the hive and eat the bees. Skunks visit the apiary in the evening and dark hours. Raising the hive 16–18 inches above the ground and using of wire netting around the hive are effective to prevent the skunk.

Fig. 4.27 Skunks

Fig. 4.28 Rodents

4.3.5.3 Rodents

Rodents such as mice and rats (Fig. 4.28) are common pests to the beekeeping. They build nest in hive boxes, destroy comb in frames and make holes in equipment. In addition, they leave droppings all over the place. Rats can be a serious problem in storage areas where bee equipment are kept. To keep mice out of hives, a mousetrap can be placed on the entrance of the hive. Reducing the entrance of the hive to one-fourth of an inch, bees will be able to come and go, but mice will not be able to enter. Bait traps can also be used for both rats and mice:

(a) *Raccoons*: The raccoons belong to family Procyonidae (Fig. 4.29). They harm the apiary by pulling the frames from the hive and build a nest to raise their young ones. They will even remove top or inner covers from stacked supers. The honey containers are taken away from the hives, and they drink the honey. Raccoons can be trapped and killed.
(b) *Pine marten*: They belong to family Mustelidae (Fig. 4.30). It is a tree-dwelling member of the weasel family. Pine martens destroy the nests of domesticated as well as wild bees for honey in hilly areas. Fencing of apiary may reduce the attack of pine marten.

Hawk moth, Indian meal moth, coleopteran beetles, dipteran flies, spiders, amphibians (toads and frogs), monkeys and langurs are minor enemies of honeybee colonies and may sometime cause nuisance to beekeepers. Among the carnivorous mammals, the jackal and some weasels are occasional enemies of bees.

Fig. 4.29 Raccoons

Fig. 4.30 Pine marten

4.4 Conclusions

From the above, it is clear that honeybee diseases, pests and parasites create problems for the beekeeper and affect colony health and productivity; therefore regular monitoring and surveillance are essential for early diagnosis of problem because care and management of diseases are easier in early stages of infection or attack. Beekeepers have traditionally utilized different approaches in dealing with these problems, some of which have involved the use of chemical treatments. However, the use of chemical treatments can significantly affect colony health, as well as create potential problems with the contamination of honey and beeswax. Therefore, there is a need to change management approaches to IPM or integrated hive management (IHM) practices whenever possible. An IHM approach involves the use of all available tactics in the design of a programme to manage bee pest populations so that economic damage and harmful side effects are minimized. An important component of this approach is the use of sustainable management practices with as little reliance as possible on chemical treatments. Always maintain the strong colony as it not only protects the honeybees from enemies and diseases but also results into higher honey production. In addition to IPM approach, high ethics and national interest can only help in safeguarding our beekeeping leading a road to prosperity of beekeepers and our rural India.

References

Abrol DP (1994) Ecology, behaviour and management of social wasp, *Vespa velutina* Smith (Hymenoptera: Vespidae), attacking honeybee colonies. Korean J Apic 9:5–10

Abrol DP (1997) Bees and beekeeping. Kalyani Publishers, Ludhiana, 450 pp

Abrol DP, Kakroo SK (1998) Studies on seasonal activity and control of predatory wasps attacking honey bee colonies. Indian Bee J 60:15–19

Abrol DP, Kumar R (2000) Impact of pesticides on the brood rearing activity and foragers of honeybee. *Apis mellifera* L. visiting *Bressica campestris* L. var. Torria. Indian Bee J 62:19–27

Adlakha RL, Sharma OP (1975) *Apis mellifera* vs. *Apis indica*. Glean Bee Culture 103(5):160

Ahmad R, Muzaffer N, Ali Q (1983) Biological control of wax moths, *Achroia grisella* (F.) and *Galleria mellonella* (L.) (Lepidoptera: Galleridae) by augmentation of the parasite. *A. galleriae* Wilk. (Hym: Braconidae). Apiacta 18(1):15–20

Akre RD, Davis HG (1978) Biology and pest status of venomous wasps. Annu Rev Entomol 23:215–238

Ali AD, Abdellatif MA, Bakry NM, El-Sawaf SK (1973a) Studies on Biological control of the greater wax moth, *Galleria mellonella* (L.). I. Susceptibility of wax moth, larvae and adult honey bee workers to *Bacillus thuringiensis*. J Apic Res 12(2):117–123

Ali AD, Abdellatif MA, Bakry NM, K. EL-SS (1973b) Studies on the biological control of the greater wax moth, *Galleria mellonella* II. Impregnation of comb foundation with Thuricide-HP as a method of control. J Apic Res 12:125–130

Allegret P (1975) Effect of an early and prolonged deprivation of protein on the ovary of the Pyralid moth *Galleria mellonella* L., reduction in the number of ovarioles. C R Seances Soc Biol 169(2):399–403

Allen M, Ball BV (1996) The incidence and world distribution of the honey bee viruses. Bee World 77:141–162

American Foulbrood | Texas Apiary Inspection Service (TAIS) (2016) Txbeeinspection.tamu.edu. N.p., 2016. Web. 21 Dec

Aronstein KA, Murray KD (2010) Chalkbrood disease in honey bees. J Invertebr Pathol 103:20–29

Asha GR, Sharma SK (2012) Efficacy of formic acid against *Varroa destructor* in *Apis mellifera* colonies. Int J Agric Sci 3(1):29–33

Atakishiev TA (1971) The bee louse (*Braula coeca* Nitzsch). Pchelovodstvo 2:16–17

Atwal AS, Goyal NP (1971) Infestations of honeybee colonies with *Tropilaelaps*, and its control. J Apic Res 10:137–142

Baggio A, Gallina A, Dainese N, Manzinello C, Mutinelli F, Serra G, Colombo R, Carpana E, Sabatini AG, Walner K, Piro R, Sangiorgi E (2005) Gamma radiation: a sanitating treatment of AFB-contaminated beekeeping equipment. Apiacta 40:22–27

Bailey L (1961) European foulbrood. Am Bee J 101:89–92

Bailey L (1983) *Melissococcus pluton*. The cause of European foulbrood of honeybees (*Apis* ssp.). J Appl Bacteriol 55:65–69

Bailey L, Gibbs AJ, Woods RD (1964) Sacbrood virus of the larval honey bee (*Apis mellifera* Linnaeus). Virology 23:425–429

Bailey L, Ball BV, Woods RD (1976) An iridovirus from bees. J Gen Virol 31:459–461

Bailey L, Carpenter JM, Woods RD (1979) Egypt bee virus and Australian isolates of Kashmir bee virus. J Gen Virol 43:641–647

Battu GS, Singh K (1977) On the possibility of using *Bacillus thuringiensis* Berliner against *Galleria mellonella* Linn. in bee hives. Indian Bee J 39(1–4):27–29

Bhalla, O.P and Dhaliwal, H.S. (1980). Potentialities and progress of beekeeping in India.In: Proc. 2nd int. conf. Apic. Trop. Climates, New Delhi (29th February- 4th March 980), pp 100–109

Bogdanov S, Kilchenmann V, Seiler K, Pfefferli H, Frey TH, Roux B, Wenk P, Noser J (2004) Residues of para-dichlorobenzene in honey and beeswax. J Apic Res 43(1):14–16

Bogus M, Cymborowski B (1977) Effect of cooling stress on growth and developmental rhythms in *Galleria mellonella* (L.). Bull Acad Sci Pologue 4:257–260

Borchert A (1974) Schadigungen der Bienenzucht durch Krankheiten, Vergifiungen und Schadlingeder Honigbiene. S. Hirzel, Leipzig

Borum AE, Ulgen M (2008) Chalkbrood (*Ascosphaera apis*) infection and fungal agents of honey bees in north-west Turkey. J Apic Res 47(2):170–171

Bradbear NJ (1988) World distribution of major honeybee diseases and pests. Bee World 69:15–39

Brar HS, Brar BS, Gatoria GS, Jhajj HS (1996) Biology of greater wax moth, *Galleria mellonella* L. infesting *Apis mellifera* L. colonies in Punjab. J Insect Sci 9:12–14

Brodsgaard C, Ritter W, Hansen H (1998) Response of in vitro reared honey bee larvae to various doses of *Paenibacillus* larvae spores. Apidologie 29:569–578

Burges HD, Bailey L (1968) Control of the greater wax moth and lesser wax moths with *Bacillus thuringiensis*. J Invertebr Pathol 2:184–195

Burgett M, Akratanakul P, Morse RA (1983) *Tropilaelapsclareae*: a parasite of honeybees in South-East Asia. Bee World 64:25–28

Burkett BW (1962) Temperature block of spinning and development in *Galleria mellonella* (Lepidoptera). Am Zool 2:396

Buys B (1990) Relationships between Argentine ants and honey bees in South Africa. In: Vandermeer RK et al (eds) Applied myrmecology: a world perspective. Westview Press Inc., Boulder, pp 519–524

Cantwell GE, Shieh TA (1981) Certan- a new bacterial insecticide against greater wax moth, *Galleria mellonella* L. Am Bee J 121:424-426–430-431

Caron DE (1990) In: Morse RA, Nowogrodzki R (eds) Honeybee pests, predators and disease. Cornell University Press, Ithaca/New York, pp 171–172

Casanova Ostos RA (1992) Three methods for control of the wax moth, *Galleria mellonella* (L.), in stored honeycombs. Revista Cientifica UNET 6(1):5–16

Chantawannakul P, Dancer BN (2001) American foulbrood in honey bees. Bee World 82:168–180

Chauvin G, Chauvin J (1985) The influence of relative humidity on larval development and energy content of *Galleria mellonella* (L.) (Lepidoptera, Pyrelidae). J Stored Prod Res 21:79–82

Chhuneja PK, Yadav S (2009) Evaluation of various bait traps for the management of wax moths in *Apis mellifera* Linnaeus apiaries. Insect Environ 15(2):63–66

Chhuneja PK, Yadav S (2011) Parasitization by *Apanteles galleriae* (Hymenoptera: Braconidae) – a limiting factor in population build-up of *Galleria mellonella* (Lepidoptera: Pyralidae) during its peak activity under the Punjab conditions. Indian J Ecol 38(1):140–141

Cobb PK (1979) Honey buzzard at wasp nests. Br Birds 72:59–64

Dave ML (1943) Waspa. Ind Bee J 5:127–131

de Miranda JR, Gauthier L, Ribiere M, Chen YP (2012) Honey bee viruses and their effect on bee and colony health. In: Sammataro D, Yoder J (eds) Honey bee colony health: challenges and sustainable solutions. CRC Press, Boca Raton, pp 71–102

Deka MK, Das PK, Majumdar A (2010) Seasonal incidence and management of wax moth (*Galleria mellonella*) in Assam. Pestology 34(12):36–38

Deosi, H.K. and Chhuneja, P.K. (2013). *Varroa destructor* Anderson & Trueman mite fall in *Apis mellifera* Linnaeus colonies following non-chemical measures during monsoon season. In: The proc. of 4th int. cong. on insect science at Bangalore, pp 67

Dubois L, Collart E (1950) L'Apiculture on *Cong Beige etan* Ruandaurundi. Deptartment of Agriculture, Brussels, 67 pp

Dyer M, Fry CH (1980) Acta 17th int. on cong., Berlin, Velrag, D.O-G, pp 44–46

Eckert JE, Bess HA (1952) Fundamentals of beekeeping in Hawaii. Hawaii Coop Ext Bull 55

Eckert JE, Shaw FR (1960) Beekeeping. Macmillan, New York

Eischen FA, Westervelt D, Randall C (1999) Does the small hive beetle have alternate food sources? Am Bee J 139:125

Ennik F (1973) Abastement of yellow jackets using encapsulated formulations of Diazinon and Rabon. J Econ Entomol 66:1097–1098

Flores JM, Ruiz JA, Ruz JM, Puerta F, Bustos M, Padilla F, Campano F (1996) Effect of temperature and humidity of sealed brood on chalkbrood development under controlled conditions. Apidologie 27:185–192

Flores J, Spivak M, Gutierrez I (2005a) Spores of Ascosphaera apis contained in wax foundation can infect honeybee brood. Vet Microbiol 108:141–144

Flores JM, Gutierrez I, Espejo R (2005b) The role of pollen in chalkbrood disease in *Apis mellifera*: transmission and predisposing conditions. Mycologia 97:1171–1176

Frazier M, Mullin C, Frazier J, Ashcraft S (2008) What have pesticides got to do with it? Am Bee J 148:521–523

Garg, R. and Kashyap, N.P. (1998). Mites and other enemies of honeybees in India. In: Mishra RC (ed) Perspective in India apiculture. Published by Agro Bikaner, pp 264–303

Garg R, Sharma OP, Dogra GS (1984) Formic acid: an effective acaricide against *Tropilaelaps clareae* Delfinado & Baker (Laelapidae: Acarina) and its effect on the brood and longevity of honey bees. Am Bee J 124:736–738

Genersch E (2008) *Paenibacillus larvae* and American foulbrood–long since known and still surprising. J Verbr Lebensm 3:429–434

Genersch E, Aubert M (2010) Emerging and re-emerging viruses of the honey bee (*Apis mellifera* L.). Vet Res 41(6):54

Genersch E, Forsgren E, Pentikäinen J, Ashiralieva A, Rauch S, Kilwinski J, Fries I (2006) Reclassification of *Paenibacillus* larvae subsp. *Pulvifaciens* and *Paenibacillus* larvae subsp. larvae as Paenibacillus larvae without subspecies differentiation. Int J System Evol Microbiol 56:501–511

Gochnauer, T.A., Furgala, B. and Shimamuki, H. (1975). Diseases and enemies of the honeybee. In: The hive and the honey bee (rev ed) Dadant Hamilton, pp 615–621

Gulati R, Kaushik HD (2004) Enemies of honeybees and their management- a review. Agric Rev 25(3):189–200

Gupta JK, Thakur RK, Sharma HK (1998) Control of predatory wasps of honeybees – a review. Ind Bee J 60(4):210–214

Hassanein MH, Abd E-Salaam AL (1962) Bull de la Societe Entomologique d'Egypte 46:87–95

Hirschfelder H (1952) Control of the bee killer wasps. Germany Anz Shadlingsk 25:122–123

Jyothi JVA et al (1990) In: Veeresh GK et al (eds) Social insects – an Indian perspective. Int. Union for Study of Social Insects Indian chapter, Bangalore, pp 219–223

Jyothi JVA, Reddy CC (1993) Influence of different comb parts as diets on the life cycle of greater wax moth (*Galleria mellonella* L.). Ind Bee J 55:29–35

Jyothi JVA, Reddy CC (1994) Rate of Oviposition and Egg Laying Pattern in Greater Wax Moth *Galleria mellonella* L. Ind Bee J 56(3–4):142–144

Kapil RP, Sihag RC (1983) Wax moth and its control. Ind Bee J 45(2/3):47–49

Khan MS (1996) Bird predators of honeybees. In: Manual on recent advances in apiculture. Department of Entomology, CCS HAU, Hisar, pp 193–210

Khan MS, Srivastava P (2010) Enemies of honeybees (other than diseases and mites) and their management. In: Ramani S, Kaushik HD (eds) AICRP on honeybees and pollinators. ICAR, New Delhi, 32pp

Khanbash MS, Oshan HS (1997) Biological study on greater wax moth, *Galleria mellonella* (L.) in Lahj region, Yemen. Arab J Pl Prot 15(2):80–83

Kshirsagar KK (1971) A hornet trap to control predatory hornet in apiary. Ind Bee J 33:56–58

Kshirsagar KK, Mahindre DB (1975) Some notes on bee predatory wasps in India. Ind Bee J 37(1–4):4–9

Kumar, K. (2000). Studies on the biology of greater wax moth *Galleria mellonella* L., and its control by *Bacillus thuringiensis* Berliner M.Sc. Thesis, CCS Haryana Agricultural University, Hisar, 83 pp

Lundie AE (1940) The small hive beetle *Aethinatumida*, Science bulletin 220. Dep. Agr. Forestry, Government Printer, Pretoria

Malonova D, Titera D (1995) Acrinathrin, an effective varroacide and its residues in stores, honey and wax. Apidologie 26:321–322

Matasuura M, Sakagami SF (1973) A bionomic sketch of the giant hornet, *vespa mandarinia*, a serious pest for japanese apiculture. J Faculty Sci Hokkaido Univ Series VI Zool 19:125–162

May AF (1961) Beekeeping. Haum, Pretoria

McGovern TP, Davis HG, Beroza M, Ingangi JC, Eddy GW (1970) Esters Highly Attractive to *Vespula* spp. J Econ Entomol 63(5):1534–1536

Menon, A. (1992). PhD. Thesis, Dept. of Zool., Bangalore Univ., Bangalore, India, 85 pp

Milum V, Geuther HW (1935) Observations on the biology of the greater wax moth, *Galleria mellonella* L. J Econ Entomol 28(3):576–578

Mishra RC, Kumar J, Gupta JK (1989) A new approach to the control of predatory wasps (*Vespa* spp.) of the honeybee (*Apis mellifera* L.). J Apic Res 28(3):126–130

Morhe RA (ed) (1999) *Honeybee pests, predators* and diseases. Cornell University Press, Ithaca/New York

Morse RA, Nowogrodzki R (eds) (1990) Honeybee pests, predators and diseases, 2nd edn. Comstock Publishing Associates/Cornell University Press, Ithaca

Nani M, Bisht DS (1972) A simple method of disinfections of honey bees combs against wax moth. Ind Bee J 34(3–4):70–77

Newton LV (1917) The domestication of the Indian honey bee. Agric J 12:44–57

Nielson RA, Brister D (1977) The greater wax moth: adult behavior. Ann Entomol Soc Am 70(1):101–103

Okumura GT (1966) The dried fruit moth (*Vitulaedmandsae serratilineella* Ragonot), pest of dried fruits and honeycombs. Calif Dept Agric Bull 55:180–186

Phillips, E.F. (1925). The bee louse *B. coeca* in US. US Dept Agric Circular 334

Ramachandaran S, Mahadevan V (1951) The wax moths and their control. Ind Bee J 13(7/8):85–88

Rana BS, Goyal NP (1991) Field toxicity of methyl demeton and dimethoate to the foragers of honeybee (*Apis cerana indica* Fab.) visitors to *Brassica chinensis* L. Ind Bee J 53:73–77

Rana BS, Kumar A, Gupta JK (2000) Influence of environmental factors on the population of predatory wasps on honeybees and evaluation of methods for controlling the wasps. Ind Bee J 62(1–2):47–54

Ratnieks FLW (1992) American foulbrood: the spread and control of an important disease of the honeybee. Bee World 73:177–191

Riessberger-Galle U, Von Der Ohe W, Crailsheim K (2001) Adult honeybee's resistance against Paenibacillus larvae larvae, the causative agent of the American foulbrood. J Invertebr Pathol 77:231–236

Ritter W, Akratanakul P (2006) Honey bee diseases and pests: a practical guide. Agricultural and food engineering technical reports. FAO, Food and Agriculture Organization of the United Nations, Rome

Ritter W, Schneider-Ritter U (1988) Differences in biology and means of controlling *Varroa jacobsoni* and *Tropilaelaps clareae*, two novel parasitic mites of *Apis mellifera*. In: Needham GR, Page Jr RE, Delfinado-Baker M, Bowman CE (eds) Africanized honey bees and bee mites. Ellis Horwood, Chichester, pp 387–395

Sapna K, Rana BS, Sharma HK, Rana K (2010) Comparative efficacy of herbal products against *Varroa destructor* Anderson & Trueman, infestation in *Apis mellifera* L. colonies. Pest Mang Econ Zool 18(1/2):26–30

Schmid E, Berg GL (1969) Oral and parenteral toxicity of *Bacillus thuringiensis* 'exotoxin', and its inactivation in larvae of *Galleria mellonella*. Experientia 25(1):96–98

Scott CD, Winston ML, Slessor KN, King GGS, Grant GG (1984) The biology and pheromone-based monitoring of the dried fruit moth, *Vitulaedmandsae serratilineella* (Lepidoptera: Pyralidae). Can Entomol 116(7):1007–1013

Shah FA, Shah TA (1991) Vespa velutina, a serious pest of honey bees in Kashmir. Bee World 72:161–164

Sharma PP, Deshraj R (1985) Diversity of Bee Flora in Kangra Shivaliks and its impact on bee keeping. Indian Bee J 47(1–4):21–23

Sharma SK, Khan MS (1995) In: Dhinds MS et al (eds) Birds in agricultural ecosystem. Pub. by Soc. For Appl. Orintho., India, pp 137

Sharma OP, Thakur AK, Garg R (1985) Control of hornets attacking bee colonies. Ind Bee J 47:27–29

Sihag RC (1992) The yellow banded brown wasp, *Vespa orientalis*, population density, bee capture efficiency and predation rate on honey bees (*Apis mellifera* L.). Korean J Apic 7(1):35–38

Singh S (1962) Beekeeping in India. ICAR, New Delhi, pp. 1–214

Singh PB, Nairn M (1994) In: Veeresh GK, Viraktamath CA (eds) Proc. 1st Nat. Sem. Cum Workshop on Honeybee Diseases, Their Diagnosis and Management. Oxford and IBH Pub. Ludhiana, India, pp 455

Singh R, Levitt AL, Rajotte EG, Holmes EC, Ostiguy N et al (2010) RNA viruses in hymenopteran pollinators: evidence of inter-taxa virus transmission via pollen and potential impact on non-*Apis* hymenopteran species. PLoS One 5(12)

Smith JIB, Caron DM (1984) Distribution of the bee louse *braula coeca* Nitzsch in honeybee colonies and its preferences among workers, queens and drones. J Apic Res 23:171–176

Spiltoir CF, Olive LS (1955) A reclassification of the genus *Pericystis betts*. Mycologia 47:238–244

Stanghellini MS, Armbrose JT, Hopkins DI (2000) Bumble bee colonies as potential alternative hosts for the small hive beetle (*Aethina tumida* Murray). Am Bee J 140:71–75

Subbiah MS, Mahadevan V (1957) *Vespa cincta* (Fab.) – a predator of hive bees and its control. Indian J Vet Sci 27(4):153–154

Szabo TI, Heikel DT (1987) Fumigation with So_2 to control dried fruit moth in honeybee combs. Bee World 68:37–38

Taber S (1999) The small hive beetle, as described by A.E. Lundie in 1940. Am Bee J 139:450–451

Thakur ML (1991) Honey and Honeybees. ICFRE, Dehradun, 301 pp

Thakur AK, Sharma OP (1984) The spider as bee enemy. J Bombay Nat Hist Soc 81:208–211

Thakur AK, Garg R, Sharma OP (1981) Vertebrate enemies of honey bees and their control in Dhauladhar (H.P.) India. Ind Bee J 43:112–113

Todd JH, de Miranda JR, Ball BV (2007) Incidence and molecular characterization of viruses found in dying New Zealand honey bee (*Apis mellifera*) colonies infested with *Varroa destructor*. Apidologie 38:354–367

Vankova J, Leskova AJ (1972) Changes in the hemolymph of *Galleria mellonella* (L.) caterpillars infected by the exotoxin of *Bacillus thuringiensis*. Acta Entomol Bohemoslov 69:297–304

Verma SK (1995) Studies on the control of greater wax moth, *Galleria mellonella* in *Apis cerana* colonies with biological insecticide, Dipel. Indian Bee J 57:121–123

Viraktamath S (1989) Incidence of greater wax moth *Galleria mellonella* (L.) in three species of honey bees. Ind Bee J 51:139–140

Wagner RE, Reirson DA (1969) Yellow jacket control by baiting, influence of toxicants and attractants on bait acceptance. J Econ Entomol 62(5):1192–1197

Walshaw ER (1967) The raiders. Am Bee J 107:14–15

White GF (1917) Sacbrood. US Dep Agric Bull 431:1–55

Whitefoot LO, Detroy BF (1968) Pollen milling and storing. Am Bee J 108:138–140

Wilkins S, Brown M, Andrew A, Cuthbertson GS (2007) The incidence of honey bee pests and diseases in England and Wales. Pest Manag Sci 63:1062–1068

Wilson WT, Brewer JW (1974) Beekeeping in the Rocky Mountain Region. Coop. Ext. Ser. Fort. Collins*:* Colorado State University

Wood M (1998) Microbes help bees battle chalkbrood. J Econ Entomol 46(8):16–17

Woodward D, Jones W (1991) Ant control in apiaries. Bee Briefs 8(2):2–5

Woyke J (1985) Further investigations into control of the parasite bee mite *Tropilaelaps clareae* without medication. J Apic Res 24(4):250–254

Woyke J (1993) Practical control method of the parasitic bee mite *Tropilaelaps clareae*. Am Bee J 133:510–511

Yadav S, Singh J, Chunneja PK (2012) Management of *Galleria mellonella* Linnaeus (Lepidoptera: Pyralidae) for comb protection during storage. Indian J Appl Ent 26(2):140–142

Zaghloul OA, Mourad AK, El Kady MB, Nemat FM, Morsy ME (2005) Assessment of losses in honey yield due to the chalkbrood disease, with reference to the determination of its economic injury levels in Egypt. Commun Agri Appl Biol Sci 70(4):703–714

Lac Culture

5

Y.D. Mishra and P. Kumar

Contents

5.1 Introduction

Lac, an important versatile commercial resin of wide utility, is the natural heritage of India. It is secreted by tiny gregarious scale insects as a protective covering around their body. Thriving on succulent shoots of a number of plant species, these insects form a thick encrustation around the twigs, which are collected and scraped for obtaining raw lac or sticklac. Passing through several steps of processing, sticklac is converted into various commercial forms as seedlac, shellac, button lac, bleached lac, dewaxed decolorized lac, etc. Several valuable by-products, like lac dye and lac wax, obtained while processing, are also of great industrial utility. Thus, lac culture can be defined as "systematic rearing of lac insects on host plants for sustainable production of commercial lac (raw lac)".

The lac culture has grown over times in the country. Centuries are witness of our traditional indigenous knowledge about lac and its uses in various spheres of life. It

Y.D. Mishra (deceased)

P. Kumar (✉)
Indian Institute of Natural Resins and Gums, Namkum, Ranchi, Jharkhand, India
e-mail: pranayak96@gmail.com

Omkar (ed.), *Industrial Entomology*, DOI 10.1007/978-981-10-3304-9_5

was known in India as *lakchha* in Sanskrit, and its mention in the holy *Atharva Veda* (Dave 1950) is evidence in favour of our claim. Its use as building material for construction of a *yatu griha* by Purochan (the *Mahabharata*), as cosmetics (*Kumarasambhavam* and *Meghdootam* by Kalidas, 200 AD) and as medicine for the preparation of *Lakshadi taila* and *Lakshadi guggulu* by Charaka and others (Ayurvedic literature) may also be quoted for reference. Its several commercial uses have also been mentioned in Ain-i-Akbari. Foreign writers (Periplus, 80 AD, to Nicolo Contil, 1560 AD) had also mentioned about lac and its uses at several places, which was well described by Watt (1908). Lac was first recognized as an export commodity in seventeenth century and was exported to Europe in 1607. It is important to note that its demand was not only for resin but also for crimson dye obtained from washing of raw lac. The demand of resin was surpassed by the high demand of lac dye during the seventeenth to nineteenth centuries. However, the demand of lac dye crashed during the later part of nineteenth century with the discovery of synthetic aniline dye. The consumption of lac as a raw material for various industries, such as gramophone, surface coating and polishing, has tremendously boosted the demand of lac resin in foreign markets during twentieth century.

The continuous growth of lac trade and its diversified uses in every sphere of life has lead to various scientific investigations since eighteenth century. Father Tachard in 1710 sent a memoir from Pondicherry to France in which the distribution of lac, its physico-chemical properties and methods of processing have been enumerated. Roxburghe (1789) made significant contribution by his study of the life cycle of lac insect. Anderson (1791) published a monograph on lac. Several workers in India and abroad, namely, Carter (1861), Stebbing (1908), Misra (1923), Imms and Chaterjee (1915) and others have also contributed towards different aspects of lac insects.

Owing to increased consumption of lac resin in electric, defence and gramophone industries, high international demand was seen during the First World War, which was followed by several scientific investigations in the field. However, drastic downfall in price of lac was recorded during the post-war scenario. A high-power committee headed by Lindsay and Harlow who submitted the report in 1921was set up by the then Government of India mainly for looking after the threats due to synthetic resin and fluctuation in production, price and quality of lac. An Indian Lac Association for Research was constituted to execute recommendations of the committee, and the Indian Lac Research Institute was established in 1924 at Namkum, Ranchi. The Institute was handed over to newly formed Lac Cess Committee in 1930 and then to the Indian Council of Agricultural Research, New Delhi, on 31 March 1966, after the dissolution of the former. Fully devoted to strategic research and development of lac, this Institute is only of its kind in the world. Studies carried out on the entomological aspects of lac insects at the Indian Lac Research Institute by Misra, Glover, Negi, Krishnaswami, Mehra, Varshney, Chauhan and many others have also provided sufficient information about the lac insects. Since the production of lac is the function of cutaneous lac glands of these insects, their knowledge thus forms the priority for those who are associated with it. However, several untold stories, still hidden within the resinous cell protecting these insects all around, are required to be searched with thorough scientific investigations.

Export of lac touched a peak of 42,367.7 MT during 1936–1937 for Rs. 234.21 lakh on increased international demand. This peak was never reched again mainly due to threats from synthetic resins, hard competition from Thailand in lac export and finally because lac was replaced by PVC in the gramophone industry. Bad days of lac actually started from 1972 to 1973 when the quantum of export went down below five figures (7563.5 MT) for the first time, after which it occasionally crossed 10,000 MT. During the last decade of the twentieth century, a turn around in lac export could be noticed as the lac resin has again been glorified due to its eco-friendly and biodegradable properties in an environmentally conscious world. In addition to the traditional areas of its utilization, some new areas have been identified of which food industries, coating of medicinal pills, cosmetics, fine chemicals (PGRs, pesticides, nematicides, pheromones, etc.) and perfumeries are worth mentioning. The internal consumption, which was once limited to 15% of the national production only, has now reached the level of 30–35%. With the advent of new sustainable and profitable lac production technologies mentioned herein and gradually increasing national and international demand of lac, this industry is expected to grow by 4–5% annually. This will help to provide bread and butter to nearly five million downtrodden people of the county, mostly tribes, inhabiting geographically handicapped regions where the green revolution is yet to reach.

5.2 The Lac Insect

5.2.1 Systematics

Lac insects are placed under the order Hemiptera, superfamily Coccoidea and family Tachardiidae. The present knowledge of lac insect taxonomy is mainly based on the systematic monograph by Chamberlin (1923a, b), its supplement (Chamberlin 1925a, b) including seven species under two genera. Catalogues by Kapur (1958) and Varshney (1966, 1977, 1990a, b, 1993) are of immense importance for updating the list of lac insects (Fig. 5.1). Although adult male and larval characters are not less important, these have not been considered due to the short duration of the life stages. However, only characters of adult female have been considered for characterizing insects of this family.

A female lac insect spends its whole life under a convex protective covering known as lac cell, which is secreted by cutaneous glands of the insect. Besides the mouth, the female has three openings at the aboral side. Of these pores, one is anal opening on anal tubercle covered by anal plate, and the two are branchial openings each on distinct branchia covered with branchial plates. Only two pairs of thoracic spirals are present, the anterior one in association with branchia and the posterior one near the rostrum. Antennae are vestigial; appendages and eyes are absent or rudimentary. The anal plate consists of chitinous fringe and ten anal setae.

Lac insects (family Tachardiidae) of the world are now represented by 87 species under nine genera distributed in two subfamilies: (i) Tachardiinae, including true lac

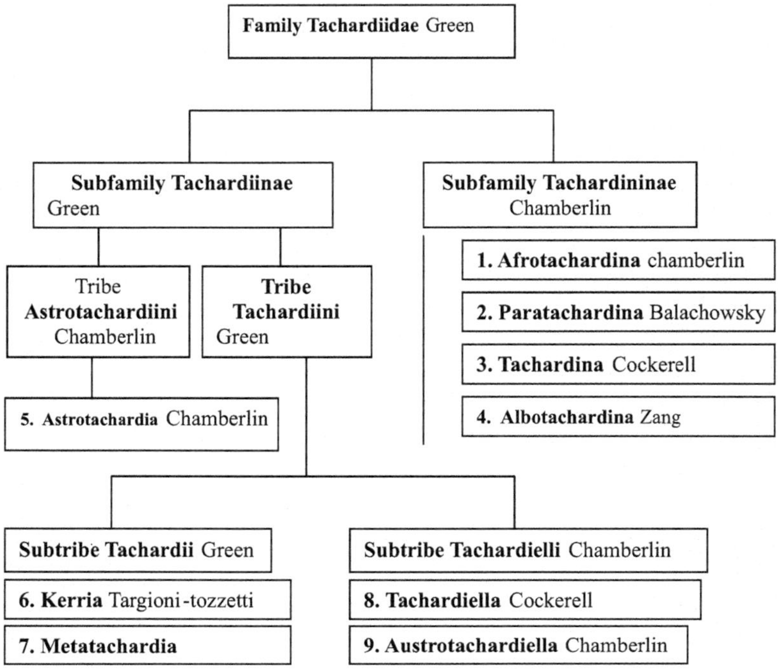

Fig. 5.1 Systemic position of lac insects genera

insects of resinous cell, and (ii) Tachardininae, including pseudo-lac insects of non-resinous cell.

(a) *Indian lac insect*

It was only 73 years after the dispatch of a memoir on lac by Father Tachard that Kerr in 1782 was able to identify the causative organism rightly as a scale insect and named as *Coccus lacca*. Fabricius (1787) described the Indian lac insect as *C. ficus* and Targioni-Tozzetti as *Kerria lacca* in 1884. Blanchard (1886) proposed the name *Tachardia* that was synonimized with *Kerria*, the valid generic name on the basis of priority by Varsaney (1966). A separate identity of these insects, proposed by Oken in 1815 placing the insect under the genus *Laccifer*, was also discarded on the basis of "*Laccifer*" being a non-Latin word. However, several genera, species, subspecies and strains have been defined without considering reproductive isolation between them. Possibly such differences may not actually exist between many of these species as most of the taxonomic characters are not related to reproductive organs or genitalia of the adult female, which are represented by the anal tubercle only. However, 16 species under the genus *Kerria* and 6 under the genus *Paratachardina* have been recorded so far from the Indian subcontinent. A recent key to the species under *Kerria* from the Indian subcontinent has been reported by Mishra and Sushil (2000) and has been shown in Table 5.1.

Table 5.1 Revised key to the Indian species of *Kerria*

Couplet	Leads to / species
1. Anal tubercle (supra-anal plate) large, elongate, longer than broad	2
– Anal tubercle (supra-anal plate) broader than long or subequal or abbreviated	9
2. Branchial plate borne on distinct elevated branchia	3
– Branchial plate sessile	*ebrachiata* (Chamberlin, 1923a, b)
3. Body long; dorsal spine smaller in comparison to other parts and body size	4
– Medium-sized globular body; dorsal spine well developed and pedicel also large	5
4. Canellar bands or chitinous trailings below anterior spiracles very long; anterior spiracle far from branchial plate	*chinensis* (Mahdihassan, 1923a, b, c)
– Canellar bands or chitinous trailings below anterior spiracles shorter; anterior spiracle comparatively nearer to branchial plate	*nepalensis* (Varshney, 1977)
5. Pedical of spine thickly sclerotized; fringe with longer lobes	*nagoliensis* (Mahdihassan, 1923a, b, c)
– Pedicel well developed, but not thickly sclerotized anal fringe lobes short	6
6. Branchia less elevated; small-size females	*communis* (Mahdihassan, 1923a, b, c)
– Branchia much elevated; large females	7
7. Number of dimples on branchial plate 9–12	*lacca* (Kerr, 1782)
– Number of dimples on branchial plate less than 9, generally 6–7	8
8. Dorsal spine half as long as diameter of branchial crater: antennae tipped with 4 setae	*indicola* (Kapur, 1958)
Dorsal spine two-thirds in length to width of branchial plate; antennae tipped with 3 setae	*chamberlin* (Varshney, 1966)
9. Marginal duct clusters with less than 30 ducts	10
– Marginal duct clusters with approximately 40 ducts	11
10. Marginal duct clusters with 10–15 ducts, anterior spiracles touching the branchial plate	*albizzae* (Green, 1911)
– Marginal duct clusters with 24–30 ducts, anterior spiracle away from the branchial plate	*sharda* (Mishra and Sushil, 2000)
11. Branchial plate and supra-anal plate subequal; area; star pores near mouth parts absent	12
– Branchial plate shorter in area than supra-anal plate; star pores present near mouth parts	15
12. Branchial crater not in centre of plate and somewhat open; dimples smaller and obscure	*sindica* (Mahdihassan, 1923a)
– Branchial crater in centre of large branchial plate, crater rim not open: dimples larger and distinct	13
13. Branchia very short (but branchial plate large and wide); pedicel of spine abbreviated	*fici* (Green, 1903)
– Branchia prominently elevated; pedicel of spine not abbreviated	14
14. Branchia club-shaped; antenna minute	*brancheata* (Varshney, 1966)
– Branchia cylindrical tube-shaped; branchial crater with 5–6 dimples; antenna apparently of 3–4 segments	*pusana*
15. Branchia nearly sessile; branchial crater well defined, but dimples weakly marked a small	*javana* (Chamberlin, 1925a, b)
– Branchia distinctly present; branchial crater not well defined, but dimples strongly marked	*rangoonensis* (Chamberlin, 1925a, b)

(b) *Strains of* Kerria lacca

Glover (1937a, b, c) has described two strains of *Kerria lacca*, the *kusmi* and the *rangeeni*, on the basis of host preference and life cycle. The former survives on *Schleichera oleosa* locally known as kusum or nagoli and has two generations in a year of almost equal duration. The latter, on the other hand, is not endowed with the mechanism to survive on kusum and has two generations of unequal duration in a year. He also submitted that no clear morphological differences could be recorded between them, hence may be regarded as strains of the same species. The species *Kerria nagoliensis*, proposed by Mahdihassan (1923a, b, c), which is still on the records as species, is the same that was regarded as *kusmi* strain by Glover. Recent evidences submitted on morphometric differences (Mishra et al. 1997) and cross-bred progeny behaviour (Mishra et al. 2001) are sufficient to consider them as separate taxa.

(c) *Trivoltine lac insects*

Mahdihassan (1923a, b, c) recorded a lac insect from Mysore (Karnataka) on jalari (*Shorea talura*), which was known to complete three generations in 13 months, and named it as *Lakshadia mysorensis*, which was later on regarded as a subspecies of *K. lacca* by Varshney (1977). Mishra and Sushil (2000) recorded a new species, *K. sharada* from the Eastern Ghats on kusum, completing three generations in 350 days at their native locality.

5.2.2 Habit and Habitat

Lac insects are parasitic to a number of plant species, called as lac host plants. These are sedentary insects excepting the newly hatched crawling stages of nymphs which, soon after coming out of the mother cell, settle on succulent shoots of the host plant where it is fixed for the whole life.

Being a plant parasite, its habitat is strictly associated with the micro-ecosystem of the host plant. These are, however, reported to occur in tropical region where temperature records between 04 and 48 °C. These insects have been reported from very low rainfall area of the desert of Thar to very high rainfall zone of north-eastern states, but certainly on different host plant species. Most preferable altitude for these insects ranges from sea level to 150 m.

5.2.3 Distribution

Although, lac insects are found throughout the tropical and subtropical regions of the world, the tropical forest of India, Thailand, China and Vietnam are the places of their occurrence in abundance. European countries are, however, devoid of these insects due to very cool climate during winter. However, as per reports, lac insects were cultured in glass houses in erstwhile USSR during winter, and the broodlac thus obtained was used for raising summer crops in the open condition on *Ficus*

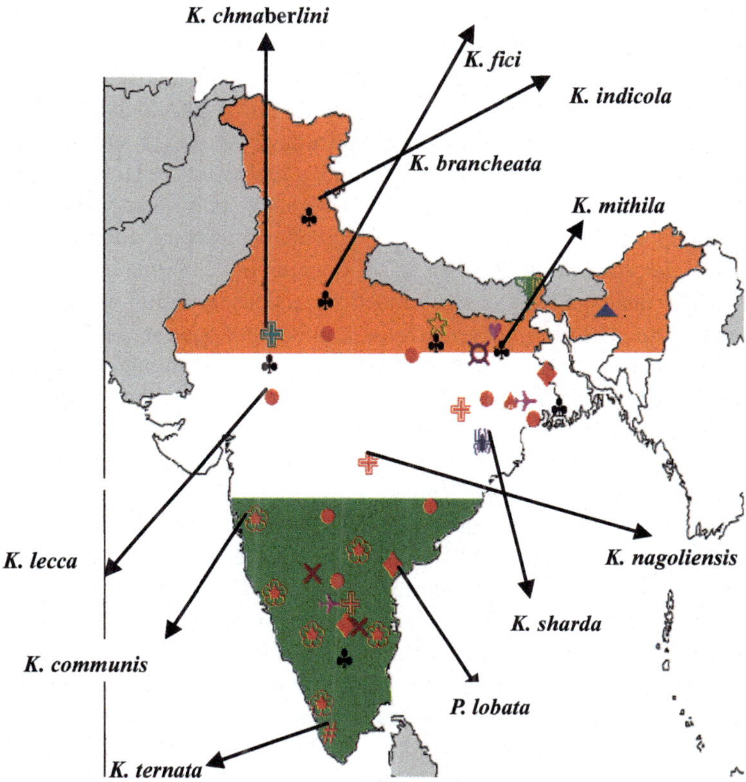

Fig. 5.2 Distribution of different lac insect species in Indian subcontinent

carica and other hosts (Malhotra 1976). Distribution of different species of *Kerria* in the Indian subcontinent has been presented in Fig. 5.2.

5.2.4 Morphology

The lac insects exhibit a unique case of sexual dimorphism and retrogressive metamorphosis. Most of the typical insect characters present in the crawler stage are lost during metamorphosis in adult females. Males, however, retain most of the insect characters after metamorphosis. Since the species, *Kerria lacca* (Kerr), is responsible for the bulk of the commercial lac production of the country, the present description accounts for this species as a type only for the sake of convenience.

(a) *Female lac insect*

The female lac insect is pear-shaped when allowed to grow in isolation and somewhat elongated in encrustation without body segmentation or demarcation between the head, thorax and abdomen. It consists of a pair of rudimentary vestigial indistinctly segmented antennae carrying a number of setae and two pairs of spiracles displaced due to torsion. One pair of these spiracles, called as posterior spiracles, lies on the oral side of the body, and the other pair that remains enclosed in branchial process or the bronchium at the aboral end of the female is called as the anterior spiracles. Absence of thoracic appendages from the entire adult life of female is also an important feature of the family Tachardiidae. However, rare occurrence of adult female (at crop maturity) with undeveloping legs and eyes has also been recorded (Mishra et al. 2002). At the aboral end, a prominent anal tubercle can be seen which terminates in the form of the opening of the alimentary canal, the anal opening. The anal opening is surrounded by a densely chitinous plate called as supra-anal plate. The anal fringe surrounding the anal opening is decorated with a circle of ten anal setae springing out of the pores in the supra-anal plate. Branchia is positioned anterior to the anal tubercle. The branchia is covered with branchial plate, which is perforated with numerous pores arranged on semilunar elevated structures known as dimples. Between the branchia and anal tubercle, a conspicuous structure known as the dorsal spine is present, which is positioned on a cuticular outgrowth, the pedicel. The dorsal spine is a unique character of the family Tachardiidae, and its function in females has not yet been defined. The size and shape of the pedicel is also an important character. The body of the female is covered with the cuticle perforated at various places by openings of cutaneous glands. The pattern of which is important for taxonomical point of view. These are (i) perivaginal pore clusters located around the anal tubercle, (ii) marginal duct clusters arranged in different patterns known as serpentine area, (iii) ventral duct clusters and (iv) lac glands spread over randomly on the surface of the body. The piercing and sucking mouthparts of the female lac insect consist of two pairs of maxillae and mandibles modified into four thin stylets. The labium is also modified into a cone-like rostrum and through this passes the tubular structure constituted by the stylets. The tube is retractile and when retracted is lodged in a crumen lying in the body cavity. All the four stylets terminate internally into four broad plates and are attached to protractor and retractor muscles for their movement (Imms and Chaterjee 1915; (Fig. 5.3).

(b) *Male lac insect*

Males, on the other hand, are long with distinct head, thorax and abdomen. The head consists of a pair of long segmented antennae, two eyes and a non-functional mouth. The thorax possesses three pairs of appendages and two pairs of spiracles. Depending on the season, species/strain, photoperiod and the locality, the adult male may be apterous or alate. The abdomen is provided with a pineal stylus at the apical aboral end serving the purpose of copulatory organ and a pair of caudal process with long waxy filaments on the dorsal side of the abdomen.

(c) *Lac and wax glands*

It was believed that lac is secreted through anal tubercle and branchial process. Misra (1930a, b) has demonstrated that the cuticle is perforated by numerous pores, which are the openings of several cutaneous glands lying under the skin. These

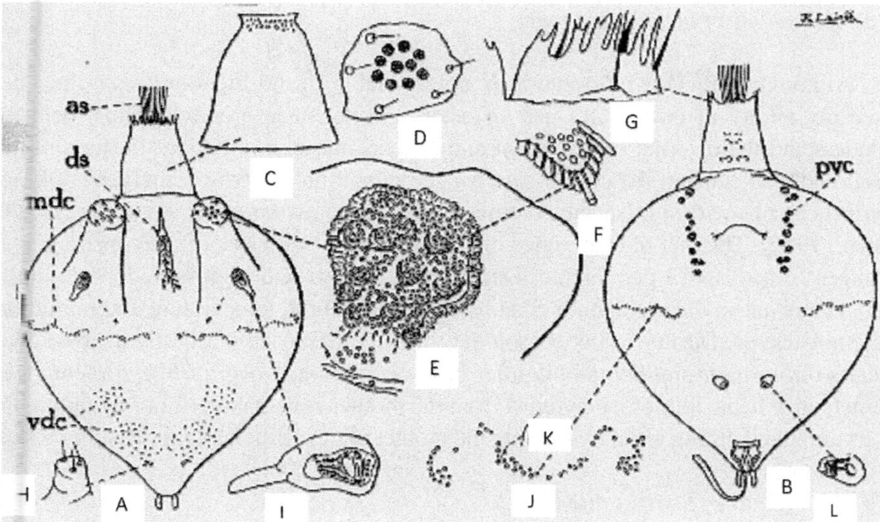

Fig. 5.3 Morphology of female lac insect. (**A**) Dorsal aspect, (**B**) ventral aspect, (**C**) supra-anal plate, (**D**) brachial plate, (**E**) the same enlarged, (**F**) cross-section through dimple to show pore and nuclear duct, (**G**) anal fringe, (**H**) antenna, (**I**) anterior spiracle, (**J**) marginal duct cluster, (**K**) individual duct, (**L**) posterior spiracle, *as* anal ring setae, *ds* dorsal spine, *mdc* marginal duct cluster, *pvc* ventral duct cluster

pores are arranged at different position in definite pattern and accordingly named as perivaginal pore, marginal duct cluster, anal and ventral duct clusters as stated earlier. Besides the above-mentioned openings, numerous glands are also seen distributed all over the body. He concluded that these glands are responsible for the secretion of lac. He further recognized three types of glands, viz. (1) minute unicellular glands, which are small and numerous, (2) large unicellular glands found associated with small glands in marginal duct clusters and (3) aggregated pluricellular glands.

5.2.5 Bionomics

The voyage of life of lac insects begins with the completion of embryonic development within the body of the mother when eggs change their position in the ovariole. Eggs then travel through the oviducts and come out of the vagina into the specially built "incubation chamber" formed by making sufficient gap between the body and resinous cell. Hatching of eggs mostly takes place before reaching the incubation chamber. Imms and Chatterjea (1915) have termed this as viviparity, an uncommon phenomenon in insects. Negi (1934) described the formation of vacant space inside the lac cell, the "incubation chamber" and the secretion of wax filaments into it. He narrated the process of egg laying as well as the emergence of larvae and called the phenomenon as ovoviviparity.

(a) *Behaviour of larval emergence*

Although sufficient information is not available in the literature regarding the sensing ability of crawlers, a pair of ocelli and antennae probably helps them to understand the external environment and proper time of coming out of the mother cell and also sensing the correct site for settlement on the twig. Emergence of lac insects begins with sunrise and continued till all the crawlers come out of the cell (A Rep., 1986). The rate of emergence of crawlers varies with size of anal opening and ranges from 3 to 13 per minute. During cloudy days and high humid conditions, crawlers wait in the incubation chamber till proper light, temperature and humidity are not sensed. During winter season the time of emergence is slightly delayed and starts during morning hours only after foggy condition is over. While studying the emergence behaviour of lac nymphs from the mother cell, emergence of nymphs can be stimulated during night by keeping the mother under artificial light (Mishra 1931).

(b) *Behaviour of larval settlement*

The crawler nymph after coming out of the incubation chamber looks ovate in outline, slightly more pointed posteriorly, soft-bodied, crimson in colour and very small in size, being usually 0.4–0.6 mm long (excluding the antennae and caudal setae) and 0.15–.2 mm wide across the thorax. During the swarming period, lakhs of such tiny insects are seen crawling over the surface of lac encrustation. The lac insect is, thus, probably named as "Laksha" in Sanskrit on this basis. Crawlers move at an average speed of 2 meter per hour on the shoot in search of adequate site for settlement (A Rep., 1986). The movement is to and fro, and the resultant distance is thus only few meters. They also pierce the bark of the host plant probably with the view of sensing the succulence of the shoot, and as soon as the correct site is obtained, they settle there with spread legs penetrating the proboscis inside the proper tissue within the bark. It has been seen that the site of settlement of these insects is always away from direct sunlight. Nearly after 2 h or so, they withdraw their legs, provided no external stimulation forces them to change the position. Finally they fix themselves at the site for the remaining period.

Lac insects are gregarious in nature. This behaviour can also be seen at the time of settlement of young crawlers. The density of settlement of lac nymphs on twigs of host plants depends on (i) host plant factors, (ii) lac insect species, (iii) different biotypes of a species and (iv) cultural practices adopted. However, the range of larval density has been found from 60 to 200 larvae per sq. centimetre.

5.2.6 Metamorphosis

(a) *The first instar larva*

A day or so after settling, the larvae start secreting lac from almost all over the body except near the rostrum and the brachial plates. Thus it gets encased in a cell of lac, which keeps on increasing with the increase in the size of the insect. The insect moults thrice before reaching maturity, the duration of each instar being dependent on several environmental factors, such as temperature, humidity, host plants, etc. Mahdihassan (1930a, b, c) has claimed the distinction between male and female nymphs at the time of settlement but due to very inconspicuous and minute differences, it is not practicable to identify the sexes at this stage on the basis of morphological characters stated by him.

(b) *The second and the third instar*

After the first moult, the male as well as the female larvae loose their legs, antennae and eyes. From this stage onwards, sexual dimorphism becomes pronounced. The lac cell of the male assumes a slipper- or cigar-shaped appearance at the final stage and has a loose operculum at the rear end. While still inside the cell, the larvae cast off the second moult, which is pushed out of the cell from the rear end. Subsequently, the larvae pass through the prepupal and pupal stages when appendages which eventually develop into legs, antennae, eyes, wings, (except in apterous males), penial stylus, etc. are easily seen. However, during the last stage, the male insect stops feeding, and the mouthparts become atrophied. The male emerges with the hind end of the body first by pushing the operculum. They may be winged or wingless, the relative number of the two forms varying considerably in different seasons of the lac crops. They copulate with the females, which remain enclosed in the lac cell, and by the time of the emergence of males, generally become pear-shaped.

The female larvae also cast off the antennae, eyes, legs, etc., after the first moult, but unlike the male, they do not develop these organs again except the rudimentary antennae instead of certain other organs, which are peculiar to the female to become conspicuous. The size of brachial plates and the number of openings in them increase; the number of setae from the anal ring plates also increases from six to ten. Openings of the marginal and perivaginal ducts are seen in clusters for the first time, but the branchia, the anal tubercle and the dorsal spine are yet undeveloped, though the upturned terminal segments of the abdomen may be regarded to mark the beginning of the anal tubercle. Two further moults are cast off and the cast skins in each case being pushed out of the lac cell at the rear end. During the second and subsequent third instar, the larva becomes more swollen and loses all traces of segmentation. Besides the increased rate of growth along the vertical axis, the terminal segments of the abdomen are directed upward. Areas around the brachial plates are demarcated and constitute what may be called the beginning of the anal and brachial lobe. Changes in the position of the organs, such as shifting of anterior pair of the spiracles towards aboral end and orientation of the alimentary canal, nervous and tracheal system, etc., also take place during this instar. The insect, thus, assumes generally the appearance of a pear-shaped structure or roundish bag completely occupying the space inside the lac cell. Openings in the clusters of ducts of the ventral, marginal and perivaginal glands also make their appearance.

(c) *Adult female*

After the final (third) moult, the dorsal spine, which is borne on an elongate tubercle, appears in the centre of the triangular lying between the two brachial and one anal tubercle. At this stage the female is sexually matured and is copulated with the male, the emergence of which synchronizes with this stage in the development of the female. The males die within few days of their emergence and copulation. From this time onwards, lac is secreted at a fast rate, and the size of the female insects and of the enveloping lac cells increases at a faster rate than in the case during the earlier stage and reaches a size several times more than that of the male lac cell. This state of activity lasts for a varying number of weeks depending upon the season, place and host plants. The female lac insects that live for a relatively longer period after mating are the chief sources of lac secretion. As the lac insects are usually situated close together, the lac secretion from adjacent cells coalesces with each other and forms a continuous encrustation on the branches. The rate of secretion of waxy filaments, which protrude out of the anal and the brachial pores, and of the excretion of honey dew also increases during this period. The cottony appearance of certain healthy encrustation of lac is due to the long filaments of wax, while the shoots appearance on the leaves of trees bearing lac is due to the growth of certain black fungi (*Capnodium* and *Fumago* species) on the honeydew that falls on the leaves.

Sticks with the lac encrustation containing gravid females are called "broodlac" stick, which are generally tied together for the purpose of infecting other trees for the succeeding crop. The duration of life cycle and numbers of generation per year depend on various factors, such as the species/strain of the lac insect, the season of development and climatic condition of the area. The duration taken for different metamorphic stages of Indian lac insects is given in Table 5.2 and life cycle in Fig. 5.4.

5.2.7 Reproduction

Reproduction in these insects was a matter of controversy till Teotia and Chauhan (1963) ruled out the possibility of facultative parthenogenesis, confirming only sexual mode of reproduction. Chauhan and Mishra (1970a) provided the evidence of multiple coitus and the effective role of males, not only in fertilization but also in maintaining the variability by copulating with a number of females and vice versa. Adult males, on emergence from the cell, start crawling around, and after finding a sexually mature female, it mounts over the body of the female and inserts the pineal stylus into the vagina by repeated attempts. Sperms then find their way to the seminal vesicles. After the completion of the process of mating, the male moves to other sexually mature female for copulation. This process of visiting females continues till the death of the male. A female can also copulate more than once with different males or with the same male. It has been confirmed that each mating contributes effectively to the progeny.

Table 5.2 List of the important pests attacking the major lac hosts

Name of the pest	Host plant	Period of incidence	Plant part affected
1. *Aonidiella orientalis* Newst. (Homoptera: Coccidae)	*kusum and ber*	February–March May–June October–November	Stem and twigs causing retarded growth
2. *Tessaratoma javanica* Thunb. (Hemiptera: Pentatomidae)	*kusum*	March and July–August	Damage the tender and growing shoots
3. *Serinetha augur* Fab. (Hemiptera: Coreidae)	*kusum*	April–May to July–August	Serious pest of fresh and dry fruits as well as seeds
4. *Coptosoma ostensum* Dist. (Hemiptera: Pentatomidae)	*palas, arhar alternate host*	February–June/July	Tender shoots, leaves and inflorescence
5. *Odontotermes obesus* Rambur (Isoptera: Termitidae)	*kusum, ber, palas, khair*	June–July	Construct closed earthen tunnels over branches
6. *Sathrophylla rugosa* L (Orthoptera: Tettigoniidae)	*palas, kusum*	April–August	Feed on tender leaves of palas. Large holes in the leaves due to feeding
7. *Teratodus monticollis* Gray (Orthoptera: Acrididae)	*palas, ber, ghont*	May–August	Midrib of leaf attacked and leaf stalk cut
8. *Myllocerus cardoni* Marshl (Coleoptera: Curculionidae)	*kusum, ber, palas, ghont*	April–November	Feed on the leaves in small semi-circles
9. *Thiacidas postica* WIK (Lepidoptera: Lymantriidae)	*ber, ghont*	January and September	Defoliate the trees
10. *Diacrisia obliqua* WIK (Lepidoptera: Arctiidae)	*kusum*	July–October	Feed on young leaves and tender seedlings voraciously
11. *Parasa lepida* Cram	*kusum, ber and palas*	July–September	Defoliates the hosts by feeding on the leaves
12. *Natada nararia* Moore			
13. *Belippa laleana* Moore (Lepidoptera: Limacodidae)			
14. *Semiothisa fidoniata* Guen. (Lepidoptera: Geometridae)	*khair*	November–December	Feed on leaves
15. *Indarbela tetraonis* Moore (Lepidoptera: Indarbelidae)	*kusum, ber*	May–July	Feed on the bark and bore into the stems of the plants
16. *Hieromantis ioxysta* Meyr. (Lepidoptera: Schreckensteiniidae)	*kusum*	April–October	Roll the edges of tender leaves and feed on them
17. *Pachyonyx quadridens* Chevr. (Coleoptera: Curculionidae)	*palas*	April–July	Producing galls on leaf petiole

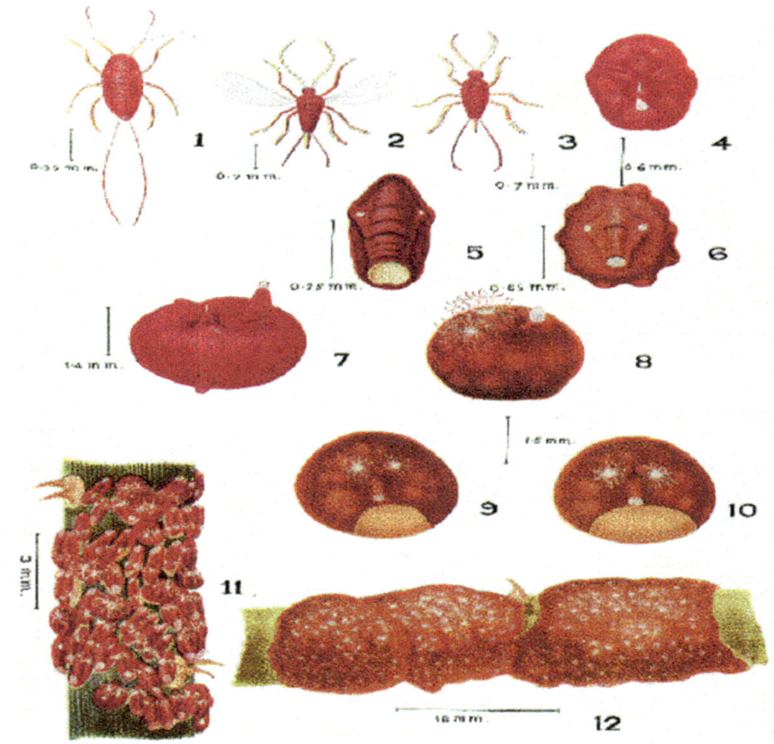

Fig. 5.4 Life stages of lac insect

5.3 Biotic Associates of Lac Insect

The lac insect ecosystem is endowed with a very rich biotic complex, which comprises a number of flora and fauna representing almost all types of biological association. Important associates are dealt below.

5.3.1 Lac Host Plants

Since lac insects draw nutrients by sucking the phloem sap (Krishnaswami et al. 1964) of host plants, these are the nearest biotic associates of lac insects and stand at the first trophic level. Thus, the occurrence of these insects cannot be assumed in the absence of host plants. Lac insects are very specific in host selection. On the other hand, survival of lac insect on a host plant also depends on several plant and soil factors, viz. succulence and age of shoots, growth period, edaphic conditions, nutrients and moisture, etc.

(a) *Categories of lac host plants*

Since the first list of host plants that included only four species (Kerr 1782), many additions have been made so far, and the global list of lac insect host plants reached up to nearly 400 species. Those referred herein are host plant species of commercial Indian lac insects (*Kerria*: Tachardiidae). In this regard, worth mentioning are the records and catalogues by Watt (1901), Roonwal et al. (1958), Srinivasan (1956), Varshney (1968), Sharma et al. (1997), etc. Lac host plants are classified mainly into three categories as per their distribution, suitability for lac cultivation and contribution to the national production of lac. These are (1) common lac hosts, (2) occasional hosts and (3) rare hosts (Roonwal et al. 1958). Out of the common host plants (17 species), three species, viz. palas or dhak (*Butea monosperma*), kusum or nagoli (*Schleichera oleosa*) and ber (*Ziziphus mauritiana*), are of all-India importance contributing nearly 95 % of the national production of lac. Hence, these are also known as conventional hosts (Fig. 5.5).

Some quick-growing lac hosts, viz. (i) *Flemingia macrophylla*, (ii) *Flemingia semialata*, (iii) *Cajanus cajan*, (iv) *Ziziphus mauritiana*, (v) *Acacia auriculiformis* (a kashmani), (vi) *Acacia nilotica* and (vii) *A. catechu*, are capable of sustaining lac crop in only one generation and hence depend on conventional host plants for maintenance of broodlac crop or on some other hosts in alternation.

Thus, due to their use for lac cultivation specifically during rainy or winter crop season, these are known as alternate hosts. Some of the common host plants known as summer broodlac preservers are used to conserve broodlac during summer season for feeding the broodlac requirement of other alternate hosts. These are (i) *Albizia lucida*, (ii) *A. saman*, (iii) *Ficus cunia* (Porho: Moraceae), (iv) *F. lacor* (pakar), (v) *F. religiosa* (peepal), (vi) *F. benghalensis* (bargad), (vii) *F. racemosa* (gular), (viii) *Leea crispa*, (ix) *L. robusta* and (x) *Ougeinia oojeinensis* (sandan). Some host species are only of regional importance, viz. *Kydia calycina*, *Leea crispa*, *L. robusta*, *Grewia serrulata* and *Cajanus cajan* (arhar) for NEH region, *Ziziphus zylopyra* (ghont) for Madhya Pradesh (MP), *Shorea talura* (jalari) for Karnataka and *Acacia nilotica* (babool) for Rajasthan. Occasional host plants and rare host plant species do not always sustain lac crop; hence, these are important for reference only and also for preserving some rare lac insect species/biotypes occurring in nature.

(b) *Susceptibility of host plants to lac insects*

Susceptibility of a host plant to lac insect depends on various factors, viz. physical factors, physiological factors and biochemical factors, which include factors associated with nutrition and antibiosis. Owing to these factors, lac insect either fail to settle on twigs or may die after settlement at any stage of development before the time of producing young ones. The number of deaths of lac insects in a colony may thus vary from a few to cent per cent. Amongst the physical factors, succulence of bark, age of shoots, smoothness of shoots, pubescence on shoots as well as colour and texture of shoots and leaves are the determining factors. The succulence of the bark is very much important for inserting the proboscis by larvae into the specific region of the shoot, failing of which, they cannot survive. Since almost all common

Fig. 5.5 Conventional lac hosts

hosts have pubescence on twigs, this character has also been considered as an important one. Srinivasan (1956) described the charka variety of kusum and palas on the basis of colour and texture of shoots and leaves and stated that the former does not yield lac. Physiological factors are also equally responsible for the growth of lac insect on a host plant. Thakur (1932) has described that sap density and sap reaction are important factors responsible for the selection of host plants. However, there could be several unidentified factors that limit lac insect growth on many plant species falling within the range of suitable sap density (0.14–0.1728) and pH (5.8–6.2). Also, physiological interactions (Chauhan and Mishra 1970b) of host plants and the lac insect have been found to determine the status of the host plant in a locality or in a particular season. There are many good hosts, which are known to

Table 5.3 List of pests attacking the bhalia (*Flemingia macrophylla*)

Name of the pest	Period of infestation	Mode of damage
1. *Hypena iconicalis* Walker	May–October	Larvae perforate the lamina. Serious defoliator of fresh leaves
2. *Hemithea tritornaria*	June–February	Looper defoliator, feed on soft portion of the lamina leaving the veins
3. *Nephoteryx leucophaella* Zell		Defoliator-cum-leaf binder/leaf skeletonizer-cum-binder
4. *Platypeplus aprobola* Meyr.		Leaf binder-cum-defoliator
5. *Dasychira mendosa* Hubn. (Tussock moth caterpiller)	July–December	Tender leaves of bhalia seedlings
6. *Prodenia litura* Boisd.	October–December. Peak June–November	Polyphagous pest

produce lac during one season only, but fail to sustain in the other season, as they are not in a right physiological state to sustain lac crop. Nutritional factors include the presence of essential lac insect nutrition in the phloem sap and essential plant nutrient in the soil. Lac insects of rangeeni strain do not survive on kusum, but the kusmi lac insects survive during the same climatic condition. Also, the *Kerria lacca* does not thrive on *Grewia serrulata*, but *K. chinensis* do so. It appears that these lac insects differ in requirement of essential nutrient, and they do not grow on host plants that do not have these particular nutrients. Mortality of lac larvae soon after settlement is the example of antibiosis. Some secondary substances/metabolic products are found to produce such effects at different developmental stages.

(c) *Intraspecific variation in lac insect susceptibility*

Host plants of the same species show great variation in susceptibility and thereby in productivity. There are two varieties of kusum and two of palas, one of each known as karia which are good host plants and the *charka* are bad hosts. In this regard, adaptive nature of lac insect is also responsible for the performance of lac insect on a particular host plant. When lac insects are introduced to different host plants, high mortality is recorded, but after some generations, it is reduced to a great extent (Srivastava et al. 1994).

(d) *Pests of host plants*

Since the productivity of lac depends on the availability of healthy shoots on host plants, damage done by host pests has direct bearing on it. It has been seen that pests of hosts, particularly sucking pests, borers, defoliators and termites, sometimes cause severe damage to host plants. List of host plant pests is given in the Table 5.2. In addition to the pests of major tree host plant species, a number of insect pests have also been recorded from the important bushy host plant bhalia (*Flemingia macrophylla*), which is depicted in Table 5.3. In addition to the above pests, more than 60 minor insect pests have also been recorded on more than one lac host plant.

5.3.2 Associated Inimical Biota

Since lac insects are sedentary in habit during its whole life span after settlement, these are susceptible to a horde of enemies, as they become an easy prey for them. These include predators, parasites and diseases. The predators of lac insect include both vertebrate and invertebrate species, while the parasites are all insects. The loss caused to the lac crop by the insect predators and parasites amounts to 35–40% annually on an average. As per available literature, about 72 species of insects are known to be parasites of the lac insects and its predators.

(i) *Predators*

Although 16 predators of lac insects have been recorded, only four of them as stated below are key predator species of commercial lac insects and are of regular occurrence.

1. *Eublemma amabilis*, Lepidoptera: Noctuidae
2. *Pseudohypatopa (Holcocera) pulverea*, Lepidoptera: Blastobasidae
3. *Chrysopa madestes*, Neuroptera: Chrysopidae
4. *C. lacciperda*, Neuroptera: Chrysopidae

5.3.2.1 Brief Life History of Major Predators

(A) *Eublemma amabilis* Moore, the white moth

The predator is widely distributed in all the major lac-growing regions of the country and is considered as the most destructive. The moth lays greyish-white, flat round eggs, depressed in the centre measuring about 0.35–0.37 mm across the centre and blessed with beautiful sculptured chorion. Eggs are generally laid singly on the lac encrustation and turn white prior to the emergence of larvae. The larva hatches out of the egg shell by gnawing a hole at the side of the chorion. The first instar larva measures 0.51–0.54 mm in length. Under laboratory conditions, egg laying of the moth can easily be obtained on plain paper by confining the adult moths in battery jars.

The newly hatched larva gets at the lac insect either through one of the openings in the test or by tunneling a hole through the encrustation. The attacked lac cell becomes hollow inside containing pink coloured discs of excreta and can be easily differentiated from the healthy cell. A single larva generally destroys 40–60 cells during various instars before preparation. The predator causes the most damage during the katki aghani lac crops, i.e. during the rainy season in comparison to the other two crops. It has been observed that the predator has six generations in a year. The durations of the generations from July onwards are about 37, 45, 42, 125, 80 and 40 days, respectively. Hibernating larvae have been observed during the winter generation, which covers about 125 days. However, some of the adults emerge during November–December, while the rest hibernate and emerge during January to March (Fig. 5.6).

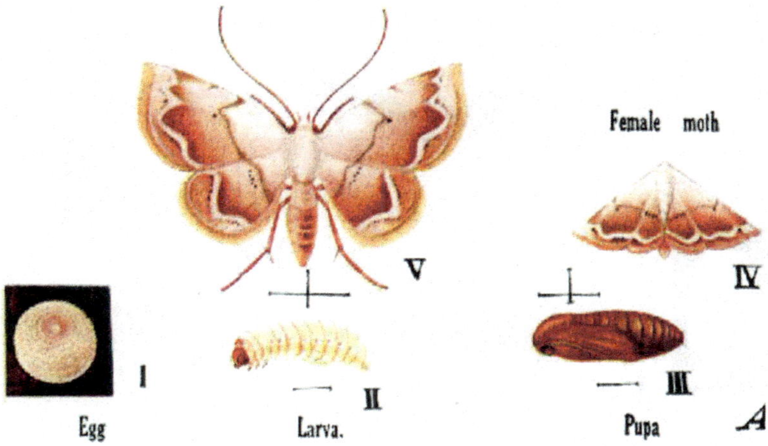

Fig. 5.6 Life cycle of *Eublema amabilis*

(B) *Pseudohypatopa pulverea* Meyr, the white moth

This predator is also widely distributed and found in all the lac-growing areas of the country. The adult moths are blackish in colour and smaller in size than *E. amabilis*. The predator feeds on the live and dead lac insects. It is also abundantly found in stored lac. The eggs are oval and laid singly on the cell of the lac insect. Measuring about 0.5 mm × 0.3 mm, the freshly laid egg is colourless, which later on turns to deep pink with growth of the embryo. Under laboratory conditions, the females can lay eggs on emery and sand paper strips in dark background. The newly hatched larva is about 1.35 mm in length, which, after passing through several instars (5–9) depending on the season, measures about 10–12 mm in length and 2 mm in breadth, before pupation. The predator completes five generations in about 381 days. It is now possible to rear the larvae under laboratory conditions on artificial diets (Bhattacharya et al. 1998). The predator feeds on the lac larvae and spins a loose web. Since a single predator is capable of destroying 45–50 mature lac cells, it is considered to be very important in view of inflicting damage to the standing lac crop and qualitative and quantitative deterioration of the stored lac (Fig. 5.7).

(C) *Chrysopa sp.*, the greenfly

Preying upon various stages of the lac insect, the larvae of the lacewings are considered to be sporadic pest. Females lay long-stalked light-green eggs either on the lac encrustation or on any part of the host trees, which then turned to brownish at the time of hatching. The larvae emerge out of the eggshell with the aid of an "egg

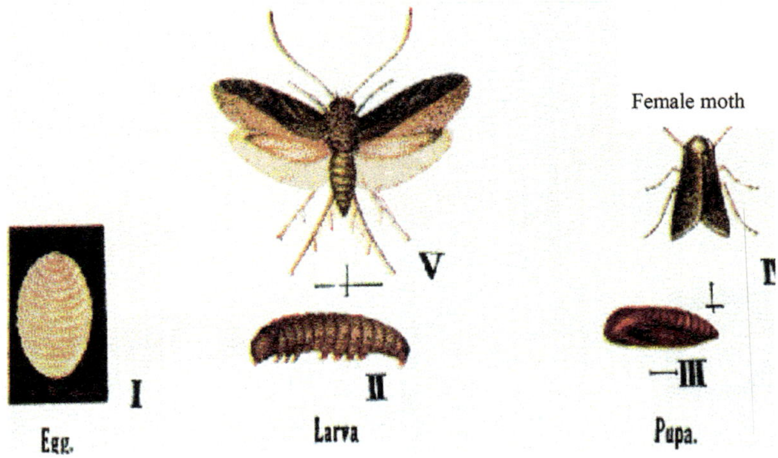

Fig. 5.7 Life cycle of *Pseudohypatopa pulverea*

buster", climb down the stalk and start feeding immediately on the insect by inserting its long pair of mandibles. It does not tunnel its way through the lac encrustation but move freely in the lac colony with body under a heap of wax filaments and insect debris. The larva passes through three instars. The larva pupates outside the lac encrustation. The predator takes 21–27 days to complete its life cycle during the rainy season and about 54 days during the cold weather.

Out of the total list of 24 parasites of lac insects, about eight parasitoids, viz. *Coccophagus tschirchii*, *Erencyrtus dewitzi*, *Eupelmus tachardiae*, *Parechthrodryinus clavicornis*, *Tachardiaephagus tachardiae*, *Marietta javensis*, *Tachardiaephagus somervillei* and *Tetrastichus purpureus*, are of regular occurrence in the lac ecosystem. Amongst these, *Tachardiaephagus tachardiae* and *Tetrastichus purpureus* are the most abundant and responsible for 6–8% damage to the lac crop. However, at times, cent per cent crop loss has been reported only due to one parasitoid.

(ii) *Beneficial parasitoids (parasitoids of insect predators)*

Besides the predators and parasitoids of lac insect, a large number of important parasitoids of lac predators have been reported in the lac ecosystem. These beneficial parasites play a vital role in the natural control of the major lac insect predators (Table 5.4).

Table 5.4 A list of major beneficial parasitoids is presented below along with host

S. No.	Name of the parasite	Host
1.	*Agathis bischoffi* Fabr (Braconidae)	*Psudohypatopa pulverea*
2.	*Agathis coryphe* Nixon (Braconidae)	*Psudohypatopa pulverea*
3.	*Agathis festiva* Muesebeck (Braconidae)	*Psudohypatopa pulverea*
4.	*Anagyrus greeni* Howard (Encyrtidae)	Cocoons of *C. madestes*
5	*Apanteles fakhrulhajiae* Mahd. (Braconidae)	*P. pulverea*
6	*Apanteles tachardiae* Cam. (Braconidae)	*P. pulverea*
7.	*Aphrastobracon flavipennis* Ashm. (Braconidae)	*Eublemma* spp. *Coccidiphaga scitula*
8.	*Brachycyrtus eublemmae* (Rao) (Ichneumonidae)	Cocoons of *Crysopids*
9.	*Brachymeria tachardiae* (Cam.) (Chalcididae)	*E. amabilis, P. pulverea*
10.	*Bracon greeni* Ashm. (Braconidae)	*E. amabilis*
11.	*Bracon hebetor* Say (Braconidae)	*E. amabilis, Coccidiphaga scitula* and *P. pulverea*
12.	*Bracon tachardiae* Cam. (Braconidae)	*E. amabilis*
13.	*Chelonus cyclopyrus* Franz. (Braconidae)	*P. pulverea*
14.	*Elasmus claripennis* (Cam.) (Elasmidae)	*E. amabilis*
15.	*Elasmus colemani* Mahd. (Elasmidae)	*E. amabilis*
16.	*Eupelmus tachardiae* (Howard) (Eupelmidae)	*E. amabilis, P. pulverea*
17.	*Eurytoma pallidiscapus* Cam. (Eurytomidae)	*P. pulverea*
18.	*Perisierola pulverea* Kurian (Bethylidae)	*P. pulverea*
19.	*Pristomerus sulci* Mahd. and Kolub. (Ichneumonidae)	*P. pulverea*
20.	*Telenomus* sp. (Scelionidae)	*C. madestes* and *C. lacciperda* eggs
21.	*Trichogramma* sp. (Trichogrammatidae)	*E. amabilis*
22.	*Trichogrammatoidea nana* Zehnt. (Trichogrammatidae)	*E. amabilis, P. pulverea* eggs

5.4 Effect of Abiotic Factors on Lac Productivity

Owing to its sedentary nature, the growth and development of lac insect is widely influenced also from abiotic factors of the surroundings. The abiotic or physiographic factors emanate from the environment and comprise of climatic or geographical and local influences. Fluctuations in lac crop yield observed in various regions of the country can be attributed, to a certain extent, to the vagaries of the weather. The productivity has been observed to vary season-wise in different localities. Important abiotic factors are (i) temperature, (ii) humidity, (iii) wind, (iv) frost and hail, (v) moisture, (vi) rainfall, (vii) soil condition, (viii) elevation and (ix) air. These factors may influence lac productivity directly, i.e. influencing growth and development of lac insects, or indirectly, i.e. by influencing other biotic or abiotic factors as depicted in Fig. 5.8. Interactions between these are very complex, and it is difficult to ascertain how they are interacting in nature. However, abiotic factors are mainly responsible for breaking the ecological balance in nature to affect the productivity either way.

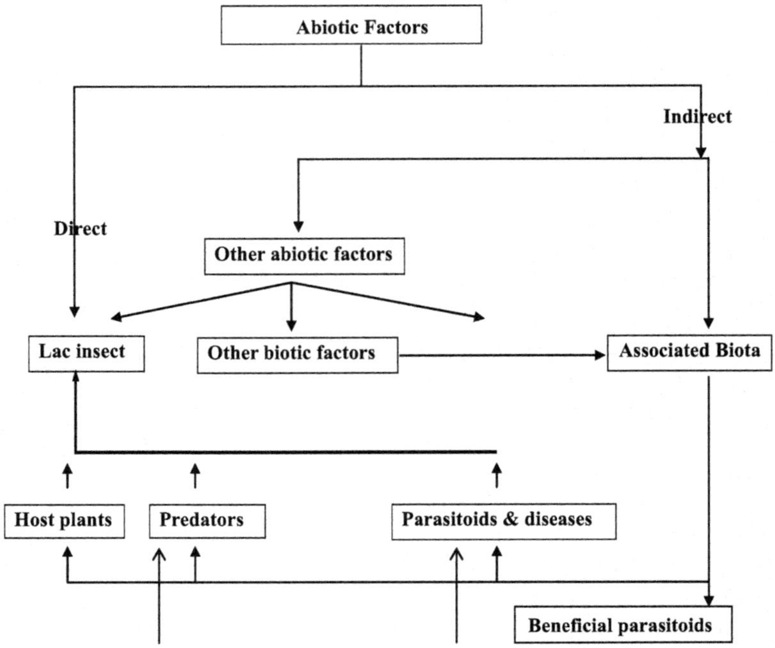

Fig. 5.8 Effect of abiotic factors on lac productivity

5.4.1 Temperature

Probably no other single factor has a greater effect upon the geographical distribution of organisms on the earth or upon the period of their activities during the annual cycle than the temperature. There is always a minimum and maximum limit of effective temperature, known as "temperature threshold", within which the organism becomes active. Beyond this limit, the organism becomes dormant, and still beyond, the dormancy is terminated into death.

Lindsay and Harlow (1921) have considered heat to be injurious to lac insect, but it can be contradicted by the fact that lac is cultured in hot areas like Sindh and Rajasthan. Some authors have considered the mortality of lac insects due to softening of resin by heat, resulting in blocking of breathing pores, but others assign the reason for this to be due to physiological activity of the host plant, which is adversely affected by the drought condition. Attempts to maintain the continuity of brood crop during Baisakhi on palas in hot areas by reducing insect stress to subnormal crowding levels have given encouraging results. This implies that the scarcity of food and moisture stress is the reason of such a heavy mortality of lac insects and indicating an indirect effect of temperature rather than direct effect. The melting of resin and blocking of hollow begin soon after the death of insects. The moisture content of the body protects the resinous cells from melting at higher temperature.

The effect of temperature on gravid female lac insects has indicated that egg laying and emergence of nymphs are controlled by climatic conditions. The nymphs become inactive below the temperature 20 °C, and oviposition virtually ceases below 17 °C in summer and 15 °C during winter. It is advisable that the broodlac of winter season should be stored in room temperature at 24 °C for emergence of nymphs, if they fail to emerge naturally. The swarming period is found to be prolonged beyond 2 months during extreme cold creating difficulty in harvesting and artificial inoculation of the lac crop. An optimum temperature range of 24–27 °C has been found the best for lac culture.

5.4.2 Humidity

The humidity of the atmospheric air is also an important factor in determining the limits of quality of lac culture (Watt 1901; Nicholson 1925). Humidity is also believed to affect the sex ratio of the lac insects (Mahdihassan 1930a, b, c, 1931a, b, c, d, e, f).

5.4.3 Wind

Strong and dusty winds are likely to cause serious loss at the critical periods of larval and male emergence. Hot winds are believed to affect the female lac insects by drying up the body fluid contents, leading to splitting up of resinous cells in the centre and change of their colour to pale brown. Similar observations were also noticed at the time of drought during Baisakhi and jethwi crops seasons after summer (Watt 1901) and due to the exposure to direct sun and wind (Malhotra 1976). Jethwi crop on ber and *Flemingia semialata* suffers male mortality during the month of April due to direct sun and hot wind causing desiccation of nonfeeding males during the third instar.

5.4.4 Frost and Hail

The force of pellets hit the lac encrustation (Watt 1901, Stebbing 1908–1910; Misra 1931a, b, c, d, e). The frost may cause an increase in relative humidity which is congenial for the growth of fungus which may cause the mortality of lac insects (Nicholson 1925).

5.4.5 Moisture

Moisture is of prime importance for the successful development of the insects, but the excess moisture affects the crop adversely. Dry and arid zones are considered unsuitable for propagating lac culture due to less moisture contents.

5.4.6 Rainfall

A range of rainfall between 1270 mm and 1520 mm has been considered most suitable for lac culture. Heavy rainfall during winter is unfavourable for lac crop due to decrease in temperature, but rain in summer has beneficial effect as it washes away the honeydew and gives a relief to insects by clearing the breathing pores. Rain is necessary for keeping the soil sufficiently moist and helps the host trees grow normally (Nicholson 1925).

5.4.7 Soil Condition

Soil rich in nitrogen and phosphorus was found helpful to host trees for lac culture. The soil should have sufficient moisture for the vegetative growth of the host plant and activity of the insects. It has been observed that the clay soils which do not allow water to percolate thorough to reach the root system and soil of dry localities where its moisture contents fall too low unless supplemented by frequent showers are unsuitable for broodlac production (Nicholson 1925).

5.4.8 Elevation

Elevation involves two important factors: (1) temperature and (2) subsoil moisture. For every 91-metre increase in altitude, the temperature drops 0.55 °C. Lac has been recorded to grow at an altitude ranging from 245 to 910 m (Lindsay and Harlow 1921; Hautefeuille 1924). It was observed that broodlac from dry locality is better than that of plains which has great subsoil moisture. Elevation as such is a physiographic factor of not much importance and has to be considered along with temperature and humidity (Nicholson 1925).

5.4.9 Air

For healthy growth of lac crops, free circulation of air is an important factor as it probably helps in reducing the high humidity. It has been observed that the trees, which are sparsely located in the forest, usually produce a good lac crop, as compared to in a dense forest.

5.4.10 Light

Light is the single source of energy for biotic system. Through photosynthesis, energy is entrapped and stored in the form of bonded chemical energy which is in turn used by biotic system. Lac insects are heliotropic by nature (Lindsay and Harlow 1921). The lac insects develop better in branches growing in full sunlight in comparison to less sunlight.

5.4.11 Fog

Fog, if appears, frequently causes the severe mortality to developing lac insects. The honeydew secreted by the lac insects invites fungal diseases due to fog. The fungus spreads and blocks the breathing pores with the ultimate result of crop failure.

5.4.12 Forest Fire

The forest fire is exceedingly destructive where lac host trees are under trials. The smokes create suffocation and lac insects die (Watt 1901).

5.5 Lac Cultivation in India

Earlier lac was not cultivated, but collected from naturally occurring infested host plants in forest or personal holdings and marketed by a particular section of tribal known as lahi/laha/lodha till the end of eighteenth century. During those days no need was felt for systematic propagation and culture of lac insects on new healthy host plants. The beginning of lac export thereby increased price realization; the artificial inoculation of healthy host trees using crude methods was adopted during the nineteenth century. During the golden days of lac in the beginning of the twentieth century, lac host plantations were raised even in the plains of India to cater the growing demand of lac. However, inconsistency in lac production by traditional methods was noticed which affected the lac trade and industry adversely. Technology intervention in the field of lac production came into force only after the establishment of the Indian Lac Research Institute, Namkum, Ranchi. Glover (1937a, b, c) published the first authentic book *Lac Cultivation in India* and suggested proper use of lac insects and host plants for profitable and sustainable production of lac. Since Glover, several improvements have been made in methods of lac culture, which mainly includes host plant management, lac crop management and lac pest management. These are separately described herein (Fig. 5.9).

5.5.1 Host Plant Management

Since commercial lac production is confined to conventional tree host species, mostly confined to natural forest, their management has been a challenging problem to the lac growers due to their scattered nature of occurrence. However, the main emphasis has been towards sustainable production of lac by investigating the optimum condition on which lac insects and their host plants interact to produce lac of commerce.

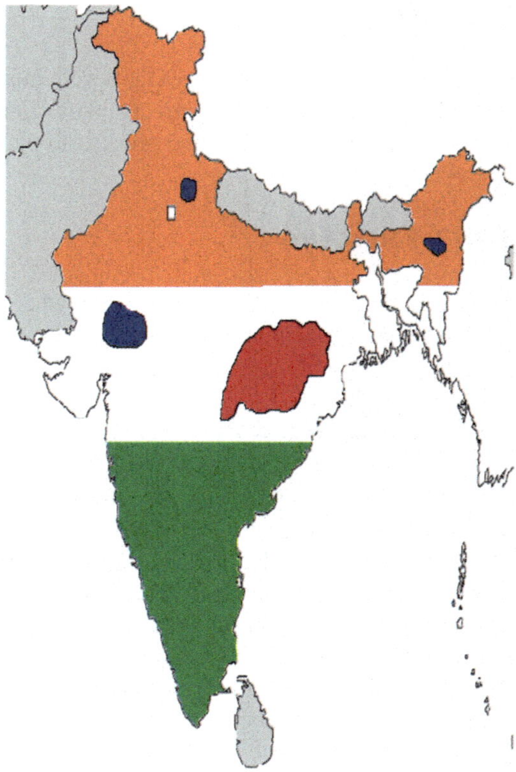

Fig. 5.9 Map of India showing lac cultivation in the country

5.5.2 Propagation of Lac Hosts

Since there was no dearth of lac host trees earlier in the forest, work conducted on propagation of tree host is very scanty in literature. Heavy deforestation and indiscriminate cutting of trees including lac hosts have made it necessary to propagate trees and bushy lac hosts not only to revive lac production but also to protect biodiversity of the lac insect and their host plants. Methods adopted for propagation of these plants are given below.

(i) *Direct sowing*

Pits of 30 cm diameter and 15–20 cm depth are prepared with thorough digging and filled up with the soil and farmyard manure along with insecticides. Lime is to be applied after investigating the pH of the soil. At the onset of monsoon, the seeds of kusum and khair are sown in the pits. Interculturing operation is done for early growth and establishment of the plants.

(ii) *Raising of plants through seedlings*

(a) *Nursery raising*: Nursery beds of 1 × 3 m are prepared by repeatedly digging up to the depth of 20–37.5 cm. Farmyard manure is mixed properly with insecticide (chlorpyriphos) to protect the seedlings from termites and white grubs. Seeds are sown during the month of April, and the beds are irrigated twice daily (morning and evening) till the onset of monsoon. Seeds of bhalia having hard seed coat can be treated with concentrated sulphuric acid for 10 minutes to break the dormancy and enhance the germination process without any adverse effect on plant attributes (Kumar and Purkayastha 1970).

(b) *Transplanting of seedlings*: Pits (30 cm diameter and 15–20 cm depth) are prepared during May–June, properly filled up with the soil and farmyard manure along with insecticides, and seedlings raised in nursery beds are transplanted in the pits. Nitrogenous fertilizers are applied in two split doses (basal and top dressing). Interculturing and irrigation are done as and when necessary for healthy growth of the plants.

(iii) *Vegetative propagation*

With a view to get true-to-type plant of desired characters, vegetative propagation has been tried through stem cuttings using plant growth regulators in various concentrations and different months. Success has been achieved for raising of galwang during March and bhalia, rain tree (*Albizzia saman*) and *Grewia serrulata* during June (Purkayastha and Kumar 1977). Success has also been achieved in clonal propagation of palas (Kumar 1989), kusum and galwang (Srivastava et al. 1994 and Srivastava et al. 1997), through air layering, reducing the gestation period by 4–5 years (Fig. 5.10b). *Ficus* spp. can be raised directly from stem cuttings.

Training of Tree Hosts into Bushes Studies made into kusum, ber, palas, rain tree and galwang in institute plantation have shown that ber and galwang are best suited for training into suitable bushes, and desired type of plant for lac culture can be raised within 4 years of planting. Lac cultivation trials on the bushes of galwang for jethwi and ber for raising aghani crops showed sticklac yield of 4–5 q/ha as compared to 2–2.5 q/ha on trees (A Rep., 1974–1976).

Crop Geometry The spacing requirement for the major lac hosts has been recommended as 3.6 × 3.6 m for palas, 6 × 6 m for kusum and 4.5 × 4.5 m for ber, khair and ghont. Equilateral triangular system of planting has also been recommended for plantation raising of these lac hosts. Under this system alternate rows can be removed after 30–35 years after sufficient growth of trees to promote free passage of light and air for better growth of the lac insects. Double-hedge system of planting for bhalia has been found best accommodating 15,000–16,000 plants/ha with a yield of 10–12 q/ha (A Rep., 1986).

Manuring and Fertilizer Application Basal dosages recommended for pits (30 cm radius and 15–20 cm depth) are 28.4 g ammonium sulphate and 450 g single superphosphate before sowing of seeds/transplanting of seedlings with irrigation once or twice in a week till the break of monsoon. The use of FYM at 90 q/ha in bhalia after 2 years of planting showed highest lac yield as manuring benefits to

Fig. 5.10 Apical/light pruning

lateral root system. Fertilizer trials on lac hosts in government forest areas both at Kundri (Palamau district of Jharkhand) and Kechki (Orissa) showed the highest lac yield in the plots with manurial treatment.

The use of sodium and calcium (micronutrients) is necessary for the growth of lac insects, but if calcium is applied to the plants without magnesium, preponderance of males results due to unbalanced action of calcium. Application of N had the lowest mortality. However, potash (K) had highest mortality. N- has tendency to reduce the mortality and the K to increase it. Spacing requirement and fertilizer doses for raising plantations of important lac hosts are given in Table 5.5.

Effect of PGR on the Growth of the Host Plants

Treatment of seeds of bhalia with 20 ppm NAA before sowing enhances the growth of plant. Pretreatment of seeds with NAA (80 ppm) showed increasing shoot length. Foliar spray of GA_3 (40 ppm) enhances the growth of seedling of bhalia, galwang and rain tree, thereby reducing the time period to be taken for the growth to reach plants up to lac inoculation stage (Purkayastha and Kumar 1981). Foliar spray of GA_3 (80 ppm.) and urea (1%) on newly appeared sprouts of kusum and palas after pruning enhanced the growth more than double (A Rep., 1974–1976).

Table 5.5 Spacing and fertilizer doses for raising plantations of various lac hosts

Host plants	Spacing (m)	Plant density/ha	Fertilizer dose/plant (g)			Sticklac yield	
			N	P_2O_5	K_2O	Kg/plant	q/ha yr
Tree hosts							
Palas (*Butea monosperma*)	3.6 × 3.6	772	100	250	75	1.25	4.8
Kusum (*Schleichera oleosa*)	6 × 6	278	100	250	75	30.00	20.8
Tree hosts to be trained into bushes							
Ber (*Ziziphus mauritiana*)	4.5 × 4.5	494	100	250	75	5.0	12.35
Rain tree (*Albizia saman*)	2 × 2	2500	50	100	20	1.5	18.75
Galwang (*Albizia lucida*)	1.8 × 1.8	3086	20	15	10	0.80	12.34
Akashmoni (*Acacia auriculiformis*)	2 × 1.8	2778	50	100	20	0.8	11.1
Bushy hosts							
Bhalia (*F. macrophylla*)	1.2 × 1.2	6944	10	10	5	0.25	8.68
F. semialata	1 × 1	10,000	10	10	5	0.40	20.0

Pruning

Pruning of lac hosts is an important operation to provide suitable space for the lac insects to feed and thrive upon them. The appearance of maximum number of shoots of suitable age for lac inoculation depends upon proper pruning. Pruning should be done with sharp instruments such as *dauli*/tree pruner/secateurs, in such a manner that branches or twigs remain free from splitting or deep scratching. Broken or damaged parts of shoots/branches should be cut neatly just below that point. Two types of pruning/coppicing, (i) apical (light) and (ii) basal (heavy), have been recommended for lac culture (Table 5.7).

(i) *Apical/light pruning*: Branches less than 2.5 cm diameter should be cut from the base, and branches more than 2.5 cm diameter should be sharply cut leaving a stump of 40–50 cm from the base. Diseased and dead portion of branches should be removed completely. Light pruning is recommended for slow-growing conventional tree host species, like palas, kusum and ber (Fig. 5.10).

(ii) *Basal/heavy pruning*: Branches having less than 7 cm thickness should be removed from the base, whereas thicker branches should be cut at a place where it has a diameter of 7 cm. In quick-growing bushy hosts, pruning should be done at a height of 10–15 cm from the ground level. For kusmi lac cultivation on mixed plantation of bhalia and galwang, plants of galwang should be first coppiced after 3 years of transplanting at a height of 15–20 cm during February–March and at a length of 5–8 cm from the origin of secondary branches in subsequent years up to 5 years. As a result of experimentation under the conditions prevailing in Chota Nagpur, the following pruning times for different lac hosts have been found suitable for lac culture.

Kusum – Pruning should be done in either February or July. However, the former month is better.

Khair – Pruning is to be done in March. However, harvesting of lac crop during February may be used to serve as pruning also.

Ber – Pruning should be done in February for inoculation in July and April–May for inoculation in October–November. For kusmi lac crop, ber should be pruned 15 months before inoculation. However, recent observations have shown that harvesting of aghani crop during January–February may also serve as pruning for inoculation in June–July.

Palas – Pruning should be done in February for lac inoculation in July and in April for inoculation in October–November.

Ficus spp. – Pruning is to be done in April for inoculation in July and in May for inoculation in October.

Period of Rest and Age of Shoot Minimum period of rest required between the time of cropping and time of reinoculation has been determined as 12 months for kusum and 6 months for palas, ber and khair. The optimum age of shoot for the time of pruning to that of lac inoculation has been determined as 18 months for kusum and 6 months for palas, ber and khair. Optimum time of pruning and period of rest have been shown as follows (Table 5.6). Broodlac requirement for quick-growing species is fulfilled from conventional/brood preserver hosts for sustained production round the year.

5.5.3 Lac Crop Management

(i) *Inoculation*

The lac crop management begins from inoculation of host plants with broodlac. Broodlac means "healthy lac encrustation consisting of gravid females about to produce young ones". The status of broodlac is similar to seeds of agricultural crops. Inoculation is of two types, i.e. artificial inoculation and self-inoculation.

Artificial Inoculation This is done when fresh host plant is inoculated from broodlac obtained from another host plant. For this, broodlac twigs consisting of good encrustation of gravid living healthy females are selected and placed into 60 mesh synthetic net containers and tied with the help of plastic *sutali* at both ends, top and bottom. The broodlac bundles are placed on host plants at a convenient place from where lac larvae can reach to the succulent twigs easily. Nearly 100 g of broodlac may be accommodated in a synthetic net container (30 × 10 cm). After a week, the broodlac bundles may be rotated and be placed where more space is vacant.

Table 5.6 Schedule of different lac culture operations for various host plants

Host	Crop season	Period of rest (months)	Pruning time	Crop duration
Conventional slow-growing tree hosts * (gestation period more than 10 years)				
Kusum	Winter cum summer	18	January/February	June/July to June/July
	Summer cum winter	18	June/July	Jan./Feb. to Jan./Feb.
Palas	Summer cum rainy season	6	April/May	Oct./Nov. Oct./Nov.
	Summer (*ari*)	6	April/May	Oct./Nov. to April
Quick growing tree host species capable of being trained into bushed ** (gestation period 4–5 years)				
Ber	Summer cum rainy season	6	April/May	Oct./Nov. Oct./Nov.
	Winter season	13	April/May	June/July to Jan./Feb.
	Summar season (*ari*)	6	April/May	Oct./Nov. to April
Galwang	Summer season	6	June/July	Jan./Feb. to June/July
Akashmoni	Winter season	18	Jan./Feb.	June/July to Jan./Feb.
Akashmoni	Rainy season	18	Oct./Nov.	June/July to Oct./Nov.
Quick-growing bushy hosts ** (gestation period 1 year)				
Bhalia	Winter season	13	April/May	June/July to Jan./Feb.
Flemingia semialata	Winter season	13	April/May	June/July to Jan./Feb.

*Hosts under this category are lightly pruned ** Hosts to be heavily pruned

Self-Inoculation When broodlac produced on a host plant is left on the same plant to inoculate vacant twigs of the same plant.

Phunki Removal As the emergence of lac larvae becomes practically over after 3 weeks of inoculation, *phunki* (empty) broodlac bundles are removed immediately.

(ii) *Harvesting*

Complete harvesting of lac crop is done only on maturity of the lac crop, but sometimes, immature crop (*ari*) is also harvested along with pruning. Three types of harvesting, i.e. partial harvesting, complete harvesting and *ari* harvesting, are done in lac culture.

Partial Harvesting When self-inoculation is contemplated, crop is harvested partially, and some quantity of broodlac is left over on the host plant for inoculation of next crop.

Complete Harvesting Lac crop is completely harvested along with coppicing of the host. Broodlac is separated for inoculation of the next crop or selling in the market, and rejected lac sticks are scraped and disposed off.

Ari Harvesting *Ari* harvesting is done when a time of coppicing/pruning does not coincide with the maturity of lac crop. Thus, the immature lac crop is harvested along with coppicing/pruning of the host plant as in case of palas during April/May.

(iii) *Pest management*

The losses due to pests in lac culture are known to be far greater than what usually occur in other agricultural crops. A horde of pests destroy approximately two-thirds of the probable crop, leaving only one-third for the cultivator to reap. In certain seasons, partial to complete lac crop failure has been attributed to this factor alone, which has also a serious bottleneck in introducing lac culture in new localities. The pests are also a source of menace in the stored lac, wherein inflicting quantitative loss is also responsible for qualitative deterioration by contaminating the produce with webs, frass and dead bodies. Since the predators are responsible for majority of the total losses, almost all the works on this aspect have been concentrated towards the control and management of the two key lepidopteran predators *Eublemma amabilis* and *Pseudohypatopa pulverea*. The entire management strategy has, therefore, been targeted to find out effective measures for the control of lac predators keeping in view the safety of the lac insect. Earlier findings were limited to preventive measures, mechanical control and identification of natural enemies for biological control based on the various reports. On the basis of subsequent findings, many other techniques were also developed. The different techniques for lac culture developed so far are mentioned below.

I. *Preventive measures*

- Healthy and mature broodlac, free from predator and parasites, should be used as far as possible.
- *Phunki* (empty broodlac sticks) should be removed from the inoculated trees within 2–3 weeks after the inoculation.
- Scraping or fumigation of excess lac stick after inoculation and also after *phunki* removal at ones should be done.
- Removal of scraped lac from the vicinity of lac inoculated trees and processing as early as possible should be made.
- As a general rule cultivation of kusumi strain should be avoided in a predominantly rangeeni area and vice versa.

II. *Cultural control*

Early Inoculation Inoculation of the lac crop earlier by 10–15 days than the normal crop has been found to have higher incidence of predators. This technique can be exploited in combination with heavy inoculation (i.e. higher brood rate) of brood.

Trap Crop Taking advantage of the oviposition behaviour of the predators in highly dense population, some of the host plants are inoculated with higher brood rate to attract the predators for egg laying. These plants serve as trap crops and have a greater incidence of predators and can be harvested as *ari* (immature lac).

Intercropping The larval ectoparasites *Bracon greeni* and *Elasmus* sp. have been reported to parasitize pink and spotted bollworm larvae attacking cotton. Intercropping of cotton and okra with lac can increase the population of natural parasites common to the crops.

III. *Mechanical control*

Sixty mesh synthetic net container bags may be used for inoculation of broodlac. The emerging lac larvae easily crawl out from the minute pores of the bag and settle on the twigs of the lac host plants, while the emerging adult enemy predator moths get entrapped within the net. This innovation can check the egg laying by the adult moths on the new crop and reduce pest infestation.

IV. *Chemical control*

Initially, chemical control was restricted to fumigation of scraped lac with carbon bisulphide as preventive measure only, but at a later stage, different kinds of chemical control have been tried.

Use of Insecticides Systemic work on laboratory screening of specific insecticides safe against the lac nymphs but effective against the predators was initiated at LARI during 1970 and at IINRG during 1972, though field screening of stomach and contact insecticides was initiated at IINRG during 1963–1969.

More than 35 insecticides belonging to various groups were studied of which endosulfan has been identified to be the safest to the first instar lac nymphs, which is of prime importance for field application. It has also been found to be highly detrimental to the larval stages of the two predators. A concentration of 0.05 % has been identified as the most effective dose without any adverse effect on the economic attributes of the lac insect. Dipping of broodlac in 0.05 % emulsion of endosulfan, prior to inoculation, results in a highly significant suppression of the predatory population. Similarly dichlorvos at 0.03 % concentration has been found to be very effective in the management of *Chrysopa* sp. as it is highly effective against all the larval stages.

Use of Other Chemicals Experiments carried out with the third-generation insecticide diflubenzuron (Dimilin) have exhibited the desired selectivity. The chitin inhibitor at 0.05 % concentration has also been recommended for effective management of the lac insect predators.

V. *Microbial control*

Effective control of *Eublemma amabilis* and *Psedohypatora pulverea* was achieved by the use of biopesticide *Bacillus thuringiensis* Berliner under field conditions.

VI. *Biological control*

Various workers from different parts of the country have recorded over 35 parasites of the lac predators. Amongst the lepidopteran predators, *P. pulverea* and *E. amabilis* have 14 and 9 parasites, respectively, and 4 are common to both, whereas *Chrysopa* spp. have 7 parasites. It has been recorded that 9 parasites are most important and promising for biological control of the predators under field condition.

Two ant predators, viz. *Camponotus compressus* and *Solenopsis geminata rufa*, have been found to be very effective in controlling the lac predators by attacking the larvae and pupae of both *E. amabilis* and *P. pulverea*.

Of late several egg parasitoids have been evaluated for management of lac predators, of which *Trichogramma pretiosum*, *T. chilonis*, *T. brasiliensis*, *Trichogrammatoidea bactrae* and *Telenomus remus* hold great promise for biological control of the predators.

VII. *Integrated pest management*

Integrated pest management is possible by integration of the following techniques:

(a) Mechanical control with 60 mesh synthetic net as brood container
(b) Cultural control with (i) trap crop through high-density settlement using higher brood rate and (ii) varying the date of inoculation
(c) Chemical control by applying Thiodan 35EC spray

Combination of the above components has been found to reduce the pest population to the extent of 60–65 % and increased yield to 140–175% over control.

Before implementation of any IPM programme, points of major considerations are: (i) current and future status of the pest, (ii) economics of the pest management, and (iii) environmental sustainability of different control measures. In IPM, the pest population is maintained at a manageable limit for which a decision-making process for action or no action is involved and has a vital role to play. Although the pest problem and their control in case of lac insects is different from those encountered in case of other agricultural crops, the basic principle of decision-making remains the same. The information regarding economic injury level (EIL), economic threshold level (ETL), population dynamics, biorational approaches, assessment of diseases, impact of intercrops, etc. are required for further strengthening and successful implementation of IPM.

5.6 Lac Production Technologies

A good number of lac culture technologies suitable for various host plant species, agroclimatic zones and consumers have been successfully evolved to fit in the strategic lac production plan.

5.6.1 Technologies for *Kusmi* Lac Production

Advanced Technology for Broodlac Production on Kusum

Lac production technologies on kusum (*S. oleosa*; Lour; Oken.) under 4/5 coupe system for brood and sticklac production offer a capital growth of 352% annually at 40% exploitation of kusum trees (Mishra et al. 1999) (Fig. 5.10).

Package of Practices of *Kusmi* Lac Production on *Akashmoni* (*Acacia auriculaeformis*) in Alternation with Kusum

Akashmoni (*Acacia auriculaeformis* A. Cunn.), a new fast-growing species, when used in alternation with kusum under four coupe system in the ratio of 40 Akashmoni and 1 kusum tree can provide a capital growth of 289% per year (Choudhary et al. 1999) (Fig. 5.11).

Package of Practices for Kusmi Lac Cultivation on Mixed Plantation of *Bhalia* and *Galwang*

Kusmi lac culture on mixed plantation of bhalia (*Flemingia macrophylla* (Willd.) O. Ktze) and galwang (*Albizzia lucida* Benth) accommodating 3600 bhalia and 1200 galwang bushes in one hectare along with intercrops can generate 156% capital growth per year (Kumar et al. 1996) (Fig. 5.12).

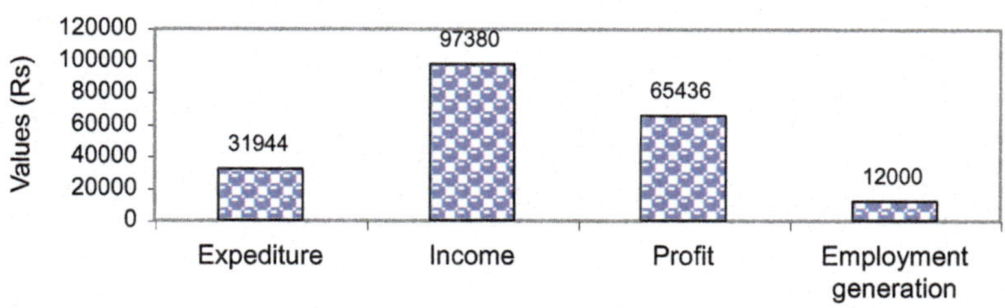

Fig. 5.11 Economics of broodlac production on kusum

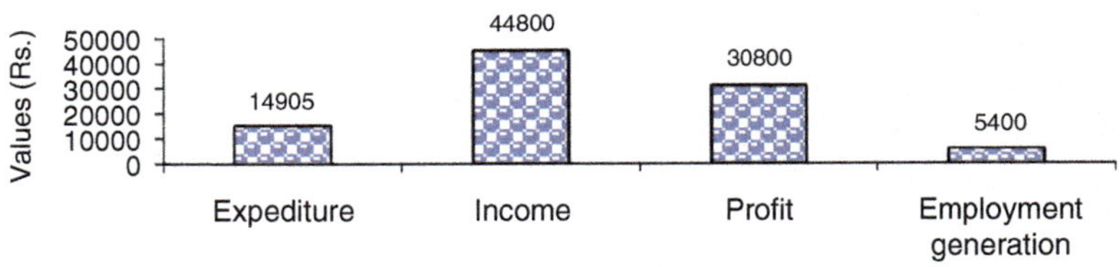

Fig. 5.12 Economics of broodlac production on Akashmoni in alternation with kusum

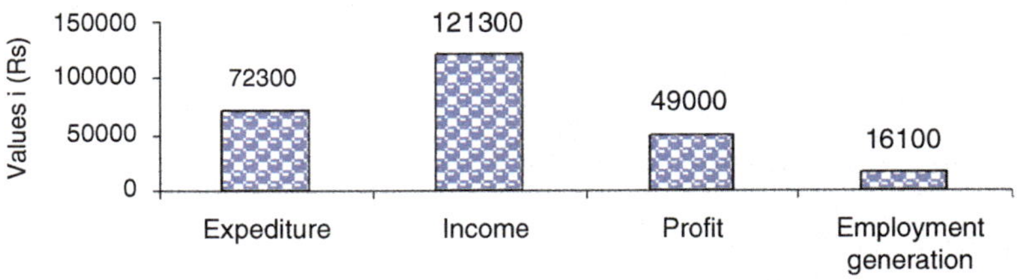

Fig. 5.13 Economics of broodlac production on Akashmoni in alternation with kusum

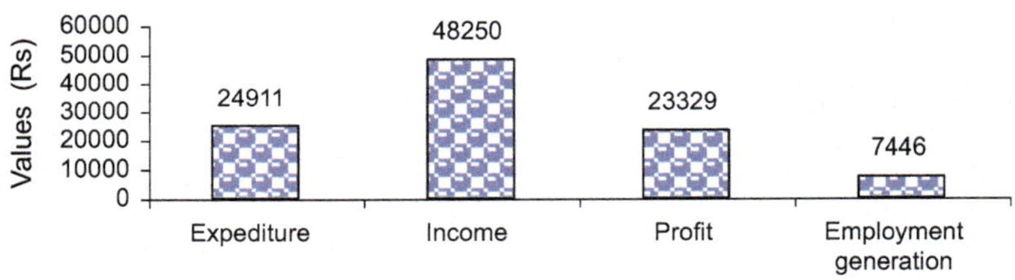

Fig. 5.14 Economics of broodlac production on palas

5.6.2 Technologies for *Rangeeni* Lac Production

Advanced Technologies for Lac Production on *Palas* in Hot Areas

Three coupe system of palas (*Butea monosperma* (Lam.) (Taub)) can provide a capital growth of 198% per year and ensure self-sufficiency in broodlac (Mishra et al. 1999a, b) (Fig. 5.13).

Technology for *Rangeeni* Lac Production on *Palas* and *Ber*

The use of palas and ber trees in the ratio of 5:1 for broodlac and sticklac production, respectively, provides a capital growth of 169% per annum. Profitability of this technology has been tested and demonstrated in the farmers field (Mishra et al. 2001) (Fig. 5.14).

Package of Practices of Rangeeni Lac Production on Akashmoni (*Acacia auriculiformis*) in Alternation with *Palas*

Akashmoni (*Acacia auriculiformis* A. Cunn.), when used in alternation with palas in the ratio of 4:1 under four coupe system, offers a capital growth of 235%

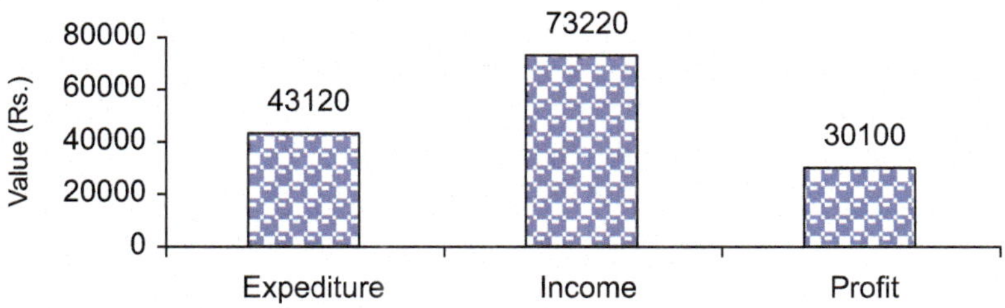

Fig. 5.15 Economics of lac production on palas in alteration with *ber*

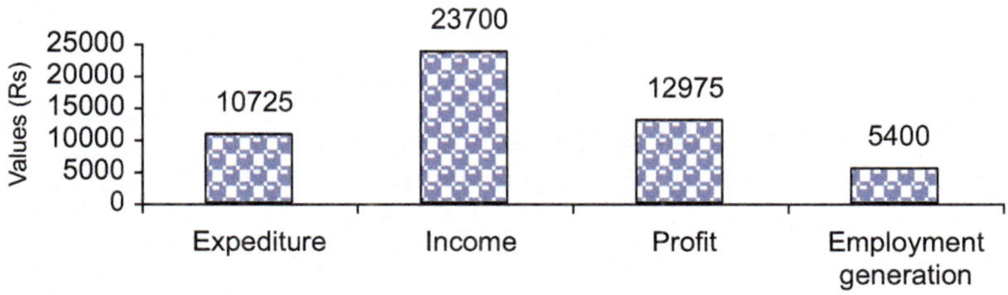

Fig. 5.16 Economics of broodlac production on *Akashmoni* in alteration with palas

per annum. In addition to the above-mentioned technologies, several other technologies are being tried and tested if it will be possible to grow *kusmi lac on palas*. Some new lac host species like *Flemingia* sp. and few others are likely to prove potential and serve the lac cultivator in increasing production of lac in the twenty-first century (Figs. 5.15 and 5.16).

5.7 Modelling and Forecasting in Lac Crop

Lac being a biological produce, its production depends on several biotic and abiotic factors described earlier. The lac cultivation has a history of extreme fluctuation in crop yield including periodic failure and bumper crop production causing uncertainty to the trade. The price of lac (like other agricultural crops) depends heavily on the quantum of production and demand in the market. The need to evolve a method of forecasting the crop production in the country was felt very much necessary for making strategic planning for sustained trade as well as causing alertness for supply of broodlac to regions where crop failure takes place. The efforts made at the IINRG for developing a model for forecasting lac production are described below.

Univariate Regression Model for Lac Production

For predicting sticklac production from summer crop of *rangeeni* lac in India and also in the states, viz. Bihar, Madhya Pradesh and West Bengal, 74, 53, 65 and 53% variation were accounted, respectively, by the production of preceding rainy season crop. The amount of variation in production of winter crop was 65, 60 and 33% for whole of India and in the states of Bihar and MP, respectively. The models for predicting summer crop of *rangeeni* (*Yrs.*) and winter crop of *kusmi* (*Ykw*) based on level of production of rainy season crop of *rangeeni* ($Y''_{(}t_{-1})$) and summer crop of *kusmi* $Ks_{(}t_{-1})$ in India were as follows:

$$\hat{Y}_{rs(t)} = 4.10 + 2.232 \quad Y_{''(t\text{-}1)}$$
$$(\pm 0.25)$$

$$\hat{Y}_{kw(t)} = 0.294 + 2.608 \quad Y_{ks(t\text{-}1)}$$
$$(\pm 0.36)$$

Econometric Model for Sticklac Production in India

Econometric multiple regression model with the variables like lagged production, world demand, rise in export price index fluctuation, sticklac price in the country and lac production in Thailand was developed. The model with variables like production in the preceding year ($t-1$) in the country (x_1) and Thai production (x_2) explained 67% variation in sticklac production in India. The model is as follows:

$$\hat{Y}_t = 9.294 + 0.797\,x_1 - 0.487\,x_2$$

Autoregression Model for Sticklac Production in the Country

Correlation study revealed that the sticklac production in any year would have spill-over effect in the following years also. The fourth-order autoregression model revealing this phenomenon was as follows:

$$Y_t = 4.142 + 0.706\,Y_{t-1} - 0.168\,Y_{t-2} - 0.006\,Y_{t-5} + 0.239\,Y_{t-4}\left(R^2 = 0.65\right)$$

Forecasting of Sticklac Yield from Standing Lac Crop on *Palas*

Yield forecast model from lac culture growing on palas tree has been developed for immature summer season crop (commercial crop of *rangeeni* lac) and rainy season crop in Palamau district of Jharkhand, known to be a hot area for lac cultivation. The yield of immature summer season crop and rainy season crop can be forecasted 20–22 and 10 weeks before the harvesting, respectively. For field immature summer crop (*Yx*), the parameters selected include quantity of broodlac used (x_1), number of host shoots with lac culture per tree (x_5), length of lac insect settled per shoot (x_6) and number of living lac insect per cm^2 (x_7), explaining 51% variation in yield. The model is as follows:

$$\hat{Y}_x = -0.773 + 0.238\,x_1 + 0.004\,x_5 + 0.014\,x_6 + 0.014\,x_7 + 0.006\,x_8$$

For yield of rainy season crop (*Yk*), the parameters selected include quantity of broodlac used (x_1), number of shoots with lac culture (x_5) and length of lac insect settled per shoot (x_6), explaining 57 % variation in yield. The model is as follows:

$$\hat{Y}_k = 0.541 + 0.171x_1 + 0.005x_5 + 0.015x_6$$

5.8 Lac Processing and Product Development

The main activities on these aspects were (i) to bring improvement in the lac processing techniques, (ii) to develop new uses of lac, (iii) to bring improvement in the existing lac-based formulations through modifications, (iv) to cater the changing demands in isolation of the chemicals from lac and (v) to synthesize diversified products from constituent acids of lac, upscaling developed technologies through pilot plant studies and facilitating their transfer to interested entrepreneurs.

To achieve the above, a clear understanding of the lac molecule was necessary. The institute, in the beginning, devoted mainly towards basic researches, which included physico-chemical characterization of the material, detailed in depth study on the chemistry of lac molecule and its structure facilitating to undertake modification reactions. Study on applied aspects was taken up side by side, especially on the use of lac in varieties of applications, suggestion of many of which came from the industry. Various problems associated with the processing of lac, including the problems posed by the lac processing industry, were also taken up. These led to the development of many products and processes. The Institute also changed priority of research according to the changing consuming pattern in these eight decades.

Many of the technologies predeveloped have been demonstrated at credible scale and have been transferred to users/entrepreneurs as and when demanded. Many important technologies are also emerging from the recent researches done at the Institute.

5.9 The Lac Resin

5.9.1 Processing

Raw lac (sticklac/scraped lac) as produced by farmers and marketed by traders in village *mandis* consists of a number of non-resinous materials or impurities. Some of them are undesired materials, and others are valuable non-resinous materials found as by-products, which may find use elsewhere. The process of removal of these impurities, as in practice in India, involves a number of steps, which are described below.

Sieving and Crushing After sufficiently drying, sticklac is sieved successively through one large (8–12 mesh) and one fine (30–40 mesh) wire nets to separate big lumps and fine dusts. Lumps obtained are subjected to crushers, intermediates are taken for making regular seedlac and fine particles for washing to obtain dust. Crushing of the sieved sticklac is done by *dhenki/janta chakki* with the help of sophisticated device or crusher.

5.9.2 Physico-chemical Nature

Lac resin consists of polyesters derived from certain hydroxy acids of aliphatic and sesquiterpenic series. It is a physical mixture of ether-insoluble hard resin (70%) and ether-soluble gummy mass (30%) known as soft resin. The former is composed of four molecules of aleuritic (9,10,16-trihydroxy palmitic) acid and four molecules of sesquiterpenic acids linked with ester and ether linkages, while the latter is composed of four pure esters comprising one molecule of aleuritic acid and one molecule of jalaric/lacci jalaric acid. Many pure component acids have been isolated and characterized from shellac and its two fractions. These are 9,10,16-trihydroxy palmitic, 6-hydroxy tetradecanoic (butolic) and 10,16-dihydroxy sesquiterpenic acids, like jalaric, lacci jalaric, shellolic, epi-shellolic, laksholic and epi-laksholic acids. These form nearly 90% of the components of lac resin. Analytical methods used for estimation have shown that at least 30% of vicinal hydroxy (aleuritic) and aldehydic (jalaric/laccijalaric) acids are present in free state.

Knowledge of different physical properties and chemical constants is very important for defining as well as characterization of lac resin. The lac resin is soluble in many organic solvents, inorganic alkalis or organic bases (Kamath 1962). Of these important solvents are alcohol, acetic acid, caustic soda, sodium carbonate and borax solution. It is partially soluble in ether, ethyl acetate, chloroform, carbon disulphide and acetone. Solubility of lac resin in alcohol organic acids and ketones confirms the presence of hydroxyl, carboxyl and carbonyl groups. Different physico-chemical properties have been presented in the Table 5.7.

5.9.3 Washing: Washing Is Done Manually as well as Mechanically

Manual Washing

Cup-shaped cemented vats called *nands* (2.5′ diameter and 2.5′ deep) are grouted keeping 3–4′ above the ground surface, and floor is cemented. A cemented rectangular water tank is also made just beside the *nand*. Twenty kilograms is put into the *nand* with sufficient water, and a skilled labor gets down in the *nand* and stamps the moistened lac with his feet for 10–15 min. After rubbing, plenty of water is poured and lac briskly churned by hand. The floated matter is taken out by sieve or cloth. This contains water-soluble lac dye, some minute particles of lac and lac wax. Washing is repeated till the washed water is clear. Two-three fractions depending upon the lac grain size are obtained. The fraction containing insect bodies and dark lac particles is known as patti (lac content 5–20%), and the other fraction containing fine lac grains is known as *rathi* or *molamma* (lac content 20–21%).

Table 5.7 Different physical constants of different forms of lac

Property	Sticklac	Seedlac	Shellac	Soft resin	Hard resin	Shellac wax
Specific gravity	1.0	1.13–1.15	1.139 (TN)			
			1.196–1.217 (bleached)			
Refractive index		1.52 (*kusmi*)	1.5235 (palas and kusmi)			
			1.5228 (dewaxed)			
			1.5295 (garnet)			
			1.5272 (TN)			
Specific heat (Cal/s/°C		0.4 (*kusum*)	0.37 (kusum)	0.48	0.34	0.43
		0.41 (*khair*)	0.36 (palas)			
			0.37 (khair)			
Melt viscosity (poise)			22,505 (at 80 °C)	117,912 (at 45 °C)	27,230 (at 85 °C)	
			2154 (at 90 °C)	10,405 (at 60 °C)	1316 (at 100 °C)	
Thermal conductivity (mw/cm/°C)			2.5 (at 30 °C)			
Molecular weight			1000	500	1900	
			949 (bleached)			
Softening temp. (°C)	65–70		75–80			
Glass transition temp. (°C)			40	10	50	
Melting temp. (°C)			77–90			
Time of polymerisation at 150 °C (min.)			30–120			
Flow Westinghouse (sec.)			55–480			
V-tube (mm.)			84–17			
ASTM (mm)			100–45			

Barrel Washing

Horizontal stationary hand fitted with the exiles caring blades and three openings on its curved surface is used. The barrel is partially filled with water and then charged with crushed lac to the full capacity. Churning the lac by revolving the blade is allowed to continue up to 1.5–3 h depending upon the quantity of the dryness of sticklac and age. Water is fed into the barrel during the operation. After the over of washing lac is discharged into the cistern, insect parts and lighter particles float on the surface and are cooped with filter cloth.

Processing

The washed lac, known as seedlac, is further processed to shellac, either by *bhatta* process at cottage industries or through mechanical process in lac factories. Processing for more specialized forms, viz. dewaxed decolorized lac, button lac, etc., is done by adopting specific methods.

5.9.3.1 Utilization

Composition and Applications of Lac

Being a versatile resin of wide utility, lac is mainly used in the surface coating industry, pharmaceutical industry, varnish and printing ink industry, adhehsive industry, electrical industry and leather industry. In addition to these, by-products like lac dye and lac wax are also used extensively.

Composition and Applications of Lac

(A natural resin of insect origin with unmatched versatality)

Shellac is nontoxic and environmentally safe.

Lac Dye

(Anthraquinone derivative)

- Dyeing of wool and silk
- Soft drink formulations
- Pill coating
- Sausage finishing
- Confectionery and chocolate coating
- Finished food coating

Lac Wax

(Mixture of higher alcohols, acids, esters and hydrocarbons)

- Shoe automobile and floor polishes
- Bottle sealer
- Tailor's chalk
- Crayons (especiallyfor writing on glass)
- Electrical potting compounds
- Lipstick
- Fruit coating

Lac Resin

(An ester complex of long chain hydroxy fatty acids and sesquiterpenic acids)

Cosmetic Industry

- Hair spray, lacquer
- Eye shadows
- Microencapsulated perfumes
- Lipstick, nail polish, mascara, etc.

Food Industry

- Coating of fruits
- Coating of chocolates, lozenges,coffee beans
- Coating of aluminium foils for making pizzas
- Nontoxic ink of marking food stuff
- Internal can coating

Electrical Industry

- Air-drying and baking-type insulating varnishes
- Coating of isolators, PCBs, etc.
- Micanites (sheets, tubes and moulded insulators)
- Cement for sockets of electrical lamps
- Shellac bond powder
- Anti-tracking insulating varnish
- Coating of spark plugs

Pharmaceutical Industry

- Coating of tablets
- Microencapsulation of vitamins
- Coating of medicines for slow-release/delayed action
- Marking ink for capsules, tabs, etc.

Leather Industry

- Adhering coating of leather with metallic and plastic foils
- Topdressing material

Varnish and Printing Ink Industry

- Furniture polish (French polish)
- Floor polish
- Sealers
- Pattern paints for wood
- Flexible and fast-drying agent for printing ink
- Heat- and waterproof varnishes
- Water-soluble lac for earthenware
- Shellac etch primer
- Metal lacquers and dry lacquering of wood
- Use in agriculture and health hazards
- Lac-coated urea
- Nematicidal activity
- Lac-coated weedicides
- Application in roach control

- Synthesisis of sex phenomones, PGRS
- Application in mosquito control

Adhesive Industry

- Gasket cement
- Optical cement
- Sealing wax
- Grinding sheel
- Hotmelt adhesive
- Polymer adhesives
- Adhesives for chips and solar cells

Miscellaneous

- Diamond and crystal cutting
- Jewellery
- Polishing stone
- High-efficiency igniter cum-fuel
- Slow-release fertilizer, pesticide, etc.
- Proofing and stiffening in hat making

Acknowledgements The authors are grateful to Dr. S.K. Yadav and Sri R.K. Singh, Scientists L.P. Division, for their help in documenting and computer work in preparing the manuscript.

References

Anderson J (1791) A letter from Dr. Anderson to Sir William Jones. Asiatic Researches, 2:359, Quoted pp. 181–384 in Watt. G. 1901. Agr Ledger Enter, Calcutta No. 9

Blanchard R (1886) Zoologic Medical 1:1

Bhattacharya A, Naqvi AH, Sen AK, Mishra YD (1998) Artificial rearing of *Pseudohypatopa pulverea* Meyr- a predator of lac insect, *Kerria lacca* (Kerr). J Ent Res 22(1):83–87

Carter HJ (1861) On the natural history of the lac Insect (*Coccus lacca*). Ann Mag Nat Hist Lond 3(7):1–10, 363–364

Chamberlin JC (1923a) A systematic Monograph of the Tachardiinae or lac insects (Coccidae). Bull Ent Res Lond 14(2):147–212

Chamberlin JC (1923b). Rev Appl Ent Lond (A)11:550

Chamberlin JC (1925a) Supplement to a monograph of the Lacciferidae (*Tachardiinae*) or the lac insects (*Homopt. Coccidae*). Bull Ent Res Lond 16(1):31–41

Chamberlin JC (1925b) 1925. Rev appl Ent Lond 3(4):420

Choudhary SG, Naqvi AH, Sen AK, Mishra YD, Kumar P (1999) Package of practices for lac cultivation on *Akashmani (Acacia auriculaeformis)*. Indian Lac Research Institute, (ICAR). Bulletin 2:1–8

Chauhan NS, Mishra YD (1970a) Multiple coitus in the lac insect, *Kerria lacca* (Kerr). Indian J Ent New Delhi 32(1):102–103

Chauhan NS, Mishra YD (1970b) Genetic evidence of nutritional difference in lac insects. Indian J Ent New Delhi 32(4):390–392

Dave KN (1950) Lac and the lac insect in the Atharva veda. Int Acad Ind Cult Nagpur, 16 pp

Fabricius JH (1787) Mantissa Insectorum 2:318–319
Green EE (1903) Remarks on Indian scale insects (Coccidae) with description of new species. Part II. Indian Mus Notes Calcutta 5(3):93–103
Green EE (1911) On some coccidae affecting rubber trees in Ceylon with description of new species. J Econ Biol Lond 6:27–37
Glover PM (1937a) Lac cultivation in India. ILRI, Namkum, 147 pp
Glover PM (1937b) Biol Abstr 13:1409
Glover PM (1937c) Sci Progress Wash, Washington 33:799 (Review)
Hautefeuille M(1924) Report on lac and its industrial treatment. Department of Industries and commerce. H.E.H. the Nizoom's Govt., Hyderabad Deccan (Translated by Syed Mahdihassan)
Imms AD, Chaterjee NC (1915) The structure and biology *of Tachardia lacca, kerr,* with observations on certain insects predaceous or parasitic on it. Indian For Mem (Zool) Calcutta 3(1):42pp
Kamath NR (1962) Chemistry of lac in lac- a monograph. Indian Lac Research Institute, Namkum, Ranchi, pp 197–220
Kapur AP (1958) A catalogue of lac insects (*Lacciferidae : Hemiptera*). Indian Lac Cess Committee, Ranchi 47 pp
Kerr J (of Patna) (1782) Natural History of the Insect which produces the gum lacca, *Coccus lacca,* Phil Trans R Soc London 17(2):374–382
Krishnaswami S, Majumdar M, Krishnan YS (1964) Studies on lac insects- host plants relation-mode of feeding of lac insect. Indian J Ent 26(4):453–457
Kumar P, Purkayastha BK (1970) A note on the hastening of seed germination in *Moghania macrophylla* (Willd) O. Ktze and *Moghania Chapper* Kntze. Sci Cult, Calcutta 38:347–348
Kumar P (1989) Vegetative propagation in *palas (Butea monosperma)* through air layering. Indian J Forestry 12(3):188–190
Kumar P,Singh BP, Srivastava SC(1996) Package of practices for *kusmi* lac cultivation on mixed plantation of ber and *galwang*. Indian Lac Research Institute, Namkum, Ranchi. (ICAR), pp 1–7
Lindsay HAF, Harlow GM (1921) Report on lac and shellac. Indian For Res, Delhi 8(1):162
Mahdihassan S (1923a) Classification of lac insects from a physiological stand point. J Sci Assoc Maharaj. College, Vizianagram, Madras 1(2&3):47–99
Mahdihassan S (1923b) 1924, Rev Appl Ent (A) London 12:61
Mahdihassan S (1923c) Observations on the life cycle of south Indian Lac insects. Indian For, Dehra Dun 49(12):653–663
Mahdihassan S (1930a) Authorised translation of M. Pidance's Report on Lac-Refining. Osmania Univ. Press, Hyderabad, Deccan: 10 pp
Mahdihassan S (1930b) The first ecdysis of the lac insects. Dt Ent Z, Berlin 4: 223–234
Mahdihassan S (1930c) The probable occurrence of sex reversal among lac insects. Z Angew Ent, Berlin 16(3):527–545
Mahdihassan S (1931a) The last two months of the lac insect. Tijdschr Ent Amsterdam 74:230–244
Mahdihassan S (1931b) Rev Appl Ent (4) London, 19
Mahdihassan S (1931c) The metamorphosis of the male lac insects., casopsis. Cas Osl Spol Ent, Praha (Acta Soc Ent, Prague) 28(1–2):5–16
Mahdihassan S (1931d) The males of lac and pseudo- lac insects. Z Wiss Zool, Leipzig 138(3):371–385
Mahdihassan S (1931e) Early sexual and other dimorphisms among lac insects. Biol Gen, Vienna & New York 7:517–530
Mahdihassan S (1931f) Das Zweite larvenstadium der weiblichen lackschildlaus (The second larval stage of female lac insect). Zool Auz, Leipzing 94(11–12):289–304
Malhotra CP (1976) Package of practices for lac production and protection (Cyclostyled). Seminar on lac production and utilisation, Indian Lac Research Institute, Namkum, Ranchi, 7 p
Misra CS (1923) A list Coccidae in the Pusa collection. Rep Proc Vth Ent Meet Pusa Calcutta, pp 345–351

Misra AB (1930a) On the post embryonic development of the female lac insect, *Laccifer lacca* Kerr (Homoptera: Coccidae) Bull Ent Res, London. 21(4):455–467

Misra AB (1930b) On a collection of lac insects from Northern India. Bull Ent Res Lond 21:161–164

Misra AB (1931a) Bio Abs, Philidelphia 9(1):239

Misra AB (1931b) Biol Abstr 5(10):2577

Misra AB (1931c) On the anatomy of the larvae of *Laccifer lacca Kerr* (*Hom: Coccidae).* Bull Res, London 22:297–306

Misra AB (1931d) On the internal anatomy of the female lac insect, *Laccifer lacca Kerr* (*Homoptera: Coccidae*). Proc Zool Soc Lond (1):297–323

Misra AB (1931e) On the internal anatomy of the male lac insect, *Laccifer lacca Kerr* (*Homoptera: Coccidae*). Proc Zool Soc Pt. IV:1359–1381

Mishra YD, Bhattacharya A, Sushil SN (1997) Effect of some systemic fungicides on the nymphs of Indian Lac Insect, *Kerria lacca* (Kerr) for their protection against fungal infection. J Ent Res 21(3):291

Mishra YD, Bhattacharya A, Choudhary SG, Kumar P (1999a) Advanced brood lac production technology on kusum. Indian Lac Research Institute, Namkum, Ranchi. Bulletin 1:1–8

Mishra YD, Choudhary SG, Bhattacharya A, Bhardwaj RL, Kumar P (1999b) Advanced technologies for brood lac production on *palas* in hot area. Indian Lac Research Institute, Namkum, Ranchi. Bulletin 3:1–6

Mishra YD, Sushil SN (2000) A new trivoltine species of genous *Kerria* Targioni- Tozzeti (*Homoptera : Tachardiidae*) thriving on *Schleichera oleosa* (Lour) Oken from eastern *India Oriental Insects*, 34:215–220

Mishra YD, Choudhary SG, Kumarr P, and Sushil SN (2001) Package of practices for rangeeni lac production on *palas* and ber Indian Lac Research Institute, Namkum, Ranchi, Bull. No. 41-4p

Mishra YD, Sushil SN, Chakrabarty PK, Bhattachraya A, Kumar S (2002) Occurrence of unorthodox female of *Kerria nagoliensis* (Mahdihassan). Shashpa 9(1):83–84

Nicholson JW (1925) Some notes on lac cultivation. Indian Forester 51(10):483–498

Negi PS (1934) A simple method for the forecast of emergence of lac larvae. Pamphlet ILRI, Namkum

Oken L (1815) Lehrbuch der Naturgeschichte Sect. 3 (Zool.1): 430

Purkayastha BK, Kumar P (1977) Note on propagation of lac hosts with growth regulators. Indian J Agric Sci 47(2):108–110

Purkayastha BK, Kumar P (1981) Effect of foliage treatment of Gibberellic acid on lac host plants. Agric Sci Digest 1(1):27–28

Roonwal ML, Raizada MB, Chatterjee RN, Singh B (1958) Descriptive account of the host plants of the Lac insect Laccifer lacca (Kerr) and the allied plants in the Indian region. Part I & II *Indian Lac Cess Committee*, Ranchi: 140 pp

Roxburghe W (1789) On the lac insect. Asiatic Researches Trans Asiatic Sec Bengal 2:360–366

Stebbing EP (1908) A manual of elementary. Forest Zool India 71–72:165–166

Srinivasan MM (1956) Lac, its cultivation and Manufacture in India Chapter IV- Host plants of the lac insect. Ind For. Dehra Dun. 82(4):180–193

Sharma KK, Ramani R, Mishra YD (1997) An additional list of the host -plants of lac insects. *Kerria* sp. (*Tachardiidae : Homoptera*). J Non-timber For Prod 4(3/4):151–155

Srivastava SC, Kumar P, Mishra YD (1994) Air layering in *galwang* (*Albizzia lucida* Benth) a lac host plant. The Indian Forester. 120(6):524–528

Srivastava SC, Kumar P, Singh BP (1997) Air layering in kusum "*Schleichera oleosa*" (Lour) Oken. J Non-Timber For Prod 4(3/4):180

Tachard Father P (1710) Sur La Laque. Historic de I: Adacemic des Sciences:44–46

Targioni T (1884) Annali Agric Rome (Minist Agric Ind Comm Italy 86–89:383–414

Teotia TPS, Chauhan NS (1963) Mode of reproduction in lac insect (Abstract). Agr Res (India) 3(2):134

Thakur AK (1932) Comparative study of lac-hosts with special reference to *Acacia catechu* and *Cassia florida*.(Abs.) Proc. nineteenth Indian Sci. Congr., Calcutta, Section I: 85

Varshney RK (1966) The Indian Lac Insect- a change in nomenclature (*Homoptera : Coccoidea*). Ind J Ent New Delhi 28(1):116–118

Varshney RK (1968) Further data on host plants of lac insects. *(Tachardiidea -Homoptera)*. J Bombay Nat Hist Soc Bombay 65(1):249–251

Varshney RK (1977) Taxonomic studies on lac insects of India Oriental insects New Delhi. Supplement, No. 5, 97 p

Varshneny RK (1990a) A synoptic updated catalogue of lac insects (*Homoptera: Coccoidea : Tachardiidae*) Proc. VIth internat. symp. Scale Insect Studies (ISSIS- VI), Krakow 1990, Part II: 71–74

Varsheny RK (1990b) Abnormal segmentation in the antenna of an apterous male lac insect, *Kerria lacca* (Kerr). Rec Zool Survey India 87(3):223–226

Varshney RK (1993) World list of lac insects. Indian Shellac: 13–15 (Annual Number 1992–1993; March, 1993)

Varshney RK (n.d.) Res. Notes Indian Lac Res Inst, Ranchi (Ent.) No. 11, 2 pp

Watt G (1901) *Tachardia (Carteria) lacca (Kerr)* (Syn. *Coccus lacca*). Lac (Lakh) and the Lac Industries. Agric Ledger (Ent Ser) 9:181–348

Watt G (1908) The commercial Products of India, being an abridgement of The Tachard (Father) P. 1710 Sur La Laque. Historic de I: Academic des sciences:44–46

Lac Insects and Host Plants

6

K.K. Sharma

Contents

6.1 Introduction

Being a natural heritage of the country, lac has been associated with tribal and backward people, providing regular income to them in the absence of other cash crops. Due to shrinking resources of livelihood and poor economic condition of the tribal, the lac cultivation has become an alternative occupation for utilising host plants available in the forests and the personal holdings not only for harvesting the lac but also for generating self-employment at their doorstep.

The lac insect ecosystem is a complex multi-trophic web of flora and fauna. It represents a rich biodiversity which includes, besides lac insects, lac host plants, several predators of lac insects, beneficial parasites, harmful parasites, microbes and a variety of host plants' pests. The lac host plants constitute the first trophic level, pests of host plants and the lac insects make the second, predators along with primary parasitoids make the third and parasitoids of lac predators constitute the fourth (Fig. 6.1).

K.K. Sharma (✉)
Lac Production Division, Indian Institute of Natural Resins and Gums,
Namkum, Ranchi 834010, Jharkhand, India
e-mail: kewalkks@gmail.com

Omkar (ed.), *Industrial Entomology*, DOI 10.1007/978-981-10-3304-9_6

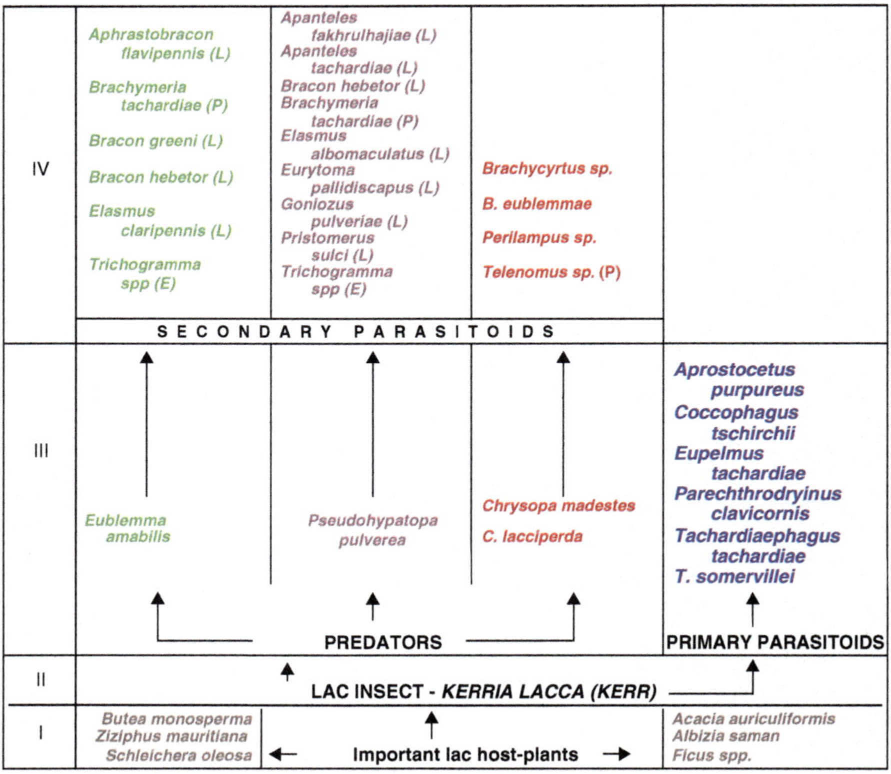

Fig. 6.1 Trophic levels (I–IV) of lac insect complex showing important species

6.2 Diversity in Lac Insect Ecosystem

6.2.1 Lac Insects and Host Plants

Lac insects depend on plants as hosts for survival. More than 400 lac hosts have been reported throughout the world to carry lac insects (Roonwal et al. 1958; Varshney and Teotia 1967; Varshney 1968; Sharma et al. 1997). Of the nine genera and 99 species of lac insects reported from the world, two genera and 26 species are found in our country, representing 26.3 % diversity of the known lac insect species. Species belonging to *Paratachardina* genus produce a hard, horny substance which is insoluble in alcohol. These are univoltine and generally treated as parasites of economically important plants such as tea and sandal. However, recently *Paratachardina* spp. have been found to be very potential biocontrol agents for managing weeds and need to be nurtured as such (Campbell et al. 1994). Lac insect species under the genus *Kerria* are generally bivoltine. However, *K. lacca mysorensis* found on jalari (*Shorea talura* Roxb.) and *K. sharda* found on kusum, *Schleichera oleosa* (Lour.) Oken. (Syn. *S. trijuga* Willd.) (Mishra and Sushil 2000) and on rain tree, *Albizia saman (Samanea saman* (Jacq.) Merr. (Leguminosae:

Mimosoidae) [Syn. *Enterolobium saman* Prain ex King and *Pithecellobium saman* Benth.] are trivoltine. Indian lac insect, *Kerria lacca* (Kerr), the most important and widely exploited insect for lac cultivation, can further be distinguished into two strains or infra-sub-species forms, the *rangeeni* and *kusmi*, on the basis of difference in life cycle, host preference and quality of lac produced. *Rangeeni* strain is characterised by unequal duration of bivoltine life cycle and non-preference of kusum as a host and *kusmi* strain by more or less equi-durational life cycle preferring kusum as a host. The quality of the resin produced by *kusmi* lac insect is superior in comparison to *rangeeni* lac insect. A significant quantitative and qualitative variation in various biological attributes of the lac insect, *viz*. yield of resin, fecundity, sex ratio and body colour, has also been reported (Chauhan and Teotia 1973; Chauhan and Mishra 1977; Varshney 1977; Mishra et al. 2000). The pigment present in the lac insect haemolymph (laccaic acid or lac dye) is non-toxic and finds numerous applications in textile, pharmaceutical and food industry. Qualitative and quantitative variations in lac insect dye have been recorded showing crimson, yellow, albino and cream body colour.

Similarly, the quantity of lac dye present in the resin (erythrolaccin) also varies depending upon the lac insect and the host plant on which the insect is reared. Considerable interspecific and intraspecific lac host variabilities *vis-a-vis* lac insect like density of settlement, initial mortality, sex ratio, size and weight of the insect cell have also been observed (Mishra et al. 1999; Srinivasan 1956).

Lac host plants can be divided into various categories based on the degree of the preference of the lac insects for the various hosts and on the abundance and quality of lac obtained by infection upon the host. It is possible to divide hosts into further categories. First, there are species, which are excellent hosts throughout the year and wherever they occur. Secondly, there are those species which are good hosts in certain restricted regions of the country, whereas in other regions these species are either indifferent or do not take lac cultivation at all. Thirdly, there are hosts which are major hosts for certain specific purposes in certain specified seasons. Again, while one sole or variety of host species is a good host, another (which may not be botanically distinguishable from the first) may be a non-host. Several examples from this type are well-known lac hosts, e.g. kusum, palas [*Butea monosperma* (Lam.) Taub. (Papilionaceae) (*Butea frondosa* Koen. ex Roxb. (Leguminosae: Papilionatae)] and *Ficus* spp. (Roonwal et al. 1958).

6.2.2 Lac Insect-Associated Fauna

The soft-bodied lac insects produce a resinous secretion, which protects them from adverse environment. In spite of this protective covering, the insects are always a sitting prey to predators and parasitoids due to their sedentary habit. Twenty-two lac insect predators, 30 primary parasitoids, 45 secondary parasitoids (Varshney 1976; Das 1990) and several fungal pathogens of lac insects as well as lac hosts (Shaoji 1993; Sharma et al. 2001), besides several other associated insects, represent a rich

Table 6.1 Number of flora and fauna in lac insect ecosystem

Sr. no.	Lac insects and associated fauna and flora	Trophic level	No. of species reported
1.	Lac host plants	I	>400
2.	Lac insects	II	99
3.	Lac predators	III	22
4.	Primary parasites	III	30
5.	Secondary parasites	IV	45
6.	Insects visiting lac insects for honeydew	–	48
7.	Unknown relationship	–	19
8.	Doubtful relationship	–	4

biodiversity of this ecosystem. Moreover, natural lac insect complex maintains a variety of other tree flora, macrofauna and soil microorganisms (Table 6.1).

6.3 Lac Insect

6.3.1 Taxonomy and Distribution

Lac insects belonging to the family Tachardiidae (Kerriidae) constitute a specialised and isolated group in the superfamily Coccoidea of the order Homoptera. At present lac insects of the world are represented by nine genera and 99 species, of which two genera and 26 species are reported from India.

Lac insects are distributed in all zoogeographical regions except Palaearctic. They, however, are concentrated in tropical and subtropical regions between 40° latitude above and below the equator on both hemispheres (Varshney 1977).

6.3.2 Life Cycle of Lac Insect

Most common Indian lac insect of commercial importance is *Kerria lacca* (Kerr). *Rangeeni* and *kusmi* are its two strains. Each of these produces two crops in a year (bivoltine). *Kusmi* insects grow well mainly on *kusum* and also on a few other trees but not on palas, whereas *rangeeni* strain grows well mainly on palas and also on a few other trees but not on kusum. The *rangeeni* insect matures once in October–November and thereafter in June–July, whereas *kusmi* matures in January–February and then in June–July. In coastal regions of Orissa and West Bengal, however, three crops (March–April, June–July and October–November) in a year are produced by *K. sharda* (trivoltine) on kusum, ber (*Ziziphus mauritiana* Lam.) and rain tree.

6.3.2.1 Crawler

The life cycle of lac insect starts with its first instar stage, the crawlers. The crawlers have soft, oval-shaped body, tapering on posterior side of the abdomen. It is 0.6 mm

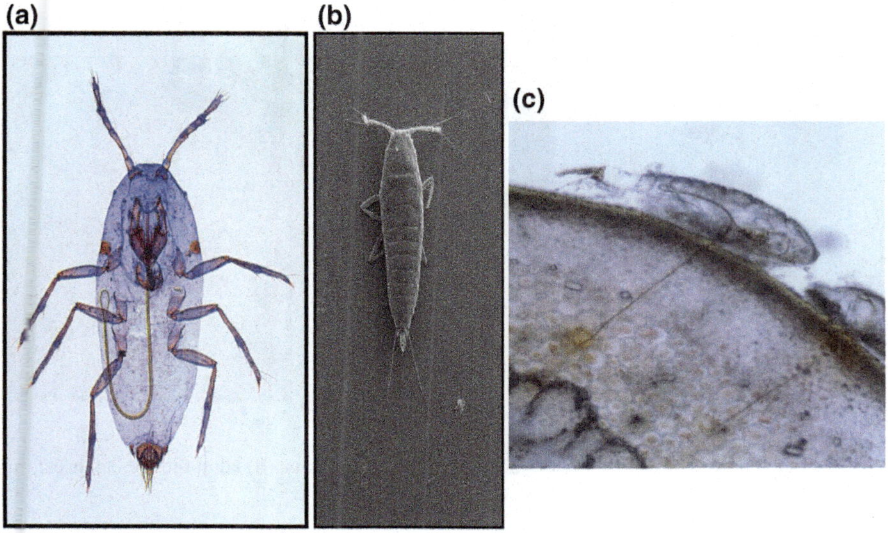

Fig. 6.2 First instar nymph. (**a**) Nymph (ventral side), (**b**) nymph (dorsal side), (**c**) nymph feeding on phloem sap

long from the head to the abdomen and 0.25 mm broad at the thorax. There is no pronounced demarcation between the head, thorax and abdomen. The head carries a pair of antennae, a pair of ocelli and ventrally the mouthparts. The oral stylets are extremely long and lie looped occupying a great deal of the abdominal cavity. The antenna is six segmented including the basal one. On the ventral side of the thorax are situated paired pro- and mesothoracic spiracles. The leg consists of the coxa, trochanter, femur, tarsus and claw. The abdomen is 8–9 segmented and the last few are telescopic (Ramani and Sharma 2016). The last segment of the abdomen forms the anal ring carrying a pair of greatly elongated setae, in addition to three shorter pairs arranged around the anal opening (Fig. 6.2a, b).

The first instar is not a sedentary stage of lac insect, but it crawls over the shoot of host plants. It has piercing and sucking type of mouthparts. Hence, it takes phloem sap by its proboscis, after piercing it into the phloem region of shoots (Sharma and Ramani 2014) (Fig. 6.2c). The lac crawlers start secreting resin in minute quantity after 2–3 days of settlement. Except the mouth, anal tubercle and respiratory pores, the lac insect covers itself by its secretion, the lac resin. To avoid covering these holes by resin, the lac insect secretes wax also, which is a white threadlike structure. Normally, 200–300 young lac insect crawlers settle on one square inch (Fig. 6.3a). The male and female insect cannot be distinguished at this stage.

6.3.2.2 Metamorphosis

Male and female insects differ in respect of its type of metamorphosis. Complete metamorphosis occurs in male insect with a resting stage, pupa between young lac insect and adult stage, whereas incomplete metamorphosis occurs in female insect

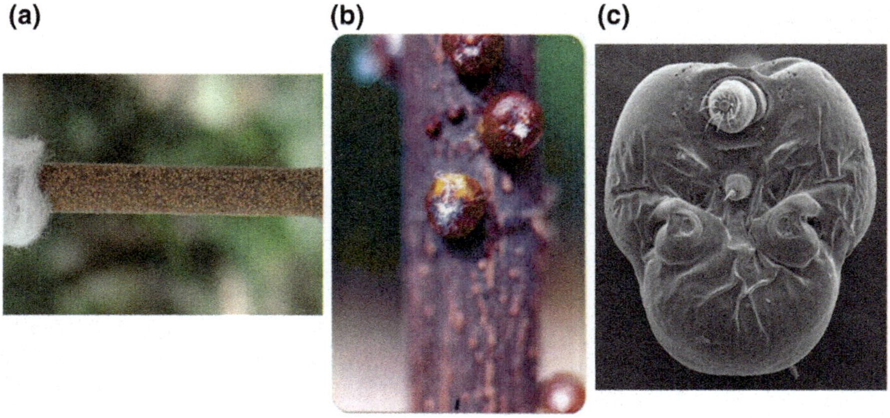

Fig. 6.3 Different stages of lac insect. (**a**) Just settled lac nymphs, (**b**) adult female lac insect, (**c**) SEM of female insect after removing resin

without any pupal stage. Hence, the structure and habit of young lac insects differ significantly from its adult stage in the case of male insect.

Before reaching the adult stage, the lac insect moults thrice. The duration of each stage depends on the host-plant species on which it feeds and the prevailing environmental condition. The nymph emerges from mature female cell and moults to reach the second instar. The male and female lac insects can be distinguished at the second instar. The male is elongated cigar-shaped, while female is round in shape. The male insect transforms into prepupa which later metamorphosises into pupa. The pupal stage is resting stage and it does not feed. Mouthparts of adult male insect are nonfunctional; hence, it is nonfeeding stage. Adult male insects emerge out of covering by removing its operculum. Within hours it starts mating with mature female insect. In the second instar of the female insect, the legs become atrophied and antennae reduce in size. The number of anal ring plates which are six in the first instar becomes ten in the second instar.

6.3.2.3 Female

At sexual maturity, the female lac insect is disc-like with wavy subcircular outline. It grows several-folds, during the post-sexual period into a pyriform or globular structure (Fig. 6.3b), measuring ~5.0 mm in length and ~3.0 mm in width. In adult stage, the female does not show segmentation (Fig. 6.3c), and due to distorted growth, anterior and posterior ends cannot be clearly distinguished. Therefore, the side attached to the host plant is known as 'oral', and the opposite side bearing the branchia, dorsal spine and anal tubercle is called 'aboral' side. The oral side of the body is covered by thick resinous covering which fixes the insect firmly onto the plant. The resinous covering bears three orifices – two for the branchia (respiratory openings) and one for the anal tubercle which bears the excretory and reproductive openings. The legs are absent and antennae can be seen as vestiges on oral aspect.

Table 6.2 Duration of different stages of bivoltine lac insects and their life span in days[a]

		Young ones			Male		Female	
Strain	Crop	I	II	III	Adult	Total	Adult	Total
Rangeeni	*Katki*	20	14	8	2	44	67	109
	(Rainy season)	(3)	(2)	(1)		(6–7)	(10)	(16)
	Baisakhi	50	40	15	3	108	145	250
	(Summer season)	(7)	(6)	(2)		(15–16)	(21)	(36)
Kusmi	*Aghani*	20	15	14	2	51	150	199
	(Winter season)	(3)	(2)	(2)		(7–8)	(21)	(28)
	Jethwi	32	25	12	2	71	90	159
	(Summer season)	(5)	(4)	(2)		(11–12)	(13)	(24)

[a]Figures in parentheses indicate time in weeks

The mouthparts consist of a long stylet comprising of four stylets (maxillae and mandibles) which project through a short conical labium. The stylets can be retracted forming a loop inside a saclike structure called crumen.

6.3.2.4 Male

The adult male is the only other stage, besides the crawler of lac insect which has resemblance to a typical insect. It measures about 1.5 mm in length. The males may be alate or apterous with a pair of wings arising from the mesothorax. The expanse of the wing is about 1.4 mm. The alate males hardly fly even though they are capable of active flight. The head bears two pairs of ocelli, one pair on the dorsal and another on the ventral side.

6.3.2.5 Duration of Different Stages

Duration of different stages depends on the strain and the season of crop besides the host plant and the location (Jaiswal and Sharma 2011) (Table 6.2). Sex ratio in lac insect population depends largely on lac crop and its season. Similarly, the type of adult male insect, alate (winged) and apterous (wingless), also depends largely on seasons. Normally, the ratio of male to female insect varies from 1:2 to 1:3. The alate (winged) male insects emerge more in summer season.

6.3.2.6 Fertilisation and Oviposition

Sexually mature male and female insect mate. It has been found that one male insect is capable of mating with 40–45 females and vice versa (multiple coitus). The life span of adult male insect is only 2–4 days after its emergence. It dies after mating, while female insects survive up to crop maturity stage.

After fertilisation, the female insect prepares itself for oviposition inside the lac cell. A mature female insect oviposits on an average of 300–400 eggs. When female is ready to oviposit, it contracts its body, thus creating space between its body and the resin cover. The eggs are laid in this space. The lac insect is ovoviviparous, i.e. it lays eggs, which hatch immediately, but the nymphs remain inside the lac resinous cell till the environmental conditions outside are congenial. The nymphs emerge

Fig. 6.4 Colour mutants of *K. lacca* (**a**) *crimson* – body colour is *crimson*, resin colour is *yellowish orange*; (**b**) *yellow* – body colour changes to *yellow*, and resin colour remains *yellowish orange*; (**c**) *cream* – both body and resin colour are creamish

through a pore one by one in large numbers over the lac encrustation. The crawling of first instar nymphs in large number is called swarming. From each female insect, nymphal emergence continues for about a fortnight.

6.3.2.7 Colour Mutants of *K. Lacca* (Sharma et al. 2006; Ramani and Sharma 1991; Ramani and Sharma 2011)

6.3.2.7.1 Wild Insect

The body is crimson in colour (Fig. 6.4a) due to the presence of a complex of water-soluble colouring pigments collectively called as lac dye. The colouring pigments are polyhydroxyanthraquinones, closely resembling in chemical structure. At least five different components, termed as laccaic acids A, B, C, D and E, have been identified in lac insects. The protective resinous covering of the lac insect from which the commercial shellac is derived appears yellowish. The colour of the resin is due to certain alcohol-soluble anthraquinone pigments related to laccaic acids, *viz.* desoxyerythrolaccin, erythrolaccin and isoerythrolaccin. The resin colour of the *kusmi* insect is lighter than that of *rangeeni* insect, in general, and this difference has to be genetic.

The biosynthetic pathway proposed for the derivation of these colouring pigments shows that both body and resin colour pigments are derived through a common precursor, the laccaic acid D, which is a component of the body colour pigments. It is recessive to the wild-type allele (Ramani and Sharma 2011).

6.3.2.7.2 White (w)

This is a mutant devoid of both body and resin colour pigments. It is due to a genetic block which prevents the formation of any of the colouring pigments. It also behaves as a recessive to the wild-type allele and is non-allelic to yellow.

6.3.2.7.3 Yellow (y)

This mutant has yellow body colour, whereas the resin colour remains normal (Fig. 6.4b). Yellow mutants are common and have been reported in both *rangeeni* and *kusmi* insects of *Kerria lacca, K. albizziae* and *K. fici.* This mutant supports the scheme of the common precursor (laccaic acid D) for both resin and body colour pigments in the biosynthesis.

6.3.2.7.4 Cream (cr)

The cream-coloured insect shows a very light resin colour and a very light yellow body colour (Fig. 6.4c). This mutant is recessive to the wild-type crimson allele. It has also been shown that it is non-allelic to yellow.

6.4 Lac Host Plants

On the basis of preference in use for lac cultivation and distribution in the country, the lac host plants are placed under three categories (Srivastava 2011), *viz.* (a) common or major host plants, (b) occasional host plants and (c) rare host plants (Table 6.3). The common or major hosts include 14 species in which three, namely, palas (*B. monosperma*), kusum (*S. oleosa*) and ber (*Z. mauritiana*), are of all-India importance since these are excellent hosts wherever they occur in the country. Other five species, namely, babool (*Acacia arabica* Willd.), banchalata (*Leea crispa* Linn. and *Leea robusta* Roxb.), arhar (*Cajanus cajan* (Linn.) Millsp.) and ghont (*Ziziphus xylopyrus* Willd.), are excellent host plants in their respective regions, i.e. in the states where they exist in abundance. Of the remaining six species, one, i.e. *khair* (*Acacia catechu* Willd.), and five species of *Ficus* are excellent host plants of all-India importance but for specific purpose only, i.e. for preservation of broodlac, the former one for growing winter *kusmi* crop and the latter five for summer *baisakhi* crop.

The largest number of host plants occurs in the family 'Leguminosae' in India, while the family 'Moraceae' (Urticaceae) with single genus *Ficus* comes next in importance in lac cultivation (Roonwal et al. 1958). The species belonging to this family find an important place for cultivating *baisakhi* lac crop as they are not deciduous for long periods during hot months.

Good to fair lac host plants are found in seven families, namely, Leguminosae, Moraceae, Sapindaceae, Rhamnaceae, Vitaceae, Tiliaceae and Dipterocarpaceae, of which Tiliaceae and Vitaceae contain regional host plants of Assam and Dipterocarpaceae contains an important regional host plant, *S. talura*, of Karnataka. Family Leguminosae comprises nine host plants of importance, out of 56 recorded within it, and the Moraceae contains ten host plants. Families Rhamnaceae and Sapindaceae each comprise unique host plants of importance.

In lac culture, the performance of host plants shows wide individual differences within the type species, thereby affecting the production of lac. Several examples of this type from well-known lac host plants, namely, *S. oleosa*, *B. monosperma*, *Ficus* spp. and others, have been mentioned. Two varieties or races have been mentioned

Table 6.3 Categorisation of host plants based on their utility

A.	**Host plants of commercial importance**	
		Palas – *Butea monosperma* (Lam.) Taub. (Papilionaceae) [*Butea frondosa* Koen. Ex Roxb. (Leguminosae: Papilionatae)]
		Kusum – *Schleichera oleosa* (Sapindaceae) (Lour.) Oken. (syn. *S. trijuga* Willd.)
		Ber – *Ziziphus mauritiana* Lam. (Rhamnaceae)
B.	**Host plants of specific importance**	
	1.	*Summer brood preservers*
		Galwang – *Albizia lucida* Benth. (Leguminosae: Mimosoideae)
		Jharkhali, Khunia – *Ficus cunia* Buch.-Ham. (Urticaceae) [syn. *Ficus conglomerata* Roxb.]
		Pakur, Pilkhan – *F. lacor* Buch.-Ham. (Urticaceae) [syn. *Ficus infectoria* Roxb. non Willd.]
		Pipal – *F. religiosa* Linn. (Urticaceae)
		Sandan – *Ougeinia dalbergioides* Benth. (*Sandan* – Leguminosae: Papilionatae) [=*Ougeinia oojeinensis* (Roxb.) Ohashe (Leguminosae)] [syn. *Dalbergia oojeinensis* Roxb. and *O. dalbergioides* Benth. (*Ougeinia oojeinensis*)]
	2.	*Alternate kusmi hosts*
		Khair – *Acacia catechu* Willd. (Leguminosae: Mimosoideae)
		Galwang – *Albizia lucida* Benth. (Leguminosae: Mimosoideae)
	3.	*Winter brood preserver*
		Bhalia – *Flemingia macrophylla* (Willd.) (O. Ktze ex Merrill) (Papilionaceae/Fabaceae)
		Semialata – *F. semialata* Roxb. (Papilionaceae/Fabaceae)
C.	**Host plants of regional importance**	
		Babool – *Acacia arabica* Willd. (Leguminosae: Mimosoideae)
		Bhawal – *Grewia didyma* or *laevigata* (Tiliaceae) [=*G. glabra* Bl. *G. disperma* Rottl.]
		Pansaura – *G. serrulata* DC. (Tiliaceae) [syn. *Grewia multiflora* Mast., F.B.I., non Juss.]
		Leea aspera Wall. Non Edg. (Vitaceae)
		Banchalta – *L. crispa* Linn. (Vitaceae)
		Galeni, Gangma – *L. robusta* Roxb. (Vitaceae)
		Jalari – *Shorea talura* Roxb. (Dipterocarpaceae)
		Ghont – *Ziziphus xylopyrus* Willd. (Rhamnaceae)
D.	**Host plants for plantation use**	
		Galwang – *Albizia lucida* Benth. (Leguminosae: Mimosoideae)
		Arhar – *Cajanus cajan* (Linn.) Millsp. (Leguminosae: Papilionatae). [Syn. *C. indicus* Spr.]
		Bhalia – *Flemingia macrophylla* (Willd.) O. Ktze ex Merrill (Papilionaceae/Fabaceae)
		Semialata – *Flemingia semialata* Roxb. (Papilionaceae/Fabaceae)
		Ber – *Ziziphus mauritiana* Lam. (Rhamnaceae)

Table 6.4 Spacing requirement and lac yield from major lac host plants

Host plant	Spacing (m)	Plant density/ha	Lac yield	
			Kg/plant	q/ha/year
A. Tree hosts				
Palas (*B. monosperma*)	3.6 × 3.6	772	2.50	9.65
Kusum (*S. oleosa*)	12 × 12	70	50.00	14.00
B. Tree hosts to be trained into bushes				
Ber (*Z. mauritiana*)	4 × 4	625	5.0	31.25
Rain tree (*A. saman*)	8 × 8	156	8.0	12.48
Galwang (*A. lucida*)	1.8 × 1.8	3086	0.80	12.34
Akashmani (*Acacia auriculiformis*)	2 × 1.8	2778	0.8	11.09
C. Bushy hosts				
Bhalia (*F. macrophylla*)	1.2 × 1.2	6944	0.25	8.68
Semialata (*F. semialata*)	1 × 1	10,000	0.40	20.0

in the literature of kusum and palas that are botanically hardly distinguishable, yet biologically easily separable by means of preference of the lac insect.

Presently, the number of host plants of the Indian lac insect, *Kerria lacca* (Kerr.), is 129 in the Indian region, of which 19 are good-quality lac host plants of commercial and other specific importance. The good host plants are classified based on the findings of a series of experiments conducted on them towards various aspects of lac cultivation.

6.4.1 Host Plant Management for Lac Cultivation

Lac host plants are the important biotic associates of the lac insects, as these insects draw their nourishment from the plants for their survival and growth. Though the insects prefer specific plants, they have shown different degrees of preference for other host plant species also. Due to scattered nature of occurrence, their management has been a challenging problem to the lac growers. The main emphasis has been towards the qualitative production of lac by investigating the optimum condition on which lac insect and their host plants interact to produce the lac of commerce.

6.4.1.1 Crop Geometry

The spacing requirement (Table 6.4) for the major lac hosts has been recommended as 3.6 × 3.6 m for palas, 12.0 × 12.0 m for *kusum* and 4.0 × 4.0 m for *ber*, *khair* and *ghont*. Equilateral triangular system of planting has also been recommended for plantation raising of these host plants to promote sufficient light and air for better growth of the lac insects. Double hedge system of planting for *bhalia* has been found best, accommodating 8000–10,000 plants/ha with a yield of 10–12 q/ha.

Management of lac crop is essential as yield of lac is influenced by various biotic and abiotic factors including edaphic factors. Management practices vary from crop to crop and host tree species on which lac culture is raised. It will be convenient and more meaningful if management strategy is explained with respect to the lac cultivation operations.

6.5 Lac Production Management

6.5.1 Availability of Succulent and Tender Shoots

Pruning of trees is an essential step for lac cultivation in order to have sufficient number of tender, succulent and healthy shoots for feeding at the time of infestation of trees with lac insect. The first instar nymph is very delicate and requires tender shoot for feeding on phloem sap as it is unable to insert proboscis in older and hard-barked shoots. Dead, cracked and broken branches, which are not suitable, are removed during pruning that also helps the tree to maintain its vigour. Pruning time and techniques for various trees have been standardised.

The trees are pruned with axe or *dauli* in such a way that (a) volume of the newly grown host tree crown is not reduced and (b) one can get access to lac-encrusted shoot at the time of harvesting. Normally, the twigs more than 2.5 cm in diameter should not be cut and those less than 1.5 cm in diameter should be cut from their point of origin. The twigs of 1.5–2.5 cm in diameter should be cut, leaving 1–1.5 ft. length from point of origin. Shoots of *Flemingia* spp. are cut 4–6 in. above the ground. However, when trees are pruned for the first time, thick branches may also be pruned to give a desired shape to the crown and make the branches accessible to crop harvesting. During lac cultivation, harvesting normally serves the purpose of pruning also.

6.5.2 Pruning Time *vis-a-vis* Crop Harvesting

Based on the lac crop maturity and time of pruning, the lac host trees can be divided into two groups: (a) tree species whose pruning time and lac crop maturity coincide, for example, *S. oleosa*, *Flemingia* spp., *Albizia* spp., etc., and (b) whose pruning time does not coincide with lac crop maturity as in the case of *B. monosperma.* Pruning of *S. oleosa* is recommended either in January–February or in June–July when crop also matures. In case of *B. monosperma*, pruning is recommended either in February or in April–May depending on the lac crop to be raised. However, *rangeeni* lac crop does not mature during these months. In such cases, lac crop is harvested in April–May to coincide it with time of pruning, or self-inoculation is practised to get broodlac in October–November.

6.5.3 Partial Pruning to Enhance Survival of Lac Insect During Summer

For protection against high temperature during summer, the palas tree having lac crop is partially pruned in the month of February. This induces early sprouting of leaves which protects lac culture against direct heat of the sun during summer season. Pruning also helps increase chances of lac insect survival in summer season as sap availability to lac insect is more in succulent shoots. Similarly, partial defoliation of palas tree at the time of inoculation in October–November particularly of unpruned trees helps in reducing the lac crop loss caused by leaf fall in February–March.

6.5.4 Inoculation of Host Tree with Broodlac

Inoculation of host trees is done by placing the broodlac (lac stick with mature gravid living mother cell) on host trees, so as to facilitate emerging crawlers to settle on shoots of the host plant. Leaving a part of mature lac crop on the tree at the time of harvesting so that the lac insect swarms from left over encrustation settles on the same tree is called **self inoculation**. Whereas, process of inoculation using broodlac harvested from other tree is referred to as **artificial inoculation**.

About 20–25 g of broodlac per metre of succulent shoots is required. However, for summer crop, lesser quantity of broodlac is utilised to avoid excessive load of lac insects to the host plants which are already stressed due to adverse environment. Broodlac is sorted out before inoculation to select out sticks with sparse settlement and dead and virgin lac cells. Selected broodlac sticks are kept preferably in bamboo basket for allowing free circulation of air to the living broodlac. Rejected broodlac sticks are scraped. The sorted broodlac is made into bundles by cutting the broodlac sticks into small pieces of 6–8″. Bundling is usually done in the afternoon when nymphal emergence has slowed down. Normally, for *kusmi* broodlac bundles weighing around 100 g each are made and for *rangeeni* broodlac bundles of 50 g weight are preferred as being light in weight its volume is more. Broodlac bundles should be placed in bags made from 60 mesh nylon nets to prevent emergence of pests from the broodlac. Small pieces of broodlac can be used by using net bag and then tying onto the host tree. The broodlac is tied at lower/inside of canopy for summer crop and outer of canopy for winter crop. This is so done to avoid heat mortality, and during rainy season sufficient aeration is not available inside the canopy which may lead to fungal infection especially on the host plant with dense foliage.

After 4–5 days of inoculation, the tender shoots are observed for settlement of lac nymphs. If adequate settlement has happened, the bundles may be removed and tied at other branches where lac nymphs have not settled in sufficient numbers. This helps in the uniform distribution of lac crawlers on tree and optimum utilisation of broodlac.

6.5.5 Removal of Used-Up Broodlac (*Phunki*) Sticks

After complete emergence of lac nymphs from broodlac, the used-up bundles (*phunki* bundles) are removed from the tree to minimise attack of predators and parasitoids to new lac crop and to avoid wastage of lac detaching from *phunki* sticks on drying.

The used-up broodlac harbours large number of insect predators and parasitoids that emerge over a period and affect the new crop; therefore, the *phunki* lac is removed as soon as emergence of lac nymphs is over (not more than 3 weeks after infestation) and is scraped and disposed of.

6.5.6 Management of Lac Crop

Lac crop under field condition is monitored regularly for possible infestation of enemy insects, predatory animals and fungi. Pest management practices depend on crop season, host tree and also environmental conditions.

- Healthy broodlac free from enemy insect infestation is used for inoculation/raising of new crops.
- Inoculated broodlac is kept on the plant for a period till larval emergence is complete and ordinarily not beyond 3 weeks from the date of inoculation.
- Sixty mesh synthetic net bags (27 × 12 cm) are used as brood containers for inoculation to trap the enemy insects. This practice is especially advisable when lac is to be introduced for the first time in a new locality.
- About 10 % of trees are inoculated heavily with broodlac (twice the normal rate of broodlac) to attract/act as trap for enemy insects. Such trees are harvested before crop maturity or are sprayed with recommended insecticides.
- Self-colonisation more than once of lac culture should be avoided.
- Harvested broodlac in excess of requirement, rejected lac and all *phunki* lac (used-up broodlac sticks) should be scraped and dried at the earliest before emergence of pests.
- Since natural population of beneficial parasitoids (parasitoids of lac predators) of lac insect is low, insect sorting device developed by IINRG should be used to retrieve beneficial insects from rejected broodlac sticks or *phunki* lac sticks or lac obtained from heavily inoculated trees.
- Cultivation of the *kusmi* strain of lac in predominantly *rangeeni* areas and vice versa should be avoided to break the life cycle of enemy insects.

6.5.7 Management of Lac Crop Harvesting

The lac crop is harvested either at crop maturity or in immature stage from a tree along with host twigs (Fig. 6.5a, b) depending upon the season, host used and purpose of utilisation. Maturity of the crop is assessed by post-embryonic

Fig. 6.5 Broodlac harvesting at crop maturity. (**a**) Broodlac being harvested from a tree and (**b**) broodlac stick

developmental stages of lac nymphs inside the mother cell. Crop is harvested when young ones are ready to hatch. However, the immature lac can be cut at any time. Harvesting of immature lac is avoided as it results in shortage of broodlac for raising the next crop. The immature harvested lac is scraped and sold.

6.5.8 Coupe System (Grouping) of Host Trees

The broodlac is harvested with secateurs to minimise loss of lac which normally gets separated from the twigs if harvested with conventional axe or *dauli*. However, if shoots are thick as in case of fast-growing host plants like *Z. mauritiana*, *A. lucida*, etc., conventional tools may be used. Normally, complete harvesting is done, but in special cases, some of the lac encrustation is left on the host for self-inoculation and the next crop is harvested completely. Harvested twigs with broodlac are also sorted and transported in bamboo basket for proper aeration before being used for raising the next crop.

The continuous exploitation of host tree for lac cultivation leads to qualitative and quantitative loss of output as a result of (a) reduction in vigour of the host tree due to sucking of sap by the insects and (b) higher incidence of insect predators and parasitoids of lac insects due to continuous self-colonisation of the host. To overcome these problems, trees are divided into sets or groups and used in rotation to give rest to the trees. Dividing of host plants into different coupes largely depends on the habit of the host species and type of insect used.

Basically, host plants can be divided into two categories (a) on which both winter/rainy and summer season crop can be cultivated, for example, *S. oleosa*, *B. monosperma*, *A. lucida*, etc. and (b) on which only one crop can be raised under normal conditions like *F. semialata*, *A. catechu*, etc. Some of the trees which are suitable for both crops are utilised as brood preservers during summer season, for example, *Ficus* and *Albizia* spp.

Table 6.5 Lac cultivation schedule on *kusum* – *Schleichera oleosa* (*five-coupe system*)

Year	Month	Set I	Set II	Set III	Set IV	Set V
First	January–February	Pruning				
	June–July		Pruning			
Second	January–February			Pruning		
	June–July	Inoculation			Pruning	
Third	January–February	Partial ➔ harvesting	Inoculation			Pruning
	June–July	Harvesting and pruning	Partial➔ harvesting	Inoculation		
Fourth	January–February		Harvesting and pruning	Partial➔ harvesting	Inoculation	
	June–July			Harvesting and pruning	Partial➔ harvesting	Inoculation
Fifth	January–February	Inoculation ←	←	←	Harvesting and pruning	Partial harvesting
	June–July	Partial harvesting	Inoculation ←	←	←	Harvesting and pruning

6.5.8.1 Host Trees Which Can Be Utilised for Both Crops Without Alternation with Other Species

6.5.8.1.1 *Butea monosperma* (Palas)

This plant species is utilised for *rangeeni* lac cultivation and is suited for both seasons, *viz.* summer (October/November–June/July) and rainy season (June/July–October/November) crop. The available trees are divided into three groups with equal numbers in each. Two sets of trees are utilised for broodlac and the remaining one for immature *baisakhi* (*ari*) crop production. The healthy trees are utilised for broodlac production, and trees on which there is a possibility of lac culture mortality due to direct exposure to high temperature during summer (normally located on rocks or where groundwater level is very low) are utilised either for immature *baisakhi* crop or *katki* crop only.

6.5.8.1.2 *Schleichera oleosa* (Kusum)

The trees of *S. oleosa* are slow growing, and the wait period between crop harvesting and re-inoculation varies from tree to tree ranging between 12 and 18 months depending upon its pruning response. Trees are divided into four or five sets when one-and-a-half-year rest is required and three sets when only 1-year rest is given (Table 6.5).

Fig. 6.6 Lac cultivation on plantation scale on *Flemingia semialata*. (**a**) Plantation of *F. semialata* for intensive lac cultivation and (**b**) lac crop on *F. semialata*

6.5.8.2 Lac Host Plants Which Require Another Species for Alternation of Crop

6.5.8.2.1 *Flemingia semialata* (Semialata)

This plant of about 6′ height can be suitably utilised for winter crop of *kusmi* (Fig. 6.6a, b) and alternated with *S. oleosa* or *A. lucida*.

6.5.8.2.2 *Acacia catechu* (Khair)

The species may be used for raising crops of both strains during winter/rainy season, but winter (*aghani*) crop of *kusmi* lac in alternation with *S. oleosa* or *A. lucida* gives better results.

6.5.8.2.3 *Ziziphus mauritiana* (Ber) for *Kusmi* Lac Cultivation

For *kusmi* lac cultivation, trees of *Ziziphus mauritiana* are utilised for winter season crop in alternation with *S. oleosa* for summer crop.

6.5.8.2.4 *Ziziphus mauritiana* (Ber) for *Rangeeni* Lac Cultivation

The trees of *Butea monosperma* (palas) are suitable for both summer and rainy seasons. Normally, it is utilised for producing broodlac while *Z. mauritiana* (ber) for immature summer crop *baisakhi* (*ari*) that is harvested in April–May. The trees are not divided into coupes.

6.5.9 Forecast of Nymphal Emergence and Crop Harvesting

If lac crop is harvested well before the crop maturity, it directly affects the quality of broodlac because the food supply is cut off to the developing embryo in the mother lac insect. As a result the emerging nymphs either remain very weak and die soon or are underdeveloped. Similarly, if broodlac is harvested after the nymphal emergence begins, especially in summer season, the emergence takes place so quickly that the growers do not have sufficient time to infest other trees, and bulk of nymphs are wasted. Moreover, treatment of broodlac with insecticide for control of enemy insects also becomes impossible, as the emerged nymphs will die. The knowledge of crop maturity and forecast of larval emergence, therefore, have got special importance in lac cultivation. The common methods used for forecast of nymphal emergence are (a) appearance of granulated material in haemolymph of mother cell, (b) appearance of cracks on lac encrustation, (c) appearance of yellow spot and (d) stage of embryonic development.

6.6 Pests and Diseases

The losses in lac cultivation due to various insect pests and diseases (Fig. 6.7) are known to be far greater than what is usually met in other agricultural crops. A number of major and minor pests at times almost destroy the lac crop, thereby not only reducing the yield drastically but also affecting the quality of the lac.

6.6.1 Predators

About 22 predators have been reported to be closely associated with lac insects, of which three are major predators, viz. *Eublemma amabilis* Moore, *Pseudohypatopa pulverea* Meyr. and *Chrysopa* spp. The first two cause on an average 35–40 % damage to lac crop. These lepidopterous predators cut a hole in the lac and feed on the insect from inside by making a tunnel. *Chrysopa*, though a sporadic predator, sometimes causes havoc particularly in the culture of *kusmi* strain.

(a) (b) (c)

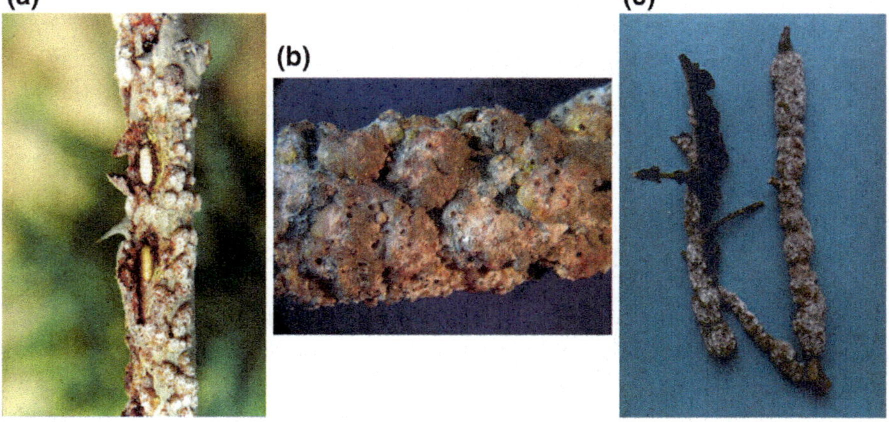

Fig. 6.7 Infested lac culture: (**a**) larval and pupal stages of predators inside the lac encrustation, (**b**) parasite-infected lac and (**c**) sooty mould on lac

Table 6.6 Important insect pests of lac ecosystem

Sr. no.	Insect	Family
A. *Predators*		
	Eublemma amabilis Moore	Noctuidae
	Pseudohypatopa pulverea Meyr.	Blastobasidae
	Chrysopa spp.	Chrysopidae
B. *Primary parasitoids*		
	Aprostocetus (syn. *Tetrastichus*) *purpureus* Cam.	Eulophidae
	Coccophagus tschirchii Mahd.	Aphelinidae
	Erencyrtus dewitzi Mahd.	Encyrtidae
	Eupelmus tachardiae How.	Eupelmidae
	Marietta javensis How. (=*M. leoperdina*)	Aphelinidae
	Parechthrodryinus clavicornis Cam.	Encyrtidae
	Tachardiaephagus tachardiae How.	Encyrtidae
	T. somervillei Mahd.	Encyrtidae
C *Secondary parasitoids*		
	Apanteles tachardiae Cam.	Braconidae
	A. fakhrulhajiae Mahd.	Braconidae
	Aphrastobracon flavipennis Ashm.	Braconidae
	Brachymeria tachardiae Cam.	Chalcididae
	Chelonella cyclopyra Franz.	Braconidae
	Elasmus albomaculatus Gahan	Elasmidae
	E. claripennis Cam.	Elasmidae
	Eurytoma pallidiscapus Cam.	Eurytomidae
	Perisierola (=*Goniozus*) *pulveriae* Kuerten	Bethylidae
	Pristomerus sulci Mahd. and Kolub.	Ichneumonidae
	Trichogrammatoidea nana Zehnt.	Trichogrammatidae

6.6.2 Inimical Parasitoids

Thirty different parasitoids of lac insect have been reported by Varshney (1976). They lay eggs into the lac cell through the anal tubercle in/on the body of lac insect. The larva that hatches feeds only on lac insect. The extent of parasitisation and relative/seasonal abundance of parasites associated with lac insect have been studied by various workers. Important predators and parasitoids are listed in Table 6.6.

Of all the parasites associated with lac insect, eight parasites, namely, *Coccophagus tschirchii*, *Erencyrtus dewitzi*, *Eupelmus tachardiae*, *Parechthrodryinus clavicornis*, *Tachardiaephagus tachardiae*, *Marietta javensis*, *T. somervillei* and *Aprostocetus purpureus*, are of regular occurrence in the lac ecosystem. Amongst these *Tachardiaephagus tachardiae* and *A. purpureus* are the most abundant lac-associated parasites.

6.6.3 Diseases

In addition to the damage caused by insect pests, lac crop yield suffers significant losses due to other biotic agents particularly fungi. The earliest recorded diseases are that of black mould species of *Capnodium* and *Fumago* caused by honeydew that drips from colonies of lac insects on the twigs of host trees (Lindsay and Harlow 1921). Avoidable losses due to fungi alone have been observed to be 40.9–59.8 % in *kusmi* strain of lac insect (Das et al. 1986). Similarly, significant reduction (75.05–88.41 %) in lac yield has been observed in *kusmi* strain of lac insect (Mishra et al. 1997). The presence of pathogenic fungi, *Pythium* sp., in female cells causes a heavy mortality in the nymphs which fail to eclose satisfactorily and lie dead in clusters within the female resinous cell. Two species of fungi, namely, *Conidiocarpus* (syn. *Podoxyphium conidioxyphium*) and *Polychaeton* sp., causing 30–40 % damage to lac insect in Vietnam and 11 species of saprophytic pathogenic fungi causing dark mildew on lac insect have been reported from China (Shaoji 1993). Recently, three species of fungi belonging to family Eurotiaceae/Aspergillaceae, viz. *Aspergillus awamori* Nakazawa, *Aspergillus terricola* Marchal and *Penicillium citrinum* Thom (syn. *P. aurifluum* Biourge), causing severe damage to lac culture have been reported. Fungal infection in lac cultivation causes losses in lac yield by (a) inhibiting respiration, (b) hindering mating process, (c) blocking larval emergence and (d) affecting lac host plant efficiency.

6.7 Pest Management

6.7.1 Cultural Control

6.7.1.1 Early Inoculation

Inoculation of the lac crop earlier by 10–15 days than the normal crop has been found to have higher incidence of predators. This technique can be exploited in combination with heavy inoculation (i.e. higher brood rate) of host plants.

6.7.1.2 Trap Crop

Taking advantage of the egg laying behaviour of the predators in highly dense population, some of the host plants are inoculated with higher brood rate to attract the predators for egg laying. These plants serve as trap crops and have a greater incidence of predators and can be harvested as *ari* lac (immature lac).

6.7.1.3 Intercropping

The larval ectoparasites *Bracon greeni* and *Elasmus* sp. have been reported to parasitise pink and spotted bollworm larvae attacking cotton. Intercropping of cotton and okra with lac can increase the population of natural parasitoids common to the crops.

6.7.2 Mechanical Control

Utilisation of 60 mesh synthetic net container bags for inoculation of broodlac: The emerging lac nymphs easily crawl out from the minute pores and settle on the twigs of the lac host plants, while the emerging adult enemy predator moths get entrapped within the net. This innovation can check the egg laying by the adult moths on the new crop and reduce pest infestation.

6.7.3 Chemical Control

Initially, chemical control was restricted to fumigation of scraped lac with carbon bisulphide as preventive measure only, but at a later stage, different kinds of insecticides have been tried. More than 35 insecticides belonging to various groups have been studied of which endosulfan was found to be safer against the lac insects. It is the safest to the first instar lac nymphs which is of prime importance for field application. It has also been found to be highly detrimental to the larval stages of the two predators. 0.05 % endosulfan has been identified as the most effective dose without any adverse effect on the economic attributes of the lac insect. Dipping of broodlac in 0.05 % emulsion of endosulfan prior to inoculation results in a highly significant suppression of the predatory population. Similarly, 0.03 % dichlorvos has been found to be very effective in the management of *Chrysopa* spp. as it is highly effective against all the larval stages. Recently, 0.02 % etofenprox has been found to be very effective against *Chrysopa* spp.

Experiments carried out with the third-generation insecticide diflubenzuron (Dimilin) have exhibited the desired selectivity. The chitin inhibitor at 0.05 % concentration has also been recommended for effective management of the lac insect predators.

6.7.4 Microbial Control

Effective control of *E. amabilis* and *P. pulverea* was achieved by the use of a commercial preparation of the biopesticide *Bacillus thuringiensis* Berliner (Thuricide W.P.) under field conditions (Malhotra and Choudhary 1968).

6.7.5 Biological Control (Sharma et al. 1999)

Various workers from different parts of the country have recorded over 35 parasitoids of the lac predators. Amongst the lepidopterous predators, *P. pulverea* and *E. amabilis* have 14 and 9 parasitoids, respectively, and four are common to both, whereas *Chrysopa* spp. have seven parasitoids. It has been recorded that nine parasitoids are most important and promising for biocontrol of the predators under field conditions. Two ant predators, *viz. Camponotus compressus* and *Solenopsis geminata rufa*, have been found to be attacking the larvae and pupae of both *E. amabilis* and *P. pulverea*.

Of late, several egg parasitoids have been evaluated for the management of lac insect predators of which *Trichogramma pretiosum*, *T. chilonis*, *T. brasiliensis*, *Trichogrammatoidea bactrae* and *Telenomus remus* hold great promise for biocontrol of the predators.

Integrated pest management is possible by the integration of the following techniques:

1. Mechanical control with 60 mesh synthetic nets as brood container
2. Cultural control with (a) trap crop by higher brood rate and (b) varying the date of inoculation
3. Chemical control by applying endosulfan/dichlorvos/ethofenprox spray

The combination of the above components has been found to reduce the populations of natural enemies of lac insect to the extent of 60–65 % and increased yield of 140–175 % over control.

6.8 Conclusion

Most of the lac host trees are confined in forest areas, thus making them the storehouse of the lac insect diversity. Cultivation of lac not only provides livelihood to millions of lac growers but also helps in conserving the vast stretches of forests, lac insects and associated biota as most of the lac host plants are in forest areas, and farmers resist felling of these trees and protect them for lac cultivation. Though classified as minor forest produce, on certain considerations, lac might rank as a potential source of revenue. Growing lac host plants for timber and fuel yields revenue in cycles of long years, whereas lac cultivation on these trees gives a return almost every year. Thus, lac growers give more importance to regular income from cultivation of lac over the years to one-time income from timber or fuel. Thereby, lac culture plays a vital role in the protection of our bioresources. Lac also provides sustained high economic returns, generates employment opportunities and has potential to pave a strong foundation for lac-based rural cottage industries. About one million man-days are generated in the existing lac processing factories. With increasing demand for natural products, e.g. in fruit and vegetable coating and as food colour, the time is ripe to introduce lac culture in farming system and on idle lac host trees in the forests. It is evident from the foregoing account that promoting and encouraging lac culture will not only check environmental degradation and help rebuild the ecological balance but also conserve endangered lac insects and associated fauna and flora for posterity.

References

Campbell MH, Holtcamp RH, McCormick LH, Wykes PJ, Donaldson JF, Gullan PJ, Gillespie PS (1994) Biological control of native shrubs, *Cassinia* spp. using native scale insects, *Austrotachardia* and *Paratachardina* sp. (Hemiptera: Kerriidae) in New South Wales. Plant Protect Quart 9:64–68

Chauhan NS, Mishra YD (1977) A new colour locus in *Kerria lacca* (Kerr). Curr Sci 46:272–273

Chauhan NS, Teotia TPS (1973) A new variant in the lac insect, *Kerria lacca* (Kerr). Entomol Newslett 3(5):33–34

Das BB (1990) Present status of entomological research on lac in India and future strategies. Part 1 Present status of knowledge. Indian Shellac (Annual Number):9–16

Das BB, Naqvi AH, Choudhary SG, Sen AK (1986) Survey of pathogenic microorganisms and their effects on development of lac insects. Annual progress report, Indian Lac Research Institute, 1986:24–25

Jaiswal AK, Sharma KK (2011) Lac insect, its development and life cycle, pp. 22–26. In: Sharma KK, Ramani R (eds) Recent advances in lac culture. ILRI, Ranchi, pp i–x, 1–319

Lindsay HAF, Harlow CM (1920 (Pub 1921)) Report on lac and shellac. Indian For Rec 8 (Part 1):162 pp

Malhotra CP, Choudhary SG (1968) Control of *Eublemma amabilis* moore (Noctuidae: Lepidoptera) and *Holcocera pulverea* Meyr (blastobasidae; Lepidoptera), predators of the lac insect *Kerria lacca* (Kerr) by *Bacillus thuringiensis* Berliner. J Invertebr Pathol 11(3):429–439

Mishra YD, Bhattacharya A, Sushil SN (1997) Effect of some systemic fungicides on the nymphs of Indian lac insect, *Kerria lacca* (Kerr), for their protection against fungal infection. J Ent Res 21(3):291–293

Mishra YD, Sushil SN (2000) A new trivoltine species of genus *Kerria* Targioni-Tozzetti (Homoptera: Tachardiidae) thriving on *Schleichera oleosa* (lour.) Oken from eastern India. Orient Insects 34:215–220

Mishra YD, Sushil SN, Bhattacharya A, Kumar S, Mallick A, Sharma KK (1999) Intraspecific variation in host plants affecting productivity of Indian lac insect, *Kerria lacca* (Kerr). J Non-Timber For Prod 6(3/4):114–117

Mishra YD, Sushil SN, Kumar S, Bhattacharya A (2000) Variability of lac productivity and related attributes of *Kerria* spp. (Homoptera : Tachardiidae) on ber (*Ziziphus mauritiana*). J Entomol Res 24(1):19–26

Ramani R, Sharma KK (1991) A review of some genetical aspects of lac insects. Ann Entomol 9(2):47–53

Ramani R, Sharma KK (2011) Lac Insect Genetics, pp 52–56. In: Sharma KK, Ramani R (eds) Recent advances in lac culture. IINRG, Ranchi, i–x + 319 pp

Ramani R, Sharma KK (2016) Lac insects, pp 53–65. In: Sharma KK, Monobrullah Md, Mohanasundaram A, Ramani R (eds) Beneficial insect farming: benefits and livelihood generation. ICAR-Indian Institute of Natural Resins & Gums, Ranchi, v + 194 pp (ISBN 978–93–5254-624-4)

Roonwal ML, Raizada MB, Chatterjee RN, Singh B (1958) Descriptive account of the host plants of the lac insect, Laccifer lacca (Kerr) and the allied plants in the Indian Region (Part 1 & 2). Indian Lac Cess Committee, Ranchi

Shaoji G (1993) An investigation on the pathogens of both lac insects and their host plants. For Res 6(6):711–713

Sharma KK, Jaiswal AK, Kumar KK (1999) Biological control in lac cultivation – limitations and prospects. J Insect Sci 12(2):95–99

Sharma KK, Jaiswal AK, Kumar KK (2001) New record of fungi associated with Indian lac insect, *Kerria lacca* (Kerr). Indian J Entomol 63(3):369–371

Sharma KK, Jaiswal AK, Kumar KK (2006) Role of lac culture in biodiversity conservation: issues at stake and conservation strategy. Curr Sci 91(7):894–898

Sharma KK, Ramani R (2014) To understand the nature of diversity in lac insects of *Kerria* spp. in India and the nature of insect x host interaction. Final report of NAIP sub-project, 136 pp. Indian Institute of Natural Resins and Gums, Ranchi

Sharma KK, Ramani R, Mishra YD (1997) An additional list of the host plants of lac insects, *Kerria* spp. (Tachardidae: Homoptera). *J*. Non-Timber For Prod 4(3/4):151–155

Srinivasan MM (1956) Lac, its cultivation and manufacture in India, Chapter IV – host plants of the lac insect. Indian Forester 82(4):180–193

Srivastava SC (2011) Lac host plants – current status and distribution. In: Sharma KK, Ramani R (eds) Recent advances in lac culture. ILRI, Ranchi, pp. i–x, pp 73–82, 1–319

Varshney RK (1968) Further data on host-plants of lac insects (Tachardiidae: Homoptera). J Bombay Nat Hist Soc 65(1):249–251

Varshney RK (1976) A check list of insect parasites associated with lac. Orient Insects 10(1):55–78

Varshney RK (1977) Taxonomic studies on lac insects of India. Oriental Insects Suppl 5 (1976), 97 pp

Varshney RK, Teotia TPS (1967) A supplementary list of the host plants of lac insects. J Bombay Nat Hist Soc 64(3):488–511

Lac Crop Harvesting and Processing 7

K.K. Sharma

Contents

7.1 Introduction

Lac production involves pruning, inoculation, *phunki* removal, harvesting and lac scraping (Mukhopadhyay and Muthana 1962). Farmers perform these operations manually using traditional equipments. The traditional equipments need modification to increase the efficiency and ultimately the production.

7.2 Improved Lac Cultivation

7.2.1 Pruning

Lac insect thrives best on tender shoots rather than on old and woody ones. In order to provide a suitable ground for the insect to feed well and thrive upon, the host plants must be receptive and sustainable. For young plants no particular preparation is required to receive their first infection since there is an abundance of tender shoots.

K.K. Sharma (✉)
Lac Production Division, Indian Institute of Natural Resins and Gums, Namkum, Ranchi 834010, Jharkhand, India
e-mail: kewalkks@gmail.com

Omkar (ed.), *Industrial Entomology*, DOI 10.1007/978-981-10-3304-9_7

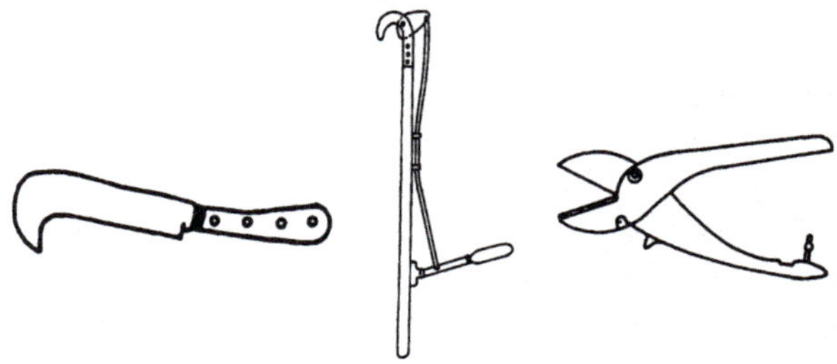

Fig. 7.1 Pruning instruments

Fig. 7.2 (**a**) Lac-bearing branches and (**b**) raw lac, seedlac and shellac

For older plants, however, a process of pruning is required prior to infection in order to stimulate the production of fresh and succulent branches. Ordinarily, branches less than an inch in diameter should only be pruned. The majority of farmers do pruning with axe, and this may explain to some extent the thickness of the branches cut. Proper pruning cannot be done with an axe. If branches of the thickness recommended are cut with axes, they will either break or split. Some farmers use sharp-edged knife for pruning which is better than the axe for pruning, but this also lacks perfection.

The ideal pruning instruments (Prasad 2011a, 2015) are long-handle tree pruner and secateurs (Fig. 7.1). The long-handle tree pruner is more suitable for pruning. This type of pruner is simple in construction and could easily be made by village blacksmith if he has a sample to copy. There are two types of secateurs – roll-cut secateurs and the French secateurs. The roll-cut secateurs are better and easier to use.

7.2.2 Inoculation

The method by which the lac insects are introduced onto a lac host plant is known as inoculation. It involves the simple process of cutting lac-bearing twigs (Fig. 7.2a) from an infected tree a few days before the emergence of the nymphs. A bundle of such twigs known as broodlac is tied to an uninfected tree on which tender new shoots are plentiful. Nymphs emerge from the broodlac which settle down on young branches of the tree. In existing practice, inoculation is done manually by climbing on the tree on suitable branches. Thus, the process is slow and laborious. A broodlac placement tool has been developed by the Indian Institute of Natural Resins and Gums (IINRG), Ranchi, to make the process faster and less cumbersome.

7.2.3 *Phunki* Removal

Broodlac left on the tree after the emergence of nymphs is popularly known as *phunki* lac. When broodlac is allowed to remain on trees even after the emergence of nymphs, harmful insects carried along with broodlac might invade new crop. Therefore, timely removal of *phunki* is necessary to prevent carryover of pests to new crops. In existing practice, *phunki* is removed from the tree mostly manually which again requires climbing on the tree. Some farmers use inverted J-shaped cutting hook mounted on a bamboo pole for pulling down the bundles from ground level, thus avoiding the climbing on the tree. As bundles fall on the ground, lac encrustation gets separated from sticks partially due to impact from the ground and is likely to be lost. A *phunki* hook with bundle collection arrangement has been developed by the IINRG.

7.2.4 Lac Harvesting

The removal of mature lac-bearing branch from the tree by cutting is known as lac harvesting. The farmers harvest crop with the help of an axe or a sharp-edged knife. The use of axe for harvesting is unsuitable as it leads to cracking or splitting of the branches. Further, to use either axe or knife, the farmer has to climb on the tree which makes harvesting process slow and laborious. Secateurs and long-handle tree pruning are better equipments for harvesting lac crop.

7.2.5 Lac Scraping

Lac scraping involves removal of lac encrustation from lac stick. Farmers scrape lac from lac sticks using traditional knives. The process is very tedious and slow. It involves sitting on the ground in a group and scraping by means of special type of knives. Lot of impurities such as sand, dirt, stick pieces, etc. also find their way with scraped/raw lac (Fig. 7.2b). The process has been mechanised by developing the following machines (Prasad and Nath 2016) for better efficiency.

7.2.5.1 Hand-Operated Disc-Type Lac Scraping Machine

The machine scrapes 5 kg lac sticks in an hour and separates about 93.7 % dry lac from lac stick. One person is required to operate the machine.

7.2.5.2 Hand-Operated Roller-Type Lac Scraping Machine

Scraping rollers are the main components of machine and comprise of two corrugated mild steel rollers each having a diameter of 125 and 200 mm long. One of the rollers is fixed and another one is spring loaded and thus adjustable. The rollers rotate in opposite direction at a speed differential between them.

7.2.5.3 Motor-Operated Lac Scraping-Cum-Grading Machine

The machine is capable of scraping a large amount of lac in a short time from the lac sticks of varying diameters. Further, machine also does crushing and grading.

7.3 Processing of Lac

The process of purification of lac (Goswami and Sarkar 2010a; Pandey et al. 2015; Prasad 2011b), in general, consists of two steps. The first step involves washing of the sticklac or raw lac to remove sand, woodchips, etc., followed by drying of the resin. This semi-refined form is called seedlac (Fig. 7.2b). The pure resin, shellac (Fig. 7.2b), is then obtained by hot filtration, either in country (*bhatta*) process or in mechanised factories. Shellac obtained/processed in mechanised factories is called machine-made shellac.

Sticklac is crushed, washed and dried (Fig. 7.3) to obtain seedlac that is further refined to get shellac either as flakes or in the shape of a button. Other refined products are bleached lac, dewaxed decolourised lac, etc. Making of the seedlac involves the following steps.

7.3.1 Crushing and Screening

Dried sticklac is crushed and sieved through 8–12 mesh wire net to obtain smaller particles. The clods bigger than the 8–12 mesh size are crushed again till they become of desired size.

Fig. 7.3 Processing machines for refining lac

(a) (b)

Fig. 7.4 Washing of sticklac (**a**) manual and (**b**) mechanical

7.3.2 Washing

Washing is done in the following ways:

7.3.2.1 Manual Washing

In manual washing (Fig. 7.4a), cup-shaped stone or cemented vats called *nands* are used. These vats are dug into a cemented floor such that the upper level of *nand* is 3–4″ above the ground level. A cemented rectangular water tank of suitable capacity is also made just beside the vat. Now, crushed lac is put into the vat and sufficient water is added to moisten the charge completely. Then, a worker gets into the vat to stamp and rub the moist lac with his feet against its rough wall for 10–15 min. After rubbing over, plenty of water is poured into the vat and the lac briskly churned by hand. This causes the lac grains to settle at the bottom, while the lighter particles made up of uncrushed lac cells (*ghongi*), wood particles and other extraneous matters float onto the surface. This is taken out with the help of either sieve or a piece of cloth. The waste water is removed by filtering through a cloth.

7.3.2.2 Mechanical Washing

Washing (Fig. 7.4b) is carried out in horizontal stationary barrel fitted with an axel on which some blades are fixed. The barrels are closed at both ends and have three openings on its carved surface, of which one is at the top for charging lac and water and another vertically below which is closed with a lid. The third one is in a horizontal row in midway between the top and the bottom hole and is fitted with fine wire net of either 60 or 80 mesh through which washed water is discharged out. There is a cistern below, which is sufficiently large enough to hold the entire charge plus the water.

7.3.2.3 Drying

After washing, lac grains known as seedlac (*chauri*), are dried by spreading them on a cemented platform, preferably in the shade. The wet seedlac is spread in very thin layers and turned over from time to time with wooden rake or hoe or by feet.

7.3.2.4 Winnowing

Dried seedlac is winnowed to separate out the particles of sand, wood, etc. (collectively called *agila*). After winnowing, it is sieved in different grain sizes. The different grain sizes are blended or mixed in desired proportions to make various commercial grades. *Molamma* or *retti* obtained in the process is either sold separately or used along with seedlac for making inferior quality of shellac.

7.3.2.5 Grading

The seedlac is graded into different categories according to its colour and content of alcohol insolubles. IS specifications (which are also the basis for ISO standards) are available, but yet, the lac put into the market is mostly collected from widely scattered small production units and blended. Its quality is, therefore, not consistent and varies widely.

Both the yield and the quality of seedlac obtained from sticklac depend upon (a) the host plant on which lac is cultivated, (b) season of cultivation, (c) type of sticklac such as *ari* or *phunki* and (d) amount of impurities and method and extent of washing.

7.3.3 Manufacture of Shellac

The seedlac contains impurities embedded in the resin which have to be removed. The principle involved is to melt the seedlac by application of heat and filter the molten mass through filter cloth. The filtered mass is converted into shellac which is in thin flakes or in the shape of button.

7.3.3.1 Handmade Shellac

The seedlac or a mixture of seedlac and *molamma* depending upon the quality of shellac required is packed into a cloth bag about 40 ft long 2.5 in. in diameter. The bag is heated portion-wise but uniformly by rotating slowly in front of a wood charcoal fire burning in a specially designed oven known as *bhatta* (Fig. 7.5). The bag is made of thick woven markin cloth.

7.3.3.2 Machine-Made Shellac

Machine-made shellac may be obtained by two processes: (1) heat process and (2) solvent process. In heat process (Fig. 5b), shellac is manufactured from seedlac, whereas in solvent process, the raw material may be *kiri* or sticklac or seedlac.

7.4 Chemistry and Composition of Lac

Lac is obtained from the tree either as *ari* (immature lac) or *phunki* (used-up broodlac), *i.e.* either before or after completion of the insect's life cycle. Lac is scraped off the lac-bearing branches with knife to obtain raw lac also called scraped lac or sticklac.

(a) (b)

Fig. 7.5 (**a**) *Bhatta* process of refining lac. (**b**) Machine-made shellac

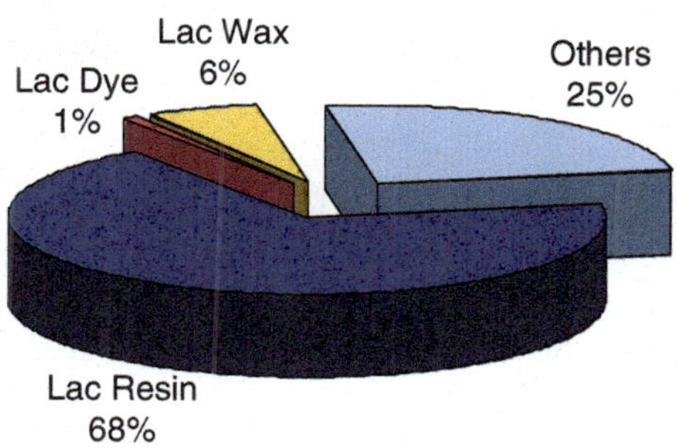

Fig. 7.6 Composition of lac

7.4.1 Composition of Lac

Raw lac contains resin, wax and lac dye (Bose et al. 1963) besides other impurities such as wood particles, sand, insect bodies, etc. (Fig. 7.6).

7.4.1.1 Lac Resin

It is a polyester type of material, comprising basically of long-chain and sesquiterpenic fatty acids. With a molecular weight ranging from 660 to 1000, the term *oligomer* is better suited for the resin. It is unique in the sense that even with its low molecular weight, it has the property of forming films on a wide variety of surfaces; these films possess outstanding adhesion, gloss, scratch hardness and resistance to petrochemical solvents. The resin can also be hydrolysed to its component acids – some of these acids serve as important starting materials in the synthesis of a wide

Fig. 7.7 Lac dye

Fig. 7.8 Lac wax

variety of compounds for the industry. The resin can be moulded into various shapes and amazingly, it has adhesive properties also. It is highly inflammable but a good electrical insulator. Indeed, it may be difficult to find another material, natural or synthetic, exhibiting such a wide range of utilities.

7.4.1.2 Lac Dye

It is a polyhydroxy anthraquinone derivative (Srivastava et al. 2010) and has brilliance and fastness, especially when used for dying silk and wool. In the early nineteenth century, lac dye (Fig. 7.7) (like indigo) was more widely used than lac resin. Lac wax, though not as versatile as carnauba wax, is an important component in many formulations for floor polishing, cosmetics, etc.

7.4.1.3 Lac Wax

It is a constituent of natural resin lac. It is secreted by the lac insect along with the resin, in the form of thin white filaments. It is generally found to the extent of nearly 4–5 % in seedlac, 3–5 % in shellac and slightly higher percentage in sticklac (5–6 %). Lac wax (Fig. 7.8) is recovered directly during processing of sticklac and also during the preparation of solvent-based dewaxed shellac generally called shellac wax.

7.5 Quality of Lac

Raw lac is not of uniform quality and not used as such and is to be subjected to a variety of refining processes to obtain a more purified end product. Some of the important quality parameters of lac are:

7.5.1 Colour Index

The colour index is determined comparing the colour of a standard solution of iodine with a solution of the sample in ethyl alcohol by diluting the sample solution progressively with alcohol until a close match is obtained. In case of seedlac, colour index ranges from 8 to 30. In case of hand-made shellac, the colour index ranges from 6 to 30. In case of machine-made shellac, the colour index varies from 8 to 50. In case of dewaxed shellac, the value ranges from 9 to 50, and in case of dewaxed decolourised shellac, it ranges from 0.9 to 5.0.

7.5.2 Flow

The method consists of melting a sample of ground shellac/seedlac (20 mesh/40 mesh) in a test tube vertically for 3 min in a thermostatically maintained bath at 100 ± 1 °C and tilting the test tube to an angle of 15° for 12 min in order to permit the sample to flow down the tube and measuring the distance in millimetre. The average flow of *rangeeni* varieties ranges from 35 to 55 mm and *kusmi* varieties ranges from 60 to 100 mm.

7.5.3 Life (Polymerisation Time)

The method consists of heating the sample of seedlac/shellac under specified condition in a test tube at 150 ± 1 °C and observing the time required for it to attain a rubbery state as indicated by the "spring back" of glass rod when twisted through full circle in the molten resin. The average life of *rangeeni* varieties ranges from 30 to 50 min, and those of *kusmi* varieties ranges from 45 to 70 min.

7.5.4 Bleach Index or Bleach Ability Test

The number of millilitres of bleach liquor with 3 % available chlorine required to bleach 30 g of seedlac in Na_2CO_3 is the bleach index of seedlac. When all the bleach liquor is consumed, the colour of the sample is to be matched with the standard iodine solution in a colorimeter.

7.5.5 Wax

Wax is a natural constituent of lac resin, varying from about 2.5–5.5 % depending on the type of lac.

7.5.6 Ash Content

A known weight of sample is heated on a silica or platinum crucible at a low heat, not exceeding dull redness, until free from carbon, and then finally heating the crucible in a muffle furnace at a temperature of 650–750 °C for 4 h. The amount of ash is computed as a percentage on the lac taken for test. In case of hand-made shellac, the ash percentage ranges from 0.5 to 1.0 %. In case of machine-made shellac, it should not exceed 0.3 %.

7.5.7 Water-Soluble Matter

A known mass of powdered sample is digested with water, made up to a known volume and filtered. The matter that goes into the solution is determined by evaporating an aliquot portion of the filtrate to constant mass and calculating for the whole solution. In case of seedlac, it should not exceed 1 %. In case of hand-made/machine-made shellac, it should not exceed 0.5 %.

7.6 Applications of Lac

Lac resin and lac dye have been used since time immemorial, and there are various references to such usages in literature (Goswami and Sarkar 2010b). Over a period of time, these uses have either changed or disappeared altogether (Sarkar 2011). However, due to its biodegradable, nontoxic, tasteless, ecofriendly and self-sustaining nature, coupled with the rising worldwide interest in "green" technologies, lac and its by-products have staged a comeback.

Lac resin is a polyester type of material, comprising basically of long-chain and sesquiterpenic fatty acids. Similarly, lac dye, which is a polyhydroxy anthraquinone derivative, has brilliance and fastness, especially when used for dying silk and wool.

7.6.1 Lac in the Past

Perhaps, one of the most infamous references of lac in the past is that of the *lakshagriha* in the epic, Mahabharata. An inflammable mansion, cunningly constructed for the *Kauravas* by the architect *Purochan*, had the sole purpose of vanquishing the *Pandavas* by burning them down. The highly inflammable and moulding property of lac was utilized in this particular instance. Before the advent of synthetic plastics, lac was an important component in the moulding industry. Due to its outstanding

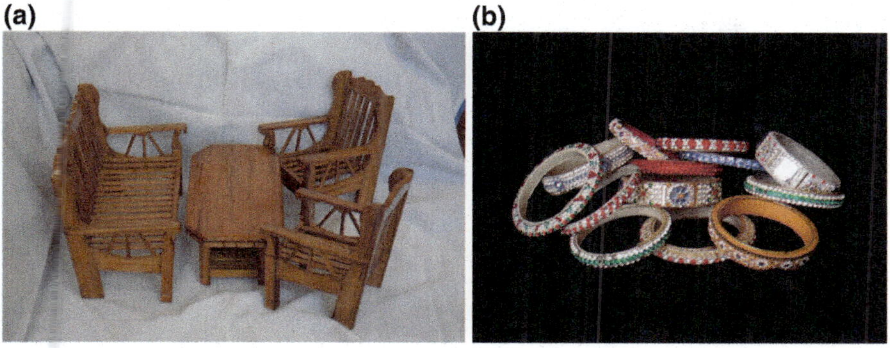

Fig. 7.9 (a) Lac varnish coating on wood and (b) shellac used in bangles

surface coating properties, the largest consumption of lac was in coating compositions for wood (Fig. 7.9a), metal, leather, etc. Traditionally, lac has been used for metal-lacquering ornamental brassware for protection against tarnishing and also in bangle making (Fig. 7.9b) and coating of earthenware.

7.6.2 Lac in the Present

Lac-based surface coating compositions were subsequently modified to meet the new challenges of the market. Shellac has been combined with nitrocellulose, urea, thiourea, etc. for improving its drying characteristics, brushability, etc. Lac-melamine (formaldehyde compositions termed as Melfolac) has proved to be an excellent varnish for the wood varnishing industry due to its improved gloss, water resistance and drying characteristics. More recent compositions are even better since they are non-spirit based and can be applied by spray also. Lac-based etch primers, antifouling paints (for the shipping industry) and strippable formulations (for the packaging industry) are also available.

Several improved insulating varnishes (baking and air-drying types) have been developed for coating of armatures/coils of electric motors, transformers, etc. Similarly, compositions for internal and external coating of tin cans used for the packaging industry have been formulated. Lac-based coating compositions are available for enteric coating of medicinal tablets (Fig. 7.10a), glazing of coffee beans and chocolates, etc. In such compositions, lac is used either as a solution or in the form of an emulsion. Lac functions as a moisture barrier for protecting the core ingredient; it also controls disintegration and serves as a finishing coat over wax, prior to printing of the trademark on the medicinal tablet. Lac dye is used for dying of silken and woollen clothes (Fig. 7.10b).

There have been concerted efforts, with some degree of success, to use lac as a binder resin in the manufacture of particle boards, as an extender in FRP (fibre-reinforced plastic) sheets and as an ecofriendly alternative to acrylic emulsions for coating cement/masonry surfaces.

Aleuritic acid, the major component acid of lac resin, has proved to be an important starting material for the synthesis of a wide variety of fine chemicals and

(a) (b)

Fig. 7.10 (**a**) Lac formulation coating on medicines and fruits; (**b**) lac dye is used for dying silken and woollen clothes

bioactive agents, including perfumery compounds, plant growth regulators, fungicidal agents, mosquito repellents, insect pheromones, etc.

7.6.3 Lac-Based Cottage Industries

Lac has tremendous potential for application in various areas which are important for mankind. It has got some nonconventional applications where synthetic resin cannot compete. It provides livelihood to a large number of people, and as such efforts need to be made to create awareness about its utilisation.

Lac as such has got utilisation in various areas, but the modification of its functional groups with some chemicals and synthetic resins has resulted in its innumerable utilisation in the area of surface coating, adhesive, electrical, food, leather, pharmaceutical and cosmetic industries.

7.6.4 Surface Coating Industries

7.6.4.1 Heat- and Waterproof French Polish (Melfolac)

A French polish composition based on shellac and synthetic resin, popularly known as Melfolac, has been developed which may be applied either by brush or spray. The varnish is not only heat- and waterproof, but also it gives a very glossy and attractive finish on wooden furnitures, radio and TV cabinets, musical instruments, etc. When pigmented, Melfolac produces quick-drying, waterproof paints suitable for application on shop windows, display panels, kitchen furniture, etc.

7.6.4.2 Picture Varnish

Pictures and paintings need protection against dust, abrasions and humid atmosphere for their long durability. Lac-based varnish composition which can be sprayed has been developed for protection of pictures and paintings.

7.6.4.3 Book Varnish

An insecticidal varnish based on lac has been developed especially to be used for binding of books and protecting them from damage caused by insects.

7.6.4.4 Water-Soluble Lac

Lac is not water soluble but it can be dissolved in water in the presence of some chemicals. The aqueous lac solution is used for coating earthen pots to prevent porosity, imparting glaze to surface and preventing against humidity. It may also be used by small-scale industry entrepreneurs in leather finishes and in photoengraving, replacing the imported glue. This is also suitable for making emulsion paints and inks.

7.6.4.5 Water-Thinned Shellac Red Oxide Primer

The modern trend in the world of paint industry is to develop as far as possible water-based primer. It has utility, particularly, in automobile industry for being economic and free from fire and health hazards. The notable features of these primers are (a) outstanding adhesion to steel surface; (b) durable adhesion to oil paints, synthetic enamels and nitrocellulose lacquers; and (c) application by any conventional method.

7.6.4.6 Shellac Etch Primer

An anticorrosive primer for the exposed surface of metals like aluminium, zinc, etc., has been developed which possesses outstanding adhesion property and gives smooth adherent film with hard mat appearance.

7.6.4.7 Fruit and Vegetable Coating

It is a very fast-expanding application. Lac wax/shellac formulations are used for coating fruits, vegetables, etc. to increase their shelf life and extend nutritional value.

7.6.5 Adhesive Industries

7.6.5.1 Gasket Shellac Compound

Shellac possesses excellent resistance to petroleum and hydrocarbon solvents, and a shellac-based composition named gasket shellac compound is used for the repair and maintenance of automobile engines.

7.6.5.2 Sealing Wax

Shellac-based sealing wax formulations possess excellent adhesion and gloss properties along with superb ability to protect and retain delicately engraved seals.

7.6.5.3 Lac Glue

A glue composition of seedlac/shellac and hydrolysed lac is suitable for pasting paper to paper and paper to other surfaces such as glass, metal, plastic wood and cloth.

7.6.5.4 Particle Board

Particle board especially for use as partition wall or false ceiling has been developed by using refuse lac and waste material of plants.

7.6.6 Electrical Industries

7.6.6.1 Insulating Varnish

Shellac possesses good electrical properties like dielectric strength, anti-surface carbonisation and arc resistance apart from its ability to adhere to a variety of surfaces. Both air-drying and baking-type insulating varnishes for coating of coils/armatures of electric motors and transformers and also for manufacture of laminated products have been developed.

7.6.7 Making Seedlac, Shellac and By-Products

The lac obtained after cultivation is processed for removal of impurities to get semi-refined product (seedlac) and refined product (shellac). A processing unit on small scale may be established to get the profitable price by sale of processed material.

7.6.7.1 Lac Dye

A water-soluble natural dye known as lac dye is obtained from the wash water of lac factory. About 150 tons of lac dye may be recovered from the wash water obtained from approximately 20,000 tons of sticklac processed annually. It has potential for dyeing wool and silk cloths. It gives different attractive shades with different mordants fast to light and wash. Moreover, the pure lac dye being natural and nontoxic has got demand in food industries for making soft drink and in pill coating, sausage finishing, confectionery and chocolate coating.

7.6.7.2 Lac Wax

Lac contains about 4–5 % wax which is a mixture of higher alcohol (in major), acids and hydrocarbons. Lac wax is a high-quality wax and possesses comparable properties with carnauba wax in regard to hardness and solvent retention. Lac wax is used for making tailor's chalk, crayons, electrical potting compound, lipstick and formulations for other cosmetic products.

7.7 Miscellaneous Applications

7.7.1 Making of Bangles

A number of small-scale industries for making of attractive bangles of lac are running in different parts of the country. It has got a wide scope for its extension. Due to its moulding and adhesive properties, lac finds applications in making of jewellery and ornaments from valuable coloured stones.

7.7.2 Other Uses

It is also used for polishing and sharpening stones, making of impact and water-resistant coal block, coating of fertiliser for slow release and proofing and stiffening of hat making (Fig. 7.11).

Fig. 7.11 Composition and applications of lac (Agarwal et al. 1998)

References

Agarwal SC, Kumar P, Goswami DN (1998) Safari of ILRI through seven decades. Indian Lac Research Institute, Ranchi, 84 pp

Bose PK, Sankaranarayanan Y, Sen Gupta SC (1963) Chemistry of lac. Indian Lac Research Institute, Ranchi, 225 pp

Goswami DN, Sarkar PC (2010a) Processing of lac. In: Baboo B, Goswami DN (eds) Processing, chemistry and applications of lac. Directorate of Information and Publications of Agriculture, Indian Council, Indian Council of Agricultural Research, New Delhi, pp 12–28

Goswami DN, Sarkar PC (2010b) Applications of lac. In: Baboo B, Goswami DN (eds) Processing, chemistry and applications of lac. Directorate of Information and Publications of Agriculture, Indian Council, Indian Council of Agricultural Research, New Delhi, pp 118–153

Mukhopadhayay B, Muthana MS (1962) A monograph on lac. Indian Lac Research Institute, Ranchi

Pandey SK, Prasad N, Sharma SC (2015) Processing of lac – seedlac, shellac, bleached lac and lac dye. In: Kumar A, Jaiswal AK, Singh AK, Yogi RK (eds) Advances in lac production, processing, product development and value addition. ICAR-Indian Institute of Natural Resins and Gums, Ranchi, pp 118–130

Prasad KM (2011a) Technology of lac processing. In: Sharma KK, Ramani R (eds) Recent advances in lac culture. Indian Institute of Natural Resins and Gums, Ranchi, pp 237–251

Prasad N (2011b) Improved equipments for lac cultivation. In: Sharma KK, Ramani R (eds) Recent advances in lac culture. Indian Institute of Natural Resins and Gums, Ranchi, pp 200–204

Prasad N (2015) Implements for lac cultivation. In: Kumar Alok, Jaiswal AK, Singh AK, Yogi RK (eds) Advances in lac production, processing, product development and value addition. ICAR-Indian Institute of Natural Resins and Gums, Ranchi, pp 140–147

Prasad N, Nath Alok (2016) Processing of natural resins and gums – an overview. In: Sharma KK, Monobrullah Md, Mohanasundaram A, Ramani R (eds) Beneficial insect farming-benefits and livelihood generation. Indian Institute of Natural Resins and Gums, Ranchi, pp 120–132

Sarkar PC (2011) Applications of lac – past, present and emerging trends. In: Sharma KK, Ramani R (eds) Recent advances in lac culture. Indian Institute of Natural Resins and Gums, Ranchi, pp 252–258

Srivastava S, Ansari MF, Goswami DN (2010) Chemistry of lac dye and wax. In: Baboo B, Goswami DN (eds) Processing, chemistry and applications of lac. Directorate of Information and Publications of Agriculture, Indian Council, Indian Council of Agricultural Research, New Delhi, pp 43–61

Mulberry Sericulture

8

T.P.S. Chauhan and Mukesh K. Tayal

Contents

8.1 Introduction

The word sericulture is derived from the Greek word 'sericos' meaning silk and the English word 'culture' meaning rearing. Sericulture is a unique combination of art and science of raising silkworms which form a protective covering called cocoon. Cocoons are boiled to get continuous filament called the silk. Sericulture is a science which deals with various aspects of silkworms. It is an agro-based rural cottage industry, and the end product of which is silk. There are three main branches of sericulture industry; moriculture (mulberry cultivation) deals with host plant cultivation, silkworm rearing that covers rearing of silkworms for cocoon production

T.P.S. Chauhan
Ex-Scientist, Central Silk Board, 21-A, Deshmeshpuri, Indira Gandhi Marg, Niranjanpur, Dehradun 248171, Uttarakhand, India
e-mail: tpschauhan.1956@gmail.com

M.K. Tayal (✉)
Regional Sericultural Research Station, Central Silk Board, Miransahib, Jammu 181101, Jammu and Kashmir, India
e-mail: drmukeshtayal@gmail.com

Omkar (ed.), *Industrial Entomology*, DOI 10.1007/978-981-10-3304-9_8

and reeling of cocoons or reeling industry which deals with raw silk production (Krishnaswami et al. 1987a).

Silk is the most elegant textile fibre in the world. Silk, the 'queen of textiles', has unparallel quality of natural sheen, inherent affinity for dyes, high absorbance, light weight, soft touch and high durability. Silk is nature's gift to mankind and a commercial fibre of animal origin other than wool. Being an ecofriendly and biodegradable material; silk has assumed special relevance in present age. Promotion of sericulture can help in ecosystem development as well as high economic returns.

Silk industry provides livelihood opportunity for millions owing to high employment generation, low capital intensive and remunerative nature of its production. The industry with its rural-based on-farm and off-farm activities and enormous employment potential has attracted the policymakers to recognize the industry among one of the most appropriate avenues for socio-economic development of a largely agrarian economy like India.

The world raw silk production reached to 178057.62 MT in 2014–2015 (Source: International Silk Association, net-generated information). China led in the world raw silk production with 1,46,000 MT and 82.00 % of the produce. India is the second largest producer of silk with 28,708 MT in 2014–2015 in the world and has 16.12 % share in global raw silk production. All the countries except China and India have been witnessing a declining trend in raw silk production in the last two decades (Tazima 1991; Machii 2015; Shetty et al. 2015).

8.1.1 Sericulture in Indian Perspective

India has a rich and complex history in silk production, and its silk trade dates back to the fifteenth century. Sericulture industry provides employment to approximately 8.03 million persons in rural and semiurban areas in India. A sizeable number of workers belong to the economically weaker sections of the society, including the women. Women constitute over 60 % of those employed in downstream activities of sericulture in the country. This is possible because sericulture activities starting from mulberry garden management, leaf harvesting and silkworm rearing are more effectively taken up by the women folk. Even the silk-reeling industry including weaving is largely supported by them.

The Indian traditional and cultural affinities in the domestic market and geographic specificity have helped the industry to achieve a leading position in the country. India has the unique distinction of being the only country producing all the five known commercial silks, namely, mulberry silk, tropical tasar silk, temperate/oak tasar silk, eri silk and muga silk, of which muga silk with its golden-yellow glitter is unique and prerogative of India.

Mulberry sericulture is mainly practised in five states, namely, Karnataka, Andhra Pradesh, West Bengal, Tamil Nadu and Jammu and Kashmir, which jointly account for about 97 % of the total mulberry silk production in the country. Indian sericulture industry has provided employment to about 80.3 lakh persons and has mulberry plantation to the tune of 2.2 lakh hectare during 2014–2015.

Table 8.1 Raw silk production in India during 2014–2015

Raw silk production (MT)							
Mulberry raw silk			Non-mulberry raw silk				Total raw silk
Bivoltine	Cross-breed (multi.x bi.)	Total mulberry raw silk	Tasar	Eri	Muga	Total non-mulberry raw silk	
3870	17,520	21,390	2434	4726	158	7318	28,708

The total raw silk production has been 28,708 MT during 2014–2015, out of which 21,390 MT belong to mulberry silk, while the remaining 7318 MT aggregated to tasar, eri and muga silks (Table 8.1). India has a target to produce 32,000 MT raw silk by 2016–2017.

India has additional requirement of around 5000 MT of silk (particularly bivoltine mulberry silk of international standard) which is imported mainly from China. The Indian silk goods are being exported to the traditional major markets, like the USA, European countries and small markets of Asia. The silk goods export earnings have been Rs. 2829.87 crore during 2014–2015. Central Silk Board, a union government organization in India, is popularizing 'Silk Mark' certification for purity of silk products through the Silk Mark Organisation of India (SMOI). Silk Mark is an assurance label that protects the interests of consumers, who are being cheated by traders by selling spurious products in the name of silk.

8.1.2 Mulberry Silkworm

The insect producing mulberry silk is a domesticated variety of silkworms, which has been exploited for over thousands of years. All strains reared at present belong to the species *Bombyx mori* L. that is believed to be derived from *Bombyx mandarina* Moore. China is the native place of this silkworm, but now it has been introduced in all the silk-producing countries, like Japan, India, Korea, Italy, France and Russia.

The races of mulberry silkworm may be identified on the basis of geographical distribution as Japanese, Chinese, European or Indian origin; or as univoltine, bivoltine or multivoltine depending upon the number of generations produced in a year under natural conditions; or as tri-, tetra- and penta-moulters according to the number of moults that occur during larval growth; or as pure strain and hybrid variety according to genetic recombination.

8.1.3 Taxonomy of Silkworm *Bombyx mori* L.

Mulberry silkworm, *Bombyx mori* L., belongs to the order Lepidoptera of the class Insecta. There are different kinds of silkworms under superfamily Bombycoidea which has eight families. All wild silkworms belong to the family Saturniidae, while mulberry silkworm, *Bombyx mori* L., belongs to the family Bombycidae.

Phylum Arthropoda
Class Insecta
Order Lepidoptera
Family Bombycidae

There are several species of silkworms that are being used in commercial silk production. These are:

Mulberry silkworm
Bombyx mori L. (Bombycidae)

Tasar silkworm

Antheraea mylitta Drury (Saturniidae)
Antheraea pernyi G.M. (Saturniidae)
Antheraea yamamai G.M. (Saturniidae)
Antheraea royeli M.r. (Saturniidae)
Antheraea proylei Jolly (Saturniidae)

Muga silkworm

Antheraea assamensis W.w. (Saturniidae)
Eri silkworm

Philosamia ricini Boisduval (Saturniidae) (=*Samia cynthia ricini*)

8.1.4 History of Silk

'Silk' has a long colourful history unknown to most people. For centuries, the West knew very little about silk. Historical evidences suggest that silk was discovered in China about 2700 BCE. Chinese kept the secret of silk altogether to themselves for more than 2000 years. Chinese legend has stated that Princess Si-ling-chi, wife of great Prince Hoang-ti, discovered the means of raising silkworms and reeling the silk from cocoons for making of silk garments. Princess Si-ling-chi was later honoured with the name Seine-Than or 'The Goddess of Silkworms'. Sericulture spread to other parts of the world from China, and silk became a precious commodity highly sought by other countries (Chang 1960; Krishnaswami et al. 1987a; Tazima 1991).

Many evidences of silk have been found at the sites of Yangshao culture in Xia County, Shanxi Province, in Northern China, suggesting the origin of silk much earlier than 2700 BCE. A half-cut silk cocoon was found from the site, dating back to between 2600 and 2300 BCE. The species was identified as *Bombyx mori*, the domesticated silkworm. Fragments of primitive loom were also unearthed from the sites of Hemudu culture in Yuyao, Zhejiang, dated to about 4000 BCE. The earliest evidence of silk fabric was found during 3630 BCE, where the fabric was used for wrapping the body of a child. The fabric was discovered from a Yangshao site in Qingtaicun at Rongyang, Henan. Scraps of silk were found in a Liangzhu culture site at Qianshanyang in Huzhou, Zhejiang, dating back to 2700 BCE. Other fragments have been recovered from royal tombs in the Shang Dynasty (Chang 1960; Kuhn 1982; Nunome 1992).

8.1.4.1 Silk Development in China

With the discovery of silk in China, it was reserved exclusively for the ruler. It was permitted only to the emperor, his close relations and the very highest of his dignitaries. Gradually the various classes of society began wearing tunics of silk, and silk came into more general use like clothing, decoration, in musical instruments, bonds of all kinds and even rag paper, the world's first luxury paper. Eventually even the common people were able to wear garments of silk (Chang 1960; Boulnois 1966).

During the Han Dynasty, silk ceased to be a mere industrial material and became an absolute value in itself. Farmers paid their taxes in grain and silk (Boulnois 1966). Silk began to be used for paying civil servants and rewarding subjects for outstanding services. Values were calculated in lengths of silk as they had been calculated in pounds of gold. This use of silk continued during the Tang Dynasty as well. It is possible that this added importance was the result of a major increase in production.

8.1.4.2 Secret of Silk Out to the World

During the later epoch, the Chinese lost their secret to the Koreans, the Japanese and, later, to the Indians. Scholars believe that starting in the second century BCE, the Chinese established a commercial network aimed at exporting silk to the West. Silk was used, for example, by the Persian court and its king, Darius III, when Alexander the Great conquered the empire. Even though silk spread rapidly across Eurasia, with the possible exception of Japan, its production remained exclusively Chinese for three millennia.

In 139 BCE, the world's longest highway was opened and stretched from Eastern China to the Mediterranean Sea. This road was the historically famous 'Silk Road', named after its most important commodity. The sericulture reached Korea around 200 BCE, when waves of Chinese immigrants arrived there. Silk reached to West through a number of different channels (Boulnois 1966; Barber 1991). Shortly after 300 CE, sericulture travelled westward from China. The emperor Justinian gained the secrets of sericulture for the Roman Empire in 522 AD, with the smuggling of the silkworm eggs from China by Persian monks. The monopoly of China on sericulture was broken, and silk import from China became smaller and smaller. In 877 AD, the rebel chief Biachu captured Canfu, the centre of foreign silk trade, put to death all its inhabitants and destroyed all of the mulberry trees and silkworms. These actions stopped foreign commerce in China for more than 60 years. However, by this time, silk production was so well established in Western Asia and Eastern

Europe that this wholesale destruction hardly affected the price of silk in the rest of the world (Boulnois 1966).

Around 440 CE, a prince of Khotan (present – Hetian), married a Chinese princess. The princess smuggled out silkworm eggs by hiding them in her voluminous hairpiece. The Khotan kept the secret of silk production too. Silk cultivation spread to Japan around 300 CE. The Byzantines managed to obtain silkworm eggs by 522 CE and were able to begin silkworm cultivation. The Arabs also began to manufacture silk during this same time. As a result of the spread of sericulture, Chinese silk exports became less important, although they still maintained dominance over the luxury silk market (Boulnois 1966).

8.1.4.3 Innovation in Sericulture Industry

The Industrial Revolution changed much of Europe's silk industry. Due to innovations in spinning cotton, cotton became much cheaper to manufacture and therefore caused more expensive silk production to become less mainstream. New weaving technologies and introduction of Jacquard loom, however, increased the efficiency of silk production. The silk industry in the Italian state of Lucca was on its peak in the eleventh and twelth centuries. The cities of Lucca, Genoa, Venice and Florence were exporting silk to all of Europe. In 1472 there were 84 workshops and at least 7000 craftsmen in Florence alone (Boulnois 1966; Krishnaswami et al. 1987a; Tazima 1991).

During the eighteenth and nineteenth centuries, Europeans produced several major advancements in silk production. By the eighteenth century, England led Europe in silk manufacturing because of English innovations in the textile industry. These innovations included improved silk-weaving looms, power looms and roller printing. In 1801, a Frenchman named Joseph Jacquard exhibited his new machine for silk weaving, which gradually spread through the industry (Krishnaswami et al. 1987a; Tazima 1991).

An epidemic of several silkworm diseases caused production to fall, especially in France, where the industry never recovered. The great French scientist, Louis Pasteur, rescued the silk industry in 1870 by showing that the then epidemic Pebrine disease of silkworms could be controlled by prevention through simple microscopic examination of adult moths (Krishnaswami et al. 1987a). These advances set the trend for a more mechanized and scientific approach to silk production than existed previously. Sericulture has also been attempted in the USA, but these endeavours have been sporadic and largely unsuccessful.

8.1.4.4 Origin of Indian Silk

There is belief that a Chinese princess got married to a Tibet king in around 140 BCE who brought silkworm egg hiding in her hairpin. The silk production might have been reached to India through Brahmaputra and the Ganga Valley. Indian historians believe that silk was a known textile fibre in India even earlier than 140 BCE on the skirts of Mount Everest and gradually spread in warmer places. It is also said that during the period of Emperor Kanishka, there was trade among India, Rome and Greece (Watt 1893; Mukharjee 1899, 1907).

Silk in India was considered a symbol of royalty and prestige, a 'pure fabric' used for all religious, ritual and ceremonial occasions. The first Indian silk fabric was produced as early as 1725 BCE from non-mulberry silkworm (tasar, eri and muga). The silk trade flourished in India during the medieval period. Silk manufactured in Kashmir and Bengal was exported to Europe in the fourteenth and fifteenth centuries (Nanavati 1965). Silk industry in Assam flourished during Ahom's kings in the sixteenth century (Varadarajan 1988).

In India, silk fabrics were made in the cottage industrial sector. Mughal emperors were very fond of silk cloths and patronized the industry in Bengal and Kashmir. The first silk textile mill, on modern lines, was started in the year 1832 by the East India Company at Haora. Later on silk factories were also started in Karnataka and Kashmir during 1845 and1892, respectively.

The industry suffered a setback between 1875 and 1915 due to the occurrence of the Pebrine disease and loss of the raw silkworm crops. However, it got boost up after the tariff protection granted in 1934. During the period of the First World War in 1922, the Indian silk industry saw many ups and downs till 1948, the period of the Second World War (Krishnaswami et al. 1987a). After independence, there has been significant increase in the production of silk textiles in the country.

8.1.5 Silk Road

8.1.5.1 Silk Road and Han Dynasty of China

In the second century BCE to the end of the fourteenth century CE, a great trade route originated from Chang'an (now Xian) in the East and ended at the Mediterranean in the West, linking China with the Roman Empire. The Silk Road was a network of trade from East to West, formally established during the Han Dynasty of China. The Silk Road was opened up by Zhang Qian in the Western Han Dynasty, and the routes were gradually formed throughout the Han Dynasty (Nanavati 1965; Tazima 1991).

The Silk Road was not a single trade route from East to West; hence, historians termed this as 'Silk Routes', though 'Silk Road' remained the more common and recognized name even today. Both the terms for this network of roads were coined by the German geographer and traveller, Ferdinand von Richthofen, in 1877 CE, who designated them 'Seidenstrasse' (silk road) or 'Seidenstrassen' (silk routes). This ancient route not only circulated goods but also exchanged the splendid cultures of China, India, Persia, Arabia, Greece and Rome. The Silk Road was used regularly from 130 BCE, when the Han officially opened trade with the West until the Ottoman Empire boycotted trade with the West and closed the routes during 1453 CE (Nanavati 1965; Krishnaswami et al. 1987a).

8.1.5.2 Silk Road in the Tang (618–907) and Yuan (1271–1368) Dynasties of China

This trade route gradually grew up in Han dynasty, but reached to its most prosperous stage in history during the Tang Dynasty (618–907). Before the Anshi Rebellion

(755–762) in the Tang Dynasty, this world-famous road experienced its 'golden age' of development from the business point of view.

During the Yuan Dynasty, the Silk Road experienced its last flourishing period. Along with the growth of the Mongolian Empire and the establishment of the Yuan Dynasty, the route regained its vigour and became prosperous once again. It enjoyed the last glorious era during this period. In 1271, the great Mongolian ruler Kublai Khan established a powerful Mongol Empire – Yuan Dynasty (1271–1368) – at Dadu (the present Beijing). The territory of the giant empire was the largest one in Chinese history, which stretched as far as Mongolia and Siberia in the north, South China Sea in the south, Tibet and Yunnan in the south-west, Stanovoi Range (Outer Khingan) and Okhotsk in the north-east and Xinjiang and Central Asia in the north-west. Even West Asia and Russia were under the control of this empire.

The Mongolian emperors welcomed the travellers of the West with open arms and appointed some foreigners on high positions. Kublai Khan gave Marco Polo a hospitable welcome and appointed him a high post in his court. At that time, the Mongolian emperor issued a special VIP passport known as 'golden tablet' which entitled holders to receive food, horses and guides throughout the Khan's dominion. The holders were able to travel freely and carried out trade between East and West directly in the realm of the Mongol Empire. Although maritime transport had an influence on the route, many Westerners, Chinese envoys and caravans travelled along this ancient trade route. However, the historically important route could not contend with expansion in the field of navigation which assisted its demise.

8.1.5.3 Persian Royal Road

The Persian Royal Road, which served as one of the main arteries of the Silk Road, was established during the Achaemenid Empire (500–330 BCE). The Persian Royal Road ran from Susa, in North Persia (modern-day Iran), to the Mediterranean Sea in Asia Minor (modern-day Turkey) and featured postal stations along the route with fresh horses for envoys to quickly deliver messages throughout the empire. The Persians maintained the Royal Road carefully and expanded it through smaller side roads. These paths eventually crossed down into the Indian subcontinent, across Mesopotamia and over into Egypt.

Alexander the Great conquered the Persians and established the city of Alexandria Eschate in 339 BCE in the Fergana Valley of Neb (modern Tajikstan). Leaving behind his wounded veterans in the city, Alexander moved on. In time, these Macedonian warriors intermarried with the indigenous populace creating the Greco-Bactrian culture which flourished under the Seleucid Empire following Alexander's death. Under the Greco-Bactrian king Euthydemus I (260–195 BCE), the Greco-Bactrians had extended their holdings. According to the Greek historian Strabo (63–24 CE), the Greeks 'extended their empire as far as the Seres'. 'Seres' was the name by which the Greeks and Romans knew China, meaning 'the land where silk came from'. It is thought, then, that the first contact between China and the West came around the year 200 BCE.

In 138 BCE, Emperor Wu, in the Han Dynasty of China, sent his emissary Zhang Qian to the West for help in trade. Zhang Qian's expedition led him into contact with many different cultures and civilizations in Central Asia and, among them, those whom he designated the 'Dayuan', the 'Great Ionians', who were the Greco-Bactrians that descended from Alexander the Great's army. This journey of Zhang Qian inspired Emperor Wu to speculate on what else might be gained through trade with the West, and the Silk Road was opened in 130 BCE. During 138–129 BCE, Mesopotamia came under Parthian rule and, with it, came control of the Silk Road. The Parthians then became the central intermediaries between China and the West.

8.1.5.4 Trade Links Along the Silk Road

The Silk Road stretched from China through India, Asia Minor, up throughout Mesopotamia, to Egypt, the African continent, Greece, Rome and Britain. The northern Mesopotamian region (present-day Iran) became China's closest partner in trade. Paper and gunpowder, which had been invented by the Chinese during the Han Dynasty, had a greater impact on culture than the silk. The rich spices of the East, also, contributed more than the fashion which grew up from the silk industry. The trade between China and the West was firmly established by Roman Emperor Augustus (27 BCE–14 CE), and silk was the most sought-after commodity in Egypt, Greece and Rome (Boulnois 1966; Francis 2002).

During 161–180 CE, silk was the most valued commodity in Rome. Silk remained popular, until the fall of the Roman Empire in 476 CE. Rome was survived by its eastern half which came to be known as the Byzantine Empire and which carried on the Roman infatuation with silk. Around 60 CE, the West had become aware that silk was actually spun by silkworms.

The Byzantine emperor Justinian (527–565 CE) had sent two emissaries, disguised as monks, to China to steal silkworms and smuggle them back to the West. The plan was successful and initiated the Byzantine silk industry. When the Byzantine Empire fell to the Turks in 1453 CE, the Ottoman Empire closed the Silk Road and cut all ties with the West.

8.1.5.5 Value of the Silk Road

The greatest value of the Silk Road was exposure to different cultures, arts, religions, philosophies, technologies, languages, sciences, architectures and civilizations along with the commercial goods that merchants carried from country to country. Along with trade through a network of silk routes, different kinds of disease also travelled. The spread of the bubonic plague during 542 CE in the Byzantine Empire is thought to have arrived through Silk Road. The closing of the Silk Road forced merchants to take to the sea to ply their trade, thus initiating the age of discovery (1453–1660 CE), which led to worldwide interaction and the beginnings of a global community (Nanavati 1965; Boulnois 1966; Tazima 1991).

8.1.5.6 Memorabilia Along the Silk Road

In history, many renowned people left their traces on the most historically important trade route, which included eminent diplomats, generals and great monks. They crossed desolate deserts and the Gobi, passed murderous prairies and went over the freezing Pamir to finish their missions. Many great events happened on this ancient road, making the trade route historically important. Famous travellers like Marco Polo, Ban Chao, Xuanzang, Zhang Qian, Ban Yong and Fu Jiezi along the road were its bright pearls, making it glorious. A great number of soldiers gave their lives to protect it. These are some of the reasons the road is still a time-honoured treasure.

8.2 Biology and Life Cycle of Silkworm

8.2.1 Biology of Silkworm

The silkworm, *Bombyx mori*, is a fully domesticated insect. It is an ideal molecular genetic resource for solving a broad range of biological problems. The biology and genetics of *B. mori* is the most advanced of any lepidopteron species. All the strains reared at present belong to *B. mori*, which is believed to be derived from the *Bombyx mandarina* Moore, native of China. Races of mulberry silkworm are classified as univoltine, bivoltine and multivoltine, depending upon the number of generations in a year under natural conditions (Krishnaswami et al. 1987a; Aruga 1994).

8.2.1.1 Voltinism

There is egg hibernation in *B. mori*. The number of life cycles per year depends on the silkworm strain and the environmental conditions, particularly the temperature. Voltinism refers to the number of generations/crops in a year. Silkworm strains which go through multiple generations (5–6) in a year are termed as polyvoltines or multivoltines. These strains do not undergo egg diapause, which is an adaptation to tropical condition where there is no severe winter. Multivoltine races found in tropical areas have the shortest life cycle (Krishnaswami et al. 1987a).

In nature, univoltine races complete only one generation during the spring, and the second generation of eggs goes through a period of rest or hibernation till the next spring. In case of bivoltine races, however, the second generation of eggs does not hibernate and hatch within 11–12 days and produce second generation normally during summer. It is the third-generation eggs which undergo hibernation and hatch in the next spring, thus producing two generations in 1 year. This is an adaptation to overcome harsh winters in temperate countries. Artificially, the egg diapause can be broken after storing them at 2.5–5 °C in cold storage for a period of 90–110 days (Krishnaswami et al. 1987a; Aruga 1994). Removal of eggs from cold storage to room temperature (25 °C) results in embryonic development in eggs until hatching.

8.2.1.2 Hibernation and Acid Treatment of Eggs

Univoltine and bivoltine races of *B. mori* lay diapausing eggs which hatch after 10 and 4 months, respectively, passing through diapause period. The diapause factor in eggs can be broken either giving winter condition artificially in cold storage or by treating eggs in hydrochloric acid of specified specific gravity (Manjula and

Table 8.2 Preservation schedule of eggs in cold storage

Hibernation schedule	Temperature/days required					
	25 °C	20 °C	15 °C	10 °C	5 °C	2.5 °C
4 months	10 days	3 days	3 days	2 days	50 days	60 days
6 months	20 days	5–7 days	5–7 days	5–7 days	40 days	96–100 days
10 months	70 days	20 days	20 days	20 days	50–60 days	100 days with intermediate care at 15 °C for 1 day after 60 days

Hurkadli 1986, 1995). Hibernated eggs are stored in cold storage under 4, 6 and 10 months of hibernation schedule (Table 8.2). Central Sericultural Germplasm Resource Centre (CSGRC), Hosur, Tamil Nadu, India, has 1180 mulberry and 443 silkworm germplasms. Silkworm Genetics Division, Institute of Genetic Resource, Kyushu, Japan, has a collection of approx. 2,20,000 clones (DNA/cDNA) from 50 libraries which included clones of wild silkworm also (Iyengar 2013). Iyengar (2013) reported in 'News brief' in Indian silk magazine that cryopreservation of silkworm eggs or the frozen ovary can help in maintenance of large number of silkworm germplasms. Banno et al. (2013) have worked on long-term preservation of *B. mori* germplasm by using frozen ovaries. The method used for cryopreservation is slow cooling of donor ovary of fourth instar larvae (1 °C per min.) until −80 °C. This could be made possible by using BICELL freezing vessel.

Eggs are allowed complete hibernation in room temperature (25 °C) and put under aestivation temperature as per the schedule of hibernation/preservation in cold storage. Eggs are released from cold storage in intermediate temperature of 10 °C, 15 °C and 20 °C before placing them in incubation temperature at 25 °C.

8.2.1.3 Acid Treatment

Acid treatment of hibernating and hibernated eggs with hydrochloric acid (HCl) of 1.064–1.075 specific gravity at 46–47 °C temperature for 4–6 min is resorted to for induced hatching and to continue the generation (Manjula and Hurkadli 1986, 1995). The cold and hot acid treatments are equally effective to break diapause in eggs of *B. mori*. These methods of artificial hatching can be used as per convenience. Cold acid treatment is done with 1.1 specific gravity hydrochloric acid for 60 min at room temperature (25 °C). Eggs are thoroughly washed in running water after acid treatment till complete acid is washed away. This can be ensured by touching egg sheet or eggs on tongue tip. If there is no sour taste, the acid is completely washed away. Eggs are then dried in shade and stored as per schedule.

Hot acid treatment requires less time and can treat large quantity of eggs in short time. Acid can be heated through hot water in hot water bath tub with thermostatic control system in glass or plastic container. Hydrochloric acid should be of analytical, laboratory or commercial grade with specific gravity of 1.15–1.18. The specific gravity can be reduced by adding water in the acid and measuring it with hydrometer.

Fig. 8.1 Moth, cocoon and larva of silkworm, *Bombyx mori*

8.2.2 Life Cycle

The *B. mori* is a holometabolous insect and undergoes a complete metamorphosis (Krishnaswami et al. 1987a; Tazima 1991). The life cycle of this insect represents the most advanced form of metamorphosis and completes life cycle through serial progression of four distinct stages of development: egg, larva, pupa and adult (Fig. 8.1). Larva or caterpillar is an economically important life stage of the insect, which produces silk cocoon in about 25 days. The whole life cycle from hatching of worms till emergence of moth (adult) takes 40–45 days. Moth does not eat anything and can survive for 1–2 weeks.

8.2.2.1 Egg

Eggs may be of diapause or nondiapause type. Egg is round and white. Eggs become darker and darker as embryonic development progresses. Races producing white cocoons lay pale-yellow/creamish-white eggs (Fig. 8.2), and races producing yellow cocoons lay deep yellow eggs. Hibernating eggs of bivoltine and univoltine races are creamish white. These eggs change their colour to dark brown or purple with the deepening of serosal pigment. The diapausing eggs are laid by the silk moths inhabiting in temperate regions, whereas silk moths belonging to subtropical regions, like India, lay nondiapause type of eggs. During diapause, all the vital activities of the eggs are ceased.

8.2.2.2 Larva

The egg hatches into larva also called caterpillar in about 10 days. Caterpillars feed on mulberry leaves after hatching. The newly hatched larval body is densely covered with bristles. As the larva grows, it becomes smoother and lighter in colour due to rapid stretching of the cuticular skin during different instars of the larval stage. The larval growth is marked by four moults and five instars. Larval stage lasts for about 25 days, which includes eating and inactive moulting period.

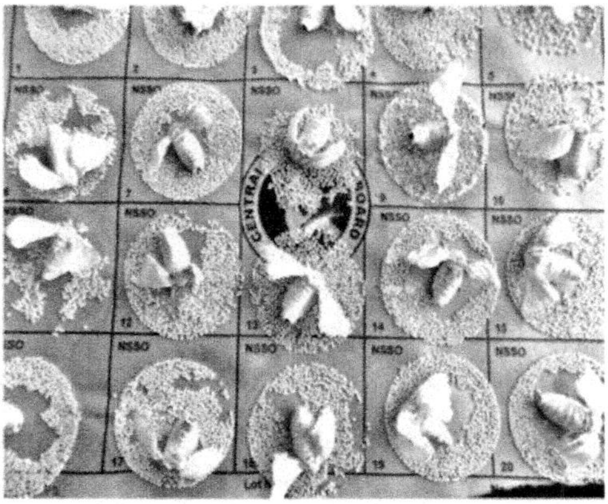

Fig. 8.2 Silkworm eggs laid by female moths

Larval skin consists of cuticle and hypodermis. Cuticle is made up of chitinous protein and is covered with a thin layer of wax, which is capable of being extended considerably to permit rapid growth of the larva during each instar. Larva bears four pairs of tubercles: subdorsal, supraspiracular, infraspiracular and basal tubercle. Each tubercle carries 3–6 setae. The larval body is composed of the head, thorax and abdomen. The head consists of six fused segments. It carries the appendages: antennae, mandibles, maxillae and labium. Six pairs of larval eyes or ocelli are located a little above the base of antennae. Five segmented antennae are used as sensory organs. The mandibles are well developed, powerful and adapted for mastication. The maxillary lobe and palpi help in discriminating the taste of food. The prementum is also chitinized, and its distal part carries a median process known as spinneret through which silk is extruded out from the silk gland. The sensory labial palpi are found on both sides of the spinneret.

The thorax has three segments: prothorax, mesothorax and metathroax. Each of the thoracic segments carries ventrolaterally one pair of true legs, which are conical in shape and carry sharp distal claws. These claws are not used for crawling, but they help in holding the leaves while feeding. The abdomen consists of 11 segments, though only 9 can be distinguished, as the last 3 are fused together to form the apparent ninth segment. Third to sixth and last abdominal segments bear a pair of abdominal legs, which are not true legs, but unjointed muscular protuberances. Eighth abdominal segment bears caudal horn on the dorsal side (Krishnaswami et al. 1987a; Tazima 1991).

The abdominal segments carry the sexual markings on ventral side, which are developed distinctly during fourth and fifth instars in the eighth and ninth segments. In females, the sexual markings appear as a pair of milky-white spots in each of the eighth and ninth segments and are called Ishiwata's fore gland and Ishiwata's hind gland, respectively. In males, a small milky-white body known as Herold's gland appears ventrally in the centre between eighth and ninth segments. Nine pairs of

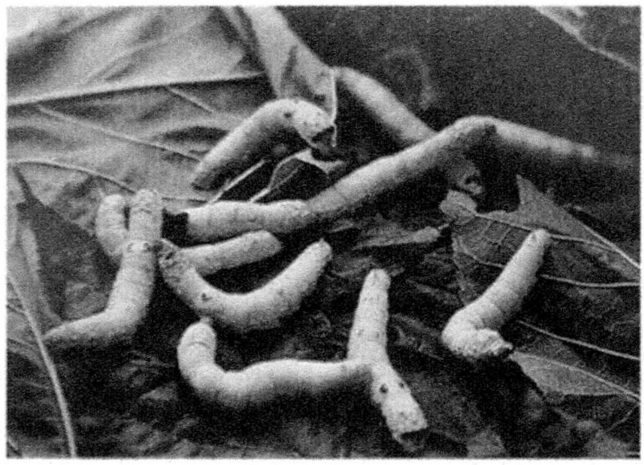

Fig. 8.3 Larvae of silkworm

spiracles are present: one pair on the first thoracic segment and eight pairs on each side of the first to eighth abdominal segments, respectively.

The full-grown caterpillar, i.e. fifth instar larva (Fig. 8.3), has a pair of well-developed silk glands, modified labial glands. Silk glands are cylindrical and divided into three segments: anterior, middle and posterior segments. The inner lining cells are characterized by the presence of large and branched nucleus. These glands secrete silk filament which has an inner tough protein, fibroin, enclosed by a water-soluble gelatinous protein, sericin. The fibrinogen which on extrusion is denatured to fibroin is secreted in the posterior segment of silk glands and forms the core of the silk filament in the form of two very thin fibres, called brins. The sericin, a hot water-soluble protein, secreted by the middle segment of the gland, holds the brins together and covers them. The duct from another small gland called Lyonnet's gland, which lubricates the tube through which the silk passes, joins the ducts of the silk glands. Finally, the silk is moulded to a thread as it passes through the silk press or spinneret. Anterior part of the silk gland does not secrete anything but acts as passage for silk material.

8.2.2.3 Pupa

It is inactive or resting stage of silkworm during which its internal organs undergo a complete change and assume the new form of adult moth. The mature silkworm passes through a short transitory stage of prepupa before becoming a pupa. During the prepupal stage, dissolution of the larval organs takes place, which is followed by the formation of adult organs. The pupa is white and soft in early stage which gradually turns brown to dark brown with harder pupal skin (Fig. 8.4).

As the pupal stage advances, a pair of large compound eyes, a pair of antennae, fore- and hindwings and the legs become visible. Ten segments are seen on the ventral side, but only nine are visible on the dorsal side. Seven pairs of spiracles are present in the abdominal region, the last pair being nonfunctional. Sex markings are prominent, and it is much easier to determine the sex of pupa. The female has a fine longitudinal line on the eighth abdominal segment, whereas such marking is absent in the male. The pupa is covered within a thick, oval, white or yellow silken case

Fig. 8.4 Pupae of silkworm

called cocoon. The pupal period may last for 10–14 days after which the adult moth emerges out slitting through the pupal skin and piercing the fibrous cocoon shell with the help of the alkaline salivary secretion that softens the tough cocoon shell.

8.2.2.4 Adult

Moth is incapable of flight because of its feeble wings and heavy body. It does not feed during its short adult life. The body of moth has general plan of insect body organization. The ocelli are absent. The antennae are conspicuous, large and bipectinate. The meso- and metathorax bear a pair of wings. The front pair overlaps the hind pair when moth is under rest. The moth is unisexual and shows sexual dimorphism. The female has comparatively smaller antennae. Its body and the abdomen are stouter and larger. The female is generally less active than the male. The male moth possesses a pair of hooks known as harpes at its caudal end. Male moth is smaller in size, active and darker in colour. Male moths copulate with females for about 2–3 h (Fig. 8.5). The female starts laying eggs when they are decoupled; this is completed within 24 h. A female lays 400–500 eggs in case of bivolitine and univoltine races. Female moths of multivoltine races lay 300–350 eggs. The eggs are laid in clusters (closely laid in group) and are covered with gelatinous secretion of the female moth (Fig. 8.6).

8.2.3 Silkworm Genetics

The biology and genetics of the silk moth is the most advanced of any lepidopteron species. Mulberry silk moth possesses 28 pairs of chromosomes. *Bombyx mandarina* (Japan) having chromosomal number of 27 pairs, *Bombyx mandarina* (China)

Fig. 8.5 Silk moth

Fig. 8.6 Female moth laying eggs

28 pairs and *Theophila religiosae* 31 pairs, are the other close ancesters of *Bombyx mori*. Sex determination in *B. mori* has been established by Japanese scientists. Tanaka (1916) was the first scientist to demonstrate inheritance of sex chromosomes. He established that a male has ZZ sex chromosomes, while female has ZW sex chromosomes. Kawaguchi (1934) confirmed the observations of Tanaka (1916) but reported the theory of heterogametic female sex determination. Hashimoto (1948) demonstrated that W chromosome monopolizes the determination of femaleness in silk moth. He found that even triploids with ZZW and tetraploids with ZZZW chromosomes determine females. Tazima (1941) demonstrated that W and Z chromosomes are always segregated, and there is no crossing over between W and Z chromosomes. The genes responsible for morphological traits are present on Z chromosome. There are no genes of morphological significance on W chromosome.

Chromosomes in *B. mori* are holocentric, i.e. they possess centromeres throughout the chromosome body. There are no recombination nodules in females. *B. mori* chromosomes are highly condensed and appear dot shaped at most meiotic and mitotic metaphase stages. Diffused centromeres and lack of special features make them difficult to identify individually.

8.2.3.1 *Bombyx mori* Genome

Japan and China agreed upon the integration of both WGS (whole genome shotgun) data sets which led to the determination of complete genomic sequence (the International Silkworm Genome Consortium, 2008). They have fully sequenced more than 11,000 full-length cDNA clones by primer walking method. The high-density SNP (single-nucleotide polymorphism) linkage map mapped 192 scaffolds on 28 chromosomes which accounted for 87 % of the genome of *B. mori*. All information including genomic sequence, physical and linkage maps and EST (express sequence tag) data were integrated into Silk DB (silkworm genome database) in China and KAIKObase (an integrated silkworm genomic database and data mining tool with three map browsers) in Japan (Mita 2012).

8.2.3.2 Industrial and Biological Advantage of Genomic Study

1. *Bombyx mori* is a model organism for Lepidoptera, containing the most disruptive agricultural pests. Silkworm genomic information will aid in searching target genes for development of new insecticides.
2. The complete genome sequence and high-quality genetic-physical map will help in positional cloning of various phenotypic mutants of this insect.
3. Silkworm transgenesis has become an important bioreactor for recombinant protein production.
4. Comparison of genomic library of *B. mori* and *B. mandarina* will lead to identification of genes responsible for the domestication of *B. mori* insect.

8.2.3.3 Molecular Characterization of Silkworm Germplasm

Molecular characterization helps to identify the genetic distinctness of a race/breed/stock and thereby helps to eliminate duplicates and reduces the cost of germplasm maintenance and volume of work. Thus, the genetic markers facilitate molecular characterization, maintenance, conservation and cost-effective management of plant and animal genetic resources (Bruford and Wayne 1993).

8.2.3.4 Faunal Diversity of *Bombyx mori* in India

India has been recognized as one among the 12 mega biodiversity rich centres of the world. India is harbouring great faunal diversity, and nearly 11.9 % of the world fauna is present in India. Floristically, India is very rich, harbouring three megacentres of endemism (ecological state of specises), i.e. Western Himalayas, Eastern Himalayas and Western Ghats. It has several diverse sericigenous flora and fauna. Wild species of *Bombyx* and other genera of *Bombycidae* exist in the great Himalayan ranges and Andaman Islands under natural habitat.

Eggs and cocoons of a wild silk moth belonging to Bombycidae were collected from wild mulberry tree *Morus serrata* near Kedarnath (30.47 °N, 79.02 °E) at an altitude of 800 metres above MSL (Tikader 2001). The eggs were incubated and

Table 8.3 Indigenous silkworm races in India

Sl. no.	Race	Voltanism	Origin	Cocoon colour and shape	Remarks
1	Pure Mysore	Polyvoltine	Karnataka	Spindle, flossy, greenish yellow	Commercially used
2	Nistari	Polyvoltine	West Bengal	Spindle, flossy, golden yellow	Commercially used
3	Saruput	Polyvoltine	Assam	Spindle, flossy, creamish white	Commercially used
4	Moria	Polyvoltine	Assam	Spindle, flossy, creamish white	Commercially used
5	C-nichi	Polyvoltine	Karnataka	White, short dumbbell	
6	Borapolu	Polyvoltine	Assam	Greenish white	
7	Chotapolu	Polyvoltine	Assam	Spindle, yellowish or creamish white	
8	Kashmir race	Univoltine	Kashmir	White or yellow, elongated oval	Extinct

reared on the mulberry plants at Central Sericultural Germplasm Resources Centre (CSGRC), Hosur, India. Cocoons and eggs of wild silk moth were very similar to *B. mori*. It is a potential genetic material with several unique characters. Such wild relatives of *Bombyx* can create additional seri-diversity and widen the genetic base of *B. mori*. India has very rich germplasm of indigenous and exotic races of *B. mori* (Tables 8.3 and 8.4). Central Sericultural Germplasm Resource Centre (CSGRC), Hosur, Tamil Nadu, India, has 1180 mulberry and 443 silkworm germplasm. Silkworm Genetics Division, Institute of Genetic Resource, Kyushu, Japan, has a collection of 2,20,000 clones (DNA/cDNA) from 50 libraries (Iyengar 2013). Some of these races have high heritability for economically important cocoon characters, which can be used in the synthesis of new genotypes with superior characters and higher survivability (Table 8.5).

8.2.4 Biochemical Genetics and Synthesis of New Genotypes

Biochemical genetics has a very important role to play in biotechnological approaches of silkworm breeding (Tazima 1954; Chauhan et al. 1997). Quantitative traits of silkworm genotypes and their correlation to the biochemical parameters help in identifying the genotypes with high quantitative characters and resistance to the silkworm diseases (Basavaraj et al. 2005). Cloning of such biochemical traits can be done through identifying the gene by RFLP mapping. Studies have been carried out by gel electrophoresis on naturally occurring polymorphism of various enzymes like esterase, amylase, phosphatase and haemolymph proteins (Doira 1978). There are positive correlations between amylase in digestive juice and

Table 8.4 Bivoltine silkworm races used commercially in India

Sl. no.	Race	Voltanism	Origin	Cocoon colour and shape	Remarks
1	KA	Bivoltine	Cross-breeding, India	Oval, white	Commercially used
2	NB4D2	Bivoltine	Cross-breeding, India	Dumbbell, white	Commercially used
3	SH6	Bivoltine	Cross-breeding	Oblongated, white	Commercially used
4	NB7	Bivoltine	Cross-breeding, India	Oval, white	Commercially used
5	NB18	Bivoltine	Cross-breeding, India	Dumbbell, white	Commercially used
6	CSR2	Bivoltine	Cross-breeding, India	Oval, white	Commercially used
7	CSR4	Bivoltine	Cross-breeding, India	Dumbbell, white	Commercially used
8	CSR6	Bivoltine	Cross-breeding, India	Dumbbell, white	Commercially used
9	CSR26	Bivoltine	Cross-breeding, India	Dumbbell, white	Commercially used
10	CSR27	Bivoltine	Cross-breeding, India	Oval, white	Commercially used

Table 8.5 Degree of heritability for different characters in silkworm *Bombyx mori* L.

Characters	Degree of heritability	Degree of heritability opinion of Japanese scientists (estimated common pool)
Hatchability	*	–
Larval duration	*	*
Pupation rate	***	**
Yield by weight	***	–
Cocoon weight	**	**
Shell weight	**	**
Shell ratio	**	***
Cocoon filament length	*	***
Filament weight	*	**
Filament size	***	***
Reelability	*	**
Raw silk percentage	***	*
Silk filament neatness	*	**

*, **>, ***> Lower to higher degree of heritability

survivability in silkworm (Chatterjee et al. 1993). Similarly, there is positive correlation between alkaline phosphatase activity in digestive juice and higher quantitative cocoon characters.

8.2.4.1 Biotechnology in Silkworm Breeding in India

Scientists of Central Sericultural Research and Training Institute, Mysore, India, have identified 13 silkworm genotypes, namely, Mysore local, Nistari, C. niche, Diazo, Sarupat, Moria, Guangnong, Hu201, J122, NB1, NB7, NB4D2 and NB18 for RFLP studies on the basis of polymorphism variation at DNA level (Dutta 1994). Two contrasting genotypes NB1 and Nistari were selected for quantitative trait linkage (QTLs) analysis through RFLPs. Two hundred recombinant clones carrying Pst 1 fragment of 0.5–2.0 KB of silkworm DNA have been isolated (Kathirvel et al. 1994). These fragments were ligated to pUC 18 vector, transformed in DH5 *Escherichia coli*, propagated and preserved for identification.

8.2.4.2 Cloning of Antibacterial and Antiviral Protein

An induced antibacterial protein was isolated and purified from haemolymph of infected silkworm, *Bombyx mori*. Attempts are being made to clone cDNA of antibacterial protein. Defence mechanism in silkworm against nuclear polyhedrosis virus (NPV) is medicated through antiviral substances in gut juice and viral inhibitory factor (VIF) produced in the haemolymph. The significance of red fluorescent protein (RFP) in antiviral activities has been established, and the prospect of identifying gene system was studied by Hayaskiya et al. (1969) and Sethuraman (1991). Scientists of CSR & TI, Mysore, India, have isolated 65 KD (NB1 x French plain) and 28 KD (Nistari) proteins from gut juice of fifth-age silkworms, which showed clear antiviral activity (Chinya and Ramesh Kumar 1994).

8.3 Mulberry Cultivation

8.3.1 Mulberry

Mulberry (*Morus* spp.) is a perennial, deciduous, deep-rooted, fast-growing and high-biomass-producing plant. Mulberry can be grown in both tropics as well as in temperate regions (Minamizawa 1997). It can be cultivated in different soil types and can be raised in both rainfed as well as in irrigated conditions. It is comparatively resistant to environmental fluctuations. Das (1983) and Rangaswami et al. (1987) described taxonomy, cytology and breeding of mulberry species. Tikader et al. (2002) gave an account of geographical distribution of Indian mulberry species. Rau (1967) and Tikader et al. (1999) have reported the oldest mulberry living tree available in the sub-Himalayan region, Joshimath, Uttarakhand, India.

There are four popular mulberry species in India, namely, *Morus alba* L., *Morus indica* L. *Morus laevigata* Wall and *Morus serrata* Royb. *Morus alba* and *Morus indica* are cultivated for silkworm rearing, while *Morus laevigata* Wall is grown as fruit-yielding tree. These four mulberry species have 28 somatic chromosome numbers. The most popular and superior mulberry varieties used in India for silkworm rearing are V1, S36, S13, S54, S146, S1635, TR10, S1 and K2 in plain subtropical area and Kosen, Ichinose, Goshoerami and KNG in hill temperate area of Kashmir and parts of Himachal Pradesh.

Temperature ranging from 24 °C to 29 °C is ideal for the growth of mulberry. Mulberry cannot sprout below 13 °C and above 38 °C. Rahaman et al. (2010) have developed cold-tolerant mulberry genotype, which has comparatively good growth during winter in West Bengal, India. An atmospheric humidity of 65–80 % is ideal for mulberry growth. Five to ten hours of daylight in the temperate region and 9–13

Fig. 8.7 Mulberry plantation

h in the tropics is considered ideal. Soil should be fertile, porous and well drained and should be with good water-holding capacity. Loamy, clay loamy or sandy loamy soils are best. Soil should be slightly acidic ranging from pH 6.2–6.8, though the level of tolerance ranges from pH 5.0–9.0.

The quality of mulberry leaf is a major contributing factor for good and successful silkworm rearing (Singhavi et al. 2000). The first important step is to select the excellent variety of mulberry for silkworm rearing. The selection of variety for cultivation should be based on the existing weather conditions, suitable for the mode of cultivation, such as tree, low bush or high bush (Fig. 8.7).

The three important characters considered for the selection of mulberry varieties for silkworm rearing are higher leaf yield, excellent quality of leaves and easy propagation of plants. Nutritionally superior varieties, which are resistant or tolerant to pests and diseases, are preferred for silkworm rearing. Nutritional status of mulberry leaves differs significantly among different varieties (Dandin 2006). V1 variety of India has 13.10 % sugar and 24.56 % crude protein which are highest among the Indian mulberry varieties. Machii and Katagiri (1991) have established the varietal difference in nutritional value of mulberry leaves for silkworm rearing. Chakraborti and Singhal (1996) have reported seasonal variation in ascorbic acid content in promising mulberry genotypes in India. Similarly, Bose and Bindroo (2001) have established the biochemical changes in leaves of seven mulberry varieties under rainfed condition in subtropics of India. All these studies support that there is a definite relationship between mulberry genotypes and nutritional profile of their leaves. The nutritional profile does differ in cultural practices and input applications. Nutritional quality of leaf acts as major contributing factor for successful rearing. Miyashita (1986) has clearly indicated the percentages of different factors responsible for successful silkworm cocoon crop as under:

1. Mulberry leaf: 38.1 %
2. Climate: 37.5 %
3. Rearing technique: 9.1 %
4. Silkworm race: 4.1 %
5. Other factors: 11.2 %

Table 8.6 Planting of mulberry plants in different spacing

Plant spacing (cm)	No. of plants/acre	No. of plants/hectare
90 cm × 90 cm	4840	12,345
(90 cm × 180 cm) × 60 cm	4840	12,345
(90 cm × 150 cm) × 60 cm	5445	13,888

Quality of mulberry leaf is a major contributing factor for successful silkworm rearing. Hence, production of higher quantity of quality mulberry leaf is given the main importance. Alternatively, the healthy condition of plants is considered as the important feature. Hence, it is necessary to select the varieties, which fulfil the objectives of cultivation and quality leaf production.

8.3.2 Plantation

8.3.2.1 Land Preparation

Generally flat lands are suitable for irrigated mulberry cultivation. If the land slope is more than 15 %, suitable land development measures, such as contour bunding, terracing, etc., should be adopted. Land should be deep ploughed up to a depth of 30–45 cm in order to loosen the soil and to bring the soil to fine particles. Weeds and stones should be removed during the preparatory stage. Basal dose of farmyard manure (FYM) at 20 tons per hectare is recommended, which has to be thoroughly incorporated into the soil.

8.3.2.2 Plantation of Mulberry

Plantation of mulberry can be taken up both by cuttings and saplings. However, saplings are always better than cuttings for quick and better establishment (Sarkar et al. 2003). Mulberry plantation can be taken up through pits of sizes 35 cm^3which are filled with FYM and soil at 1:2 ratio 1 month before the plantation. It is preferable to start plantation of saplings during the rainy season. Plant spacing can be selected as per the requirement and convenience to cultural operation or mechanical maintenance of mulberry garden (Table 8.6). Generally 6–12-month-old saplings are preferred for plantation. Saplings take about 1 year to establish as plant. Foliage harvest can be taken by next year of establishment period. Plants are trained (maintaining plant height and crown through pruning) for plant crown height after 6 months of plantation which differs for low, high-bush and tree mode of plantation.

8.3.3 Pruning of Mulberry Plants

Pruning of mulberry plants is required to improve the yield of foliage and to maintain the shape of plants for scheduled harvest of the leaf for silkworm rearing. Pruning of mulberry plant is useful in adjusting the production period to

Table 8.7 Application of chemical fertilizer in mulberry plantation

Plantation	Schedule of chemical fertilizer application	N	P	K
Rainfed				
	First year	50	25	25
	Second year onwards	100	50	50
Irrigated				
Shoot harvest	First year	100	50	50
	Second year onwards	300	120	120
Leaf harvest	First year	100	50	50
	Second year onwards	300	120	120

synchronize with the leaf requirement for silkworm rearing and leaf production in all the rearing seasons. Pruning of plants ensures maintenance of convenient height for foliage harvest, removal of dead wood and better sunlight and makes cultural operation easier. Mulberry plants are cut from a particular height to maintain plant height convenient to leaf harvest and for quality leaf production.

8.3.3.1 Method of Pruning

Method of pruning varies from place to place, according to climatic geographical location, method of silkworm rearing, type of plantation as tree, low bush or high bush, etc. Pruning is also closely related to harvesting of foliage and the type of rearing (Das et al. 1993, Ghosh et al. 1997). The method of pruning in subtropics of India is as below:

(A) Bottom pruning or (low-cut pruning at 30 cm): Branches are cut to the base during June–July months.
(B) Middle pruning (mid-trunk cut at 60–70 cm): Branches are cut at about 1 metre height during December–January months.
(C) Top pruning (high-cut pruning): Branches are cut at the top or to the soft portion.

8.3.3.2 Input Application

FYM has to be applied at 20 tons/ha/year in two split doses following pruning. Fertilizer has to be applied as per the recommended schedule. Micronutrients have to be applied wherever necessary. Foliar sprays, such as boron (1 %), urea (0.5 %), zinc sulphate (0.1 %), etc., will improve the leaf quality. The recommended fertilizer dose (kg/ha) is indicated in Table 8.7.

Among the various inputs, irrigation ranks high in giving quick and good results. Regular irrigation at an interval of 8–10 days is ideal for quality leaf production. Normally, 1.5–2 in. of water per irrigation is sufficient. In case of water scarcity, drip irrigation can be adopted.

Table 8.8 Operation schedule for irrigated mulberry garden in subtropical condition

Sl.no	Operations	Timings
1	First bottom pruning	With commencement of south-west monsoon (early June)
2	First weeding	Within a week after pruning (2nd week of June)
3	Application of FYM/compost at 10MT/ha/yr	Within a fortnight after pruning (3rd week of June)
	Sowing of green manure of sunhemp (*Crotalaria juncea*) and dhaincha (*Sesbania aculeata*)	Within a fortnight after pruning (3rd week of June) and mulching the green crop in mid of August
4	First dose of fertilizer application	Within a month after pruning (early July)
5	First harvest of leaf	During September
6	Second weeding, inter-cultivation and irrigation	Within a week of first leaf harvest (4th week of September)
7	Second harvest of leaf	Early October
8	Mid-cut pruning	During winter (December)
9	First weeding	Within a week after pruning (3rd week of December)
10	Application of FYM/compost at 10 MT/ha/yr	Within a week after pruning
11	Second weeding and inter-cultivation	During February
12	Fertilizer application	First week of February
13	Leaf harvest	During March and April

8.3.3.3 Leaf Yield

Generally, the expected annual yield is 10–20 tons of leaves per hectare in two harvests in the sub-Himalayas and 30 tons in the case of tropical region. In shoot harvest method, harvesting can be done at an interval of 70 days (five harvests in a year) in tropical condition. In case of leaf harvest method, first harvest is taken 70 days after first pruning and second and third harvests (coinciding with second bottom pruning) at an interval of 55 days followed by bottom pruning. The fourth harvest is taken 70 days after the third harvest and the last two harvests at an interval of 55 days (Table 8.8).

8.3.4 Tree Plantation

Mulberry as tree can be raised in wastelands and denuded areas and as block plantation in watershed development. It is a fast-growing tree and produces large quantity of renewable biomass in the form of branches, shoots, leaves and fruits. Accordingly, mulberry can be utilized by raising it as an ‘energy plantation’ in cultivable land, wasteland, watershed area, canal bunds, road sides, etc. under various developmental and conservation programmes. One-year-old saplings are most suitable for tree plantation of mulberry. Pit preparation of 45 cm^3 size should be taken up after the pre-monsoon showers with the spacing depending upon the topography, the land availability and the purpose for which trees are raised. It is desirable to have wider spacing for mechanical operations and to save expenditure on labour input.

8.3.5 Economic Use of Mulberry

Mulberry trees are also cultivated for harvesting of fruits. Besides the cocoons, various other products are also obtained from mulberry, like heath drink (Soumen et al. 2013), cattle feed and vitamin supplements. Mulberry plant has been used in various medicines in China and Japan to cure age-induced problems. Mulberry fruits can be used as vitamin supplements as it contains carotene, thiamin, nicotinic acid, riboflavin and ascorbic acid.

8.3.6 Foliar Diseases of Mulberry (Khan et al. 2004)

8.3.6.1 Leaf Spot

Leaf spot is a fungal (*Cercospora moricola* Cooke) disease of mulberry leaves which occurs during rainy and winter seasons. The fungus attacks on leaf surfaces, resulting in brownish necrotic irregular spots and forming shot holes on leaf surface in advanced stage. The disease is estimated to cause leaf damage of about 10–12 % (Fig. 8.8).

Fig. 8.8 Leaf spot

Fig. 8.9 Powdery mildew

It is recommended to remove and burn affected leaves and adopt wider spacing of plantation to control the fungal disease. Spray of 0.2 % Bavistin (carbendazim 50 % WP) can be taken as chemical control of disease. Maji et al. (2006) have studied ecofriendly management of foliar diseases of mulberry through botanical and biocontrol agents. They have claimed to control leaf spot (*Myrothecium*) and powdery mildew diseases to the extent of 30 % through application of botanical-based control agents in India.

8.3.6.2 Leaf Rust

Leaf rust is caused by pathogen, *Cerotelium fici* (Cast.) Arth during rainy and winter seasons. The infected leaf develops pinhead-like brown eruptive lesions on the surface. The disease is estimated to cause leaf damage of about 10–15 %. Spray of 0.2 % chlorothalonil 75 % WP can be taken as a chemical control measure to control the disease.

8.3.6.3 Powdery Mildew

Powdery mildew is caused by pathogen *Phyllactinia corylea* (Pers) Karst, during receding rainy and winter seasons. The infected leaves develop white powdery patches on the ventral surface which gradually turn to yellow-brown and black before the infected leaves fall off. The disease is estimated to cause leaf damage of about 5–10 % (Fig. 8.9). Spray of 0.2 % Karathane (dinocap 30 % EC)/Bavistin can be taken as chemical control measure to control the disease.

8.3.6.4 Fungal Leaf Blight

The fungal leaf blight is caused by pathogen, *Alternaria alternata* (Fr.) Keissl, during summer and rainy seasons. The infected leaves develop browning/blackening from the leaf tip or edge of leaf lamina. The disease is estimated to cause leaf damage to the tune of 10–12 %. 0.2 % Dithane M-45 (mancozeb 75 % WP) can be sprayed over leaves to control the disease.

8.3.6.5 Root Rot (Fungus: *Fusarium solani* Mart. Sacc)

The disease causes decay of mulberry plant roots resulting wilting and withering of leaves in plants. It is recommended to uproot and burn the affected plants. 5–10 g Dithane M-45 can be applied around the root zone to control the disease.

8.3.6.6 Root Knot (Nematode: *Meloidogyne incognita* Kofoid and White)

The disease causes poor growth in mulberry plants. Roots develop knots/galls which arrest the food supply to mulberry plants. Sharma (1998) has studied ecofriendly approach for management of root knot in India.

It is recommended to apply 400 kg of neem oil cake per acre/year in four split doses with more FYM dose.

8.3.7 Pests on Mulberry

Mulberry being a perennial plant attracts more than 300 species of arthropodes, mostly lepidopteran defoliators (Chauhan et al. 2008; Irfan et al. 2013). The other mulberry pests belong to Homoptera, Coleoptera, Thysanoptera and Diptera. Majority of these pests are polyphagous in nature, and few of them cause considerable damage to mulberry plants. Whitefly (*Dialeuropora decempuncta*) and giant African snail (order, Stylommatophora; class, Gastropoda) *Achatina fulica* Bowdich have been reported as pests on mulberry from India (Shree et al. 2006). Gopinath et al. (2013) have reported infestation of giant African snail on mulberry garden in Tamil Nadu, India, with considerable loss to leaf yield. He suggested that spray of 4 kg salt in 10 liter water can control the snail infestation in mulberry garden. Some of the lepidopteron pests are listed below:

1. *Glyphodes pyloalis* Walker (order Lepidoptera, family Arctiidae) (Figs. 8.10, 8.11)
2. *Spilosoma obliqua* Walker (order: Lepidoptera, family Arctiidae) (Fig. 8.12)
3. *Hemerophila* sp. (looper)
4. *Amata passalis* Fabr.
5. *Spodoptera* sp.

Coleopteran Pests

1. *Mimastra cyanura* Hope
2. *Episomus lacerta* Fabr (Fig. 8.13)
3. Stem borer

Homopteran Pests

1. Jassids (Homoptera)
2. Scale insects (Homoptera: Hemiptera) (Fig. 8.14)
3. Pink mealy bug *Maconellicoccus hirsutus* (Homoptera: Pseudococcidae) (Fig. 8.15)

Fig. 8.10 Single larva of *Glyphodes pyloalis*

Fig. 8.11 *G. pyloalis* larva feeding mulberry

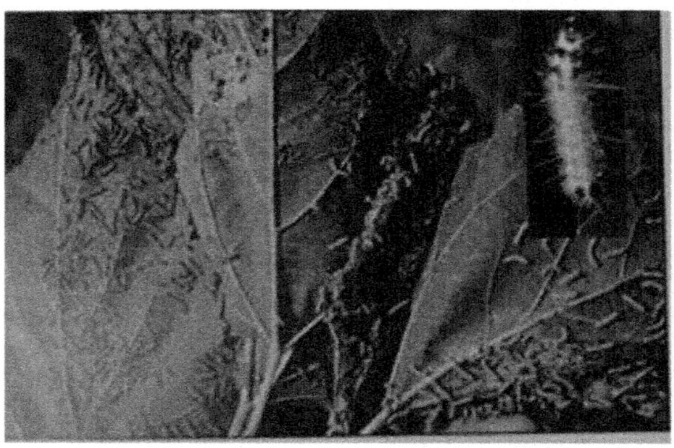

Fig. 8.12 Infestation of *Spilosoma obliqua* (Source: Singh et al. 2000)

Fig. 8.13 Adult of *Episomus lacerta*

Fig. 8.14 Scale insect

Fig. 8.15 Infestation of pink mealy bug

Fig. 8.16 Infestation of thrips

Other minor pests are thrips (*Thysonoptera*) (Fig. 8.16) and acarine pests. The most common acarine pests are red spider mite (*Tetranychus* sp.) and yellow mite, *Polyphagotarsonemus latus* Banks (Chauhan et al. 2011).

8.3.8 Leaf Harvest and Preservation

8.3.8.1 Leaf Harvest

Mulberry leaves should be harvested during cooler hours in the morning and evening to avoid moisture loss in leaves during harvest and transportation. The harvested leaves should be collected in wet gunny cloth-covered baskets and transported to a leaf chamber/room carefully in well-ventilated condition. Leaf should be preserved in a separate room or leaf chamber made up of wood with sufficient number of ventilators and covered with wet gunny cloth. This helps in maintaining leaf quality and moisture in leaves. Leaves can also be stored in a separate ventilated room with well-disinfected floor. Leaves should never be preserved on heaps. Leaves scattered on floor should be frequently tilted up and down in summer.

8.3.8.2 Shoot Harvest

Shoots of 4–5 ft height are cut with the help of sharp sickle and bundled with 10–12 kg weight, convenient to carry them to shoot preservation room (Fig. 8.17). Shoots are properly wrapped with wet gunny cloth and preserved vertically in upward direction. Floor of rearing room should be washed with 1 % bleaching solution daily before preservation of shoots. Separate footwear can be used for leaf chamber, which is disinfected daily with soap solution. While transporting from mulberry garden to preservation room, shoots should be covered with wet gunny cloth or polythene sheet to avoid water loss in leaves.

Fig. 8.17 Mulberry shoots harvest in India

8.4 Silkworm Rearing Techniques

8.4.1 Silkworm Rearing

Multivoltine races of *Bombyx mori* lay non-hibernating eggs, while bivoltine or univoltine races lay hibernating eggs. Non-hibernating eggs hatch in 9–10 days, while hibernated eggs hatch in 10–11 days after induced hatching with acid treatment or release from cold storage on completion of hibernation schedule. Hatched out larvae feed on mulberry leaves during their larval life for 22–25 days, passing through four moults and five instars. Indoor rearing of *B. mori* larvae is termed as silkworm rearing in commercial sericulture. In fact the larval stage of *B. mori* is an economically important stage which produces cocoons for raw silk production.

8.4.1.1 Silkworm Rearing Planning

Silkworm rearing is a technical process with scientific logics that require skill and resource management. This clearly indicates that management plays a very important role in silkworm rearing. The commercial silkworm rearer should make assessment of the resources, such as mulberry leaves, rearing space, rearing appliances, manpower, etc., before taking up silkworm rearing. He/she should look for good rearing season and availability of quality hybrid silkworm seed for scheduled hatching date from any certified seed-producing agency.

8.4.1.2 Rearing House and Appliances

There should be separate rearing house with good ventilation, high roof at 10–12 ft height, separate leaf preservation room and verandah all around the building. Generally, 35′ × 30′ ft floor area including verandah and leaf preservation room is required for 100 DFLs (disease-free layings) rearing in shoot rearing system. There are two systems of silkworm rearing; one is tray rearing with leaf feeding, and another is shoot rearing with shoot feeding (Rajan et al. 2003).

8.4.1.3 Disease-Free Laying (DFL)

One disease-free laying (DFL) is a cluster of eggs laid by single female moth. The female moth is crushed in 0.6 % K_2CO_3 solution and centrifuged. The smear of centrifuged sample is examined under microscope for presence of pebrine spore, the pathogen of pebrine disease. The egg laying free from pebrine spore, after female moth examination, is termed as DFL. There are about 50,000 eggs in 100 DFLs in case of new bivoltine silkworm hybrids. However, the number of eggs per laying varies between 400 and 550 in different silkworm races/hybrids.

8.4.2 Incubation of Silkworm Eggs

8.4.2.1 Required Temperature, Humidity and Photoperiod (Tazima 1972, 1978)

Generally, silkworm eggs require 10–11 days to hatch after oviposition. Silkworm eggs after oviposition or release from cold storage are disinfected with 2 % formalin solution before placing them inside incubation room. The first step of silkworm rearing is incubation of silkworm eggs in optimum condition (Figs. 8.18 and 8.19). The success of silkworm rearing is closely related with incubation of silkworm eggs under ideal environmental conditions, such as temperature, humidity and photoperiod. Temperature during incubation has direct bearing on embryonic development and voltinism. The ideal temperature for incubation of silkworm eggs is 25 °C and relative humidity between 75 and 80 % for healthy development of embryo inside

Fig. 8.18 Egg-carrying boxes

Fig. 8.19 Incubation frame

eggs and good hatching. The temperature more than 30 °C results in occurrence of dead eggs and poor irregular hatching. Similarly, incubation below 15 °C results in poor embryonic development and occurrence of nondiapause eggs during egg laying. Low humidity during incubation causes poor, irregular hatching and weak larvae. Similarly, humidity more than 95 % makes larvae weak. Photoperiod of 16 h light and 8 h dark period is ideally required during incubation (Benchamin et al. 1990; Aruga 1994).

8.4.2.2 Synchronization of Hatching (Black Boxing)

During pigmentation or early blue egg stage, the complete developed embryo (on development of mouth and abdominal parts), eggs are wrapped in tissue paper and placed under total dark condition inside the black box. The black box can be made from thick black chart paper/sheets or covering rearing tray with thick black cloth. Eggs are placed loosely inside the box to allow the air circulation. Black box checks hatching in fully developed advanced eggs and allows the embryonic development in eggs, which are lagging behind in development. This process of black boxing ensures synchronization of hatching from eggs at a time when exposed to a light source.

8.4.3 Hatching and Brushing

The process of brushing refers to the transfer of newly hatched worms from egg sheets to the rearing tray. Eggs under black box are removed and spread in single layer in rearing trays. These eggs are exposed to cold light (40 W tube light or 15–18 W CFL) in early morning for hatching. Tiny larvae are hatched out by breaking eggshells after exposure to light. High-power electric bulbs should not be used for hatching, and eggs should be placed at least one metre away from light source. This will ensure good hatching in eggs and keep newly hatched larvae away from heat radiation of light source. The process of hatching is carried out in incubation room BOD incubator at 25 °C, 75–80 % R.H.

Fig. 8.20 Hatched larvae

Fig. 8.21 Feeding of hatched larvae

Young-age silkworms require comparative higher temperature of 28–29 °C for good feeding and optimum growth of larvae. Hatched larvae are placed in the rearing room at the temperature of 28–29 °C and relative humidity more than 80 %. Small mesh cotton or nylon net is spread over hatched larvae in case of loose eggs (Fig. 8.20). Tender mulberry leaves (second and third apical leaves of branches) are plucked and chopped in small pieces (squire or long 0.5 cm width). Chopped leaves are sprinkled over hatched worms above the net and left for 20 min to allow the worms to crawl over chopped leaves (Fig. 8.21). Then the net is raised and placed on another rearing tray, the bottom of which is covered with paraffin paper. This process is called brushing (Figs. 8.22 and 8.23). These hatched/brushed larvae fed with mulberry leaves are covered with paraffin paper to maintain required temperature and humidity in the rearing bed. This is zero day of rearing of first instar. First to third instar larvae are categorized as young-age silkworms or chawki worms, while fourth and fifth instar larvae are called late-age worms.

Fig. 8.22 Separation of larvae from eggshells

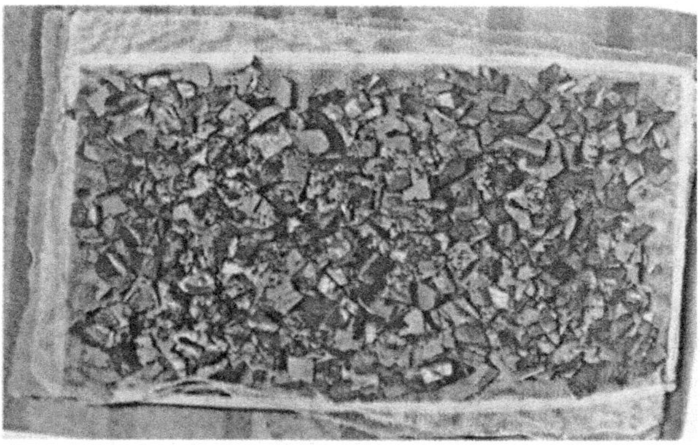

Fig. 8.23 Larvae feeding chopped mulberry leaf

8.4.4 Young Instar Silkworm Rearing (Tazima 1972, 1978)

Young instar silkworm rearing (chawki rearing) has closest bearing on the successful cocoon crop with higher productivity and quality (Krishnaswami 1988). Silkworm requires balanced nutrition for good growth and to acquire resistance against common diseases. The major parameters that determine the quality of mulberry leaves for young instar rearing are water content, protein and carbohydrates. The determinant nutritional status in mulberry leaves varies in different mulberry varieties and also with input given to mulberry plants. There should be exclusive mulberry plot for young instar silkworm rearing with good sunlight, assured irrigation with recommended organic manure and dose of chemical fertilizer. An ideal young instar silkworm rearing requires mulberry leaves with 65–70 % water content, 25 % protein and 11–14 % carbohydrate.

Young instar larvae grow very fast. The first instar larvae grow 15 times in 50 h of feeding, second instars grow five times in 45 h and third instar larvae grow three times in 60 h of feeding. Temperature and quality of feed play a vital role on the

Table 8.9 Standard rearing table for young-age silkworm of 100 DFLs

Factor	First instar	Second instar	Third instar
Temperature (0 °C)	27–28	27–28	25–26
Humidity (%)	85–90	80–85	75–80
Leaf size (sq. cm)	0.5–1.5	1.5–4.0	4.0–6.0 (entire leaf in case of shoot feeding)
Quantity of leaf (kg)	5–6	14.0–15.0	50–55
Bed area (sq. m)			
(a) In the beginning	0.36	1.35	4.05
(b) At the end	1.35	4.05	9
Bed cleaning	No cleaning	Twice	Daily

Source: Jolly (1987)

growth of larvae. Silkworm is cold blooded, and temperature has direct bearing on various physiological activities of larvae.

Rearing of young-age worms is conducted at 28–26 °C descending temperature and 75–90 % humidity (Table 8.9). The humidity should be reduced at moulting stage, while 80 % humidity should be maintained during feeding period in each instar by covering rearing bed with paraffin paper. Early stop of feeding, when 90 % larvae are settled for moult and late resumption of feeding after moult, ensures uniformity in rearing. There is no cleaning of bed during first instar. However, the first cleaning is resorted to after resumption of feeding in second instar and second cleaning before settling for second moult. Worms are spread properly with optimum spacing in third instar after resumption of feeding and cleaning of bed. Third instar larvae require almost double the quantity of mulberry leaves from second instar rearing. This is the stage of rearing when silkworms develop resistance to various diseases and harsh rearing conditions. Leaves are chopped in 0.2–0.8 mm^2 for first, 1–2 mm^2 for second and 2–4.5 mm^2 for third instar larvae.

Young-age silkworm rearing is the most important step in commercial sericulture. This can be conducted in box system, where trays are piled up one above other or shelf rearing on stands (Figs. 8.24 and 8.25). In India, young-age silkworm rearing is conducted in young-age rearing centre called Chawki Rearing Centre (CRC) under strict supervision of experienced technical staff (Sekharappa et al. 1995). During rearing, worms should not be touched with dirty hands, and cleaning can be carried out with cotton or nylon nets.

8.4.4.1 Disease Monitoring During Young-Age Rearing

Eggshell testing under microscope is recommended for detection of incidence of pebrine disease. The second moult larvae should be tested under microscope for infection of other diseases including pebrine. Rearing batch with incidence of disease should be rejected. Exclusive mulberry garden for young-age rearing can be maintained on recommended package of technology to ensure good feed to young worms.

Fig. 8.24 Young instar rearing (box system)

Fig. 8.25 Young instar rearing (tray)

8.4.5 Late Instar Rearing

Late-age rearing has utmost importance in silkworm rearing because they grow thousandfolds and eat major quantity of mulberry leaves during this age. The silk glands develop fast in third instar stage and reach maximum by the end of fifth instar. In fifth instar alone, silkworm grows about seven- to eightfolds from fourth age.

The late-age silkworm rearing, though looks easier than young-age rearing, is vulnerable to diseases, management failures and pest attack. It is, therefore, necessary to have proper technology for late-age rearing in different agroclimatic zones that suits to the rearers of all the economic groups (Tazima 1991). Late-age rearing requires comparatively lower temperature (26 °C in fourth and 24 °C in fifth instar)

Table 8.10 Requirement for late instar rearing

Factor	III Age	IV Age	V Age
Temperature (0 °C)	25–26	24–25	23–24
Relative humidity (%)	75–80	70–75	65–70
Quantity of leaf (kg)	50–55	175–200	1000–1200
Bed area (sq. m)	4.05–9.00	9.00–20.7	20.7–43.2

and less humidity (65–70 %) than the young-age rearing. Leaf consumption increases gradually and reached maximum during the fourth to seventh day of fifth instar (Table 8.10).

Silkworm rearing technique varies from place to place according to the suitability of the agroclimatic conditions of the area (Sekharappa et al. 1994). The most commonly used techniques of silkworm rearing in different area are:

1. Shoot rearing
2. Tray and shelf rearing

8.4.6 Shoot Rearing of Late-Age Worms

Commercially this method is more popular in sericulture industry. Fourth and fifth instar silkworms are voracious eaters and consume about 94 % of total leaf consumption. Harvesting of this quantum of mulberry leaf and feeding to silkworms are immensely labour-intensive. China and Japan practise shoot rearing for fourth and fifth instars to save labour input for many years (Sekharappa et al. 1997).

Shoot rearing comprises of feeding entire leaves attached on shoots (Fig. 8.30) instead of individual mulberry leaves to fourth and fifth instar silkworms (Narasimhamurthy and Subramanyam 1998). The shoot rearing has the following advantages:

1. It saves 50–60 % labour and 15–20 % leaf. Leaf cocoon ratio is lowered, and cost of rearing on equipment is reduced.
2. Shoot rearing requires minimum handling of larvae and, hence, bears minimum risk of contamination in silkworm rearing platform.
3. Leaves attached to shoots retain leaf moisture for longer time and provide quality food to silkworms with better aeration (Fig. 8.31).
4. Cocoon quality and productivity get improved in shoot rearing.

8.4.6.1 Rearing House for Shoot Rearing Racks

There is need of 520 square ft floor area for 100 DFLs rearing of bivoltine silkworms. A rearing house of 52 ft length × 10 ft width can accommodate 100 DFLs rearing. Rearing of 100 DFLs worms require about 700 square ft bed space (Rajan et al. 2003), which can be achieved by putting two rearing racks of 23′×5′×5′ ft with three tiers in 1.5–2 ft distance (Figs. 8.26 and 8.27). Shoot racks can be made up of wood, steel or locally available materials (Figs. 8.28 and 8.29). The rearing bed can

Fig. 8.26 Standard shoot rack

Fig. 8.27 Conventional shoot rack

Fig. 8.28 Fifth instar rearing on shoot rack

Fig. 8.29 Fifth instar rearing shoot rack

Fig. 8.30 Other shoot rack

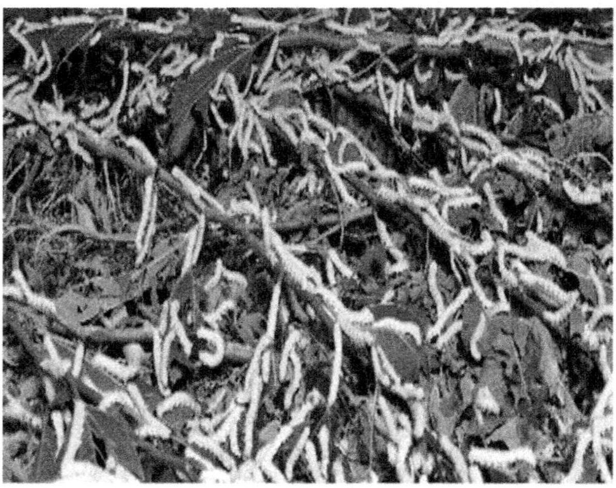

Fig. 8.31 Fifth instar larvae feeding on shoot leaves

be made up of nylon mesh or bamboo mats supported below with criss-cross steel wires. Six inch border all around in each tier is required to prevent silkworms falling from the bed.

8.4.6.2 Shoot Harvest and Preservation

Mulberry shoots of about 4 ft height are cut with sharp sickle (Fig. 8.17), pruning scissor or mechanical harvester during cooler hours and transported to preservation room under cover of wet gunny or polythene sheet. Shoot preservation room should be thoroughly cleaned with soap and 1 % bleaching powder solution everyday in the morning before preservation of shoots. Shoots are placed vertically upright during preservation.

8.4.6.3 Feeding of Silkworms

First to third instar larvae are reared on trays, and fourth instar larvae after two feedings in shelf rearing are transferred to shoot rearing racks spreading evenly. Shoots are placed over worms widthwise, bottom and top end of shoots alternatively arranged to ensure even mixing of different ages/qualities of leaves. Shoots should be fed twice in a day in winter/monsoon and thrice during summer.

Bed cleaning needs to be done once on the second day of fifth instar. This can be done either by net or rope. In the rope method, two ropes of 20 ft length are spread over the rearing rack keeping a distance of 3 ft between two ropes. Two feeds of shoots are given over the rope, and when the larvae crawl on the top of the shoots, the rope ends are pulled to the centre. The old bed is separated and removed, and the new bed is spread evenly again. In the net method, net of 5′ × 6′ with 2 cm^2 mesh size is spread over the rearing bed, and two feeds are given over the net. The net is then lifted to clean the bed. Requirement of mulberry leaf and shoot for 100 DFLs (a unit of 100 disease-free egg laying) rearing is explained in Table 8.11.

8.4.6.4 Care During Moult

The worms showing symptoms of moult develop shiny skin, and larvae stop feeding mulberry leaves. When 90 % of worms are settled for moult, feeding should be stopped and slaked lime powder can be dusted with the help of muslin cloth to reduce humidity in rearing bed. After 5–6 h of lime dusting, unsettled, undersized and diseased worms should be discarded in case of tray rearing. Feeding can be resumed when 90–95 % worms are out of moult.

Table 8.11 Estimate of mulberry leaf/shoot requirement for 100 DFLs rearing

	Time duration		Quantity of mulberry leaf (kg)	
Larval stage	Eating period	Moulting period	Leaf	Shoot
First instar	4 days	12 h	5–6 kg	00
Second instar	3 days	6 h	14–15 kg	00
Third instar	4 days	14-16 h	50–55 kg	00
Fourth instar	4 days	24-30 h	180–200 kg	460 kg
Fifth instar	8 days	0	1000–1200 kg	2800 kg
Total	23 days	2–3 days	1259–1476 or 1500 kg	3260 kg

8.4.6.5 Use of Bed Disinfectant

There are chances of spread of silkworm disease due to leaf contamination and improper handling of worms. Hence, it is necessary to use bed disinfectant to avoid any contamination in the rearing bed. There are many bed disinfectants, namely, RKO, BPL, Labex, Resham Jyothi, Ankush, Sanjivani, Vijetha and Vijetha supplement. Any recommended bed disinfectant has to be applied over the moult-out silkworms before feeding them with mulberry leaves. Powdery formulation of bed disinfectant should be dusted at 5 g/square ft area with the help of fine muslin cloth, 20–30 min before resumption of feeding in every moult. About 4 kg of these disinfectants are required for 100 DFLs rearing in case of tray rearing and 6 kg in shoot rearing. The method of application of these disinfectants is same for all the formulations.

8.4.6.6 Mounting and Harvesting

Mounting can be done by three methods: hand picking, shoot shaking (Jobarai) and self-mounting. The spinning worms are collected by hand picking and placed on mountage at 40–50 worms/square ft for spinning/cocooning. If picking of spinning worms is delayed, worms may start spinning in the bed itself. In shoot shaking method (Fig. 8.34), when 5–6 % worms are matured, they can be handpicked and mounted. Later mulberry shoots are placed on the rearing bed. After 1 or 2 h, majority of the worms crawl over the shoot. The shoots are separated from bed and then shaken it jerkily to and fro manually on vinyl sheet spread on the floor to collect the dropped worms from shoots for mounting (Figs. 8.32, 8.33, 8.35 and 8.36). Remaining larvae left on the bed can be handpicked and mounted. In this method of mounting, 38 % of time can be saved without affecting the cocoon quality and reeling characters of cocoons. In self-mounting method, plastic mountages are placed over matured worms. Majority of the spinning worms crawl over to mountages for cocooning. But 25 % of worms spin cocoons on rearing bed. Self-mounting affects adversely the quality of cocoons. The cocoons can be harvested on the 6th day after

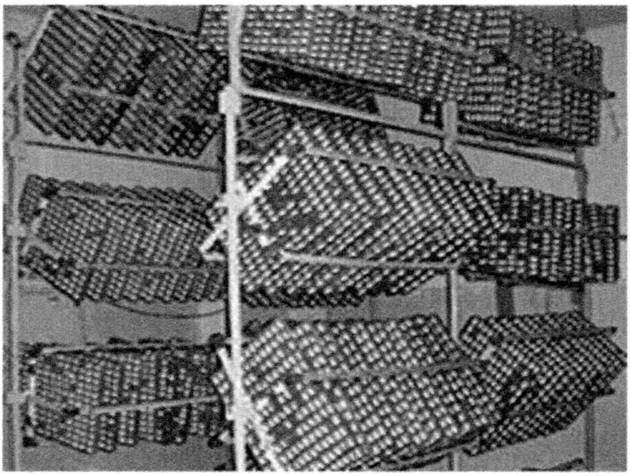

Fig. 8.32 Cocooning in rotary mountages

Fig. 8.33 Full grown 5th instar silkworm feeding mulberry leaves

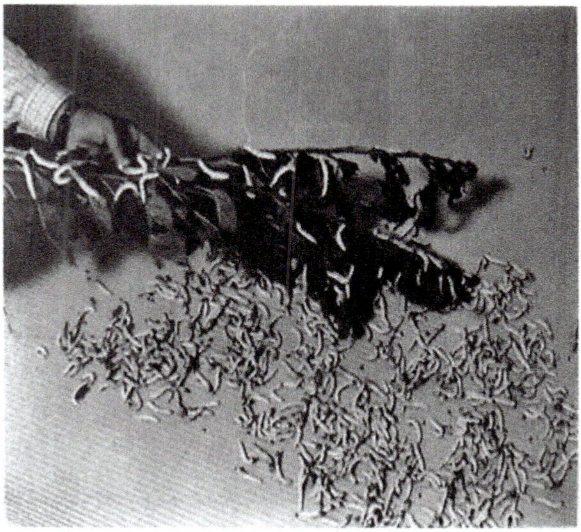

Fig. 8.34 Shoot shaking (Jobarai) method for separation of spinning worms

Fig. 8.35 Spinning larvae crawling on rotary mountage

Fig. 8.36 Cocooning in rotary mountages

mounting. Defective cocoons are sorted out before taking them to cocoon market. Transportation of cocoons should be done during cooler hours of the day in loosely packed cotton/gunny bags/plastic crates for marketing.

8.4.7 Tray Rearing

It is estimated that approximately 480 square ft space is required for the rearing of 100 DFLs worms in final instar. This estimate holds good for multivoltine x bivoltine hybrid silkworm and traditional bivoltine hybrids; however, new bivoltine hybrids require about 700 square ft space for 100 DFLs or 50,000 larvae in final instar.

Approximately 1200–1500 kg mulberry leaf is required to rear 100 DFLs worms, 94 % of which is used in fourth and fifth or final instars. In final instar alone, leaf used is about 80–85 % of the total leaf consumption of all the five instars. Rearing articles required in tray rearing method are given in Table 8.12.

8.4.7.1 Silkworm Feeding Schedule

The fourth and early fifth instars require medium tender leaves, while late fifth instar needs medium to course leaves. It is necessary that plant twigs should be 65 days old after pruning to get required succulent mulberry leaves for the silkworm. Young instars require 65–70 % moisture in leaves. The mulberry leaves for final instar should contain less moisture as compared to early instars.

Larvae can be fed three or four times in 24 h depending upon climatic conditions; however, quantity of leaf should remain the same in three or four feeding schedules. The correct amount of feeding can be estimated by calculating the amount of leaf fed and leftover leaf. The approximate quantity of leaf requirement is 50 kg in third instar, 200 kg in fourth instar and 1200 kg in fifth instar for 100 DFLs rearing of new bivoltine hybrids (Table 8.11).

Table 8.12 Requirement of rearing equipments

Sl. no.	Name of rearing articles	Quantity of rearing articles required
1	Rearing tray	120 No.
2	Rearing racks	10 No.
3	Feeding stand	4 No.
4	Basin stand	4 No.
5	Wooden leaf chopping board	1 No.
6	Leaf chopping knife	2 No.
7	Leaf carrying trolley	1 No.
8	Plastic basin (big and small)	4 No.
9	Bed cleaning nets (preferably soft nylon)	20 No. 0.5 × 0.5 cm
		100 No. 1.5 × 1.5 cm
		250 No. 2.5 × 2.5 cm
10	Disinfection pump	1 No.
11	Plastic drum (big and small)	2 No.
12	Plastic collapsible mountages	150 No.
13	Deflossing machine	1No.

Larvae after every moult may be resumed with soft leaves followed by medium leaves. Feeding of fifth instar larvae can be resumed with soft leaves for three to four feeding after moult followed by medium and course leaves till spinning. Mulberry leaves (chopped or entire) should be distributed uniformly in the bed during every feeding (Figs. 8.37 and 8.38). Leaves are chopped in 0.2–0.8 mm^2 for first, 1–2 mm^2 for second, 2–4.5 mm^2 for third, 5–10 mm^2 for fourth and 10–20 mm^2 or entire leaf for fifth instar worms. In case of low humidity in the rearing room during summer, mulberry leaves on trays can be covered with newspaper after feeding. Newspaper should be removed 1 h before the next feeding.

8.4.7.2 Cleaning

Silkworm rearing bed should be cleaned once in a day using nylon or cotton nets (Fig. 8.39) of required mesh size (0.5–2 mm for first and second, 10 mm^2 for third and 20 mm^2 for fourth and fifth instar). Diseased, unhealthy or undersized worms must be picked up and removed before start of cleaning. Diseased and unhealthy worms should be collected in 2 % formalin or 5 % bleaching powder. Cleaning nets must be washed every day with soap solution. There should be minimum handling of silkworms.

8.4.7.3 Spinning of Cocoons

Silkworm grows optimum and reaches to its peak growth period in 6–8 days of fifth instar (at 24–25 °C). During the 5th to 7th day, the fifth instar larvae eat maximum mulberry leaves to attain peak growth and production of silk protein in silk glands. Silkworms slow down feeding after 8th day and ultimately stop feeding when silk glands become mature. Matured silkworms shrink in size, body becomes translucent, and worms start searching for dry places where they can spin cocoons. This is

Fig. 8.37 Conventional bamboo tray rearing

Fig. 8.38 Silkworm rearing on plastic trays

the time to pick up the worms from rearing trays/platforms and put them on mounting frames/mountages for cocoon formation (Figs. 8.36, 8.40 and 8.41).

Chandrakala et al. (1999) and Trivedi et al. (2003) have studied the application of phytoecdysteroid for early and uniform maturation of silkworms (ready for spinning cocoons). Phytoecdysteroids are extracted from plant *Silene gallica* (common catchfly plant), Caryophyllaceae, and diluted extract is sprayed over mulberry leaves before feeding them to silkworms. Application of phytoecdysteroid can bring down fifth instar duration by 24–36 h. This saves labour and minimizes the incidence of silkworm diseases during rearing.

The worms ready for spinning should be strictly fed with chopped coarse mulberry leaves, and bed should not be covered with paper cover. Humidity in the

Fig. 8.39 Fifth instar larvae on cleaning net

Fig. 8.40 Cocooning on plastic collapsible mountage

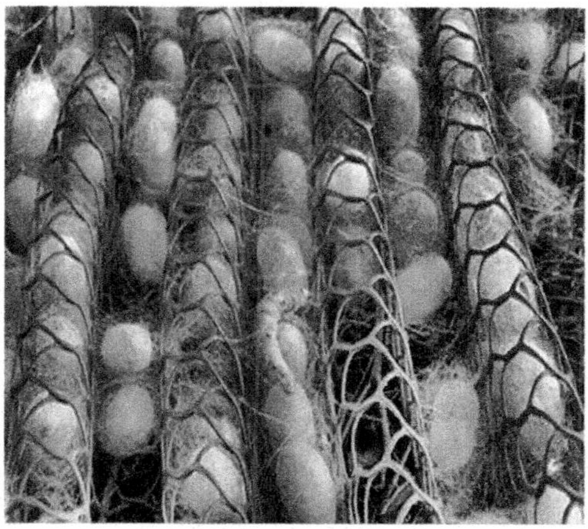

Fig. 8.41 Cocoon formation

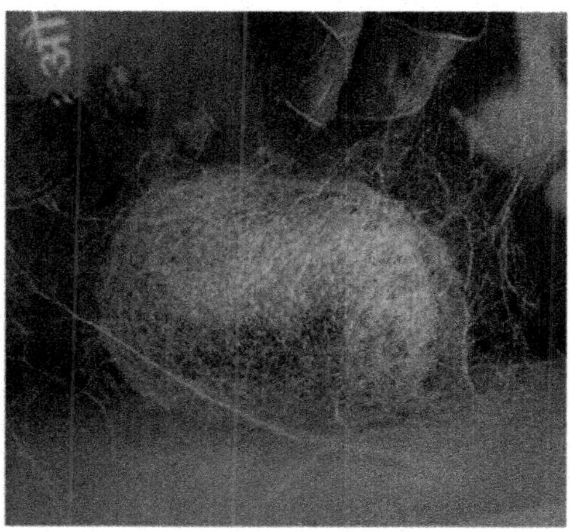

rearing bed should be less during spinning stage and free flow of air should be allowed inside the rearing room. There are chances that disease may come in the spinning stage. Hence, in the first picking, diseased larvae should be picked up and rejected. If the diseased worms are put on the mountage, it may spoil good cocoon by straining them. Only recommended number of spinning larvae should be put on mountages. Plastic collapsible mountage of 2′ × 3′ size can accommodate about 300 worms (Fig. 8.40). Overcrowding of worms on the mountage may result in double cocoon formation, and less number of worms may result in more floss on cocoons. Worms on mountages should be kept in a well-ventilated room with very less humidity. High humidity in the rearing bed and in mountage may adversely affect the reeling quality of the cocoons and the quality of the silk filament.

It is necessary to arrange required number of mountages in advance. Mountages should be cleaned and disinfected. Usually 40–50 spinning larvae can be mounted in 1 square ft area. Presently, the most widely accepted method of mounting in India is bamboo mountage (Chandrika) or plastic collapsible mountages. The plastic collapsible mountages are specially designed to place them in 2′ × 3′ size tray. Newspaper is spread below the mountage to absorb urination of spinning larvae. This newspaper placed below the plastic mountage should be removed gently without shaking mountage, once primary cocoon formation is over. This will help in reducing the humidity on mountage and improve the reelability of cocoons.

The desired temperature during spinning is 23–24 °C and humidity 60–70 %. The temperature above 28 °C and humidity more than 80 % are detrimental to the quality of cocoons. The initial 72 h of spinning is very important for the quality of cocoons, and attention must be given to keep the optimum temperature and humidity during this period.

Cocoon formation is completed by 72 h, and cocoon can be harvested on the 6th day in spring and autumn seasons, while in winter cocoon harvest may be done on the 7th day. Harvested cocoons must be spread in single layer on rearing trays to allow the air to pass through. Stained and defective cocoons must be separated and kept separately.

8.4.8 Cocoon Formation

There are different types of mounting frames used in silkworm rearing. Plastic collapsible mountage (Fig. 8.40), bottle brush (Fig. 8.42) and rotary mountage (Figs. 8.36) are used in China, Japan and India. In addition, sericulture farmers in India also use bamboo Chandrike mountage (Fig. 8.43) in Karnataka, Tamil Nadu and West Bengal. Among all the mounting frames used in sericulture industry, rotary mountage is the best for cocoon formation and quality of cocoons (Singh et al. 1998). The quality of mounting frames are assessed on the parameters, like cocooning percent, defective cocoon percent, floss percentage, reelability percentage and uniformity in cocoon shape (Table 8.13).

Fig. 8.42 Bottle brush mountage

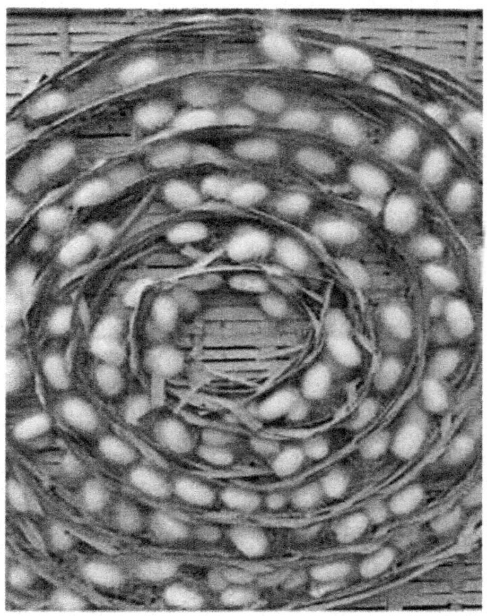

Fig. 8.43 Bamboo spiral mountage

8.4.8.1 Care During Spinning of Cocoons

The most suitable temperature and humidity during spinning of cocoons is 23–24 °C with 60–70 % RH. During spinning of cocoon, temperature more than 28 °C and humidity beyond 80 % affect adversely cocoon quality and cocooning as well. Mounting of premature worms on mounting frames adversely affects the quality of

Table 8.13 Effect of different mounting frame on cocooning

Type of mountage	Cocooning (%)	Defective cocoon (%)	Floss (%)	Reelability (%)
Bamboo chandrike	90.2	9.1	4.1	81.0
Plastic collapsible mountage	89.4	8.0	3.3	82.0
Bottle brush mountage	93.5	5.9	2.8	87.0
Rotary mountage	95.2	2.0	1.8	90.0

Source: (Muroga, 1996)

Table 8.14 Effect of mounting time on cocoon characters

Mounting time	Cocoon weight (g)	Cocoon shell weight (g)	Raw silk (%)	Cocoon filament length (m)
Control (most suitable time)	1.70	0.385	18.9	1124
Before 12 h	1.71	0.389	18.7	1159
Before 24 h	1.64	0.368	17.7	1063
Before 36 h	1.53	0.315	17.1	970
Before 48 h	1.46	0.304	16.8	969

Source: JICA publication, India

cocoons (Table 8.14) and cocooning. There should be proper ventilation in rearing room during spinning of cocoons. Doors and windows should be kept open to allow air current across the rearing room and to reduce the humidity.

8.4.8.2 Air Circulation and Cross Ventilation

Silkworm releases considerable amount of water in the atmosphere during rearing and spinning of cocoon. According to an estimate, 43.9 l of water is released in atmosphere in 50 DFLs rearing, out of which 21 l alone is released during spinning of cocoons. Air current across the rearing room helps in reducing humidity essentially required for spinning silkworms.

The mature silkworms excrete more urine during spinning of cocoons and create humid atmosphere in the mounting frame. If humidity in mounting frame remains higher during cocoon formation, the reelability of cocoons gets decreased, and chances of pupal mortality are increased. Hence, the air circulation and cross ventilation should be increased to improve reelability of cocoons (Table. 8.15). Harvested cocoons should be placed in single layer in a well-ventilated room before their marketing. Seed cocoons, if used for silkworm seed preparation, should be strictly placed on disinfected plastic trays in a single layer (Fig. 8.44).

8.4.8.3 Silkworm Rearing in India

Silkworm rearing in various parts of the country, like Karnataka, Andhra Pradesh and Tamil Nadu, is mostly carried out either in trays and on shoot rearing racks.

Table 8.15 Effect of temperature, humidity and air current on cocoon reelability

Temperature	Humidity (%)	Air current cm/s	Reelability
23°°C	65	0	92.3
		50	96.2
	90	0	53.5
		50	91.0
30 °C	65	0	85.2
		50	93.6
	90	0	28.4
		50	83.4

Source: JICA, India

Fig. 8.44 Storage of cocoons in plastic trays

Small and marginal silkworm rearers mostly prefer tray rearing (bamboo round tray, Fig. 8.37), while big rearers prefer shoot rearing. Mature silkworms are mounted on bamboo chandrika (bamboo mat with spiral ridges). Rearers also use plastic mountages and bottle brush mountages for cocooning.

Silkworm rearers of West Bengal and adjoining areas use bamboo trays called 'dala' for silkworm rearing. Silkworms are fed with mulberry leaves as shoots, and mature silkworms are mounted on bamboo chandrika for cocooning. Silkworm rearers in Maharastra also use bamboo trays for silkworm rearing. Few farmers adopt shoot feeding for silkworm rearing, and cocooning is carried out either on bamboo chandrika or plastic collapsible mountage.

Rearers of Kashmir and other parts of North India follow tray rearing, platform rearing, floor rearing and also shoot rearing in some pockets. Mature silkworms are mounted on locally available mounting materials, like dried grass, paddy stumps, mustard stumps, dried eucalyptus leaves, dried mango twigs with leaves, etc. Few farmers use plastic collapsible mountages for cocooning.

8.4.9 Maintenance of Hygiene in Silkworm Rearing

The following hygienic practices are followed during silkworm rearing:

1. Rearing room and rearing appliances are fully disinfected before start of silkworm rearing.
2. Silkworm rearers should wash their hands with soap or 500 ppm chlorine dioxide in 0.5 % slaked lime solution every time before entering into the rearing house. They should use separate footwear in the rearing house. Foot mat dusted with 5 % bleaching powder in slaked lime powder can be placed in the entry place.
3. Diseased and dead larvae in the rearing bed, if any, should be picked up using plastic forceps or chopsticks and collected into 5 % bleaching powder in 0.3 % slaked lime solution. It is advisable to burn diseased/dead larvae or bury in a pit away from the mulberry garden and rearing house.
4. It is advisable to use vinyl sheet/bag for collection for rearing waste and shifting it into waste pit. Disinfection of the vinyl sheet should be carried out after the disposal of waste by dipping in 2 % bleaching powder in 0.3 % slaked lime solution.
5. Floor of rearing house should be wiped daily with 2 % bleaching powder in 0.3 % slaked lime or 500 ppm chlorine dioxide in 0.5 % slaked lime solution. If the floor is mud floor, 5 % bleaching powder in slaked lime can be dusted at an interval of 3 days.
6. Mulberry leaves are stored in a separate room with separate entry from outside and into the rearing house. Leaf storage room/chamber should be washed periodically using 2 % bleaching powder in 0.3 % slaked lime solution.
7. Slaked lime powder should be dusted on worms settled for moult to reduce rearing bed humidity and to prevent secondary infection of disease.
8. Dusting of bed disinfectant is carred out after every moult (before resumption of feed) and on fourth day of final instar as per the schedule and quantity (Table 8.16), dusting of disinfectant powder should be avoided on silkworms settling for moult, under moult or on mulberry leaves that are to be eaten by larvae (Fig. 8.45).

In case of incidence of muscardine, a fungal disease of silkworm, during winter and rainy seasons, 2 % Dithane M-45 (75 % WP) or 2 % Captan (50 % WP) in kaolin should be dusted on the 4th day of final instar larvae at 5 g/square ft.

Table 8.16 Schedule of bed disinfectant

After first moult, before feeding	3 g/sq.ft
After second moult, before feeding	3 g/sq.ft
After third moult, before feeding	5 g/sq.ft
After fourth moult, before feeding	5 g/sq.ft
On 4th day of final instar	5 g/sq.ft

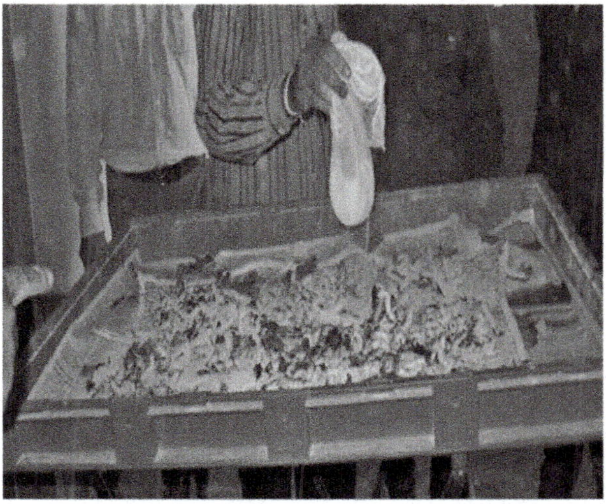

Fig. 8.45 Application of bed disinfectant

8.5 Mechanization in Sericulture

Sericulture industry has two important activities: one is pre-cocoon and another post-cocoon activity. Pre-cocoon activity involves 'on-farm activity', like mulberry cultivation, cultural operations in mulberry garden, plant protection, input application and irrigation. The other activity is 'silkworm rearing' which involves disinfection, mulberry leaves harvest for feeding silkworms, cleaning of rearing bed, mounting of mature silkworms for cocooning, silkworm seed production and other related activities (Verma and Dandin 2006). Post-cocoon activity involves cocoon harvest, sorting of cocoons, deflossing of cocoons, processing of cocoons, reeling of cocoons, twisting of silk threads, dying, weaving and fabric making. All these activities are labour-intensive and require skilled labour. After cocoon harvest, drying, sorting, etc. should be considered as post-cocoon activity.

Large-scale farming and silkworm rearing need huge manpower which makes it a less remunerative and combustive activity. The labour cost on silk production is around 60–70 %, mainly on mulberry cultivation and silkworm rearing.

Use of mechanization in farm activity and silkworm rearing can reduce labour input and increase income margin to sericulturist. Mechanization helps to reduce drudgery in sericulture and makes large-scale farming easy, remunerative and less time-consuming. It reduces production cost in silk production and increases productivity in each sector. The main components where mechanization is used are as below:

8.5.1 Mulberry Cultivation

Mulberry cultivation, especially the large-scale farming, needs small and big tools/ machineries. The important machineries and their use are listed below (Sarkar et al. 2003):

8.5.1.1 Tractor and Power Tiller

These are used for tilling/ploughing of soil in mulberry garden. Few attachments of tractor, viz. hole digger, tractor trencher and fogger, are also used in mulberry garden (Fig. 8.46).

8.5.1.2 Power Tiller Sprayer

It is used for spray of chemicals/insecticides in mulberry garden.

8.5.1.3 Shoot Harvester

This is a mini tractor fitted with mechanical harvester for the cutting of mulberry shoots.

8.5.2 Silkworm Rearing

Commercial hybrid silkworm rearing requires big labour input, majority of which is shared by women labour. Silkworm rearing has skilled and nonskilled activities, which can be performed by specialized machines. Few of these are explained below (Rajan et al. 2003):

8.5.2.1 Power Sprayer

Power sprayer is used for disinfection of rearing building and rearing appliances more effectively and efficiently. This reduces labour input on disinfection and saves time to perform the activity (Fig. 8.47).

8.5.2.2 LPG Flame Gun

This is a small device with LPG cylinder, connecting tube and flame gun. This is used for burning of floss on rearing stand, cardboard and rotary mountage and to sanitize metallic equipments (Fig. 8.48).

Fig. 8.46 Power tiller operation (*Source*: Verma and Dandin 2006)

Fig. 8.47 Power sprayer for disinfection of rearing house and rearing equipments

Fig. 8.48 LPG flame gun for burning of floss on rearing stands, floor and card board rotary mountages

8.5.2.3 Leaf Chopping Machine

This device is more useful for large-scale rearing, which not only saves manpower but also ensures more hygienic condition during rearing. Mulberry leaves can be chopped in different sizes as per the requirement of rearing (Fig. 8.49).

8.5.2.4 Cocoon Deflossing Machine

Manual removal of floss from cocoons requires considerable manpower and more time which can be reduced by using this devise. This is more useful in silkworm seed production and in reeling unit (Fig. 8.50). Selvakumar et al. (2015) manufactured a low-cost cocoon deflossing machine, which is easy to operate, gives minimum jerks to pupae and can be operated both either by electricity or manually.

8.5.2.5 Cocoon Cutting Machine

Cocoon cutting is carried out for sex separation of pupae, which is an important activity in hybrid seed production. This device, though effective, has drawback of damaging pupae with knife injury.

Fig. 8.49 Leaf chopping machine for chopping of mulberry leaves

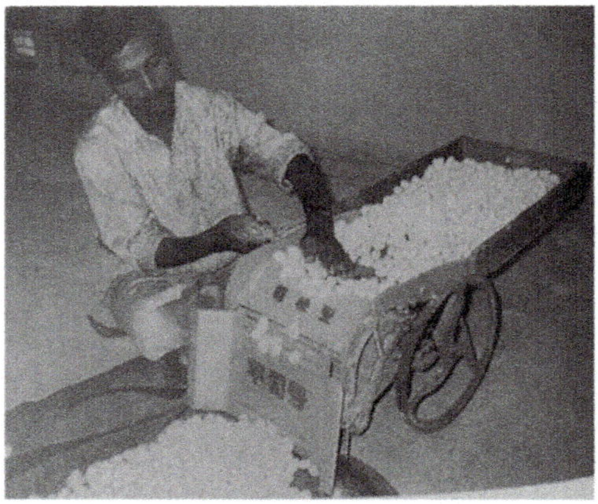

Fig. 8.50 Cocoon deflossing machine to remove silk floss from cocoon

8.6 Reeling of Cocoons

Reeling of cocoons is a process of unwinding of silk filament from cocoon shell. Cocoons are boiled in hot water to remove cementing element sericin and make silk filament free for unwinding. The silk filament is made up of two types of protein, i.e. fibroin (76 %) and sericin (22 %). The reeled silk is called raw silk. The raw silk may be of any thickness which is termed as denier in silk industry. The raw silk is twisted and used for the weaving of fabrics. Reeling process involves sorting of cocoons, cooking of cocoons, reeling of cocoons and re-reeling of silk filament

(Kawakami and Somashekar 1997). Sorting of cocoons is resorted to for the removal of defective cocoons, which are not reelable and can adversely affect quality of silk. Cocoon shell contains 72–81 % fibroin, 19–28 % sericin, 0.5–1.4 % fat/wax and 1–1.4 % colouring matter/ash.

Cocoon sorting is done to separate good/reelable cocoons and defective cocoons. Defective cocoons are classified under:

1. Double cocoons
2. Flimsy cocoons
3. Pierced cocoons
4. Stained cocoons
5. Deformed cocoons
6. Thin/open-end cocoons
7. Insect-damaged cocoons
8. Melt cocoons
9. Crushed cocoons
10. Mould-attacked cocoons
11. Unsized cocoons

8.6.1 Cooking and Reeling (Kawakami and Somashekar 1997)

Cooking process involves softening of sericin and unwinding of silk filament. The process of cooking is sequenced as below:

1. Wetting/retting (soaking of cocoons) at 70–75 °C for 60–90 s.
2. High-temperature treatment at 90–92 °C for 90–120 s.
3. Low-temperature treatment at 70–75 °C for 60–90 s. This process is also called as permeation.
4. Cooking of cocoons using steam at 97–98 °C for 90–120 s and immersing them in 97 °C water for 60 s.
5. Conditioning is carried out by sprinkling of cold water and reducing of temperature from 97 to 70 °C in about 4–6 min.
6. The cocoons are then brushed to find their free ends and put to reeling basin/machine. Cocoons are then reeled on reels/reeling machine.
7. A required number of baves from cocoons according to the size of raw silk thread to be reeled are combined and passed upwards through the guided eye, which is either a porcelain button or a jettebout with the object of primary grouping of several baves drawn and combined to form the required composite thread.
8. Each reeling basin is provided for reeling several ends in modern reeling machines (Fig. 8.51).

8.6.1.1 Re-reeling of Silk

Defects of direct reeled silk are removed by process of re-reeling of the silk already reeled onto standard reels. Re-reeling provides smoothness and strength to silk filament with minimal defects.

Fig. 8.51 Reeling of cocoons

8.6.2 Examination of Silk

Silk examination (Table 8.17) is done in a special constructed rectangular hall running east to west and having skylights of special ground glass windows on the northern side in order to obtain good defused light. Some of the silk examination tools are winding frame, seriplane, cohesion tester, denier scale and serigraph (Kawakami and Somashekar 1997).

8.6.2.1 Tactual Examination

Raw silk is evaluated under the following parameters:

1. Winding brakes (brakes)
2. Average size (d)
3. Size deviation (d)
4. Maximum deviation (d)
5. Evenness variation I (count)
6. Evenness variation II (count)
7. Evenness variation III (count)
8. Cleanness %
9. Neatness %
10. Low neatness %
11. Tenacity (g/d)
12. Elongation %
13. Cohesion (strokes)

Table 8.17 Visual examination and grading

Sl. no.	Characters	Gradation
1	Uniformity of colour, luster and hand	Good
		Fair
		Slightly inferior
		Inferior
2	General finish	
	(a) Finish	Loose/cut ends/raised ends/soiled/damaged
		(a) Good
		(b) Fair
		(c) Slightly inferior
		(d) Inferior
	(b) Gum spots	Nil/few/many
3	Nature	
	(a) Luster	Bright/medium/dull
	(b) Colour	
	White raw silk	White/greenish/brownish/grayish
	Yellow raw silk	Yellow/reddish/darkish/greenish/creamish
	(c) Hand	Ordinary/slightly rough/rough

Table 8.18 Table for raw silk classification

Thickness/size	Grades
18 and 19 denier and above	6A, 5A, 4A, 3A, 2A, A, B, C, D, E
34 denier and above	4A, 3A, 2A, A, B, C, D, E

8.6.2.2 Raw Silk Classification

The quality of silk is graded differently for different thicknesses of raw silk (Krishnaswami et al. 1987b).

Size Grades (Table 8.18)

18 denier and above 6A, 5A, 4A, 3A, 2A, A, B, C, D
19 to 33 denier 6A, 5A, 4A, 3A, 2A, A, B, C, D, E
34 denier and above 4A, 3A, 2A, A, B, C, D, E

8.6.3 Stifling of Cocoons (Krishnaswami et al. 1987b)

This is the process of killing the pupae without affecting quality of cocoon shell for reeling. There are several methods of killing live pupae inside cocoon shell. Live pupae can be killed by sun drying of cocoons, steam stifling, hot air stifling, X-ray radiation, infrared radiation and passing poisonous gas in cocoon chambers.

8.6.3.1 Hot Air Stifling (Krishnaswami et al. 1987b)

The most reliable and recommended method of stifling of cocoon is hot air stifling, which kills pupae without affecting the quality of cocoon shell. Cocoons are generally harvested from mountage on 6th or 7th day of spinning. These cocoons are deflossed (removal of floss from cocoons) and sold in cocoon market. Commercial reeling units purchase these cocoons from rearers and stifle them before reeling. Unstifled cocoons will encounter emergence of moth in 13–14 days of spinning by cutting one end of cocoons. This will turn them to pierced cocoons, which are not reelable as there will be several breaks in cocoon silk filament.

There are two methods of hot air drying of cocoons. In one method, cocoons are exposed to low to high temperature from 60 to 110 °C in 5–6 h. Another method recommends drying of cocoons in higher temperature of 100–110 °C and then subjected to lower temperature of 60–65 °C in 5–6 h. Stifled cocoons must be stored in a well-ventilated room and should be protected from insect damage.

8.6.4 Technical Terms Related to the Silk-Reeling Industry

1. Denier or size of filament: The size of filament is expressed as denier which is calculated as

$$\text{Denier} = \frac{\text{Weight of silk filament in gram}}{\text{Length of silk filament in metre}} \times 9000$$

2. Reelability: It is expressed in percent and calculated as

$$\text{Reelability\%} = \frac{\text{Total number of cocoons reeled}}{\text{Total number of castings or breaks}} \times 100$$

3. Percentage of raw silk yield:

$$\textit{Raw}\,\text{silk\%} = \frac{\text{Silk (weight)}}{\text{Cocoon weight}} \times 100$$

4. Raw silk recovery:

$$\textit{Raw}\,\text{silk recovery (\%)} = \frac{\textit{Raw}\,\text{Silk\%}}{\text{Cocoon shell\%}} \times 100$$

5. Renditta (this term is used to express quantity of cocoons required to produce 1 kg of raw silk):

$$\textit{Raw}\,\text{silk\%} = \frac{\text{Silk obtained}\,(\textit{wt.})\,\textit{By}\,\text{reeling of cocoons}}{\text{Cocoon reeled (weight)}} \times 100.$$

8.7 Conclusions

Silk is a commercial fibre of insect origin. Sericulture being ecoconservation activity provides high economic returns with low investment. Sericulture has been identified as employment generation industry which has about 60 % participation of women. Leaf harvesting and silkworm rearing are more effectively taken up by the women folk. Even silk reeling and weaving is largely supported by them. The industry provides gainful employment, economic development and improvement in the quality of life to the people in rural area. It plays an important role in antipoverty programmes and prevents migration of rural people to urban areas in search of employment. It provides job opportunity to the rural people at their doorstep. Money flows in silk industy from lower to higher state of the society. Sericulture being a cottage industry is practised primarily by rural poor people, and the final product, the silk, is consumed by the rich and effluent class of society. Majority of silk goods are purchased by the urban rich and the middle class.

Sericulture industry is the most scientifically organized low investment industry. It is highly disciplined but very sensitive to different guiding factors that contribute to the success of the industry. Silkworm rearers and traders are the highest-income gainers with 54.6 and 17.8 % profit share, respectively, while cocoon reelers, silk twisters and silk weavers get 6.6, 8.7 and 12.3 % profit share, respectively, of the total silk production activities (Table 8.19). The whole silk industry is broadly divided under cocoon production and silk production. The cocoon production involves mulberry leaf production, an on-farm activity and indoor activity of silkworm rearing. Silk production covers reeling or extraction of silk from cocoons, twisting of silk thread and conversion of fabrics from silk thread.

Cocoon production is the most vital activity in the silk industry, which is greatly influenced by climatic factors, disease-free environment in rearing house/room, good-quality nutritious mulberry leaf and managerial skill of commercial silkworm rearers. The highly productive silkworm hybrids, supply of quality silkworm hybrid seed in time, good rearing season and disease management during rearing of silkworms are the guiding factors for higher-quality cocoon production and production of superior-quality raw silk. Superior-quality cocoons should bear compactness and uniformity in shape and size, should contain less defective cocoons (double cocoons, stained with urine or disease larvae, mountage pressed, thin-layer cocoons, open-end cocoons, cocoons with dead pupae) and should have higher reelability and higher silk recovery. Production of quality cocoons can be ensured through

Table 8.19 Income distribution in the sericulture industry

Operator	Income share
Silkworm rearers	54.6 %
Cocoon reeler	6.6 %
Raw silk twister	8.7 %
Weaver	12.3 %
Trader	17.8 %

disease-free environment, higher pupation rate and use of proper mountages (plastic collapsible or rotary mountage) for spinning silkworms (cocoon formation).

Sericulture being a highly labour-intensive activity has lot of drudgery involved in it. The sericulture scientists have attempted to evolve new technologies, which are women- and old-aged-person-friendly with less drudgery involved. There are as many as six silkworm cocoon crops from the same set of mulberry plantation in the tropical area of South India and West Bengal in India, while there are only two main silkworm cocoon crops during spring and autumn seasons in temperate areas of Kashmir, Himachal Pradesh and Uttarakhand in India. Majority of cocoon production in Japan and China comes from temperate areas. In India, the major part of cocoon production comes from tropical and subtropical areas of South India. Temperate area contributes mostly for bivoltine cocoon production.

Success of silkworm rearing and cocoon crop depends largely on proper disinfection of rearing houses and rearing appliances. Silkworm eggs are disinfected with 2 % formalin solution for 10 min before incubating them in 24–25 °C temperature and 80 % humidity. The surface disinfection of silkworm eggs ensures disease freeness in eggs especially pebrine disease. The incubation of eggs in optimum conditions leads to proper development of embryo in side egg, which ultimately results in good hatching and success of silkworm rearing.

There are two recommended methods of silkworm rearing, i.e. tray rearing and shoot rearing. Tray rearing needs more floor area for rearing of 100 DFLs silkworms than shoot rearing. Tray rearing needs about 1200–1500 kg of mulberry leaves, while shoot rearing requires about 3500 kg of mulberry shoots for rearing of 100 DFLs silkworm seeds. Quantity of leaf or shoot requirement varies on the population of worms especially at fifth instar rearing, and the larval period is greatly influenced by the climatic factors particularly the temperature in the rearing room. Under shoot rearing method, silkworms are reared on trays up to third instar and then transferred on mulberry shoots (1–1.5 m length) for fourth and fifth instar rearing. Rearing bed is cleaned only once after fourth moult with the help of two ropes placing them over the bed keeping distance of 3 ft between them. Shoots are placed over these ropes keeping their apical portions alternatively arranged so that every worm gets quality mulberry leaves for feeding. Ropes are lifted alongwith shoots and worms to clean the rearing bed. Shoots are given twice in winter or spring season when environmental temperature is low and thrice during summer in 24 h.

Bombyx mori is a poikilothermic animal. Temperature more than 32 °C adversely affects the metabolic activity of larvae, while temperature below 15 °C decreases metabolic activity resulting in decrease in biological functions in larvae.

It is mandatory to apply silkworm body disinfectants (dusting) with the help of muslin cloth after every moult and on 4th day of final instar at 4 kg in tray rearing and 6 kg/100 DFLs shoot rearing to avoid secondary contamination during silkworm rearing. Dusting of disinfectant is done at 3–4 g/sq. ft in first to third instar rearing and 5–6 g/sq. ft area in fourth and final instars. In India, the most commonly used body disinfectants are Vijetha and RKO. Dusting of such bed disinfectants should not be conducted when silkworms are feeding. The most vital step in silkworm rearing is cocooning, i.e. silk cocoon formation on mounting frame by spinning worms.

Silkworms during spinning of cocoons excrete a lot of water content on mountages which increase humidity in rearing room. High humidity condition during spinning affects adversely cocoon quality; hence, it is necessary that room humidity during spinning of cocoons should be around 60–70 %. Doors and windows should be kept open to allow free air flow in room and to keep temperature down. The most suitable mountages for spinning worms are plastic collapsible mountage and rotary mountage. Usually 40–50 spinning larvae can be mounted in one square ft area. Hence, attention must be given to keep the optimum temperature and humidity during spinning of cocoons. The cocoon formation is completed by 72 h, and cocoons can be harvested on the 6th day during spring and autumn seasons and on the 7th day in winter season. Harvested cocoons are kept in single layer on rearing trays to allow the air to pass through. Stained and defective cocoons are separated and stored separately.

Good-quantity mulberry leaf has direct bearing in healthy silkworm and good cocoon crop. Silkworms do not have any established immune or defence mechanism against pathogens. Recent studies on *Bombyx mori* have established the production of red fluorescent protein (RFP), viral inhibitory factor and bacterial inhibitory substances in the midgut and haemolymph of the silkworm (Basavaraja et al. 2005). These substances inhibit the multiplication of virus and bacteria and provide a defence mechanism against the pathogens. Production of these proteins and substances depend on the quality of feed and the inorganic tissue salts present in microquantities. The presence of these micronutrients has direct effect on the growth of silkworm, economic characters of cocoons and the reproductive potential.

Mulberry plant takes about 6 months to grow as bush plant and about 2 years to grow as tree for commencement of silkworm rearing in tropical conditions. Mulberry, once planted, will go on supporting silkworm rearing year after year for 15–20 years depending on inputs and management provided. Sericulture waste from rearing can be recycled as inputs to garden. Dried mulberry twigs and branches are used as fuel in place of firewood and therefore reduce the pressure on vegetation/forest. Mulberry can also be cultivated as intercrop with numerous plantations. Mulberry being a deep-rooted perennial plant can be raised in vacant lands, hill slopes and watershed areas. Currently, only about 0.1 % of the arable land in the country is under mulberry cultivation.

Acknowledgements The authors are thankful to the Central Silk Board, Bangalore, India, for providing the opportunity to learn and understand the core issues related to sericulture industry in general and Indian sericulture in particular, which has helped us in writing this manuscript.

References

Aruga H (1994) Principles of sericulture. Oxford and IBH Publishing Co. Pvt. Ltd., New Delhi, pp. 1–362

Banno, Y. et al (2013) Cytology 66(3):283–287. Cited in Iyengar MNS (2013) Long-term preservation of *Bombyx mori* silkworm strains using frozen ovaries. Indian Silk 4(3):22–23

Barber E (1991) Prehistoric textiles. Princeton University Press, Princeton

Basavaraja HK, Aswath SK, Suresh Kumar N, Mal Reddy N, Kalpana GV (2005) A Text Book on Silkworm breeding and genetic. Published by Central Silk Board, Ministry of Textiles, Government of India, Bangalore

Benchamin KV, Gapuz JV, Jayaramraju P (1990) Influence of photoperiod on emergence, fecundity and fertility in multivoltine breeds of silkworm *Bombyx mori.* L. Indian J Seric 29(1):110–118

Bose PC, Bindroo BB (2001) A comparative biochemical study of seven promising mulberry (*Morus alba* L.) varieties under rainfed condition of sub-tropic region. Indian J Seric 40:171–173

Boulnois L (1966) The silk road. George Allen & Unwin editions, Ltd, London

Bruford MW, Wayne RK (1993) Microsatellites and their application to population genetic studies. Curr Opin Genet Dev 3:939–943

Chakraborti S, Singhal BK (1996) Seasonal variation in ascorbic acid content of promising mulberry genotypes. Indian J Plant Physiol 1:298–299

Chandrakala MV, Maribashetty VG, Jyothi HK (1999) Application of phytoecdysteroid in sericulture. Curr Sci 74:341–346

Chang KC (1960) Archaeology in China. Cambridge University Press, Cambridge

Chatterjee SN, Rao CGP, Chatterjee GK, Aswath SK, Patnaik AK (1993) Correlation between yield and biochemical parameters in the mulberry silkworm *Bombyx mori* L. Theor Appl Genet 87:385–391

Chauhan TPS, Lakshmanan V, Rajlakshmi E, Thiagarajan V (1997) Biotechnological approach in synthesizing new genotypes of silkworm *Bombyx mori* (L). pp 183–193 cited in book Mordern trends of research in Zoology. Edited by MM Prakash and PK Sandhvi. Pub. RBSA Publisher S.M.S. Highway, Jaipur, India

Chauhan TPS, Tayal MK, Dhar A, Khan MA (2008) Mulberry pest forecasting and control measures. Asian Text J 17:67–70

Chauhan TPS, Tayal MK, Rajlakshmi E, Bania HR, Dhar A, Bindroo BB, Khan MA (2011) Yellow mite (*Polyphagotarsonemus latus*): Incidence on mulberry. Indian Silk 2 (50 old) (4):8–10

Chinya PK, Ramesh Kumar S (1994) Studies on molecular basis of immunity in silkworm *Bombyx mori*: Identification of antibacterial antiviral protein. In: Annual report (1993–94), Central Sericultural Research and Training Institute, Mysore, India. p 66

Dandin SB (2006) Mulberry host plant management and improvement. In: Proceedings of regional seminar on prospects and problems of sericulture as an economic enterprise in North India. Organised by Central Silk Board, Dehradun, India. pp 1–10.

Das, BC (1983) Mulberry taxonomy, cytology and breeding. Cited in Lead paper presented in National Seminar on Silk Research and Development. Publisher Central Silk Board, Bangalore, India. pp 1–9

Das PK, Choudhury PC, Gupta YK, Ghosh A, Datta RK (1993) Effect of different levels of nitrogen, phosphorus and spacings on the growth, yield and nutritional quality of mulberry under rainfed conditions. Indian J Seric 36(2):201–207

Doira H (1978) Genetic stocks of silkworm. In: Tazima Y (ed) The silkworm: an important laboratory tool. Kodansa, Tokyo, pp 53–81

Dutta RK (1994) Biotechnology signals new phases. In: Souvenir, international conference on sericulture global silk scenario- 2001, 25–29 Oct 1994. pp 71–78

Francis W (2002) The silk road: two thousands year in the heart of Asia. University of California Press, Berkeley, pp 9, 13–33

Ghosh A, Ambika PK, Mishra RK (1997) Effect of varities, spacings and fertilizer doses on growth, yield and quality of mulberry. Indian J Seric 36(2):138–141
Gopinath OK, Rajkumar S, Balakrishna R, Qadri SMH (2013) Giant Agrican snail noticed infesting mulberry gardens in Tamil Nadu. Indian Silk 3(10,11):12
Hashimoto H (1948) Sex-limited Zebra, an X-ray mutation in the silkworm. J Seric Sci Jpn 16:62–64
Hayaskiya K, Nishida J, Matsubara F (1969) The production of antiviral substances, red florescent protein in the digestive juice of the silkworm larvae. Appl Entomol Zool 4:154–155
Irfan I, Vishal M, Ramegowda GK, Dhar A (2013) Leaf webber-a serious insect pest of mulberry in Kashmir. Indian Silk 3 (51 old) (10&11):8–10
Iyengar MNS (2013) Long-term preservation of *Bombyx mori* silkworm strains using frozen ovaries. Indian Silk 4(3):22–23
Jolly MS (1987) Appropriate sericulture techniques. Edited and Printed by Geetanjali Printers, Bangalore, pp 1–175
Kathirvel P, Chatterjee GK, Rao CGP, Chinya PK, Dutta RK (1994) RFLP mapping in *Bombyx mori* (L). Annual Report (1993–94), Central Sericultural Research and Training Institute, Mysore, India
Kawaguchi E (1934) Chromosomal behavior in tetraploid female of the silkworm. J Seric Jpn 5(2):73–79
Kawakami K, Somashekar TH (1997) Operational manual for reelers on bivoltine silk reeling technology. CSTRI, Bangalore
Khan MA, Dhar A, Zea SB, Trag AR (2004) Pest and diseases of mulberry and their management. Publisher Bishen Singh Mahendra Pal Singh, Dehradun, India pp. 1–68
Krishnaswami S (1988) Chawki- the crucial input for the success of sericulture. Indian Silk 26(10):82–87
Krishnaswami S, Narasimhanna MN, Suryanarayan SK, Kumarajan S (1987a) Sericulture manual 2 – silkworm rearing. In: FAO manuals on sericulture. United Nations, Rome, pp 10–131 (Reprinted by Central Silk Board, Bangalore)
Krishnaswami S, Madhava Rao NR, Suryanarayan SK, Sundaramurthy TS (1987b) Sericulture manual 3 – silk reeling. In: FAO manuals on sericulture. United Nations, Rome, pp 3–112 (Reprinted by Central Silk Board, Bangalore)
Kuhn D. (1982) The silk workshops of the Shang Dynasty (16th-11th century BC). In: Hu Daojing, exec. (ed) Explorations in the history of science and technology in China. Shanghai Classics Publishing House, Shanghai, pp 367–408
Machii H (2015) Current status of sericulture and prospects for silk industry in Japan. Indian Silk 5(10, 11):18–19
Machii H, Katagiri K (1991) Varietal differences in nutritional value of mulberry leaves for rearing silkworms. JARQ 25(3):202–208
Maji MD, Banerjee R, Chattopadhyay S, Saha AB, Ghosh PK, Ghosh PL (2006) Eco-friendly management of foliar diseases of mulberry through botanical and bio-control agents. In: Proceedings of regional seminar on prospects and problems of sericulture as an economic enterprise in North India. Organised by Central Silk Board, Dehradun, India, pp 74–80
Manjula A, Hurkadli H (1986) Cold acid treatment of bivoltine silkworm eggs, *Bombyx mori* for tropical countries. Séricologia 26(1):138–141
Manjula A, Hurkadli HK (1995) Chilling of silkworm eggs, *Bombyx mori*. Indian Text J 11:70–74
Minamizawa K (1997) Moriculture-science of mulberry cultivation. Oxford & IBH Publishing Co. Pvt. Ltd., New Delhi (Translated from Japanese)
Mita K (2012) Silk genomes opens new silk road for insect science and its applications. Indian Silk 2(10):4–9
Miyashita Y (1986) A report on mulberry cultivation and training methods suitable to bivoltine rearing in Karnataka, pp 1–7
Mukharjee NG (1899) Hand book of sericulture in Bengal. Calcutta
Mukharjee NG (1907) A bird eye views of Indian sericulture. Calcutta

Muroga A (1996) Comprehensive report. Central Silk Board, Bangalore

Nanavati MM (1965) Silk from grub to glamour. Central Silk Board, Bombay

Narasimhamurthy CV, Subramanyam KV (1998) Effect of shoot rearing on silkworm cocoon crop performance. Indian J Seric 27(2):156–158

Nunome J (1992) The archaeology of fiber before your eyes: a compilation of photographs of fiber artifacts. Senshoku to Seikatsusha, Kyoto

Rahaman MS, Gandhi Dass S, Ghosh MK, Bajapai AK (2010) A new cold tolerant mulberry variety genotype. Indian Silk 1(7) (old 49):8–9

Rajan RK, Singh GB, Himantharaj MT, Nataraju B, Balavenkatasubbaiah M (2003) In: Kawakami K, Yanagawa H (ed) Illustrated working process of new bivoltine silkworm rearing technology. Published under JICA, India P.E.B.S. project, CSRTI, Sripampura, Mysore-570008, India, pp 1–92

Rangaswami G, Narasimhanna MN, Kasiviswanathan K, Sastry CR, Jolly MS(1987) Sericulture manual 1-Mulberry cultivation. In: FAO manuals on sericulture. United Nations, Rome, pp 34–82 (Reprinted by Central Silk Board, Bangalore)

Rau MA (1967) The sacred mulberry of Joshimath, U.P. Indian Forester 93(8):333–335

Sarkar A, Balakrishna R, Mogli T, Thippeswamy T, Urs MKP, Yamajaki A (2003) In: Kawakami K, Yanagawa H (ed) Illustrated working process of new mulberry cultivation technology. Published under JICA, India P.E.B.S. project, CSRTI, Sripampura, Mysore, pp 1–102

Sekharappa BM, Radhakrishna PG,Raghuraman R, Dandin SB (1994) Consolidation of new rearing technologies in the field for productivity improvement by 2001. In: Proceeding of international conference on sericologia, pp 214–221

Sekharappa BM, Muniraju E, Gururaj CS (1995) Use of isolation chamber for young age (chawki) silkworm in triopics. Séricologia 35(3):525–534

Sekharappa BM, Gururaj CS, Raghuraman R, Dandin SB (1997) *Shoot feeding for late age silkworm rearing*, Technical bull no 5. KSSR&DI, Bangalore, pp. 1–27

Selvakumar T, Mallikarjuna, Shivanna MS, Shashidhar K, Angadi BS (2015) A new low cost drilling machine cocoon deflosser. Indian Silk 5(10, 11):26–27

Sethuraman BN (1991) Isolation and characterization of antiviral factor(s) in silkworm, *Bombyx mori*. Dissertation submitted to Central Sericultural Research and Training Institute, Mysore

Sharma DD (1998) Ecofriendly approach for management of root knot. Indian Silk 37(4): 15–16

Shetty KK, Dilip Kumar K, Kumaresan P (2015) Sericulture industry in India – prospects Galore. Indian Silk 5(10, 11) (old 53):14–17

Shree MP, Ravikumar K, Nagveni V (2006) Infection of giant African snail on mulberry. Indian Silk 45(3):14–16

Singh GP, Himantharaj MT, Kamble CK, Datta RK (1998) Impact of different types of mountages on cocoon and rearing parameters in silkworm *Bombyx mori* L. Séricologia 38(2):321–325

Singh RN, Samson MV, Dutta RK (2000) Pest management in sericulture. Indian Publishers Distributor, Delhi

Singhavi NR, Reddy MJM, Chakravarthy S, Rekha M, Sarkar A, Subbaswamy MR, Datta RK (2000) Leaf nutrient status of a few popular mulberry varieties. Indian J Seric 39(1):93–96

Soumen C, Rita B, Debasish B, Atul Kumar S, Bindroo BB (2013) Antioxidants from mulberry to combat degenerative chronic diseases. Indian Silk 3(10,11):4–7

Tanaka Y (1916) Genetic studies in the silkworm. J Coll Assoc Sapporo 7:129–155

Tazima Y (1941) A simple method of sex discrimination by means of larval markings in *Bombyx mori*. J Seric Sci Jpn 12:184–188

Tazima Y (1954) Mechanism of sex discrimination in the silkworm, *Bombyx mori*. Proc Int Cong Genet Cytol 6(Suppl):858–960

Tazima Y (1972) Handbook of silkworm rearing, agriculture technical manual-1. Fuji publication Co. Ltd., Tokyo, pp. 122–127

Tazima Y (1978) Silkworm Egg. Published by Central Silk Board, Bombay

Tazima Y (1991) Improvement of biological functions in the silkworm (translated from Japanese). Oxford & IBH Publishing Co. Pvt. Ltd, New Delhi

Tikader A (2001) Occurrence of wild mulberry silkworm in Uttaranchal. Indian Silk 40(3):19–20

Tikader A, Rao AA, Mukharjee P (1999) *Ex sito* conservation of oldest mulberry tree. Indian Silk 38(2):17–18

Tikader A, Rao AA, Thagavelu K (2002) Geographical distribution of Indian mulberry species. Indian J Pl Genet Resources 15(3):262–266

Trivedi K, Nair KS, Ramesh M, Gopal N, Kumar N (2003) Early and uniform maturation in silkworm *Bombyx mori* L. by phytoecdysteroid extracted from a plant of family Caryophyallaceae. Int J Indus Entomol 7:65–68

Varadarajan L (1988) Silk in Northeastern and Eastern India: The indigenous tradition. Mod Asian Stud 22(3):561–570

Verma S, Dandin SB (2006) Mechanisation in sericulture. Central Sericultural Research and Training Institute, Mysore

Watt G (1893) Dictionary of the economic products of India, vol VI, Pt. iii. W.H. Allen, London

Silkworm Diseases and Pests

9

Mukesh K. Tayal and T.P.S. Chauhan

Contents

9.1 Introduction

Silkworm *Bombyx mori* L. has very weak immune system to combat pathogens causing various diseases (Lea 1993; Samson and Chandrashekaraiah 1998; Govindhan et al. 1998). Domestication of this insect for thousands of years might have caused reduction or even loss of natural resistance/tolerance to certain pathogens. Tropical races of *B. mori* have higher degree of tolerance to pathogens than bivoltine and univoltine races. The knowledge on biology of pathogens, their incubation period and association with host tissue need to be clearly understood before reaching to the management of disease. Resistance or susceptibility to diseases is a

M.K. Tayal (✉)
Regional Sericultural Research Station, Central Silk Board, Miransahib, Jammu 181101, Jammu and Kashmir, India
e-mail: drmukeshtayal@gmail.com

T.P.S. Chauhan
Ex-Scientist, Central Silk Board, 21-A, Deshmeshpuri, Indira Gandhi Marg, Niranjanpur, Dehradun 248171, Uttrakhand, India
e-mail: tpschauhan.1956@gmail.com

Omkar (ed.), *Industrial Entomology*, DOI 10.1007/978-981-10-3304-9_9

comparative term (Aizawa 1962; Jaiswal et al. 2003; Gupta et al. 2006; Govindhan et al. 1998). Baig et al. (1991) screened 21 races of *B. mori* including the hybrids for their relative susceptibility to nuclear polyhedrosis viruses (NPV) under natural and induced conditions. Tripathi and Gaur (1991) reported higher incidence of diseases in *B. mori* during autumn than in spring crop. The extent of cocoon crop loss during autumn is reported to the tune of 35–47 %, while it is 15–20 % in spring season in north India. There is no race of *B. mori* complete by resistance to diseases and pests. There are two ways to control infection of diseases in silkworms (Samson and Chandrashekaraiah 1998; Watanabe 2002). The first and foremost tool of disease control is pathogen-free environment before and during silkworm rearing. Second step is to prevent entry of pathogens into the rearing room. It is very difficult to create zero pathogen environments in rearing room, but preventive measures can bring down the pathogens' load to a tolerable level. Kawakami (1982) isolated 395 strains of *Aspergillus* fungi from cooperative rearing house for young silkworms. Out of those 79.8 % were found virulent to silkworms. 87 % of these strains of fungi were tested resistant to formalin and 48 % showed resistance to mercuric fungicides.

It is always better to nip the disease in bud condition. The effective disinfection, preventing entry of pathogens by means of hygiene during silkworm rearing and providing optimum environment for healthy growth of silkworms, contributes to good cocoon crop. Hence, most effective disinfectant should be used before the start of rearing. Genuine commercial brand disinfectants, right kind of chemical for a particular pathogen, disinfection in appropriate time and testing of efficacy of disinfection are key factors to control diseases in silkworms during rearing.

9.2 Common Silkworm Diseases

Grasserie and flacherie are the major killer silkworm diseases prevailing in India. The incidence of silkworm diseases is higher during autumn season than spring. The high disease incidence during autumn season can be correlated with the higher pathogen load, wide range of temperature fluctuation from day to night, high humidity in late stage of rearing and bad quality of mulberry leaf (Table 9.1). According to an estimate, there is average silkworm crop loss of 15–47 % in India, 30 % in developing countries of Southeast Asia and 10–15 % in developing countries, like Japan, China and Italy. Out of total crop losses in India, 57.22 % loss is due to flacherie (bacteria), 33.88 % grasserie (virus), 2.32 % pebrine (microsporidia) and 0.487 % muscardine (fungus). Prevalence of silkworm diseases is more in autumn (35–47 %) than in spring season (15–20 %).

Seven types of silkworm bacteria have been identified, out of which five are more pathogenic causing considerable cocoon crop loss. These bacteria are *1. Streptococcus bombycis*, 2. *Staphylococcus* sp., *3. Serratia marcescens* Bizio, 4. *Micrococcus* sp., *5. Pseudomonas* sp., *6. Bacillus* sp. and *7. Bacillus thuringiensis* (*Bacillus sotto* Ishiwata).

Viruses causing disease in silkworms are *Bm* infectious flacherie virus (IFV), *Bm* densonucleus virus ($DNV_{1,}$ *Bm* $DNV_{2),}$ *Bm* nuclear polyhedrosis virus (NPV) and *Bm* cytoplasmic polyhedrosis virus (CPV).

Table 9.1 Pathogenesis and incubation period of silkworm diseases

	Silkworm diseases			
Pathogenesis and factors	Grasserie (nuclear polyhedrosis)	Flacherie	Muscardine and aspergillosis	Pebrine
Occurrence	Throughout the year but intensive during rainy and summer season	Throughout the year but intensive during rainy and summer season	During winter and rainy season. While high temperature and high humidity are congenial for aspergillosis	Any time and not season specific
Causative agent	*Bombyx mori* nuclear polyhedrosis virus (*Bm* NPV), a baculovirus	The disease is caused by *Bombyx mori* infectious flacherie virus (IFV) and *Bombyx mori* densonucleosis virus (densonucleosis) and Kenchu virus. *Staphylococcus* sp., *Streptococcus* sp., *Bacillus* sp. and *Serratia* bacteria, individually and in association, cause bacterial flacherie	White muscardine is caused by a fungus *Beauveria bassiana*. Aspergillosis is caused by *Aspergillus flavus*, *A. oryzae* and *A. tamari*	A microsporidian, *Nosema bombycis* and strains of *Nosema* sp. NIK-2r, NIK-3h and NIK-4 m
Source of infection	Secondary infection from diseased silkworm, its body fluid and contaminated silkworm rearing waste	Dead/diseased silkworms, its faecal matter, gut juice body fluid, alternate hosts, contaminated rearing house and appliances	Mummified/diseased silkworms, contaminated silkworm rearing waste, environment, house, appliances and alternate hosts (most lepidopteron pests)	Egg surface, diseased silkworm, its faecal matter, alternate hosts, contaminated rearing house, appliances and mulberry leaves are source of infection. Transovarial transmission of infection from mother moth to the progeny

(continued)

Table 9.1 (continued)

Pathogenesis and factors	Silkworm diseases			
	Grasserie (nuclear polyhedrosis)	Flacherie	Muscardine and aspergillosis	Pebrine
Infection and spread of disease	Silkworm gets infected when it feeds on contaminated mulberry leaves with polyhedra extrude from the body of diseased worms along with haemolymph. The infected larvae die in 5–7 days due to diseases	Pathogen with faeces and vomit contaminate mulberry leaf and silkworm rearing environment. Silkworms get infected on feeding the contaminated mulberry leaf.	Mummified/diseased silkworms contaminated rearing environment, house, appliances and several agriculture pests/alternate hosts form the source of pathogen to spread the disease	Silkworm gets infected on consumption of mulberry contaminated with spores of pathogen. Infected female moth lays diseased eggs. Pebrinized silkworm extrudes faecal matter and gut juice containing pathogens, which contaminates the rearing environment, appliances and mulberry. Alternate hosts may also become source of contamination
Favourable conditions	High temperature and high humidity	High temperature, fluctuation in temperature and humidity and poor quality of mulberry leaves	Low temperature and high humidity for muscardine and high temperature and high humidity for aspergillosis	There are no specified conditions
Persistence of pathogen	The pathogen persists for over 15 years under natural rearing house conditions	The pathogen of IFV persists more than 1 year under natural rearing house conditions	The conidia are known to remain viable for 1 to 5 years depending upon the temperature and humidity	Spores remain viable for a maximum period of 225 days in wet soil and 135 days in wet compost. Spores coated with tissues under humid conditions have been found to be viable for several years

(continued)

Table 9.1 (continued)

Pathogenesis and factors	Silkworm diseases Grasserie (nuclear polyhedrosis)	Flacherie	Muscardine and aspergillosis	Pebrine
Susceptibility of silkworm	Young-age larvae are more susceptible than late-age larvae. Resistance increases about tenfold with each moult. Bivoltine breeds are more susceptible than the multivoltine breeds	The early instar larvae are more susceptible to viruses (IFV, CPV, DNV) than late-age instars, while in case of bacterial infection, mortality is more in late-age instar before spinning. Bivoltine races are more susceptible than multivoltine	Fifth instar larvae are more susceptible than younger instar. The moulting stage of the larvae is more susceptible. In case of aspergillosis, first-stage larvae are more susceptible with 100 % mortality than third and fourth instars. Multivoltine breeds are less susceptible. (Chinnaswamy and Devajah 1984; Bhagyalakshmi 1994)	Japanese breed, bivoltine breeds, early instar silkworms and newly moulted worms are highly susceptible

9.2.1 Grasserie (Nuclear Polyhedrosis)

Grasserie has the lower gestation period of 5–7 days. The high temperature and high humidity favour the pathogen multiplication. The disease incidence is usually higher in late-age rearing, which can be attributed to higher pathogen load, poor hygienic conditions during rearing and poor quality of mulberry leaves.

9.2.1.1 Symptoms

At the early stage of infection, silkworms appear normal, but microscopic examination of haemolymph may indicate the presence of polyhedra within haemocytes. The larvae do not settle for moult and develop shining integument. The diseased larvae develop swollen intersegmental region and larval integument becomes fragile, which breaks easily oozing turbid white fluid containing huge numbers of polyhedra. The larvae become restless and move towards rim of the rearing tray. The infected larvae die in 5–7 days (Fig. 9.1).

9.2.1.2 Prevention/Control of Disease

Incubation period of pathogen *Bm* nuclear polyhedrosis virus (*Bm* NPV) is very short. Hence, the larvae once infected cannot be cured. The disease either can be prevented through disinfection or by preventing secondary infection to healthy larvae. Three-tier disinfection and maintenance of hygiene during silkworm rearing should be practised in every cycle of rearing. The detailed procedure of disinfection

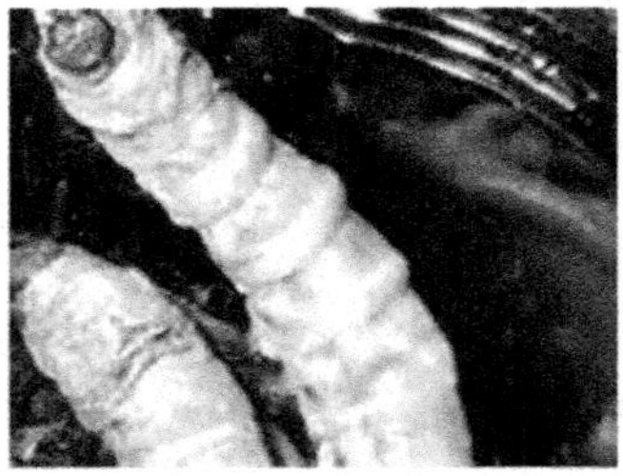

Fig. 9.1 Grasserie infected silkworm

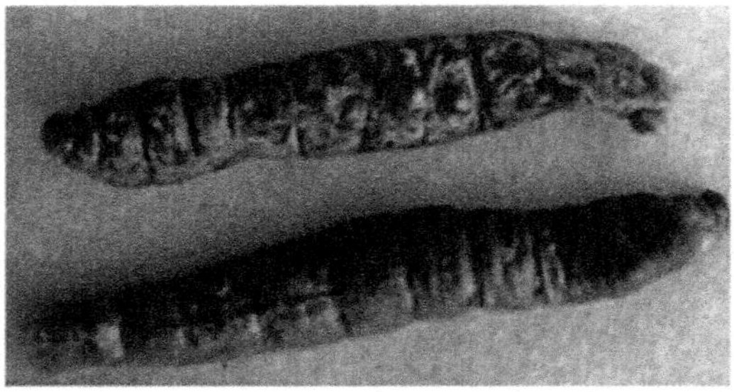

Fig. 9.2 Flacherie infected silkworms

has been described under heading "Disinfection" in this chapter. A spray of 0.3 % slaked lime solution in addition to usual disinfection of rearing house and appliances is more effective in checking the high incidence of disease. Diseased larvae should be picked out from rearing beds and destroyed.

Any bed disinfectant (Sanjeevani, Vijetha, R-K-O, Suraksha or Resham Jyothi), patented commercial formulations should be applied as per recommended schedule and quantity (at 3 g per square feet of bed area after the first and second moults and at 5 g per square feet of bed area after the third and fourth moults and fifth day of the fifth instar before half an hour of resumption of feeding). Mulberry pests as alternate hosts for the pathogens of the disease should be destroyed from the mulberry garden (Balvenkatasubbaiha et al. 2012a b). Usually 100 DFLs of rearing requires 4 kg bed disinfectant in tray rearing and 6 kg in shoot rearing.

9.2.2 Flacherie

Flacherie being the syndrome associated with various viruses (IFV, CPV, DNV and Kenchu virus) and bacteria (*Staphylococcus*, *Streptococcus*, *Serratia* and *Bacillus*) is the major disease during autumn season. The main factors responsible for the spreading of disease are fluctuation in temperature, poor quality of mulberry leaves and high humidity. Mostly this disease occurs in the late age, and the silkworms die within 10–15 days after infection. If the infection is associated with *Streptococcus* sp. of bacteria or with viruses (IFV, CPV, DNV etc.), the mortality is reported to be at higher rate with reduced incubation period of 5–7 days (Fig. 9.2).

9.2.2.1 Symptoms

Infected larvae become lethargic and loose appetite. Larvae do not crawl over mulberry leaves for feeding but hide under leaves. As the infection grows, larvae become soft and flaccid. The larval growth retards; larvae become dull, vomit gut juice and sometimes show symptoms of diarrhoea and excrete chain-type faecal bodies. The cephalothoracic region becomes translucent (common in infectious flacherie, densonucleosis, cytoplasmic polyhedrosis). The midgut becomes turbid white due to formation of polyhedra, and the faeces become whitish black (cytoplasmic polyhedrosis). The larvae on death putrefy, develop different colours and emit foul smell due to bacterial infection.

9.2.2.2 Prevention/Controls of Disease

Besides proper disinfection and hygiene maintenance during rearing, larvae should be provided with good quality of mulberry leaves, proper spacing to larvae and proper ventilation. The temperature and humidity are maintained inside rearing room and sudden change in temperature and humidity is avoided. Hygienic measures should be maintained during silkworm rearing. Diseased larvae should be picked out from rearing beds and destroyed. Bed disinfectant (Sanjeevani, Vijetha, R-K-O, Suraksha or Resham Jyothi), patented commercial formulations should be applied after every moult before half an hour of resumption of feeding (at 3 g per square feet of bed area after the first and second moults and at 5 g per square feet of bed area after the third and fourth moults and fifth day of the fifth instar before half an hour of resumption of feeding). Mulberry pests, which act as alternate hosts for the pathogens of the disease, should be destroyed from the mulberry garden (Balvenkatasubbaiha et al. 2012a, b).

9.2.3 Muscardine and Aspergillosis

Muscardine is a fungal disease and occurs under high humidity conditions. The incidence of this disease prevails during autumn season. The favourable condition for this disease is low temperature and high humidity. However, aspergillosis requires high temperature and high humidity (Chinnaswamy and Devajah 1984; Chinnaswamy et al. 1986; Bhagyalakshmi 1994).

9.2.3.1 Symptoms

The infected larvae loose appetite and become inactive. They develop oil stain-like or black marks without clear border on the body surface of worms, and they die due to aflatoxin secretion by the fungus. After death, larvae become flaccid and gradually become harder. The fungus overgrow the larvae, develops fruiting body and produces conidia on surface giving white colour in case of white muscardine and green colour in aspergillosis. Aspergillosis generally occurs in early instar silkworms. The early instar silkworms become inactive and die without clear morphological symptoms (Fig. 9.3).

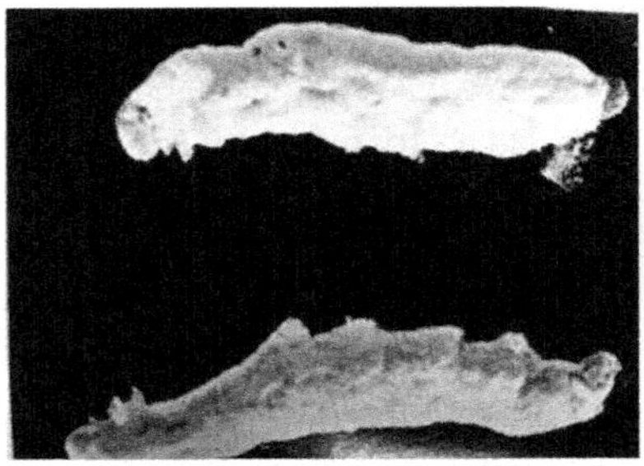

Fig. 9.3 Muscardine infected silkworms

9.2.3.2 Prevention/Control of Muscardine and Aspergillosis

Three-tier disinfection and maintenance of hygiene during silkworm rearing should be practised in rearing house. *Aspergillus* sp. is comparatively more tolerant. Kawakami (1982) reported that some of the *Aspergillus* fungi are resistant to formalin and mercuric fungicides. It is necessary to adopt a complete disinfection package (described later in this chapter) followed by disease monitoring and control during rearing. Hence, in addition to the normal disinfection with chlorine dioxide, 3 % formalin/2 % bleaching powder in 0.3 % slaked lime solution should be used for disinfection. Diseased worms should be picked up from rearing bed and destroyed at the earliest. Slaked lime powder should be dusted at every moulting stage and in between to regulate humidity in bed and keep the rearing bed dry. Good ventilation is also useful to regulate the humidity.

In addition to the bed disinfectant, it is recommended to dust 1–2 % (1 % for first and second instars, 2 % for third, fourth and final instars) of Dithane M-45 in slaked lime or 1–2 % Captan in Kaoline (China clay white-coloured clay material) on silkworm body after every moult and on the fourth day of final instar at 3–5 gm/square feet. As a specific measure, the use of sprinkle of formalin chaff (0.6 to 0.8 % formalin mixed with partly burnt paddy husk in 1:10 ratio) over worms at 3–5 g per square feet of bed area is also beneficial to check the fungal infection.

9.2.4 Pebrine

Disease is caused by a microsporidian, *Nosema bombycis* Nageli. It is a transovarian disease, which spreads from infected mother moth to its offspring. This disease may prevail in any season of rearing. The infection in early instar larvae (transovarian or in the first instar) results in death of larvae by the third instar, while in late-age infection worms spin cocoons, and mortality may occur in the pupal or moth stage.

The pebrine spores enter into silkworm digestive system through contaminated mulberry leaves with pebrine spores. The digestive juices act on the spore; the polar filament is extruded and detached from the spore (sporoplasm). The sporoplasm creeps out as an amoebula containing two nuclei which soon fuse into one to form planont. These planonts pass between epithelial cells of silkworm intestine into haemocoel and multiply by binary fission. The planonts move to various tissues of

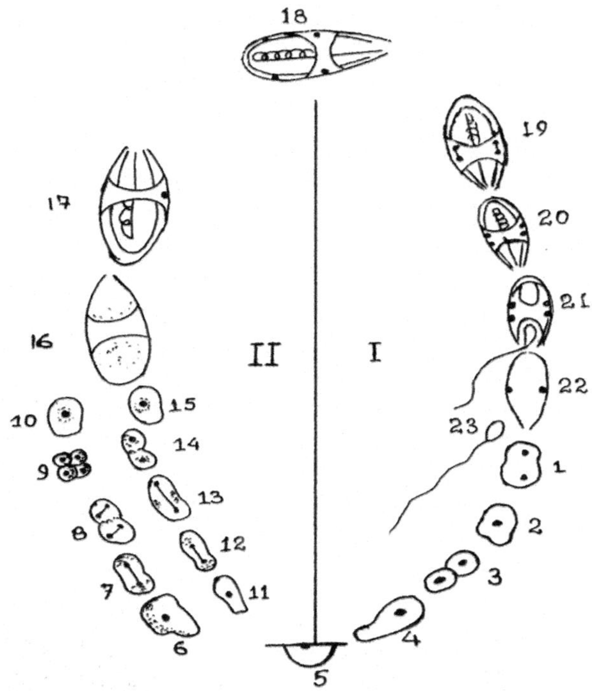

Fig. 9.4 Life cycle of *Nosema bombycis* Nageli. *I* extracellular stages, *II* intracellular stages, *1–4* planonts, *5–17* meronts, *18–20* sporulation, *21–22* spores in the midgut, *23* extrusion of polar filament, *24* amoebula leaving spore

the body including gonads, malpighian tubules, silk glands, etc. These planonts enter into cell and soon become covered with a membrane and called as schizont or meront. These schizonts undergo division by binary fission, budding or multiple division. These schizonts become sporonts and finally spores inside the cell. This stage in the life cycle of *Nosema bombycis* Nageli is called sporogamy. The life cycle of the *Nosema bombycis* Nageli is completed in about 7 days in cold climate and about 4 days in hot weather (Fig. 9.4).

9.2.4.1 Symptoms

Silkworm larvae suffering with pebrine disease do not show any symptoms until the disease is in advance stage. Generally suffering larvae have poor appetite and irregular moulting and growth. In advance stage of disease, larvae become sluggish and dull, move slowly, become paler and translucent, loose appetite and cease to moult. Unequal development of larvae is the most striking feature of pebrine disease. Black spots and wrinkles on the body are visible symptoms in advance stage of disease.

In highly infected rearings, pupae become flaccid and fail to metamorphose into adult. The infected moths show the symptoms of poor eclosion, inactiveness and improper mating following by poor egg laying. The eggs laid by the pebrine-infected mother moths are less in number with poor and irregular hatching. Pebrinized mother moths lay unfertilized and dead eggs in higher numbers.

9.2.4.2 Prevention/Control of the Disease

Scientific processing and mother moth examination under microscope for detection of pebrine spores should be strictly followed to ensure disease freeness of silkworm seed. Periodically testing of silkworm pellets as per procedure should be followed during rearing, and if spores are detected, the whole batch of silkworm crop must be rejected. Disinfection of rearing house, appliances, surroundings and surface disinfection of silkworm eggs with 2 % formalin solution for 5 min should be strictly followed. Hygienic condition should be maintained during silkworm rearing and silkworm seed production. Diseased silkworms/cocoons/eggs should be destroyed under fire immediately after detection of disease. Any bed disinfectant (Sanjeevani, Vijetha supplement, R-K-O, Suraksha, Ankush or Resham Jyothi) should be applied as per recommended schedule and quantity (3–5 gm/sq. ft.). Mulberry pests as alternate hosts for the pathogens of the disease should be destroyed from the mulberry garden (Balvenkatasubbaiha et al. 2012a, b).

9.3 Silkworm Disease Monitoring

Hitherto only preventive measures are available to control silkworm diseases during rearing. However, there have been attempts to cure these diseases in the past. Nataraju et al. (2005) have studied the feasibility of oral vaccination of silkworms against NPV. Balvenkatasubbaiha et al. (2012a) have developed an ecofriendly botanical-based formulation "Amruth" to cure grasserie and flacherie. 20 g of "Amruth" powder is dissolved in 1 litre water, and 70 ml of this solution is sprayed over 1 kg mulberry leaves to feed silkworms. The scientific group claims that it is first ever curative measures to control grasserie and flacherie diseases.

Monitoring silkworm disease incidences during rearing is the first step of disease control. These diseases are transmitted either horizontally or through vertical transmission (from parents). Horizontal transmission can be checked through proper disinfection, while vertical transmission of disease can be minimized/prevented through proper monitoring of disease and nip the pathogens at a particular stage to stop their further multiplication. Pebrine is one such dreaded disease, which is transmitted both horizontally and vertically. The possibility of infection of silkworms and transmission of pathogens are studied during disease monitoring.

9.3.1 Prerearing Monitoring

9.3.1.1 Rearing Environment

Dust and dirt from the rearing house floor, walls, rearing trays and soil in front of rearing building should be collected and mixed with 0.6 % potassium carbonate solution (four times to the weight of dust) and homogenized thoroughly. Samples can be tested under microscope to detect the presence of pathogens in the environment. This process should be repeated before and after disinfection or at any stage.

9.3.1.2 Forced Eclosion Test

This test is conducted for seed cocoons to be used for preparation of hybrid silkworm seed. Samples are collected from every cocoon lot and kept at constant temperature of 32 °C to get 1- or 2-day early emergence. The emerged moths are tested under microscope.

9.3.2 Monitoring During Rearing

9.3.2.1 Brushing Stage

Silkworm egg surface is disinfected by treating eggs in 2 % formalin for 5 min before placing them in the incubation room. The eggs/egg shells should be tested as per the procedure under microscope for the presence of pebrine spore.

Unhatched eggs, dead eggs and egg shells are collected in a mortar separately and ground thoroughly. The material is mixed with 0.6 % potassium carbonate solution (four times to the weight of material). Samples are examined under microscope.

Microscopic examination of chawki worms can be taken up before their distribution to the farmers in the third instar stage. The testing can be conducted in the third-stage larvae after the second moult before resumption of feeding to ensure the disease-free status of chawki worms. The information generated may also give an account of secondary infection of diseases, if any, in the rearing house.

9.3.2.2 Larval Stage

Silkworm larvae should be monitored at every moulting stage, by taking samples from each and every silkworm crop. Undersize/unequal, lethargic larvae unsettled for moult and dead larvae should be tested under microscope. This process of microscopic examination will help in the forecasting of possible infection or epidemic in the final stage of rearing.

9.3.2.3 Faeces Test

Faeces test should be conducted by collecting faecal pellets during the entire course of silkworm rearing. This test is most convenient for monitoring of pebrine disease. The larvae suffering with pebrine extrude pebrine spores with faecal pellets everyday, which contaminate the rearing environment.

9.3.2.4 Pupal Stage

This is more useful for seed crop rearings. The remnant gut examination of pupa is more reliable test for pebrine detection than the entire pupa. Before the cocoon lots are selected for grainage, pupae from randomly selected cocoons are subjected to

Fig. 9.5 Centrifuge machine

pebrine examination. The pupa is punctured ventrally, and the midgut region is extracted and crushed in a mortar with pestle. The material is then subjected to microscopic examination. The practice of pupal gut testing is carried out till emergence.

9.3.3 Mother Moth Examination

Silkmoths after oviposition are collected in perforated cardboard boxes/covers. The boxes are properly numbered as per egg sheets and preserved in well-ventilated room at 20–25 °C for mother moth examination. Examination is conducted under microscope.

Mother moth examination is conducted by using fresh (green) or dry moths to detect pebrine disease. For green moth testing, 0.6 % of potassium carbonate (K_2 CO_3) and for dry moth testing, 2 % potassium hydroxide (KOH) is used for crushing. Individual mother moth examination is conducted at grandparent level, where each mother moth is crushed in mortar pestle. Under cluster moth examination, group of 20 moths are homogenized in 80 ml of 0.6 % K_2 CO_3 solution (Fig. 9.6). The floating tissues are filtered through muslin cloth or absorbent cotton and centrifuged (Fig. 9.5) at 3000 rpm for 5 min to get clear sediment pellet. These pellets are dissolved in 0.6 % K_2 CO_3, and the samples are examined under microscope in 600× magnification.

9.3.3.1 Identification of Pebrine Spore

Pebrine is detected by the presence of oval or ovocylindrical and highly refractive spore of *N. bombycis* in the macerated tissue of host. The spore measures 3.8 μm in length and 2.6 μm width and exhibit characteristic Brownian movement (Fig. 9.7).

Fig. 9.6 Moth grinder

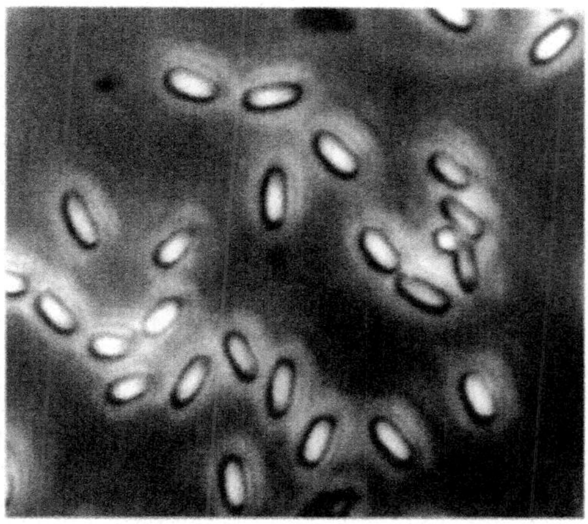

Fig. 9.7 Pebrine spores under microscope

9.3.4 Forecasting of Diseases

A bulk of data on seasonal incidence of diseases for considerable period of 5–6 years is required before arriving at any conclusion on disease calendar. Seasonal incidence of diseases and associated factors, such as temperature, humidity, rainfall, etc., are collected to make a complete calendar of silkworm disease incidence and the forecasting system. The forecasting of diseases can be addressed through strict disease monitoring system.

9.3.4.1 Disease Incidence in Relation to Ecofactors

Occurrence of silkworm diseases has definite correlations with the climate, temperature and rainfall. The spring season with a moderate temperature and low humidity in India has minimum incidence of diseases. The average temperature during the spring season in north Indian subtropic regions remains between 22 and 26 °C, and humidity varies between 55 and 65 % with hardly any rain. The low incidence of silkworm diseases in the spring season can also be correlated with low density of pathogens in the environment.

The sporadic information available on the disease incidence in north India during summer suggests that most of the crop losses are due to the grasserie disease. However, the total disease incidence and crop losses during the summer are lesser than the monsoon season. The monsoon season with high temperature and high humidity (32–36 °C and RH 90–100 %) provides a congenial atmosphere for the multiplication of bacteria and viruses, the causative agents of flacherie and grasserie diseases.

9.3.4.2 Analysis of Forecasting Data

The autumn is the second important season for the silkworm crop in north India. It has equal chances of failures due to the flacherie and grasserie diseases. At the start of silkworm rearing by the first week of September, the atmospheric temperature and humidity are higher which gradually go down and reach lowest of autumn before the onset of winter. During the last week of September and first week of October, there is hardly any rain in subtropics of north-west India. The difference between day and night temperatures is 10–15 °C, which may be the possible cause of flacherie disease in silkworms during autumn season.

The information available of the meteorological data and the seasonal disease incidence reveals that the density of the specific pathogens is the first cause of silkworm disease, whose intensity varies from season to season due to variation in temperature, humidity and rainfall.

A quality of mulberry leaves has direct impact on the health of silkworms. The silkworms develop tolerance against diseases is chawki stage. Healthy chawki worms reared on good quality mulberry leaves ensure good cocoon harvest (Tayal et al. 2003, 2004, 2006, 2007, 2014).

There are many lepidopteran pests on mulberry which are carrier/alternate hosts of pathogens of silkworm diseases, i.e. *Spilosoma obliqua* Walker, *Glyphodes pyloalis* Walker besides *Daphnia pulverulentus* Walker, etc. (Chinnaswamy et al. 1986). A high incidence of disease during autumn season can be correlated with the high density of pest during September to October with poor quality of mulberry leaves.

9.4 Disinfection

9.4.1 Disinfection of Rearing House and Appliances

Killing of pathogens to control the disease incidence is termed as disinfection. The most effective disinfection is carried out with the help of various chemicals, having capability to act against pathogens. Disinfection of rearing space, leaf storage/preservation room, mounting hall and all rearing appliances under use in silkworm rearing is of paramount importance for successful silkworm rearing. In fact disinfection before and after silkworm rearing is the key factor for good cocoon crop. The disease-causing pathogens remain hidden in various places. Hence, special attention is required to conduct thorough and complete disinfection before the commencement of rearing. It is important to use chemicals/disinfectants of desired concentration and potency. Shelf life of the chemical should be checked before its use. Due attention is needed to select the appropriate disinfectant to make the disinfection meaningful. There are four commonly used disinfectants in sericulture industry:

1. Formalin 2 % + 0.05 % detergent
2. Bleaching powder 2 % in 0.3 % slaked lime solution
3. Chlorine dioxide (Sanitech) 500 ppm in 0.5 % slaked lime
4. 0.5 % slaked lime solution

9.4.1.1 Formalin

Disinfection with formalin is more effective in close door airtight condition at temperature more than 25 °C with 70 % humidity. It is commonly used for fumigation of rearing appliances and rearing rooms. Formalin is commercially available as 36–37 % formaldehyde solution. Efficacy of formalin increases as the temperature rises. For effective use of formalin, the rearing space should be made airtight and formalin solution mixed with 0.05 % detergent solution. It has a strong irritant effect on the eyes and nasal mucous membrane. Hence, utmost care is needed during its application. It is advisable to wear gloves and gas mask whenever formalin spray is done.

Preparation of 2 % formaldehyde solution

$$\frac{\text{Required concentration of formalin} \times 1000}{\text{Available concentration formaldehyde}}$$

Example:

$$\text{For 01 litre of 2 \% formaldehyde solution} = \frac{2 \times 1000}{36} = 55.55\ \text{ml}$$

Hence, 56 ml of commercial grade formaldehyde of 36 % concentration to be added to 944 ml of water to make 1 litre of 2 % formalin solution.

OR

$$\frac{\text{Available conc. of formaldehyde} - \text{Required concentration of formaline}}{\text{Required concentration of formaline}}$$

Example

$$2\%\ \text{formaldehyde solution} = \frac{36-2}{2} = 17$$

$$= \text{One part formaldehyde solution with 17 parts of water.}$$

9.4.1.2 Bleaching Powder

It is an amorphous powder with a pungent smell of chlorine and is also called as chlorinated lime powder. Being hygroscopic, it absorbs moisture from the atmosphere and becomes ineffective. The chlorine content of bleaching powder should be 30 % for effective disinfection, and it must be ensured that it is stored in sealed bag, free from moisture. The pathogens are killed with 2 % spray of bleaching powder solution due to release of nascent oxygen and the chlorine produced, which has bactericidal action. It has slow action and, therefore, mostly used as surface disinfectant. 2 % bleaching powder mixed with 0.3 % slaked lime is more effective disinfectant.

Preparation of 2 % bleaching powder solution To obtain 2 % solution of bleaching powder, 20 g of bleaching powder is dissolved with little amount of water making it as paste and solution is made up to 1 litre water. This solution ensures 0.6 % chlorine, which is effective against many pathogens including viruses.

9.4.1.3 Slaked Lime

Slaked lime has antiviral properties, absorbs moisture and is effective in regulating the bed humidity in the rearing room. It is very useful to spray or dust lime along with bleaching powder around the rearing house and premises of rearing. For disposal of dead or diseased larvae, slaked lime containers are more useful than formalin solution.

9.4.1.4 Chlorine Dioxide Solution

This is most effective disinfectant to kill bacteria and viruses causative agents of various silkworm diseases in sericulture industry. Chlorine dioxide solution in 500 ppm is very effective disinfectant in sericulture industry and is safe to human body. To prepare chlorine dioxide solution of 500 ppm, the following materials are required.

Materials Required

(a) Chlorine dioxide solution 20,000 ppm (commercially available)
(b) Activator (commercially available formulation)
(c) Slaked lime
(d) Water

Model for Preparation of 20 litre of Solution

Solution A

50 g of activator crystals is added to 500 ml of chlorine dioxide solution (Serichlore or Sanitech brand name). Activator crystals are dissolved properly, and solution is allowed to stand for 5 min till colour of solution changes to light greenish yellow.

Fig. 9.8 (**a**) Putting activator crystals in plastic basin. (**b** and **c**) Adding chlorine dioxide into activator crystals and mixing it. (**d**) To make up required quantity by adding water. (**e** and **f**) Preparation of slaked lime solution and adding (**a** and **b**) solution

Solution B

100 g of slaked lime powder is dissolved in little water and made up to 19.5 litres with water. Then solutions A and B are mixed (Fig. 9.8) to obtain 20 litres of solution for use. The area of rearing building can be calculated as length (m) × breadth (m) of rearing room and solution at 2 litres per square metre is sprayed. In addition, requirement of additional 10 % solution for outside rearing room and 25 % for rearing appliances should be taken before the start of disinfection programme.

Fig. 9.9 Washing of rearing appliances

9.4.1.5 Disinfection of Rearing House

Preliminary or Postrearing Disinfection

Thorough disinfection of rearing building and rearing appliances without disturbing the appliances inside rearing house should be conducted with 2.5 % chlorine dioxide (500 ppm) in 0.5 % slaked lime water or 2 % bleaching powder after completion of each rearing operation.

Washing of Rearing House and Appliances

After the post rearing first round disinfection, rearing trays, cleaning nets, basins, polythene sheets, etc. should be washed with 2 % bleaching powder + 0.3 % slaked lime solution (Fig. 9.9). Appliances should be sun-dried for 10 to 20 h after washing. Washing of rearing house, incubation room and leaf preservation room with 2 % bleaching powder + 0.3 % slaked lime is required to remove any organic matter attached or left during rearing.

Final Disinfection

Spray of 2.5 % chlorine dioxide solution with the help of power sprayer or foot sprayer is conducted in such a way that all the walls, corners, windows, doors, roof and floor are drenched with solution (Fig. 9.10). Rearing appliances are placed inside the rearing room and disinfected along with verandahs, leaf chambers and surrounding of rearing buildings simultaneously. In addition, fumigation can be done with 10 % formalin solution making rooms and building airtight. Rearing house should be closed and opened 2 days prior to the commencement of brushing/start of next rearing.

9.4.1.6 Disinfection of Incubator

1. Incubation room or incubator must be disinfected with 2.5 % chlorine dioxide 1 week prior to commencement of incubation.
2. Incubator with metal sheet can be disinfected with spray of 70 % alcohol.
3. Silkworm eggs are placed in the incubation chamber after surface disinfection with 2 % formalin for 5 min. It is not advisable to disinfect eggs in advance stage of embryonic development, i.e. after pinhead pigmentation stage.

A B

Fig. 9.10 Spray of disinfectant with power sprayer. (**a**) Disinfection outside and around the rearing house (**b**) Disinfection inside rearing housealong with rearing appliances

9.4.1.7 Maintenance of Hygiene

1. Mixture of bleaching powder and slaked lime (1:19 parts) is sprinkled at the entrance of the rearing house and its surroundings 3 days before brushing to avoid secondary contamination.
2. Rearing house floor is cleaned by swapping with 2 % beaching powder + 0.3 % slaked lime solution during rearing operation.
3. Separate footwear should be used for inside rearing room.
4. The rearers are supposed to have hand wash with germicidal soap before entering into rearing house and after each feeding or cleaning operation.
5. Foot mat is sprinkled with bleaching powder and slaked lime mixture (1:19) at the entrance of rearing house.
6. Nets, chopsticks and foot wears are washed daily.

9.5 Pests of Silkworm

9.5.1 Uzi Fly, *Exorista bombycis* Louis

There are at least four Uzi flies that infest different silkworms (Tanaka 1964; Jolly 1967). These are Indian Uzi fly, *Exorista bombycis* (Louis); Japanese Uzi fly, *Crossocosmia sericariae* (Rondani); *Ctenophorocera pavida* (Meigen); and tasar Uzi fly, *Blepharipa zebina* Walker. The flies are very common in Burma, Thailand and South Korea besides India (Krishnaswami et al. 1987). The taxonomic position of Uzi flies is systematically presented by Narayanaswamy and Devaiah (1998). The biology of the fly along with external features of egg, different larval instars, puparium and adults has been described by Siddappaji (1985). The Uzi fly infestation is higher in West Bengal, Karnataka, Andhra Pradesh and Tamil Nadu than the rest of India (Sriharan et al. 1971; Kumar et al. 1993; Rao et al. 1993).

Uzi fly *Exorista bombycis* (Diptera) is a serious endolarval parasitoid of silkworm in India. This is reported to cause 10 to 20 % damage to cocoon crop in

Fig. 9.11 Uzi fly *Exorista bombycis*

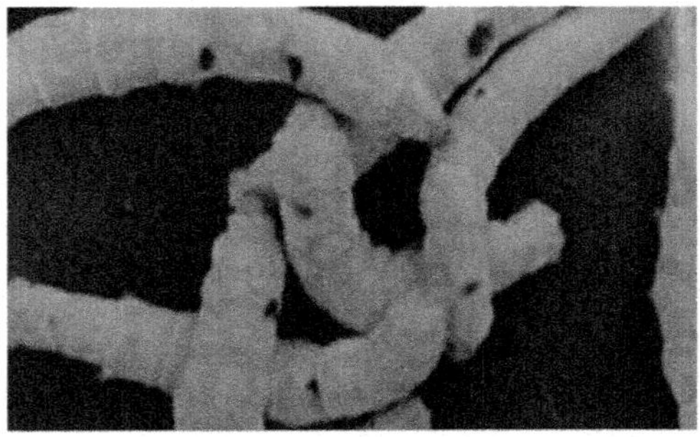

Fig. 9.12 Black colour scar mark on Uzi-infested larvae

Fig. 9.13 Cocoons damaged by Uzi fly

Karnataka, Andhra Pradesh and Tamil Nadu in India. It is a large fly with prominent black and grey strips on body. The female fly lays white, glossy, small oval eggs at intersegment space/body of fifth instar fully grown larvae. This can be identified with black colour small scar mark on ventrolateral side of the larval body (Figs. 9.11 and 9.12). These eggs hatch in 2–3 days after oviposition, and the young maggots bore its way into the body of silkworm's larvae. The maggots live in the tissue of the silkworms especially the fat body for about a week. The fully grown maggots, after completing its feeding period, pierce out from larval body by killing them. Adult flies emerge in 10–15 days (Fig. 9.13).

Wiremesh or nylon nets are placed around rearing trays and rearing rooms to prevent flies from entering into room. Uzitrap solution with yellow colour in white tray is placed inside and outside the rearing room. Uzi fly is attracted towards solution, drops inside and gets killed. A chemical formulation has been prepared with 10 g of benzoic acid crystals dissolved in 125 ml acetone and made to 1 litre by adding water. This solution can be sprayed over worms every alternate day after the second moult till spinning except moulting stage at 5–6 ml per square feet. *Nesolynx thymus*, a pupal parasitoid of Uzi fly, is released on the second day of the fifth instar and on mountages with spinning worms.

9.5.2 Dermestid Beetles

Dermestid beetles (order, Coleoptera; family, Dermestidae) are pest of dead pupae, which attack on stored stifled cocoons and make holes in them (Tewary et al. 2006). So far 30 species of Dermestidae are reported to be associated with sericulture industry world over (Vijayveer et al. 1996). Twenty-four species of dermestids are reported in India (Shekar et al. 2000). *Dermestid ater* DeGeer feeds on cocoons, silk floss, stifled cocoons, dead/live pupae, moths and eggs of *B. mori* (Ayuzawa et al. 1972; Patil and Govindan 1985; Ansari and Basalingappa 1987; Geetabai 1988; Kumar et al. 1988; Vijayveer et al. 1996; Arulmozhi and Chandramohan 1999; Shekar et al. 2000). Ansari and Basalingappa (1987) reported that early instar (first to third) grubs of *D. ater* feed only on the silk floss, whereas the later instar grubs (fourth to fifth) attack the fresh stifled cocoons and feed on the dead pupae. On an average an individual beetle during its larval period damages 5–7 cocoons. However, the extent of damage caused to pupae (dead and live) and moths of *B. mori* by *D. ater* was 16.62 and 3.57 %, respectively (Thiagarajan and Govindaiah 1987). The variation in the developmental periods of *D. ater* has been reported, which differs according to the type of feed, place of storage, age of stored cocoons and environmental conditions. Shubha (2000) reported that four late instar larvae of *D. ater* were sufficient to devour one moth in 2 days. Ten beetles were sufficient to destroy one laying within 10 h. In addition, the pest also damages silk cloth and carpet.

Fumigation with methyl bromide (0.5 kg per 283 m^3) for a day or with chloropicrin (0.5 kg per 283 m^3) for three days can prevent this pest from entering into cocoon storage room. In addition to these, ants, lizards, rats, squirrel and birds eat young larvae and pupae after cutting cocoons.

9.6 Conclusion

Silkworm cocoon crop suffers mainly from four common diseases, namely, flacherie, grasserie, muscardine and pebrine. The average silkworm crop loss in India due to diseases is to the tune of 15–47 %, while loss is 10–15 % in other countries, like Japan, China and Italy. Out of total cocoon crop loss in India, 57.22 % is on account of flacherie, 33.88 % grasserie, 2.32 % pebrine and 0.487 % muscardine. The incidence of silkworm diseases is more in autumn season (35–47 %) than spring (15–20 %) in north-west India. Bivoltine silkworms are more susceptible to diseases as compared to multivoltine silkworm breed.

Occurrence of silkworm diseases has definite correlations with the climate, temperature and rainfall. The spring season with a moderate temperature and low humidity has minimum incidence of diseases. The low incidence of silkworm diseases in the spring season can also be correlated with low density of pathogens in the environment due to low temperature and low humidity. The higher incidence diseases during autumn season can be correlated with higher density of pathogens, possibly due to higher temperature (32–35 °C) and higher humidity between 70 and 80 %.

Disease control in silkworm cocoon crop is attempted by the disinfection of rearing room and appliances, preventing entry of pathogen into rearing room or control of secondary infection (use of bed disinfectant and maintenance of hygiene), making immune system of silkworm stronger through nutritious feed and use of curative measures (Samson and Chandrashekaraiah 1998; Watanabe 2002).

It is a scientific fact that there is no disease in pathogen-free environment. But it is very difficult to create zero pathogen environments in rearing room. However, preventive measures can bring down the pathogen load to a tolerable level of silkworm. There is no race of *B. mori* totally resistant to diseases and pests. Hence, proper disinfection is required to kill the pathogens present in the rearing environment and to eliminate risk of disease transmission.

Silkworm disease management is a holistic approach of disease control in silkworm rearing. It is mandatory to follow the disinfection calendar before and after every rearing. Disinfection is more effective when it is carried out in proper time (calendar) with effective chemicals and appropriate spray pumps (power sprayers). 2 % solution of bleaching powder with minimum 30 % chlorine and 500 ppm solution of chlorine dioxide are more effective and economical disinfectants. The chlorine dioxide solution in 500 ppm can be used for spray of disinfectants on wall, roof, door, window of rearing room and all rearing equipment to be used for rearing. Rearing equipment, floor of rearing room, leaf chamber and any covered space with concrete flooring should be washed with 2 % bleaching solution in 0.3 % slaked lime (0.6 % chlorine). The spray of 2 % bleaching powder solution kills the pathogens due to release of nascent oxygen and the chlorine produced which has bactericidal action.

Hitherto, majority of silkworm disease management measures used in silkworm rearing are prophylactic; however, there has been few attempts in past and recent past to cure diseases through use of antibiotic and botanical formulations (Manimegalai and Chandramohan 2005). Scientists of Central Sericultural Research and Training Institute, Mysore, India, have developed an ecofriendly botanical-based formulation "Amruth" to cure grasserie and flacherie disease (Balvenkatasubbaiha et al. 2012a).

Any bed disinfectant (Sanjeevani, Vijetha supplement, R-K-O, Suraksha, Ankush or Resham Jyothi) should be applied as per recommended schedule (Balvenkatasubbaiha et al. 2012b) and quantity (3–5 gm/sq. ft.). It is recommended to dust 1–2 % (1 % for first and second instars, 2 % for third, fourth and final instars) of Dithane M-45 in slaked lime or 1–2 % Captan in Kaoline on silkworm body after every moult and on the fourth day of final instar at 3–5 gm/sq. ft.

There are many pests of silkworm, silk cocoons, pupae, moths and eggs. The Uzi fly *Exorista bombycis* Louis is a serious endolarval parasitoid pest of silkworm. The infestation of this fly is often very high in West Bengal, Karnataka, Andhra Pradesh and Tamil Nadu states of India. Another major pest of *B. mori* is dermestid beetle which attacks on stored stifled cocoons and makes holes to them. The early instars (first to third) grubs of *D. ater* feed only on the silk floss, whereas the later instar grubs (fourth to sixth) attack the fresh stifled cocoons and feed on the dead pupae. On an average an individual larval damages 5–7 cocoons.

Acknowledgements The authors are thankful to the Central Silk Board, Bangalore, India, for getting an opportunity to learn and understand the core issues related to sericulture industry in general and Indian Sericulture in particular, which has helped us in writing this manuscript.

References

Aizawa K (1962) Antiviral substance of gut juice of silkworm *Bombyx mori*. J Insect Pathol 4:72–76

Ansari MS, Basalingappa S (1987) Percent infestation of silkworm cocoons by dermestid beetle, *Dermestid ater* (Dermestidae:Coleoptera). Sericologia 27(4):587–592

Arulmozhi M, Chandramohan N (1999) Population level and damage potential of dermestid beetle *Dermestid ater* in Tamil Nadu. National Semi Tropical Service UAS, Bangalore, p. 148

Ayuzawa C, Sekido I, Yamakawa K, Sakurai U, Kurata S, Yaginuma Y, Takoro Y (1972) Hand book of silkworm rearing. Fuji Pub. Co. Ltd., Tokyo, p. 319

Baig M, Sharma SD, Balavenkatasubbaih, Samson MV, Sasidharan TO, Noamani MKR (1991) Relative susceptibility of different races of silkworm, *Bombyx mori* L to nuclear polyhedrosis under natural and induced conditions. Sericologia 31(3):417–420

Balavenkatasubbaih M, Chandrashekharan K, Sharma SD, Narsimha Nayak AR (2012a) Silkworm diseases and their control. Pamphlet published by CSR & TI, Central Silk Board, Mysore

Balavenkatasubbaih M, Chandrashekharan K, Sharma SD, Narsimha Nayak AR (2012b) Disinfection and hygiene in silkworm rearing. Pamphlet published by CSR & TI, Central Silk Board, Mysore

Bhagyalakshmi TA (1994) Host range and strain virulence of *Aspergillus tamari* Kita and its pathogenicity to different races of *Bombyx mori* L. M.Sc.(Seri.) thesis, University of Agriculture Sciences, Bangalore, pp 86

Chinnaswamy KP, Devajah MC (1984) Susceptibility in different races of silkworm, *Bombyx mori* Linnaeus to aspergillosis caused by *Aspergillus tamari* Kita. Sericologia 24:513–517

Chinnaswamy KP, Reddy DNR, Govindan R, Narayanaswamy TK (1986) Pathogenicity of *Aspergillus tamari* Kita to *Dysichira mendosa* Hubner (Lepidoptera: Lymantriidae), a pest of mulberry. Entomon 11:67–68

Geetabai M (1988) Pests of *Bombyx mori* (Lepidoptera: Bombycidae) in grainage and their control. Sericologia 28:379–382

Govindhan R, Narayanaswamy TK, Devaiah MC (1998) Principles of silkworm pathology. Publication Series Scientific Publishers, Bangalore, p 420

Gupta V, Jaiswal K, Gangwar SK, Dwivedi P, Srivastava A, Kumar R (2006) Disease susceptibility in different hybrid breeds of Bombyx mori L. in Uttar Pradesh climatic conditions. Proceedings of the "Regional Seminar on Prospects & problems of sericulture as an economic enterprise in north-west India", Dehradun, India, pp 286–289

Jaiswal K, Gangawar SK, Dhami SS (2003) Screening of popular bivoltine hybrides of silkworm, (*Bombyx mori* Linn.) in monsoon season of Uttar Pradesh climatic conditions. In Recent Researches on Indian Sericulture Industry, pp 108–112

Jolly MS (1967) A brief report on wild sericigena in India with special reference to tasar culture. World Meet. Silk Prodn, Mourcia, pp 1–6

Kawakami K (1982) Causal pathogens of *Aspergillus* disease of silkworm and its control. JAQR 15(3):185–190

Krishnaswami S, Narasimhanna MN, Suryanarayan SK, Kumarajan S (1987) Sericulture manual 2-Silkworm rearing in *FAO Manuals on Sericulture*. United Nations, Rome, pp 102–129 (Reprinted by Central Silk Board, Bangalore, India)

Kumar P, Jayaprakash CA, Singh BD, Sengupta K (1988) Studies on the biology of *Dermestes ater* (Dermestidae: Coleoptera), a pest of silkworm pupae and adults. Curr Sci 57(222):1253–1254

Kumar P, Manjunath D, Datta RK (1993) Biological control of uzi fly. In: Proceedings of National Seminar Uzi fly and its control. KSSRDI, Bangalore, pp 91–106

Lea HZ (1993) Principles and technique of silkworm breeding. ESCAP, United Nations, New York, p 114

Manimegalai S, Chandramohan N (2005) Botanicals for the management of bacterial flacherie of silkworm, *Bombyx mori* L. Sericologia 45(1):55–58

Nataraju B, Sathyaprasad K, Manjunath D, Aswani Kumar C (2005) Silkworm crop protection. Central Silk Board pp 61–85

Narayanaswamy KC, Devaiah MC (1998) Silkworm uzi fly. Zen Publishers, Bangalore, p. 232

Patil GM, Govindan R (1985) Beetles attacking silk cocoons and their control. Resham Krishi 7(12):15–17

Rao CHN, Veeranna G, Prasad NR (1993) Studies on the infestation of *Exorista sorbillans* on *Bombyx mori* L- Incidence during monsoon season. In: Proceedings of National Seminar on Uzi fly and its control, KSSRDI, Bangalore, pp 49–56

Samson MV, Chandrashekaraiah (1998) The concept of disease resistance and its relevance in silkworm breeding, In: Sreerama Reddy G (ed) Silkworm breeding. IBH Publishing Co. Pvt. Ltd., New Delhi, pp 149–158, 354

Shekar MA, Sen AK, Sathyaprasad K, Katiyar RL, Rajadurai S, Datta RK (2000) A package for management of dermestid beetles in pierced cocoon storage rooms of commercial seed production centres. Seminar on sericulture technique annual appraisal, CSRTI, Mysore, India, pp 163

Shubha MR (2000) Pests of silkworm grainage and their management. M.Sc. (Seri) thesis, University of Agricultural Sciences, Bangalore, pp. 72

Siddappaji C (1985) Biology and management of the Indian uzi fly, *Exorista sorbillans* (Wiedemann) (Diptera:Tachinidae) a parasite of mulberry silkworm. PhD thesis, UAS, Bangalore, p 163

Sriharan TP, Samson MV, Krishnaswami S, Datta RK (1971) Laboratory investigations on uzi fly *Tricholyga bombycis* Beck: a tachinid parasite of silkworm (Bombyx mori L) 1. Rearing and reproduction of the parasite. Indian J Seric 10:14–22

Tanaka Y (1964) Sericology. Central Silk Board, Bombay, pp 277

Tayal MK, Chauhan TPS, Siddiqui AA, Saraswat SB, Singh BD (2003) Role of dietary inorganic tissue salts to improve the disease resistance in silkworm *Bombyx mori* L.: an observation. In: Proceedings of the "National Seminar on Mulberry Sericulture research in India". KSSRDI, Bangalore, 26–28 Nov 2001, pp 810–813

Tayal MK, Chauhan TPS, Siddiqui AA, Mishra PN, Singh BD (2004) Response of inorganic salts supplementation on health status and tolerance in silkworm *Bombyx mori L.*: new approach in disease management. In: Proceedings of the "National Seminar on Disease & pest management in sericulture, Sericulture College, Chintamani, UAS, Bangalore, 12–13 Sept 2003, pp 137–142

Tayal MK, Chauhan TPS, Mishra PN, Chakrabarti S (2006) Dietary supplementation of inorganic tissue salts in *Bombyx mori* L.: a new hope to stabilise autumn crop in North West India. In: Proceeding of the regional seminar on "prospect and problems of sericulture as an Economic enterprise in North West India", 11–12 November 2006, Dehradun, pp 184–187

Tayal MK, Chauhan TPS, Mishra PN (2007) Dietary supplementation of inorganic tissue salts and disease control in silkworm *Bombyx mori* L. J Seric 11–15(1,2):55–60

Tayal MK, Chauhan TPS, Dhar A, Raina SK, Sahaf KA (2014) Strategy of disinfection and disease management in commercial rearing of silkworm *Bombyx mori L.* in sub tropics of north India. Abstract published in proceedings of 101st Indian Science Congress in section of Animal, Veterinary and fishery sciences, pp 131

Tewary P, Tayal MK, Ahmad MN, Dhar SL, Sengupta D, Chakrabarti S, Khan MA (2006) Extent of damage to stored silkworm cocoon of *Bombyx mori* L. by Dermested beetles in North west India. In: Proceeding of the regional seminar on "prospect and problems of sericulture as an Economic enterprise in North West India", Dehradun, 11–12 November 2006, pp 177–180

Thiagarajan V, Govindaiah (1987) Menance of dermestid beetle in grainages. Indian Silk 26(3):26–27

Tripathi PM, Gaur JP (1991) Intensity of different diseases in different seasons. Annual Report, Regional Sericultural Research Station, Central Silk Board, Shahaspur, Dehradun, India

Vijayveer, Negi BK, Rao KM (1996) Dermestid beetles and some insect pests associated with stored silkworm cocoons in India, including a world list of dermestid species found attacking this commodity. J Stored Prod Res 32:69–89

Watanabe H (2002) Genetic resistance of the silkworm, *Bombyx mori* to viral disease. Curr Sci 83:439–446

Tropical Tasar Sericulture 10

D.D. Barsagade

Contents

10.1 Introduction

In India, the silks are produced commercially through the cultivation of all four silk varieties, viz. mulberry and three types of non-mulberry silk, eri, muga and tasar. The tasar silk divided in to two category, tropical tasar and temperate tasar. Among non-mulberry, the tasar silk is produced by silkworm, *Antheraea mylitta* (Drury) and is preferably used for commercial exploitation in the tropical zone of India. The tropical tasar growing area is represented by the belt of humid forest placed over the central and southern plateau passing through Bihar, Jharkhand, Madhya Pradesh, Chhattisgarh, Orissa, West Bengal, part of Uttar Pradesh, Andhra Pradesh, Maharashtra and Karnataka. Maharashtra is one of the traditional tasar-producing states in India, and Vidarbha region is known in the tasar silk production (Jolly et al. 1968, 1974; Yadav et al. 1996; Alam et al. 2000; Kushwaha et al. 2004).

The characteristics of wild silkworms in India changes accordingly to the ecological conditions and food plant availability (Suryanarayana et al. 2007;

D.D. Barsagade (✉)
Department of Zoology, MJF Educational Campus, Rashtrasant Tukadoji Maharaj Nagpur University, Nagpur 440033, Maharashtra, India
e-mail: dr_ddbars@rediffmail.com

Omkar (ed.), *Industrial Entomology*, DOI 10.1007/978-981-10-3304-9_10

Srivastava and Thangavelu 2005). The 44 ecoraces of *A. mylitta* are confined to different ecological niches in India (Thangavelu 1991; Sinha and Sinha 1997; Mathur et al. 2005; Srivastava and Thangavelu 2005). These ecoraces have been identified mainly on the basis of their distribution and cocoon characteristics.

Antheraea mylitta completes its life cycle in wild condition, and larvae are exposed to diverse meteorological conditions, such as temperature, humidity, sunlight and rainfall. These environmental variations make the larvae vulnerable to microbial diseases caused by bacteria, virus, protozoan and fungi (Sen et al. 1969; Barsagade et al. 2009, 2011). Similarly, pests and predators also cause heavy losses of silk production by affecting various stages of silkworms from larvae to adult. Infestations of pests and predators on the silkworm are serious threats to the silk industry in India, particularly West Bengal and then Bihar, Assam and Maharashtra (Dasgupta 1962; Sriharan et al. 1971).

Various insects, such as *Xanthopimpla predator* (Ichneumonidae: Hymenoptera) and Uzi fly *Blepharipa zebina* (Tachinidae: Diptera) are pests, and *Canthecona furcellata* (Pentatomidae: Hemiptera), *Sycanus collaris* (Reduviidae: Hemiptera), *Hierodula bipapilla* (Mantidae: Dictyoptera), *Vespa orientalis*, *Polistes hebraeus* (Vespidae: Hymenoptera), ant *Oecophylla smaragdina*, dermestid beetle, birds, lizard, squirrel, rat, snake, etc. are among the important predators which feed on tasar silkworm (Jolly 1976; Singh 1991; Singh and Thangavelu 1991; Veer and Rao 1994; Yadav et al. 1996; Shivakumar and Shamitha 2013).

Due to the attack of numbers of insect as well as non-insect pests and predators, tropical tasar silkworm, *A. mylitta*, is in trouble. These major and minor threats of silk industry cause heavy losses to the total silk production in India, resulting into the loss of economy. Therefore, in the present chapter, the structure of eggs, post-embryonic larval development and metamorphosis, cocoon formation in the field and attack of various pests and predators on *A. mylitta* have been explained.

10.2 Host Plants

Jolly et al. (1968) stated that the tasar silkworm is polyphagous in nature and feeds on several host plants. However, it exhibits host preference. The host plants on which silkworm normally prefers are known as primary host plants. Other host plants, where the silkworms can sustain their life but normally do not prefer, are called secondary host plants. According to Jolly et al. (1968, 1974) and Yadav et al. (1996), arjun (*Terminalia arjuna*), asan/yen (*T. tomentosa*) and sal (*Shorea robusta*) are primary host plants of *A. mylitta*. Besides these plants, Suryanarayana et al. (2007) found that the larvae of tasar silkworm also feed on several other plants known as secondary host plants (Table 10.1).

Table 10.1 Primary and secondary host plants of *A. mylitta*

Name of the host plant	Common name	Family
Primary host plant		
Terminalia arjuna Bedd.	Arjun	Combretaceae
T. tomentosa W. & A.	Asan/Yen	Combretaceae
Shorea robusta Gaertn.	Sal	Dipterocarpaceae
Secondary host plants		
Anogeissus latifolia Wall.	Dhaura	Combretaceae
Bauhinia variegata L.	Kachnar	Caesalpiniaceae
Bombax ceiba L.	Semal	Bombacaceae
Careya arborea Roxb.	Kumbhi	Lecythidaceae
Carissa carandas L.	Karaunda	Apocynaceae
Dalbergia sissoo Roxb.	Seesam	Fabaceae
Hardwickia binata Roxb.	Anjan	Caesalpiniaceae
Lagerstroemia indica L.	Saoni	Lythraceae
L. parviflora Roxb.	Bendar	Lythraceae
L. speciosa Pers.	Sidha	Lythraceae
Madhuca longifolia Macbr.	Mahua	Sapotaceae
Syzygium cumini Lam.	Jamun	Myrtaceae
Terminalia chebula Retz.	Harada, Haritaki	Combretaceae
T. bellirica Roxb.	Bahera	Combretaceae
T. catappa Linn.	Jangli badam	Combretaceae
T. paniculata Roth.	Kinjal	Combretaceae
Tectona grandis Linn.	Sagwan	Verbenaceae
Ziziphus mauritiana Lamk.	Ber	Rhamnaceae
Z. jujuba Lamk.	Jangli ber	Rhamnaceae

10.3 Classification of *Antheraea mylitta*

Richards and Davies (1957) classified tasar silkworm of the tropical region as follows:

Class – Insecta
Subclass – Pterygota
Division – Endopterygota
Order – Lepidoptera
Suborder – Ditrysia
Superfamily – Bombycoidea
Family – Saturniidae
Genus – *Antheraea*
Species – *mylitta* (Drury)

Fig. 10.1 (**a**–**l**) Life cycle of tasar silkworm, *Antheraea mylitta* (D). (**a**) Natural plantation of *Terminalia tomentosa*, (**b**) eggs, (**c**) first instar larva, (**d**) second instar larva, (**e**) third instar larva, (**f**) fourth instar larva, (**g**) fifth instar larva, (**h**) hammock formation, (**i**) spinning of cocoon, (**j**) cocoon, (**k**) pupa inside cocoon and (**l**) adult moth

10.4 Life Cycle of *Antheraea mylitta*

According to Jolly et al. (1974), the life cycle of tasar silkworm comprises of the egg, larva, pupa and adult stages (Fig. 10.1).

10.4.1 Egg Laying

After the period of coupling, the mated female is placed singly in small cages (manias) for egg laying. The female moth deposits eggs in batches of 5–10. The egg laying continues up to the fourth day with an average laying up to 200 (Sen et al. 1969; Barsagade et al. 2011) (Fig. 10.2a).

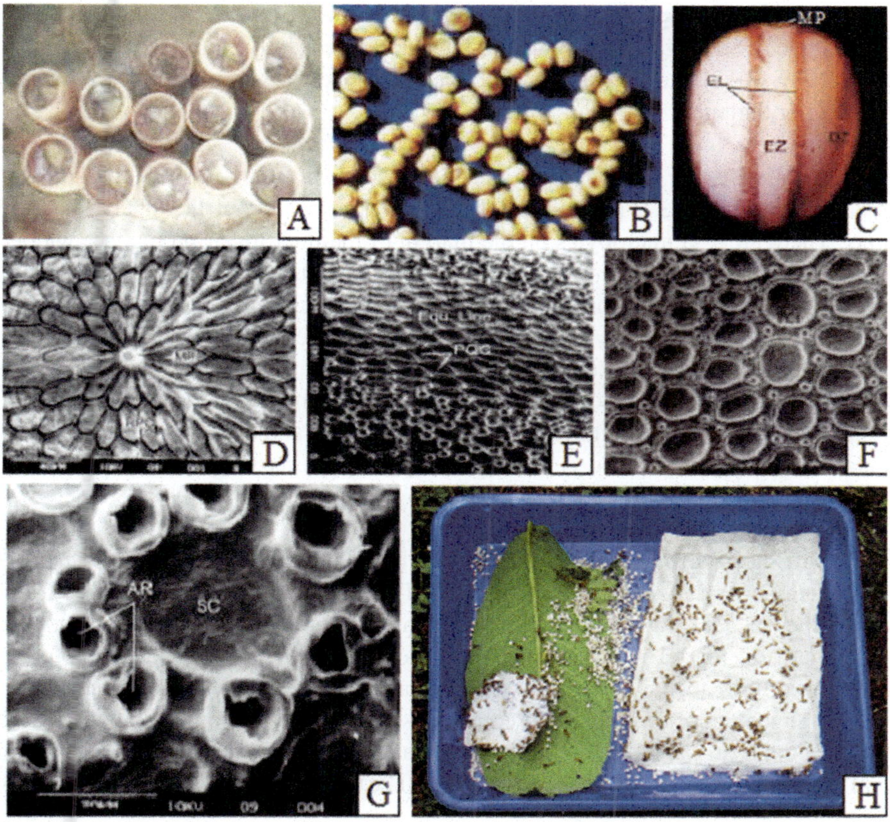

Fig. 10.2 (**a–h**): Showing structure of egg (**a**) Egg laying, (**b**) Structure of egg, (**c**) magnified view of egg equatorial zone, (**d**) scanning electron microscopic (SEM) of eggshell, showing micropyle zone, (**e**) edge zone, (**f**) disc zone, (**g**) magnified view of aeropyle and (**h**) hatching of larvae

10.4.2 Egg

Jolly et al. (1979) observed dark-brown freshly laid eggs owing to the gummy coating of meconium. They are oval, dorsoventrally flattened and bilaterally symmetrical along the anteroposterior axis. They are about 3 mm in length and 2.5 mm in diameter. Soon after oviposition, they become brown in colour and are coated with gelatinous secretion of accessory glands which facilitate the eggs to stick firmly with the substratum (Barsagade et al. 2011). After washing, they become light yellow or creamy in colour. Two brownish parallel lines run along the equatorial plane of the egg, called the equatorial lines (Fig. 10.2b, c).

10.4.2.1 Scanning Electron Microscopic (SEM) Structure of Eggshell

The surface of the egg is differentiated into four zones (Jolly et al. 1979; Barsagade et al. 2009, 2015), the micropylar zone, the edge zone, the aeropyle crown zone and the disc or flattened zone (Fig. 10.2c).

(a) The Micropylar Zone It is a deeply concealed area forming a distinct micropylar pit of about 5 ± 2 μm in diameter at the anterior pole of the anteroposterior axis. There are 7–8 micropyles on the margin of the micropylar pit and are widely separated from each other. Each micropyle is about 0.8–1.2 μm in diameter. The micropylar area is surrounded by a group of radiating petaloid micropylar cells known as primary micropylar cells. These cells are separated by protrusions and show a narrow margin proximal to the micropylar area, while distal margin is wider and rounded, forming a sort of rosette around the micropylar pit. The primary micropylar cells are surrounded by secondary, tertiary and quaternary layers of the petaloid cells. The petaloid cells of primary to quaternary layers change in size gradually and become polygonal in shape (Fig. 10.2d).

(b) The Edge Zone The edge zone lies in between two equatorial lines running from the micropylar zone to the caudal end. The polygonal cells are separated from each other by a narrow intercellular space. In the intercellular space, the aeropyles are found together. The cells are either quadrangular or pentamerous sculptures. Equatorial lines appear in between the edge and aeropyle crown zone due to compact arrangement of follicular polygonal cells. The aeropyles appear at the junction of three adjacent cells. Each middle polygonal cell is encircled by a big aeropyle and peripheral cells by five aeropyles (Fig. 10.2e).

(c) The Aeropyle Crown Zone The aeropyle crown zone lies on the exterior side of the edge zone and is distinguished into apical and distal region. According to Mazur et al. (1980), in apical region the aeropyles with fine setae-like processes encircle the polygonal cells and form a sort of crown. The paired aeropyles are lying in between adjacent cells. In distal region, the cells are spherical and are surrounded by about eight aeropyles. The aeropyle crowns are devoid of setae-like processes (Barsagade et al. 2009).

(d) The Disc Zone The disc or flattened zone occupies the major surface of the egg. The polygonal cells are separated from each other by intercellular space, and they are small and large in shapes. The aeropyles are widely separated from each other and do not form a pair. The numbers of aeropyles are from five to seven. The fine setae-like processes around the aeropyle openings are lacking in the disc zone. Mazur et al. (1980) conformed the chorion of *A. mylitta* is composed of two main layers, the trabecular layer and the lamellar layer. The trabecular layer consists of three sublayers: (1) the innermost wax layer, (2) the inner chorionic layer and (3) the outer chorionic layer. The inner and outer chorionic layers are connected to each other by vertical pillars forming a series of cavities, and as a result, the entire trabecular layer becomes a perforated structure. The cavities of the trabecular layer contain flocculent material. The trabecular layer is externally covered with several layers of helicoidally arranged fibrillar lamellae. According to Barsagade et al. (2009), the lamellar layer is perforated by the aeropyles running from the outer surface of the chorion to the inner cavities. Some of the lamellae are running obliquely and are helicoidally twisted (Fig. 10.2f, g).

10.4.3 Mother Moth Examination and Selection of Disease-Free Laying (DFL)

Jolly et al. (1979) described microscopic examination of the mother moth ensures disease-free laying. For mother moth examination, after egg laying, the abdomen of the female moth is cut, and smear is taken in a little water and is examined under a microscope at l0x × 40x. The special attention is paid to microsporidian infection because of its transovarial mode of transmission. The mother moth exhibiting even the slightest hint of microsporidiosis is noted, and eggs laid by that particular moth are rejected. Infected mother moths and their laying are burned or buried immediately (Barsagade and Kadwey 2010). If other diseases are detected, the laying is rejected only in cases of severe infection.

10.4.4 Disinfection of Eggs

As tasar silkworms eat a portion of the eggshell during and after hatching, therefore surface sterilization of the egg should be ensured so as to avoid infection through contamination. The selected laying is placed in a clean cloth or bag and dipped in water to remove the meconium. They are then rinsed with 2–3 % formalin solution for 5 min followed by washing in running water. The washed eggs are kept in the incubator for drying.

10.4.5 Egg Incubation

Egg boxes are placed in a room for incubation at 30–32 °C temperature and 70–80 % humidity. In this way 80–90 % hatching is obtained. The incubation period is about 7–8 days.

10.4.6 Hatching

According to Jolly (1976), hatching of eggs commences early in the morning. The hatching of properly incubated eggs started on the seventh or eighth day and was completed within 2–3 days. The percentage of hatching was higher in the egg laid on earlier days (Fig. 10.2h).

10.4.7 Post-embryonic Development of *A. mylitta*

The post-embryonic development has been studied first by Jolly (1976), and later Jolly et al. (1979) it was modified as follows:

(a) First Instar Larvae
On hatching, the first instar larvae appear dull brownish and yellow with black head, measuring about 7 mm in length and 1 mm in width. Jolly et al. (1979) described that the body normally turns green and the head turns brown after 48 h. Occasionally, yellow, blue and almond colour larvae are also observed. The prothoracic hood is well developed with a black, oval, dorsal spot. The anal flap and claspers bear a triangular black mark separately. The larvae contain a red midventral line extending up to the seventh abdominal segment and a dumbbell-shaped black mark on the eighth segment. Laterally, the first seven abdominal segments have a pair of black vertical and oblique lines on the eighth abdominal segment. The larvae have five types of black tubercles, dorsal, upper lateral, lateral, lower lateral and caudal, and fine white hairs irregularly distributed all over the body (Fig. 10.3a–c).

(b) Second Instar Larvae
Freshly moulted larvae are yellowish green with brown-coloured head and have an 'M'-shaped black mark on the prothoracic hood. This mark later on becomes 'V' shaped with two dots. The anal flap shows a brownish 'V'-shaped mark. A red mid-ventral line appears, extending up to the seventh abdominal segments. Black triangular mark appears on each of the claspers. The body tubercles are orange-red in colour (Fig. 10.3d–f).

(c) Third Instar Larvae
The larvae are yellowish green with brown head and a red mid-lateral line extending the length of the abdomen. A yellow lateral line extends from the second to the tenth abdominal segment. The body tubercles turned violet in colour. A yellow lateral line appears from the second to the tenth abdominal segment. Silvery white shining spots of various sizes (either oval or triangular) develop on the lateral side near the foot of the upper lateral tubercles of the second to the seventh abdominal segment (Fig. 10.3g–i).

(d) Fourth Instar Larvae
The larvae are brown headed with yellowish-green body. Prothoracic hood markings reappear as two semilunar red marks corresponding to the epicranial plates. The triangular mark on each of the claspers becomes brown in colour. The tubercles are violet, and the lateral line extending from the second to the tenth abdominal segment is bordered by a brown upper line (Jolly et al. 1979), and the tubercle setae reduce in number (Barsagade 1998) (Fig. 10.3j–l).

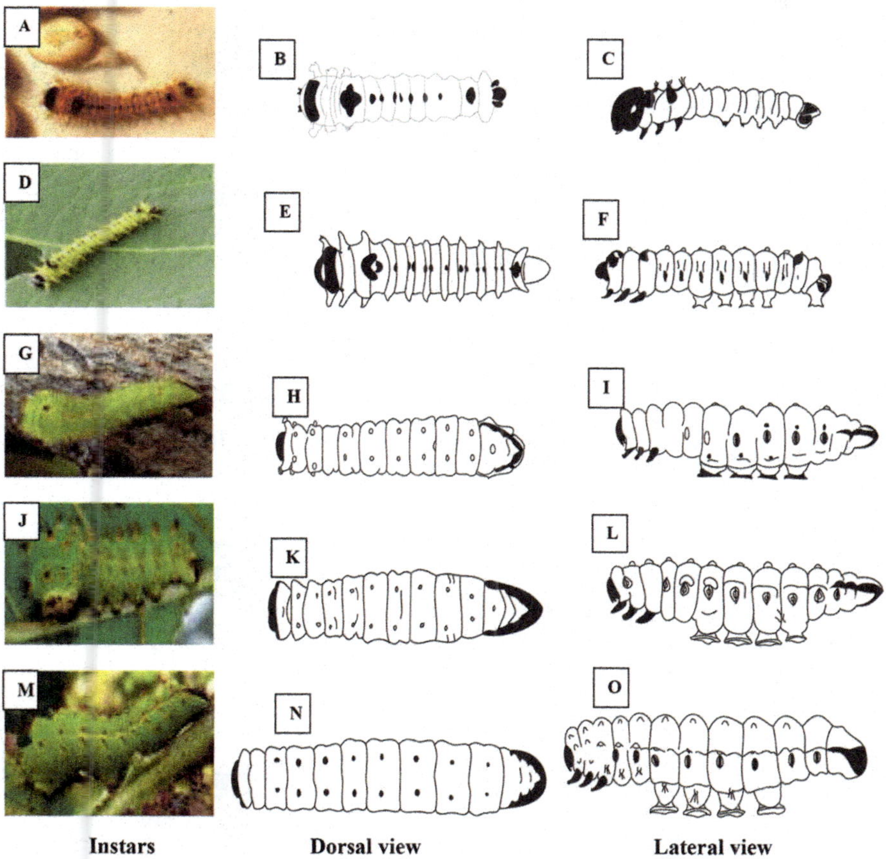

Fig. 10.3 (**a–o**) Showing morphological changes in the first to fifth instar larvae of *Antheraea mylitta* during metamorphosis: (**a–c**) first instar larva, (**d–f**) second instar larva, (**g–i**) third instar larva, (**j–l**) fourth instar larva, (**m–o**) fifth instar larva

(e) Fifth Instar Larvae

They are brown headed with green or yellowish-green body; two semilunar markings are present on the prothoracic head. At maturity the size and weight increase greatly, and the larvae measure about 13 (±1.7) × 2.1(±0.5) cm in length and width and ~50 (±5.0) g in weight (Barsagade 1998). The brown-coloured triangular marks on each clasper are well evident. The red midventral line now existed up to the tip of the abdomen. The lateral line is bordered by a brown upper line and extends from the second to the tenth abdominal segment. The sexual marking appears late in the fifth instar as milky-white spot on the ventral surface of the eighth and ninth abdominal segments. The violet-coloured body tubercles with irregular shining spots are present on the lateral tubercles (Mishra et al. 2008) (Fig. 10.3m–o).

According to Tembhare and Barsagade (2000); Thakre (2015), during the postembryonic development, morphological changes occur in larvae particularly in prothoracic hood, anal flap and claspers (summarized in Fig. 10.4).

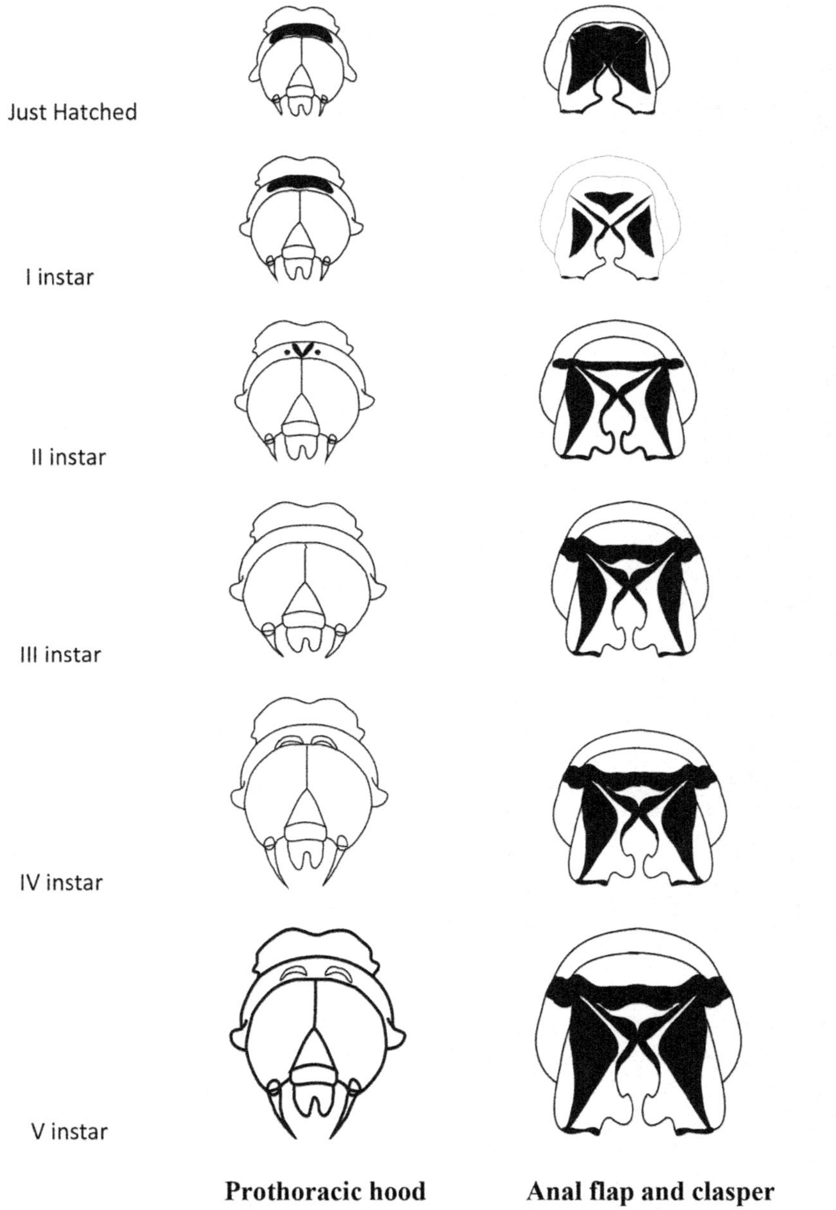

Fig. 10.4 Developmental changes in prothoracic hood and anal flap in *A. mylitta* during the first to fifth instar metamorphosis

10.4.8 Larval Behaviour

Jolly et al. (1979) studied the cocoon formation behaviour of fifth instar larva of *A. mylitta* in detail, and he described three stages of cocooning as follows:

Table 10.2 Development and feeding of larvae of *Antheraea mylitta*

Instar	Leaf consumed per larva	Instar life span (days)	Total food consumed (gm)	Instar mean weight (gm)	Length (mm)	Width (mm)
I	0.25	3.30	0.38	0.05	7.00	1.00
II	1.29	4.00	1.80	0.26	22.00	4.00
III	4.14	5.50	6.87	1.33	40.00	7.80
IV	17.12	6.80	16.10	5.90	65.00	10.00
V	120.10	15.60	98.50	23.25	100.00	18.00

(a) Feeding The larvae feed from the margin up to the midrib of leaves. The younger larvae rest on leaf margin, and the older ones attached themselves to the petiole or even small twig with the help of their abdominal feet and claspers. The leaf lamina pulled up between the mandibles by the thoracic legs. The early instar larvae relish tender leaves, and in advanced stages they prefer mature leaves. On average, a larva consumes nearly 300 g of foliage, about 20 % during the first four instars and 80 % during the fifth instar (Table 10.2). The rate of feeding is minimal before and after the moult but reaches its peak between these two periods. Every stretch of feeding is followed by a rest period of about 5 min or of longer duration in winters and in unfavourable weather. While resting, the larva assumes a typical sitting posture, by raising its head and contracting its body. The larva continues to feed in light and dark conditions and stops feeding at temperature less than 15 °C.

(b) Moult The larvae stop feeding just before the onset of a moult and become motionless. They secure a firm grip either at the underside of leaves (early stages) or the shaped part of twigs (late stages). During the moult, the anterior part of the body remains obliquely suspended, the prothoracic hood is fully stretched and the protruding dark head is bent ventrally inwards (Jolly et al. 1979). They seldom lose their grip, but once detached they cannot re-establish and moult with difficulty. During ecdysis, the old head capsule is detached intact along the vertex, after which the larvae push forward, slitting the old skin along the middorsal line. Moulting continues for 23–24 h depending on the climatic conditions. The freshly moulted larvae have tender, pale skin. They eat a portion of the cast-off skin and commence active feeding after a short rest.

(c) Larval Growth Rate Tropical tasar silkworms have a remarkable growth rate. Their length, width and weight increase during post-embryonic development up to 13 cm, 2.1 cm and 50 g, respectively, at maturity (Barsagade and Tembhare 2000).

10.4.9 Cocoon Formation

The cocoon formation process is studied in detail by Jolly et al. (1979). He described three steps as hammock formation, ring and peduncle construction and spinning during formation of cocoon. According to them, the larval feeding phase ends at maturity, marked by evacuation of end of the gut contents or 'passing of last excreta', as a green semisolid mass followed by a colourless slimy substance. Exploration for cocooning spots starts after 10–25 min of rest and may last for 1–1.5 h. A group of three or more leaves is usually chosen by larva for hammock formation. The cocoon formation is completed in the following three stages:

(a) Hammock Formation The mature V instar larva joins the leaves to form a small nest or hammock, with an opening at the top by irregularly throwing out silk filaments. It spends nearly 6 h to complete the hammock (Fig. 10.5a).

Fig. 10.5 Life stages of *A. mylitta* (**a**) Hammock formation, (**b**) Cocoon, (**c**) Male and female pupa, (**d**) Male moth, (**e**) Female moth, (**f**) Cocoons in grainage, (**g**) Coupling of moths, (**h**) Controlled indoor rearing of larvae on branches of *Terminalia tomentosa,* (**i**) Cocoon cooking for uniform softening of hard cocoon shell for reeling of silk thread, (**j**) Spinning of cooked unreelable cocoons to produce hand -drawn coarse yarn ghicha

Fig. 10.5 (continued)

(b) Ring and Peduncle Formation The larva moves out of the hammock and with its mandibles straps bark 2–3 mm wide around the twig in the form of a ring in 25–30 min; the space from within the ring to the top of the hammock with a few strands of silk is later completed by adding more strands. The usual to and fro motion of the head during these operations changes to semicircular and linear during ring and peduncle formation, respectively. These stages are completed in 10 h.

(c) Cocoon The cocoons are hard shelled provided with a black well-formed peduncle and ring. The cocoons are generally oval, with a smooth surface. At the anterior end, there is a well-formed dark-brown peduncle with a ring at the distal end. The cocoons are generally light, bright yellow to light grey in colour. The female larvae spin larger cocoon than the male. After formation of ring and peduncle, spinning of the cocoon is initiated, usually during the daylight hours. During spinning which is most regular, the movement of larval head traces horizontal figure of eight. In 24 h, the larva is completely enveloped by the cocoon; but spinning continues for another 2–3 days, during which the cocoon shell is further thickened. After spinning it places its head towards the peduncle end. Before pupation, the larva very slowly discharges a small quantity of light milky yellowish substance for 2–3 h. The discharge makes the cocoon yellow or grey. The larva converts into pupa after 4–6 days. The mature larva secretes silk which hardens into cocoon around its body and metamorphoses into a pupa (Fig. 10.5b).

(d) Pupa Jolly et al. described the structure of pupa first time in 1974 and then modified in 1979. The pupae are dark brown in colour with well-defined, segmented body. They measure about 4.5 (±0.02) cm in size and 10 (±0.8) g in weight. The anterior end of the body is broad, while the posterior end is tapering and conical. Distinct wing pads are well evident on the dorsal surface of the pupae (Fig. 10.5c). According to Mishra et al. (2008), the appendages of the pupae are firmly soldered to the body by a secretion produced during the last larval moult. The pupae are of obtect-adecticous type and dark brown in colour. On an average, each male pupa measures about 4.3 × 2.2 cm in size and weight 9.2 g, while the females are 4.6 × 2.5 cm in size and 11.00 g in weight, respectively. The body segmentation is well developed, and the exposed surface of the appendages is heavily cuticularized than those adjacent to the body. The head is represented by a vertex, a frontoclypeus, a distinct but small labrum and a pair of prominent broad bipectinate antennae in both sexes. A prominent brain window is present between the bases of antennae. However, the mandibles are reduced. Three thoracic segments are distinct on the dorsal side, but ventrally they are concealed by the appendages. The prothoracic segment is the broadest. A pair of forewing is folded ventrally and covers the first two and half of the third abdominal segment. The forewing almost entirely concealed the hindwings except a narrow strip along the dorsal margin of the latter. Ten segments are observed in the abdomen, the first seven of which bear spiracles on each side. The first four segments are fixed and immovable. A genital aperture in the form of a minute line in female on eighth to ninth and a dot in male is present ventrally on ninth segments. The sexual dimorphism is not prominent except in size, tip of the abdomen and structure and position of the genital aperture. The pupae of bivoltine and trivoltine races undergo diapauses after first and second generations, respectively (Jolly et al. 1979).

10.4.10 Adult Moths

In adult moths, sexual dimorphism is well evident. The female moth is bigger than male moth with a distended abdomen and narrow bipectinate antennae. The male moth is smaller, with a narrow abdomen and broad antennae. The female moth is polymorphic in colours, grey or yellow, whereas the male is brown. The adult moth is representing a non-feeding stage. Barsagade et al. (2009) studied surface ultrastructure of mouthparts and suggested that the mandible and labial palps are rudimentary. The terminal parts of maxillae are modified into vestigial proboscis. According to an earlier study, the forewings and a prominent postmedian red line with white line on its border are well evident. The postmedian line of hindwing is represented by a wavy margin. The entire moth is covered with conical, narrow, bristlelike scales. Two pairs of membranous wings are attached to the pterothoracic segment (Jolly et al. 1979; Mishra et al. 2008). The wings, however, show remarkable sexual dimorphism and hence have been discussed sex-wise.

(a) Male Moth The wing span of male is about 16 cm. The forewing has a concave anterior margin and slightly pointed at the tip. The postmedian line (PM) is red with an outer white border. The antero-median line (AM) is usually black or dark brown; in some cases it has white inner border. There is another black or dark-brown oblique line (OL), having an inside white lining which connects costa (C) at the base of the junction of the third medial (M) and first cubitus (Cu) veins (Mishra et al. 2008). The males are generally brown, dark brown and brick red, but grey and yellow males are also found, rarely. The prominent oval-shaped ocellus is present on both wings. The space between the ocellus (OC) and postmedian line is darker (DS). The ocellus is larger on the forewing than the hindwing. The hindwing is smaller than the forewing and its outer margin is convex (Jolly et al. 1979). The antero-median line is slightly wavy and black in colour but without any white lining. The dark space (DS) is demarcated from the postmedian line by a wavy margin. But the oblique line present in the forewing is absent. The remaining colouration of both forewing and hindwings is almost identical except for the yellow stripe of ocellus, which is broader and prominent in the hindwing (Jolly et al. 1979; Barsagade and Tembhare 2000; Mishra et al. 2008) (Fig. 10.5d).

(b) Female Moth According to Jolly (1976), the wing expanse of female moth is about 18 cm. The outer margin of the forewing is less concave than male moth. The areas of the ocellus and hyaline spot of the forewing are larger than those of the hindwing. The antero-median line of both fore- and hindwings shows variable colours, namely, grey, deep yellow, brown and black in different moths. Likewise, the background colour of the wings exhibits diversity. The yellow and grey females are of common occurrence while brown colour is rare. The dark space of the hindwing is not so sharply demarcated from the postmedian line. The colour pattern of the ocellus resembles with that of males, except for some minor structural difference (Jolly et al. 1979; Mishra et al. 2008) (Fig. 10.5e).

10.4.11 Grainage

Grainage (seed collectivity) is the establishment of healthy silkworm egg production. A systematic approach not only reduced mortality and saves labour but also improves for the progeny. The selection and preservation of seed cocoons, the preparation of disease-free laying and their disinfection and the incubation are among the important aspects of grainage described by Jolly et al. (1979) and Mishra et al. (2008) as follows (Fig. 10.5f):

(a) Selection of Seed Cocoons

The healthy and tough cocoons are selected visually for seed (eggs) production.

(b) Preservation of Seed Cocoon

The pupal stage of nondiapausing crop lasts hardly a month, during which the atmospheric temperature and humidity are kept within the optimum range (25–30 °C and 70–80 %) for their development (Jolly et al. 1979). A very simple type of seed preservation house well ventilated with false ceiling is used for seed cocoon preservation with room temperature less than 5–6 °C atmospheric temperature. The selected seed cocoon arranged in a garland hangs from the roof of a house and allows for safe emergence of moths.

(c) Emergence of Adult Moths

The development of pupae to the moth takes about 25–30 days depending on the conditions, i.e. optimum temperature (28–30 °C) and relative humidity (75–85 %) (Jolly 1976). According to Barsagade and Tembhare (2000), the indoor condition normally satisfies optimum requirement of temperature and relative humidity. Maximum emergence of the moths takes place in the afternoon but the peak period is 1900–2100 h.

Emergence in the diapausing stock generally coincides with the onset of monsoon (the last 2 weeks of June and the second and third crop taken place in August–September and November–December, respectively).

(d) Coupling

As per study of Jolly et al. (1979), after 2–3 h of emergence, the moth starts to couple. The peak period begins from midnight to 2:00 am. For the good results, equal numbers of male and female moths were kept in large bamboo basket or plastic basket for mass coupling. Tasar silk moth generally prefers to couple in darkness at comparatively lower temperature (24–26 °C) for 30 min–1 h. The coupling in the plastic basket was allowed for 4–6 h; although for ensuring normal egg laying, coupling required 10–12 h (Fig. 10.5g).

(e) Rearing of Larvae

According to Jolly et al. (1979), outdoor rearing exposes tropical tasar silkworms to unfavourable weather conditions and attack by pests and predators and occurrence of various diseases. These account for losses of 50–55 %, mainly of the early instars due to improper handling of larvae and the faulty selection of rearing sites. These losses can be substantially decreased by more rational approach to rearing. The fate of the crop largely depends on choice of rearing site and food plants, brushing, supervision and maintenance of larval population and other rearing operations. Slackness in any operations adversely affects the yield. The rearing technique suggested by Jolly et al. (1979) are given below.

10.5 Rearing and Harvesting

Before the onset of rearing, the site and surrounding area should be cleared of weeds, which might otherwise induce pest and predator attacks. Apart from removing unsuitable foliage, care should be taken to free the bushes of insects, particularly ant nests. A band of straw with little ash should be tied around the trunk to check the downward movement of larvae. The trunk base should be encircled with a thin band of Gammexane powder to prevent attack by ants and other insects.

(a) Quality of Leaf
According to Kushwaha et al. (2004), the quality of leaf in relation to larval age is a major determinant of health and vigour. The younger larvae thrive on juicy, tender leaves, while the late instars require medium to mature leaves. Outdoor rearing cannot ensure provision of the proper quality of leaves during the different instars. Nevertheless, this can be achieved to a degree through various cultural operations and selective utilization of food plants.

(b) Brushing
Suryanarayana et al. (2007) described the brushing method. Brushing is the placing of the hatching larvae on the leaves. The traditional cultivators tie the leaf cups containing the eggs on the bushes for hatching. Both the developing embryo and the newly hatched larvae are thereby exposed to fluctuating temperature and humidity, heavy rains, storms and other hazards. This results in poor hatchability and heavy losses of larvae. A small twig should be placed over each of the newly hatched larvae, which are then tied on the bushes in a uniform distribution. This operation should not be carried out in strong sunlight, heavy rain or other inclement conditions.

(c) Improved Rearing Techniques
As the *A. mylitta* is wild, there is a very heavy population loss, particularly during the first instar, due to outdoor rearing (Alam et al. 2000). According to Srivastava and Thangavelu (2005), the complete indoor rearing has been not so far successful; however, controlled rearing techniques are useful and have been developed by the Central Silk Board of India to protect the early stages larvae as follows.

(d) Controlled Indoor Rearing of Early Larvae
The first instars until their first moult are reared indoors. The fresh branches of host plants are used for controlled indoor rearing (Shivakumar and Shamitha 2013).

(e) Rearing Set
According to Jolly et al. (1979), the rearing set consists of water-filled bottle holding 3–5 twigs (about 60 cm long) with quality foliage and a cylindrical polythene/mesh net cloth enclosure with a split bamboo/wire frame. The cut end of the twig is inserted well under water. This arrangement keeps the foliage turgid for 3–4 days of feeding. Proper clustering of the foliage to form the centre of the set prevents the

larvae from crawling onto the polythene/mesh net cloth. The mouth of the bottle is plugged, to check an increase in humidity due to gradual evaporation of the water. The rearing capacity of the set can be increased considerably by using a big earthen pot or tin container (Fig. 10.5h).

Brushing is conducted according to the improved technique of using only one day's hatchings and avoiding overcrowding. The polythene/mesh net cloth enclosure, which is tied at the bottom around the neck of the bottle and fastened at the top to a support, should be opened for about 15 min daily for cleaning and for adequate aeration. Special care should be taken to remove the enclosure as soon as the larvae start settling to moult; otherwise, the ecdysis is so difficult as it causes heavy mortality. The larvae coming out of moult are allowed to crawl on fresh twigs and transferred outdoors. The temperature and humidity should be maintained up to 32 °C and 70–80 % RH.

(f) Rearing Huts

The rearing sets are placed inside a hut thatched with straw and leafy branches. It should be constructed high enough from the ground to avoid water logging. According to Yadav et al. (1996), the hut should be bordered with a thin band of Gammexane powder to keep away ants. The open side should not directly face the sun and should be closed at night and during bad weather. Rearing up to the third instar is conducted on economic plantation, preferably under nylon netting. Although the overlapping twigs allow the larvae to crawl from one bush to the other, the medium size and regularity of the plantation not only permit more efficient management, supervision and operation but also minimize losses. As soon as the larvae have completed the second or third moulting, the twigs bearing them are cut and transferred to the forest or block plantation for further rearing.

Rearing of Late Larval Population in Field

Outdoor rearing calls for dawn-to-dusk vigilance against pests and predators. Traditional cultivators brush the larval population too densely irregardless to the quantity of foliage. This high density lowers the yield owing to the higher disease mortality and also adversely affects the economic character of the cocoons.

According to Yadav et al. (1996), frequent direct handling of the larvae causes considerable injury to larval health and contaminates the population. It is therefore desirable to transfer the larvae only once or twice by cutting the small branches bearing larvae and attaching them to unused food plants. A secondary advantage of this system is light pruning of the plants. Jolly (1976) stated that the moulting and spinning larvae also need special attention; the former should not be disturbed, and the latter require enough foliage to form the hammock properly.

As per Jolly et al. (1979), dead larvae hanging on the bush or fallen on the ground should be collected every morning and evening. A sample of the dead larvae should be examined for microsporidiosis. All the dead worms should be buried outside the rearing site. The rearing equipments and the hands of field workers should be disinfected with Dettol water after every contact.

(g) Cocoon Harvesting

According to Mathur et al. (2005), the cocoons should be harvested after 6 or 7 days of spinning. The terminal braches are cut and the cocoons are pulled off from the twig. The adhering leaves are removed, and the cocoons are graded.

(h) Silk Reeling and Spinning

The filament of tropical and temperate tasar is continuous and reelable as the cocoons are of the closed type.

(i) Sorting

The traditional method of sorting simply eliminates flimsy, stained and pierced cocoons. As cocoon colour, size and compactness have a significant influence on reeling performance, sorting of those lines helps ensure uniform cooking and reeling for the production of quality raw silk. According to Jolly et al. (1979), the cocoons of *A. mylitta* are either yellow or grey with different grades of compactness. The cocoons can be graded into large, medium and small according to the size ranges.

(j) Stifling

The fresh cocoons are stifled not only to prevent moth emergence by killing the pupae but also to ensure proper preservation of reeling cocoons by eliminating the moisture content (65 %). The different methods of stifling are briefly described below given by Jolly et al. (1979):

1. *Sun drying*: Although sun drying is economical and widely practised owing to its simplicity, prolonged exposure to the sun denatures the sericin, which makes the cooking of the cocoons difficult and thereby affects the reeling performance.
2. *Steam stifling*: Sun drying is necessary after steam stifling for proper preservation of the cocoons. The storage of semi-wet cocoons causes fungus attacks and vitiates reeling efficiency. Basket and chamber steaming are two common methods of steam stifling.
3. *Basket steaming*: This very crude method is widely used in small reeling establishments. 1–2000 cocoons are put into a bamboo or cane basket having tightly woven sides to check the steam outflow and a loosely woven bottom to let in the steam. The basket is wrapped with a thick, wet cloth to ensure effective steaming and placed over a water-filled container which is heated by fire. The cocoons are steamed for 20–30 min.
4. *Chamber steaming*: This method is practised in large tasar reeling establishments. The steamer is a double-walled cylindrical chamber with a rectangular bottom for water, a small furnace and a pressure gauge. The cocoons are placed in a removable perforated cylinder. Smoke is exhausted through the space between the two walls and the 'chimney'. The cocoons are steamed for 20–30 min at 5–10 lb/in^2.

5. *Hot air drying*: The hot air oven or chamber has an electrical or fuel-operated heating source and a built-in air circulation mechanism for drying. The cocoons are spread on perforated shelves. At 80–90 °C, the pupae are stifled, and in 10–15 h, they are completely desiccated. If the quantity of cocoons to be stifled is large, partial (20 or 40 %) desiccation may be effected to shorten the operation.

This technique is superior to sun drying and steam stifling not only because it has a limited effect on both the sericin and the fibroin but also the cocoons can be stored immediately.

(k) Storage

Thangavelu (1991) stated that even fully dried cocoons are liable to mould damage during periods of high humidity. Reeling cocoons are ideally stored at 25–30 °C and 60–65 % RH. The storage room must be rats and insects proofed, ventilated and periodically disinfected with formalin. The stifled cocoons should be thinly spread, preferably on wire-mesh racks.

(l) Reeling

According to Jolly et al. (1979), reeling involves simultaneous unwinding of filaments from several cocoons, imparting twist and winding the composite raw silk thread onto a bobbin, and it is completed by the following steps:

1. *Cooking*: Uniform softening of the hard cocoon shell by cooking is a vital operation as it makes possible smooth unwinding of the filament without frequent breakages, thereby improving productivity and the raw silk quality. The technique depends on the nature and content of the sericin present (Fig. 10.5i).

   Tasar cocoon is very hard and compact. Unlike the mulberry cocoon, it cannot be softened by boiling in plain water. The poor solubility of the sericin in water and alkaline/acidic solutions, due to the presence of tanning substances, makes the cooking and degumming of tasar cocoons difficult.
2. *Traditional method*: The cocoons are individually wrapped in silk waste, especially at the peduncle end, to prevent the shell from bursting and are treated in a boiled alkaline media for 4–5 h. This technique involves much labour and the reeling results are poor.
3. *Improved technique*: Enzymatic decomposition of the sericin at low temperature renders it soft enough for reeling. Solutions of cocoonase, papain, trypsin, pepsin and Biopril-50 have been tried; the last has been found to be the most efficient because it not only provides uniform softening but also reduces the cooking/steaming period. The cocoons are first boiled in plain water for 1 min and then steamed at 15 lb/in^2 for 40–60 min after which they are left in the chamber while the pressure is gradually released. The cocoons are then loosely wrapped in a porous cloth and soaked in 0.1–0.2 % Biopril-50 solution for 20–22 h, initially at 40–50 °C and later at room temperature. After being spread out on an ash bed,

the cocoons are semidried and then deflossed for dry basin reeling. This method gives about 65 % reelability with Daba and Bogai races of *A. mylitta.*

4. *Deflossing*: The cooked cocoons are deflossed until the filament end is found. Deflossing is carried out individually by hand, as collective brushing results in opening of the peduncle ends. Little time is required for the individual hand brushing of tasar because the cocoon is large and the number consumed daily per reeler is very small.
5. *Reeling operations*: Normally, five or six deflossed cocoons of *A. mylitta* (Daba) are fed into the reeler per end to produce, respectively, 60/65 denier or 40/45 denier thread. The filaments are passed together to the delivery roller, from which they pass to the bobbin through the final thread guide. Twisting and winding take place simultaneously by means of a ring and a traveller on the ring rail. The formation of thread on the multiend reeling machine is different. The filaments are passed together through a jette boute, and then the composite raw silk thread, travelling through the thread guides, is finally wounded without twist on the wooden reels.
6. *Reeling appliances*

Various reeling appliances are described by Jolly et al. (1979) as follows:

Natwa: This appliance is a hand swift made of bamboo and wood. The reeler, sitting cross-legged, reels a number of filaments and winds them on the natwa after imparting twist to the composite yarn by hand. This is such a slow process that a reeler can only reel bout 80 cocoons in 8 h; however, because it ensures better control of the individual cocoon filaments, the resulting yarn is of good quality. Despite low productivity, it is widely used because of its simplicity and small investment cost.

Trivedi reeling machine: This pedal-driven machine has four spindles with a vertical traverse mechanism for distributing the reeled thread uniformly on the bobbin. The average daily production of raw silk is 180–200 g.

CTRS reeling machine: Designed on the spinning principle, the CTRS reeling machine is also pedal driven and four spindled and has a wooden swift (50 cm in circumference) for delivering the composite silk yarn to the fast-rotating bobbin fixed on the spindles. There is a small step pulley on the spindle-driving shaft for changing the spindle speed when necessary. Besides increasing the delivery rate, the wooden swift also reduces lapping by lessening the contact of the raw silk thread with the wood surface. All the working points are fitted with ball bearings for very smooth operation. The machine is also equipped with various other devices which help to increase its efficiency.

Modified CTRS multiend reeling machine: Tasar cocoons can also be reeled on a dry basin with a multiend machine, such as the cottage type of 4-end reeling machine with jette boute (without croisure) and button. The average output of CTRS multiend reeling machine is higher than the CTRS reeling machine when cocoons soaked in Biopril-50.

(m) Re-reeling

After reeling, the raw silk from bobbins or reels is rewound onto standard-sized reels about 150 cm in circumference.

Motor-Driven Standard Re-reeling Machine

This two-sided machine is widely used in large reeling operations. One operator can attend to about 15 ends. It has an improved traverse distributor system that forms diamonds in the hank.

The traverse mechanism consists of adjustable gears. For instance, with a gear ratio of 10:18, for every 18 revolutions of the reel, the traverse eccentric disc makes 10 revolutions, thus moving the traverse bar forward and backward ten times to produce ten webs across the face of the hank. This pattern can be changed by altering the combination of gears.

Each silk hank formed on the re-reeling machine is 150 cm in circumference and weighs 50–70 g for local transactions. The hank is twisted to form a skein, and a number of skeins are press-packed to form a 'book' or bundle for commercial use.

(n) Spinning

The filament from a cocoon is almost 60–70 % reelable; however, the remaining can be spun. According to Jolly et al. (1979), pierced and defective cocoons and those which have been spoiled by improper cooking or reeling are also utilized for spinning. Spun yarn is coarser and more irregular in size than reeled yarn.

(o) Silk Waste

Unreelable silk that is utilized for spinning falls into the following categories by Jolly et al. (1968):

1. *Cocoon waste*: Cocoons pierced by the emerging moths are the best-quality waste. Discarded cocoons which are unreelable because of certain defects (insect infestation, thin ends, doubles, etc.) can nevertheless be used for spinning.
2. *Reeling waste*: Reeling waste which constitutes the major portion of total waste includes cooking and brushing waste as well as that obtained during reeling on account of cocoon feeding and end breakages or as basin residue (innermost shell layer). Like cocoon waste, it can readily be spun by hand or on a charkha.
3. *Thread waste*: This consists of all bits of yarns obtained during knotting and cleaning in the various stages of commercial silk yarn production: re-reeling, winding, throwing and weaving. It is mechanically processed in the spinning operation.

(p) Spinning Yarns

Jolly et al. (1979) discuss the types of yarn spinning as follows:

1. *Ghicha*: Cooked tasar cocoons which are unreelable owing to opening of the peduncle end or a hard shell are used to produce the hand-drawn coarse yarn called *ghicha*. The spinner holds the cocoon in the left hand and draws out the

Fig. 10.6 Parasite of tasar silkworm *A. mylitta*, (**a**) *Xanthopimpla pedator*, (**b**) *X. pedator* infected pupa within cocoon and (**c**) Uzifly, *Blepharipa zebina*

entangled filament with the right hand, imparting a few twists by rubbing the yarn on the surface of an inverted pitcher. The spun yarn, subsequently made into hanks, is used only for weft because of its softness (Fig. 10.5j).

2. *Katia*: A lump of degummed and open tasar reeling waste is drafted by hand. Twist is given to the strand by rotating the spindle on the indigenous 'takli'. The yarn is subsequently wound on the spindle.
3. *Jhuri*: The kind of yarn spun from uncleaned and unopened tasar cocoons is known as *jhuri*.

10.6 Occurrence of Pests and Predators of Tasar Silkworm

On the basis of feeding behaviour, damage caused by predators and insect pests to *A. mylitta*, Jolly et al. (1979) described them as the larval–pupal pest tachinid fly (*Blepharipa zebina*), cocoon pest yellow fly (*Xanthopimpla predator*), and predators such as *Canthecona furcellata*, *Hierodula bipapilla*, *Vespa orientalis*, *Oecophylla smaragdina*, *Myrmicaria brunnea*, dermestid beetle, birds, lizards, squirrels, rats and termites. According to Singh and Thangavelu (1991), occurrence of these pests and predators is considerably fluctuated during each crop as per their availability of habitat and climatic condition, i.e. rainy, winter and summer season. Pests and predators directly affect the tasar silkworm and kill the larvae, pupae and sometimes adult stages of *A. mylitta* (Figs. 10.6 and 10.7).

10.6.1 Pests of *A. mylitta*

(**a**) *Xanthopimpla predator*

An ichneumonid, *Xanthopimpla predator*, is a major pupal endoparasitoid of tasar silkworm commonly known as 'yellow fly' (Jolly et al. 1979). According to Gathalkar (2014), female yellow fly searches out the suitable tasar cocoon/pupa as a host. After the host confirmation by palpating its antenna on the host cocoon, *X. predator* pierces the cocoon by 1 cm long egg-laying apparatus called ovipositor. The yellow fly lays a single egg on the developing prepupa/pupa of *A. mylitta*. Mostly, the parasitoid prefers pupa as their suitable host and rarely pierces the

Fig. 10.7 Predators feeding on tasar silkworm *A. mylitta*, (**a**) *Canthecona furcellata*, (**b**) *Hierodula bipapilla*, (**c**) *Vespa orientalis*, (**d**) *Oecophylla smaragdina*, (**e**) *Myrmicaria brunnea* and (**f**) *Monomorium* sp.

cocoons those are yet to be developed into pupa. The *X. predator* preferred sunny days for egg laying, and its infestation was noted maximum after 12:30–5:00 pm. It completes life cycle in about 16–18 days totally devouring the body content of the host pupa by the larval stages. Pupation takes place inside the tasar pupa, and adult emerges by rupturing the head end of a dead pupa and peduncle end of the cocoon by cutting it with strong mandibles. It causes heavy loss to silk production by infecting the tasar cocoons and completing its life cycle inside the pupae of silkworm. Pupal mortality was recorded about 7 %, 9 % and 11 % of the total crop. Total crop production as well as the seed cocoon is being affected harshly by this pest (Fig. 10.6a, b).

(**b**) *Blepharipa zebina*
According to Dasgupta (1962), the Uzi fly *Blepharipa zebina* (Walker) is a larval endoparasite of tasar silkworm. The gravid female of Uzi fly lays eggs on silkworm larvae from the third instar onwards. The newly hatched maggot penetrates into the body of tasar silkworm and feeds on haemolymph. The maggots moult two times and form three instars inside the host body within 10–20 days. The mature maggots of Uzi fly come out of the cocoon by making a hole and pupate outside. Larval mortality was recorded about 3 % of the total crop (Fig. 10.6c).

10.6.2 Predators of Tasar Silkworm

(**a**) *Canthecona furcellata*
The carnivorous stink bug *Canthecona furcellata* (Hemiptera: Pentatomidae) is the most harmful predatory insect found on the *A. mylitta*. Barsagade and Gathalkar (2016) describe the attack of nymphs and adults on the early stages of tasar

silkworm usually on first to third instar larvae, rarely on fourth and fifth instar. According to Barsagade and Gathalkar (2016), the rate of predation is serious during moulting. The mouthparts of the bug are sucking type and known as rostrum (proboscis). The rostrum or proboscis is pierced into larval integument and sucks haemolymph. Sometimes, the bug sucks the haemolymph from the spinning larva, through moist and thin network of silk thread of cocoon. The adult bug *C. furcellata* is about 15 mm long. The life cycle of a bug is completed in two monthly cycles, i.e. March–April, June–July and September–October in a year. Both nymph and adult feed on haemolymph of the *A. mylitta*. Serious damage to tasar larvae is caused by both the immature and adult bug *C. furcellata*, and tasar larval mortality increased up to 11 % (Fig. 10.7a).

(b) *Hierodula bipapilla*

Praying mantis *Hierodula bipapilla* is also an important predator of *A. mylitta*. The insect is recognized by foreleg, i.e. the raptorial type, a small triangular head and slender body belonging to the order Mantodea (Jolly et al. 1979). The wings are well developed. The pincer-like forelegs are modified for grasping the larvae and well-developed mouthparts for biting and chewing mode of feeding. According to Gathalkar (2014), *H. bipapilla* is active in all the seasons and causes damage to crop I, crop II and crop III. Both the nymph and adults preferably favour and feed on early instar larvae of *A. mylitta*. The female deposited its eggs in a definite pattern and glued together into egg mass, called 'ootheca', fixed to tasar food plants. The whole egg mass is built up, layer by layer, and young mantis emerges out from the eggs. Nymphs and adults are predacious at all times, and adults attack third and fourth instar larvae, and sometimes it also attacks on adult moth. Due to predating on larval stages, the damage caused is up to 3 % of larval mortality (Fig. 10.7b).

(c) Common wasps – *Vespa orientalis* **(Hymenoptera: Vespidae)**

The wasp, *V. orientalis*, is a serious predator of *A. mylitta*, and it attacks on the early larval stages of silkworm. According to Gathalkar (2014), the mouthparts of the wasp are of biting and chewing type. Wasp has strong mandibles by which they catch the first to third instar larvae of *A. mylitta*. After catching the host larvae, it cuts the larvae into pieces and feeds on them. *V. orientalis* was predacious in each crop throughout year. *V. orientalis* is a medium-sized reddish- or brown-coloured wasp with remarkably yellow bands on slender, elongated and spindle-shaped abdomen. The nests are constructed on tasar host plants and consist of a number of circular combs attached one over the other. Each comb contains numerous hexagonal membranous cells. The whole nest is enclosed in a papery envelop. These are the social wasps with queen, worker and drone. Fertile queen lays eggs in each cell of comb where the egg hatches into larva. Then, the larva spins a silk cap over the cell's opening and transforms into an adult. The adult comes out by eating the silk cap. The activity of wasp increases in the warm weather. Their young ones were fed on small larvae of *A. mylitta* brought into the nest by their parent or worker wasps. *V. orientalis* is active throughout the year, and due to the predatory activity, 5 % of total crop damage was noticed, and the larval mortality fluctuates in between 4 and 5 % (Fig. 7c).

(d) The Ants

According to Gathalkar and Barsagade (2016), among the aggressive and omnivorous ants, *Oecophylla smaragdina*, *Myrmicaria brunnea* and *Monomorium sp.* are the most abundant terrestrial insects that attack on early larval instars of *A. mylitta*, sometimes *Oecophylla smaragdina* and other ants also attack on the pupae and adult tasar silk moth. The predatory nature of these ants is described as follows:

1. *Oecophylla smaragdina*: The weaver ant *O. smaragdina* makes large nest on the host plant of *A. mylitta*, i.e. *Terminalia* species. It is a very common forager, attacking the larval stages of tasar silkworm from first to third instar, and sometimes it attacks on fourth and fifth instar larvae. The nest contains queen, workers, and drones. The workers are very aggressive and attack on the early instars especially from first to third instar of *A. mylitta*. The ant workers cut the larvae into pieces by their strong mandibles, and the pieces are carried towards their nest. During feeding the workers release irritating secretion through the mandibular glands. The sting apparatus is absent and therefore it does not stink, but it releases formic acid from the last abdominal segment causing the irritation to larval skin. Because of the predatory foraging habit of workers, it attacks in groups, and within a minute, they kill early larval stages of *A. mylitta* (Fig. 10.7d).

   The life cycle of the weaver ant *O. smaragdina* passed through four stages, i.e. egg, larva, pupa and adult. It is polymorphic and three castes of queen, male and worker are found in a colony. On the basis of size workers, they are categorized into two types, i.e. major and minor workers. Due to the predatory behaviour of ant, the larval mortality of *A. mylitta* increases up to 5 %, and the total annual crop damage was observed approximately 4–5 %.
2. *Myrmicaria brunnea* (Saunders) belonging to family Formicidae and subfamily Myrmicinae was identified as a predator of *A. mylitta*. *M. brunnea* has a distinctive downcurved abdomen and two spines on the metathorax. Workers are chestnut brown in colour with shining mandibles. The worker ants attack on the host tasar larvae in groups. Initially, the host larvae are captured by few workers and then they pricked it. After that nearby workers, attack the same prey from all sides. Predatory workers make attack with high aggressiveness over the tasar larva and cut prey in small pieces and later on transported to the nest (Fig. 10.7e). Due to the predatory behaviour of *M. brunnea*, about 2 % larval mortality were observed, and the total annual crop damage was noted about 3 %.
3. *Monomorium* sp.: These are the small ants, reddish brown in colour, and live in colony belonging to order Hymenoptera and family Formicidae. The workers attack on the tasar silkworm and affect the first instar to third instars. Sometimes they attack on the pupae of tasar silkworm, by making a small hole in the cocoon, and fed over the pupa inside the cocoon resulting into the death of the pupa. Destruction of stages of silkworm occurred, and the total tasar silk production is being affected. (Fig. 10.7f)

4. *Dermestid beetle*: Jolly et al. (1979) observed the attack of dermestid beetle on the stored tasar cocoons of *A. mylitta*, in the form of pierced cocoons, in grainage house or storage rooms. It completely destroys the developing pupa of *A. mylitta*. According to Gathalkar (2014), the female beetles lay eggs in the floss of cocoons. Due to its attacks, the pupae and quality of tasar cocoon get damaged. Larvae of dermestid beetle bore into cocoon shell and feed on pupae. Due to the attack of beetle on *A. mylitta*, 0.5–1.0 % mortality of pupae was observed. Among the non-insect predators, birds, lizards, rats and squirrels were found damaging on the tasar silkworm. Due to these predators, about 4–5 % larval and pupal mortalities were recorded.

   Jolly et al. (1979) also described some birds, reptiles and mammals as predators of *A. mylitta*.
5. *Birds*: Birds are very common in tasar field, which cause the larval mortality; among which crows, rufous treepie (*Dendrocitta vagabunda*) and common hawk-cuckoo (*Hierococcyx varius*) are very common. The birds pick up the tasar larvae directly from the host plants.
6. *Garden lizard*: Garden lizard (*Calotes versicolor*) is also reported from the field of tasar silkworm; it feeds on the larval stages and helps for loss of tasar crop.
7. *Mammalian predators*: Some of the mammalian predators, like squirrels and rats, create the serious problem in tasar sericulture. The attack by rats in the field and stored cocoons is recorded. Rat (*Rattus rattus*) attacks are very common in grainage house where the cocoons get damaged and become valueless. Similarly, the seed cocoons were lost. The attacks of squirrels are occasional, and it attacks on matured hanging cocoon on tasar host trees and causes damage to cocoon by cutting cocoon shell.

   Termites are found at tasar rearing sites and also attack on the adult tasar moth.

10.7 Conclusions

The tropical tasar silkworm, *A. mylitta*, is a wild polyphagous silkworm, feeding on three primary and dozens of secondary food plants in wild conditions. *Antheraea mylitta* contains 44 ecoraces distributed in various regions of India. Each ecorace has its specific characteristics and contributes in the development of tasar sericulture as it is natural source of income.

Acknowledgements I am thankful to Dr. D. B. Tembhare, ex-professor and head, Department of Zoology, RTM Nagpur University, Nagpur, for the support and critical evaluation of the manuscript. I am also thankful to Rajiv Gandhi Science and Technology Commission, Government of Maharashtra, Mumbai, for their financial support and Directorate of Sericulture, Government of Maharashtra, and Regional Tasar Research Station, Bhandara, India, for their field support. I am thankful to my students Dr. Mangala Kadwey, Dr. Shruti Gharde, Dr. Hemant Meshram, Dr. Manoj Thakare, Dr. Ganesh Gathalkar, Miss Rani Thakare and Mr. Jitendra Kirsan for their help.

References

Alam MO, Pandey RK, Yadav GS, Sinha BRBP, Sinha SS (2000) Studies on rearing performance of wild eco-race model of tasar silkworm, *Antheraea mylitta* (D) on deferent food plants. Int J Wild Silkmoth Silk 5:82–86

Barsagade, DD (1998) Studies on the silk gland development and silk protein synthesis in the tasar silkworm, *Antheraea mylitta* (D) (Lepidoptera: Saturniidae), PhD thesis submitted to Nagpur University, Nagpur (India)

Barsagade DD, Tembhare DB (2000) Development, structure and secretory activity of silk gland complex in the tropical tasar silkworm, *Antheraea mylitta* (Drury). (Lepidoptera: Saturniidae). Sericologia 40:231–245

Barsagade DD, Kadwey MN (2010) Ultrastructural Changes in Midgut of CPV Infected Tropical Tasar Silkworm *Antheraea mylitta* (D) Lepidoptera: Saturniidae. Int J Ind Entomol 21:117–125

Barsagade DD, Gathalkar GB (2016). First predation record of *Canthecona furcellata* (Wolff.) (Hemiptera: Pentatomidae) on spinning stage silkworm *Antheraea mylitta* (Drury) Entomological Research 46(4):236–245

Barsagade DD, Pankule SD, Tembhare DB (2009) Impact of Fungus on egg shell of tropical tasar silkworm, *Antheraea mylitta*: an ultrastructural approch. Int J Ind Entomology 18(2):77–82

Barsagade DD, Kadwey MN, Gharde SA, Thakre MP, Meshram HM, Gathalkar GB, Munir M, Andrew RJ (2011) Biology and effects of environmental factors and pathogens on the Vanya tasar silkworm, *Antheraea mylitta* (D) eco-race Bhandara. Pro Nat Con Enviro Bio Biodiver:162–169

Barsagade DD, Chamat MV, Husain IR, Meshram HM, Thakre RP (2015) Scanning Electron Microscopic (SEM) and Energy Dispersive Analysis of X-rays (EDAX) studies on the eggshell of silkworm, *Philosamia ricini* (Lepidoptera: Saturniidae). Int J Zool Res 5(4):19–26

Dasgupta KP (1962) Observation on the behaviour of uzifly maggots. Indian J Seric 1:16–18

Gathalkar, GB (2014) Studies on the pests and predators of tropical tasar silkworm *Antheraea mylitta* (D) eco-race Bhandara, with special reference to *Xanthopimpla* and *Oecophylla* Lepidoptera: Saturniidae), PhD thesis submitted to RTM Nagpur University (India)

Gathalkar GB, Barsagade (2016) Predation biology of weaver ant *Oecophylla smaragdina* (Hymenoptera: Formicidae) in the field of tasar sericulture. J Entomol Zool Stud 4(2):7–10

Jolly, MS (1976). Package of practices for tropical tasar culture, Ranchi. Central Tasar Research Station, (Central Silk Board, Bombay), pp 32

Jolly MS, Chaturvedi SM, Prasad SA (1968) Survey of tasar crops in India. Indian J Seric 7(1):56–57

Jolly MS, Sen SK, Ahsan MM (1974) Tasar culture. Ambica publisher, Bombay, p 252

Jolly MS, Sen SK, Sonwalkar TN, Prasad GK (1979) Non mulberry silk. In Agric Serv Bull FAO UN 29:1–178

Kushwaha RV, Singhvi NR, Mathur SK, Lochan R, Sinha BRRP (2004) Study on food plant preference in Bhandara eco-race of *Antheraea mylitta* (D). Plant Arch 4:467–469

Mathur SK, Singhavi NR, Kushwaha RV (2005) Ecology, commercial attributes and status review of Bhandara eco-race of Indian tropical tasar silkworm, *Antheraea mylitta* (D). Proceeding of Workshop on Strategies for Non mulberry Germplasm Maintenance, pp 143–155

Mazur GD, Regier JC, Kafatos FC (1980) The silkmoth chorion: morphogenesis of surface structures and its relation to synthesis of specific proteins. Dev Biol 76:305–321

Mishra PK, Sharan SK, Kumar D (2008) Morphology of egg, larva, pupa and moth. In: Vijaya Prakash NB et al (eds) Tasar culture – principles and practices volume –2. Tasar Silkworm Biology. Central Tasar Research & Training Institute, Piska- Nagri, Ranchi, pp 29–34

Richards OW, Davies RG (1957) Imm's general text book of entomology. Chapman and Hall, New York/London

Sen SK, Jolly MS, Jammy TR (1969) Diseases of tasar silkworm *A. mylitta* (Saturniidae). Indian J Seric 8:11–14

Shivakumar G, Shamitha G (2013) Studies on larval mortality: diseases, pest and predator menace in outdoor and indoor reared tasar silkworm, *Antheraea mylitta* Drury (Daba TV). Res J Anim Vet Fish Sci 1(4):1–7

Singh KC (1991) Controlling the insect enemies of oak tasar silkworms. Indian Silk:19–23

Singh RN, Thangavelu K (1991) Parasites and predators of tasar silkworm – *Antheraea mylitta* has many enemies. Indian Silk, pp 33–36

Sinha SS, Sinha AK (1997. Eco-race of *Antheraea mylitta* (D) and exploitation strategy through hybridization. Proceeding of Seminar on Current Technology of Non mulberry Sericulture. 1–39

Sriharan TP, Samson MV, Krishnaswami S, Dutta RK (1971) Laboratory investigation on uzifly, *T. bombycis*, a Tachiniid parasite of silkworm (*B. mori*). Indian J Seric 10:14–22

Srivastava PK, Thangavelu K (2005) Sericulture and seri – biodiversity. Associate Publishing Company, New Delhi

Suryanarayana, N., Shova Beck. & Gangwar S. K. (2007). Tasar food plants and their morphoanatomical features. In: Suryanarayana N et al. (eds) Tasar culture – Principles and practices volume – 1 tasar silkworm host plants – production, protection and improvement, Central Tasar Research & Training Institute, Piska- Nagri, Ranchi. p. 33–50

Thakre MP (2015) Studies on the metamorphic changes and egg maturation in tropical tasar silkworm, *Antheraea mylitta* (Drury) eco-race Bhandara (Lepidoptera: Saturniidae), PhD thesis submmitted to RTM Nagpur University India

Tembhare DB, Barsagade DD (2000) Cephalic neuroendocrine system in tropical tasar silkworm, *Antheraea mylitta* (Lepidoptera: Saturniidae): metamorphic and sericotropic functions. Int J Wild Silkmoth and Silk 4:1–9

Thangavelu K (1991) Wild sericigenous insects of India: a need of conservation. Wild Silkmoths 91:71–77

Veer V, Rao KM (1994) A new species of *Trogoderma* (Coleoptera: Dermestidae) found damaging store silkworm cocoon in India. Sci Dir J Store Prod Res 30:283–295

Yadav GS, Jaipal Reddy K, Singh BMK, Sinha BRRP, Sinha SS (1996) Tasar eco-race Bhandara – need for conservation. Indian Silk 34(11):31–34

Temperate Tasar Culture

11

Rakesh K. Goel

Contents

11.1 Introduction

Natural silk is broadly classified into two categories – mulberry silk and non-mulberry silk or wild silk or *vanya* silk. Each variety is produced by a different variety of silkworm though all non-mulberry silk worms belong to one family only. Jolly et al. (1975) have reported nearly 80 species/forms/variants, which produce wild silk of economic value; the majority of them are wild and polyphagous in nature, feeding upon a variety of plant species. Of all the non-mulberry sericigenous insects, the following lepidopteron are widely exploited for their commercial use.

Antheraea pernyi G.M.	-	Oak tasar of China
Antheraea mylitta D	-	Tropical tasar of India
Antheraea yamamai G. M.	-	Oak tasar of Japan
Antheraea assamensis Ww	-	Muga of India
Philosamia cynthia ricini D	-	Eri of India
Antheraea polyphemus Cram	-	Tasar of the USA
Antheraea proylei Jolly	-	Oak tasar of India

R.K. Goel (✉)
Central Silk Board, Haridwar, Uttarakhand, India
e-mail: grakeshkr@yahoo.com

Omkar (ed.), *Industrial Entomology*, DOI 10.1007/978-981-10-3304-9_11

The members of the genus *Antheraea* occupy different ecological niches, ranging from tropical to temperate with transitional zones. Out of several species of oak tasar silkworms, like *Antheraea roylei*, *Antheraea frithi*, etc., based on their primary food plant *Quercus*, *Antheraea proylei* Jolly is the most prominent, viable and commercially exploited species in India.

11.2 Ecological Distribution

Temperate or oak tasar silkworm, *Antheraea proylei* Jolly, is cultivated throughout whole sub-Himalayan belt extending from Jammu and Kashmir in the north to Manipur in the far east between 26°–34° latitude and 2000–8500′ ASL altitude. The entire sub-Himalayan belt has two distinct ecological niches – the north-eastern region comprising the states of Manipur, Nagaland, Mizoram, Arunachal Pradesh, Assam, Tripura and Meghalaya having sub-tropical climate with excessive rainfall and shorter summer and the north-western region comprising the states of Jammu and Kashmir, Himachal Pradesh and Uttarakhand having temperate climate with severe cold and longer winter.

Antheraea proylei is an interspecific hybrid of *Antheraea roylei* Moore of India (n = 30), which spins double-layered unreelable cocoons, and its Chinese counterpart *Antheraea pernyi* GM (n = 49), which spins single-layered golden yellow-coloured cocoons. This fertile hybrid has higher silk content and also eliminated outer flossy layer of its parent. This species has stable karyotype of n = 49 over the years (Jolly et al. 1969, 1973; Bhagirath et al. 1988).

Its systematic position is as follows.

Phylum	-	Arthropoda
Class	-	Insecta
Subclass	-	Pterygota
Division	-	Endopterygota
Order	-	Lepidoptera
Suborder	-	Ditrysia
Superfamily	-	Bombycoidae
Family	-	Saturniidae
Genus	-	*Antheraea*
Species	-	*proylei*

11.3 Oak Host Plants

Antheraea proylei is a phytophagous insect thriving upon a variety of oaks belonging to the beech family Fagaceae, order Fagales and genus *Quercus*. Though *Quercus,* the largest genus in Fagaceae, comprises more than 400 species (Trelease 1924) widely distributed throughout the temperate regions of the north hemisphere and extending up to tropics and subtropics of south America, high altitudes of India

Table 11.1 Main forest types of oak species in India

Region	Forest type
Western Himalaya	Ban oak forest (*Quercus leucotricophora*)
	Moru oak forest (*Quercus floribunda*)
	Kharsu oak forest (*Quercus semecarpifolia*)
	Upper oak-fir forest
	Oak scrub
	Other forest types (Himalayan chir pine forest, moist deodar forest, temperate mixed coniferous forest, moist temperate deciduous forest, low-level and high-level blue pine forest and subalpine forest)
Eastern Himalaya and Northeast India	Sub-tropical wet hill forests
	Khasi sub-tropical wet hill forest
	Buk oak forest
	High-level oak forest
	Other forest types (Naga hill temperate forest, East Himalayan mixed coniferous forest, *Abies* forest and subalpine forest)

Table 11.2 Availability and distribution of oak in sub-Himalayan belt of India

State	Area under OAK (ha)
Jammu and Kashmir	409,061
Himachal Pradesh	139,503
Uttaranchal	551,436
Manipur	40,000
Nagaland	20,000
Assam	25,000
Meghalaya	23,000
Mizoram	75,000
Arunachal Pradesh	1,225,000
Total	2,508,000

and Malaya, about 35 well-known species of oaks occur all along the north-western sub-Himalayan region of India at 1200–3500 m AMSL and in north-eastern region at 600–1800 m AMSL (Negi and Naithani 1995). Many species of oaks are gregarious in their occurrence making regular altitudinal zones mainly in western Himalaya regions. They also occur as associates of different forest types in various parts of the Himalaya and hills of east India along with conifers and mixed forests of the temperate and alpine zones (Table 11.1).

There is vast wealth of oak flora in India. It is estimated that nearly 2.5 million hectare of forest is covered by different species of natural oak plants in the entire sub-Himalayan belt of India (Table 11.2). Except for *Quercus semecarpifolia* found at high altitude, other species, viz. *Q. leucotrichophora* syn. *Q. incana*, *Q. floribunda* syn. *Q. himalayana*, *Q. serrata*, *Q. griffithii*, *Q. dealbata*, *Q. semiserrata*, *Q. ilex* and *Q. glauca*, occur below 2000 m AMSL (Goel 2000a; Goel and Rao 2004; Thangavelu 2004; Goel 2000b; Goel 2010a) (Fig. 11.1).

Fig. 11.1 Oak flora of India

Almost all oak species are sexually propagated (propagation through seeds) and grow well in rich moderately moist or dry sandy soils. Oak plants pollinate by wind in the early part of the spring season. Naturally, they reproduce by seeds, but some vegetative propagation techniques are also employed for getting speedy multiplication of plant material and maintenance of desired genotypes and phenotypes. Most common vegetative propagation technique in oak is air layering (Srivastava et al. 2000b). Propagation through *Quercus serrata* stem cuttings and budding has also been reported (Chandra and Mahendru 1977; Srivastav et al. 2000a). The nursery techniques for raising their plantation through seed have been standardized (Lal et al. 1993, 1999; Lethal et al. 1996) and package of practices for raising and maintenance of plantation developed (Srivastava and Noamani 1995; Srivastava and Singh 1999; Singh et al. 2001; Mishra et al. 2004), which includes selection of viable seeds; soaking of seeds followed by sclerotization; seed germination in

Table 11.3 Calendar of activities in raising of systematic plantation of *Q. serrata*

Month	Raising of Plantation	Plantation Maintenance
November December	Collection of *Q. serrata* seeds	
January February	Preservation of seeds	
March	Selection of seed Mechanical scarification of seeds Soaking of seeds for 48 hrs to 96 hrs in water Heap formation, covering it with gunny bags and sprinkling of water.	1st hoeing & weeding Basin formation
April	Pit digging Nursery bed preparation Filling up of polythene tubes with rooting media Seed germination starting from 9th day of heap making Putting germinated seeds in tubes Arranging tubes in nursery beds Pot watering	
May	Pit digging Pot watering Cleaning and weeding Spraying of 0.05% urea solution 1st foliar spray of 0.07% Nuvan	
June	Pit digging Maintenance of seedlings 2nd spray of urea solution 2nd spray of Nuvan Pot watering if required	Application of NPK (1st dose)
July	Transplantation of seedlings Pot watering	
August		Application of NPK (2nd dose); Cultural operations like weeding, basin formation, loosening of soil etc.
September	Loosening of soil Basin formation Weeding	
October	Application of FYM	Application of FYM

heaps; putting germinated seeds in polythene tubes or direct germination of seeds in nursery beds; seedling raising in nursery for 6 months to 1 year; transplantation of seedlings during monsoon season in the pits (2×2×2 feet size) with adequate dose of FYM/vermicompost; timely cultural operations, viz. basin formation, watering, weeding, application of FYM, watch and ward, etc.; application of chemical and/or biofertilizers; and pruning and training of plants (Table 11.3). *Quercus serrata* plantation becomes productive within 5–6 years of transplantation of seedlings following package of practices for plantation maintenance. However, the plantations of *Quercus incana* and *Quercus semecarpifolia* take longer duration for reaching the bearing stage. Consequently, a large area has been brought under systematic plantation of *Q. serrata* not only in north-east but also in north-west (Bahl and Pandey 1989; Das and Pandey 1991; Dhar et al. 2006).

11.4 Life and Crop Cycle of Oak Tasar Silkworm

Oak tasar silkworm is a semi-domesticated (completely domesticated in the north-western region) lepidopteran reared at an altitude between 2000′ and 8500′ ASL. One complete life cycle of oak tasar silkworm is called a crop. The silkworm shows weak bivoltine character amenable towards uni-/bi-voltinism, and yet two crops are not feasible at a particular location. Depending on the geographical and agroclimatic conditions, one or more crops can be taken in a year. Normally, the first crop starts from the first/second week of March during spring season (hence referred as spring crop) and the second crop from September during autumn season (referred as autumn crop) in north-east. In north-west, the second crop is taken in the last week of May or first week of June at high altitude and known as summer crop.

11.4.1 Life Cycle

Oak tasar silkworm develops through four stages, viz. egg, larva, pupa and adult moth (Fig. 11.2), out of which the larval stage is the only feeding stage, while other stages are reproductive in nature or transitional stages for tiding over adverse climatic conditions. Each cycle normally completes within 65–100 days from cocoon to cocoon; the cycle span being dependent upon the rearing season and prevailing atmospheric conditions. Its life cycle starts with the emergence of moths from the cocoons on the onset of favourable climatic conditions. Prior to start of grainage, seed cocoons are shifted from preservation site (high altitudes or cold storage) into grainage where a temperature of 25±2 °C is maintained. Moths prefer to emerge in late evening hours and continue till midnight. Wing expanse is more in adult female (16 cm) than adult male (14 cm). The forewing of male moth is light grey, and its outer margin below the postmedian is having brownish tinge, whereas the forewing of female moth is brownish, and its outer margin below the postmedian is light brown. Male and female moths couple during the late night hours or early morning hours. After 5–6 h of copulation, the moths are decoupled and females are kept for

Fig. 11.2 Different stages of oak tasar silkworm life cycle

egg laying. In about 72 h, more than 90 % eggs are laid by a female, which are considered for incubation. Optimum temperature and humidity for egg laying are 20–24 °C and 70–80% R.H., respectively. Average fecundity is quite low; on average being 120 per laying though it varies from 50 to 450 per laying. The eggs are nondiapausing in nature but can be effectively preserved at 5 ± 2 °C temperature for 17 days without any adverse effect on hatching or rearing performance. Eggs are oval in shape measuring 2.5 × 2.0 mm in length and width and weighing about 6.0–7.0 mg. They are gently detached and, on removing dried appendages from the egg mass, they are washed in soap solution and disinfected with 2% formalin solution. Eggs are dried under shade and after drying incubated at 20±2 °C temperature and 70–80 % R.H. Hatching takes place within 8–9 days of incubation.

Just after hatching, the newly hatched larvae eat up the egg chorion and then feed upon oak leaves when mounted. The larvae start feeding immediately on mounting. It starts feeding from the margin of the leaf blade and proceeds towards midrib. It eats up midrib and even consumes whole leaf petiole during final instars whenever there is scarcity of foliage. Young-age larva prefers soft, tender and succulent leaves, while late-age larva the mature leaves. In larval stage, there are three main activities – exploration (search movements for suitable feed and proper place for resting, moulting and cocoon spinning), feeding and resting (Fig. 11.3). The scale of these activities varies with the age of the larva. While exploration decreases with advancement in the age of the larva, feeding increases. However, the resting period increases up to certain level and then decreases (Table 11.4).

Oak tasar silkworm is a tetra moulter passing through four moults and five instars. Newly hatched larva weighs about 6–7 mg, and its colour is black with red head capsule. It is called first instar larva, which feeds for 6–7 days after which it sits into moult. After coming out of the first moult, the body colour of the larva

Fig. 11.3 Various behavioural activities of oak tasar larva

Table 11.4 Physical units on various behavioural activities of oak tasar silkworm larva

Instar	Total duration (hrs)	Duration of different activities (hr:min)			Moulting duration (hrs)
		Feeding	Exploration	Rest	
I	159	49: 00 (31.00)	09: 06 (5.60)	100: 54 (63.40)	26–28
II	130	39: 55 (30.00)	14: 28 (11.90)	65: 37 (57.40)	28–30
III	164	39: 45 (27.90)	03: 54 (2.40)	120: 21 (69.70)	28–32
IV	197	80: 30 (40.90)	02: 58 (1.50)	113: 32 (57.60)	36–40
V	144	69: 58 (48.60)	04: 22 (3.10)	69: 40 (48.30)	
	159	84: 06 (52.90)	03: 00 (1.90)	71: 54 (45.20)	

[Figures in parenthesis indicate percentage of different activities]

Table 11.5 Average leaf requirement by a single larva of *A. proylei* J

Instar	Leaf consumption (g)	Proportionate ratio	Consumption time (hr:min)	Feeding duration (%)
I	0.21	–	50: 24	30.0
II	0.73	3.48	37: 12	31.0
III	2.59	3.55	22: 24	31.0
IV	8.97	3.46	78: 40	41.0
V	60.33	6.73	144: 00	50.0
Total	72.83	290	332: 40	

changes to green and that of head capsule to brown. Like this, the larva sheds its skin after every moult and grows in size and weight. Average weight of mature larva is about 20 g, growing by about 3000 times during its entire larval span. Total larval period varies from 35–60 days depending upon the season, geographical region and meteorological conditions. The duration of first, second, third, fourth and fifth larval instars varies from 6–7, 4–6, 6–7, 8–10 and 12–18 days, respectively. Total leaf requirement is 72 g per larva (Rana et al. 1987; Table 11.5). In final stage, the larva feeds voraciously consuming about 80% of total leaf requirement. The suitable rearing temperature and humidity are 20–25 °C and 70–75% R.H., respectively. The temperature below 20 °C and above 25 °C is harmful for larval growth.

The fully mature larva stops feeding and passes last excreta before spinning the cocoon. The spinning process starts with tying of 3–4 leaves with the help of silk thread, which results in hammock formation. After the larva spins the peduncle, it enters into the hammock and starts spinning the cocoon shell. During cocoon formation, the larva moves its head in the form of horizontal eight (∞) and throws silk thread in regular and uniform pattern. Though, at the time of cocoon formation, the larva moves its body in different directions, on completion it sits inside the cocoon with its head towards the peduncle end. In last, the larva discharges a light milky yellowish nonviscous and nonsticky fluid or excreta, which moistens the whole cocoon. This discharge on drying helps the cocoon to harden and also imparts colour to it. The cocoon formation is completed within 4–6 days. Harvesting and cleaning of cocoons is done after the sixth day of spinning when larval-pupal transformation is almost complete. Average cocoon weight is about 6.0 g varying from 4.0 to 12.0 g and size is 4 × 2.5 cm. The pupa is brown in colour, about 5.6 g in weight and 3 × 1.8 cm in size.

In this way, the oak tasar silkworm completes its life cycle like other silkworms (Fig. 11.4). In bivoltine brood, spring crop cocoons are nondiapausing, i.e. moth emergence starts after 25 days of pupation. The cocoons harvested from autumn crop or high altitude summer crop are diapausing and are required to be preserved during adverse climatic conditions.

11.4.2 Crop Cycle

Natural abode of oak tasar silkworm is whole temperate zone sprawling throughout the sub-Himalayan belt. It has got a varied topography with high terrains and deep

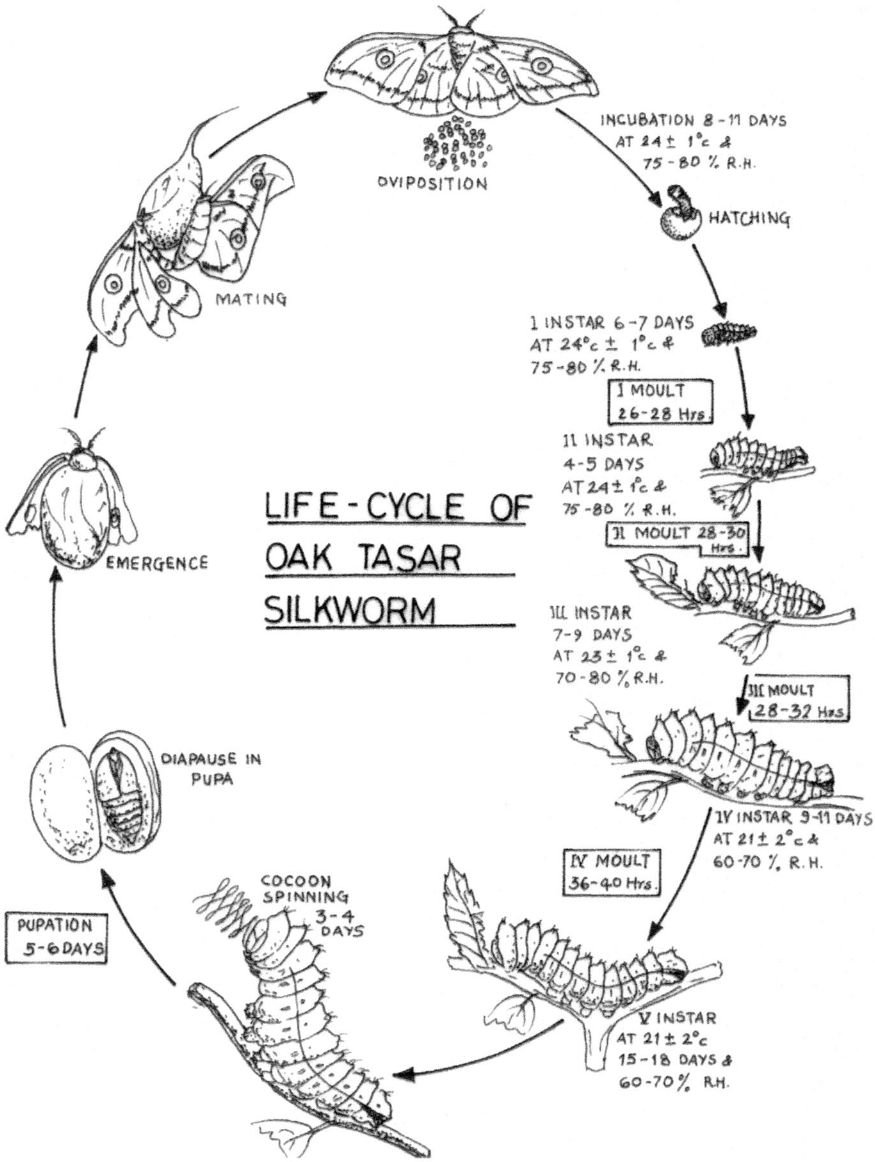

Fig. 11.4 Life cycle of oak tasar Silkworm *Antheraea proylei* J

gorges in north-west to low terrains and plain areas in north-east. Besides, there are wide disparities in their agroclimatic conditions too. While the climate of entire north-eastern region is sub-tropical with excessive rainfall and shorter summer, that of north-western region is more severe with extreme cold and low temperature in a longer winter period. As such, crop cycle of silkworm varies from one region to another.

11.4.3 Crop Pattern in North-East

In this region, the spring crop is the main crop (seed or commercial), which starts from the first week of March when either the natural sprouting takes place in *Q. serrata* plants or sprouting is manipulated by subjecting oak plants to pruning during December–January months. This crop is diapausing one, necessitating preservation of seed cocoons for 8–9 months either at low altitude in the well-ventilated rooms where the temperature remains below 20 °C from June to February or at high altitude (6000–6500′ ASL) to check erratic emergence. However, an additional subsidiary crop can also be taken up during summer season or autumn by breaking the pupal diapause through photoperiodic treatment and making suitable foliage available by giving light pruning to the plants before 35–40 days of brushing. The phase-wise pruning and grainage activities should coincide for taking second or third crops (Srivastava and Singh 1999).

11.4.4 Crop Pattern in North-West

In north-western region, availability of oak species, their sprouting behaviour and leaf maturity are entirely different than those of north-eastern region. Even the same oak species sprouts at different times at different altitudes. *Quercus semecarpifolia* is abundant at high altitudes (7000–9000′ ASL), while *Q. incana* and *Q. floribunda* are found at an altitude of 3000–7000′ ASL. *Quercus incana* is a fast-maturing species and raising of the second crop on it is very uncertain. Hence, a concept of low-middle-high altitude rearing is found viable to raise successful seed and commercial crops. Generally, three crops, viz. preponed seed crop, commercial crop and seed/commercial crops, are taken in a year during February–April, April–June and June–August (Goel and Rao 2004).

The first crop rearing on *Q. incana*/*Q. serrata* at low altitude is preponed by a fortnight so that a successful second crop may be raised on naturally sprouted *Q. semecarpifolia* foliage at high altitude by synchronizing hatching of layings produced from preponed crop cocoons with the sprouting. For preponement of the first crop on *Q. incana*, plants are subjected to defoliation during November–December and seed cocoons to photothermal treatment during January–February. The second crop starts in the last week of May or first week of June and continues up to July/August. These two crops are considered as seed crops, out of which the preponed spring crop is nondiapausing, while second summer crop is diapausing one. At middle altitudes, the silkworm rearing season is April–May when the temperature and humidity remain favourable. Initially, *Q. incana* leaves are fed to the larvae and from second instar onwards the *Q. floribunda* leaves or in some places of Uttarakhand and J&K, *Q. floribunda* leaves are fed throughout the larval span. This crop is considered as commercial crop and cocoons are used for reeling purposes (Goel 1990, 1991, 2010a, b).

11.5 Silkworm Rearing Technology

Main objective of rational rearing method is to produce quality cocoons besides reducing the cost of production through increased productivity. As the silkworm is completely domesticated in north-west and semi-domesticated in north-eastern region, the rearing technology differs from one region to another. It is therefore indispensable to adopt suitable rearing techniques for indoor and outdoor silkworm rearings.

11.5.1 Rearing Technique of North-East

Oak tasar silkworms were reared completely outdoors on oak bushes in the initial years of its origin. However, being reared outdoor, the silkworms were subjected to natural vagaries, and hence new technique of rearing was developed which resulted in better crop yields and higher productivity (Jolly 1972; Tikoo 1987; Pandey et al. 1991; Pandey 1994; Singh et al. 2001; Goel and Rao 2004).

Young-stage silkworm rearing is conducted indoors on twigs inserted in bottles, tins, pitchers, trays, etc. (Fig. 11.5). After the second moult, the worms are transferred to oak bushes outdoor, and late-age rearing is conducted completely under nylon nets (Tikoo 1987).

The concept of chawki rearing up to the second moult in chawki gardens under nylon nets followed by late-age rearing on systematic plantations of *Q. serrata* plants was developed for large-scale rearings in north-eastern region (Goel and Rao 2004).

After hatching, newly hatched larvae are allowed to crawl on the tender and succulent leaves of the freshly cut twigs, which are put on the eggs. After the larvae take firm grip on the leaves, these twigs are transferred to the plants in the chawki garden or twigs/small branches put into the bottles/tins/pitchers or the trays having small twigs bearing fresh tender leaves. The larvae are allowed to feed upon the leaves up to the second moult with change of feed as and when required. The suitable temperature and humidity for indoor chawki rearing are 22± 2 °C and 75–80% relative humidity, respectively. The proper hygienic conditions are maintained by way of cleaning, removing litter and replacing dry twigs. Overcrowding of the larvae is avoided and sizing of the worms is done at regular intervals. After the second moult, the larvae are transferred outdoors to the systematic plantation or natural plantation as the case may be.

Late-age rearing is conducted outdoors under nylon nets so as to save silkworm larvae from attack of pest and predators. The suitable temperature is 24 ± 2 °C and favourable relative humidity is 65 ± 5% R.H. The consumption of leaves during the first four instars accounts for only 20% of total consumption throughout the larval period, and the rest 80% leaves were consumed during the fifth instars alone. Hence, care is taken by adjusting the larval density on a plant that minimum handling of worms is done up to the fourth stage.

Fig. 11.5 Indoor chawki rearing on bottles

Fig. 11.6 Transfer of silkworm larvae in outdoor rearing

During the final stage, the larvae are transferred during cooler hours of the day and when they have consumed about 70–75% plant foliage. While transferring, either small defoliated branches of the same plant with larvae are cut or the larvae are manually picked up by gently lifting the caudal legs of the worm and placed them on branches of other plants and these branches are then placed in a clean bamboo tray/basket (Fig. 11.6). The branches with worms are put on new plants and the worms gradually crawl onto the plant leaving the old branches, which are removed later.

Oak tasar larva starts spinning the cocoon about 12–15 days after coming out of fourth moult. The mature larva is green and attains a weight of about 15–20 g. The larvae spin cocoons on the branches itself. Spinning starts with hammock formation followed by throwing of silk in the form of cocoon shell, and this process takes about 6–7 days after which the cocoon gets mature.

11.5.2 Rearing Technique of North-West

In north-west, oak tasar silkworm is reared indoor like *Bombyx mori*. Different rearing methods have been developed over the years and are in vogue depending upon the local conditions. Complete indoor rearing is done in dwelling houses, in cowsheds, in the houses constructed for temporary migration in high altitudes, in the rearing huts constructed temporarily with the help of nylon nets and polythene sheets or in the specially designed polyhouses, and hence the rearing method differs from place to place or even from season to season – floor, bottle, pitcher, tray or machan feeding (Bahl et al. 1985, 1987, 1987a, 1987b; Bajpai and Sinha 1993; Bhat 1994; Bhat et al. 1987; Dhar 1987; Dhar and Sinha 1996; Goel and Juyal 2009; Joshi et al. 2010; Pandey et al. 1989, 1991; Singh and Mishra 2000).

Indoor rearing technique involves centralized chawki silkworm rearing and late-age rearing at farmers' level. Before start of chawki rearing, all the rearing appliances and rearing room/hut/polyhouse are disinfected with 2% formalin solution. The larvae after hatching are brushed on oak leaves, which are placed into the trays or on the newspaper placed on the floor, bottles, pitchers as the case may be, and the larvae are allowed to feed. During the first instar, which is of 7–8 days of duration, only tender soft and succulent leaves are provided to the silkworms twice a day (first feeding in the morning and second in the evening). Quantity of feeding is regulated based on the larval population in the rearing bed. In each feeding, fresh twigs are placed on the old ones so that the larvae can migrate onto fresh leaves. The old twigs are removed after 2–3 h or during the bed cleaning. The doors of the rearing house are opened for one hour daily in morning and evening for proper aeration. The larvae undergo first moult during which no feeding or bed cleaning is done. Once the larvae come out of the first moult, fresh leaves are given to them. After 6–7 days of feeding, the larvae again undergo the moult, which is of 28–32 h and after moulting, the larvae enter into the third stage. Optimum temperature and humidity for chawki rearing is 23±2 °C and 75–80% R.H. The larval population should be optimum per unit area for proper feeding and growth.

Late-age rearing is done indoors in the rearing rooms and/or in the polythene tents (size 25′ × 20′) in the forest areas following floor or machan feeding method. Larval density is maintained as per the rearing bed size (Fig. 11.7). Medium to slightly mature leaves are fed to the larvae depending upon their age. Sufficient foliage is given to the larvae in the form of twigs twice a day in the third stage, thrice in the fourth stage and four times in the final stage. Frequency of feeding may be increased depending upon the feeding speed. The twigs are placed in the rearing bed horizontally and vertically alternately so as to make a sieve-type structure for easy movement of the larvae. The rearing beds are cleaned twice in the third and fourth stages and thrice in the final stage depending upon the accumulation of leftover foliage in the bed. Utmost hygienic conditions during rearing with the use of bleaching powder-slacked lime powder dusting.

Larval density and spacing are important aspects. Overcrowded beds lead to weak worms and disease infestation. So, proper spacing and density of worms

Indoor Floor Rearing method

Indoor Bottle/Pitcher/Tin Rearing method

Indoor Machan Rearing method

Fig. 11.7 Complete indoor rearing technique – different feeding methods found in north-west

should be ensured during late-age rearing. The rearing bed should be spread after every moult. Normally, spinning takes place in the rearing bed itself. However, the mature larvae may be picked up and placed in the suitable mountages for cocoon spinning. Twigs having dry leaves and nylon bags are also used as mountages in oak tasar.

11.6 Silkworm Seed Production Technology

Silkworm seed technology involves the processes from preservation of seed cocoons to incubation of eggs under hygienic conditions. Although the semi-domesticated to domesticated *A. proylei* race shows good moth emergence and pairing aptitude in captivity, sometimes it behaves abnormally, which can be checked by adopting proper grainage technologies. Moreover, seed production technology ensures timely production of quality seed, which in turn increases cocoon productivity and quality also.

11.6.1 Selection of Seed Cocoons

After completion of pupation in the cocoons harvested, oak tasar cocoons are subjected to visual gradation. Based on the shape, size, weight and compactness of cocoon shell, they are sorted out in two categories. The flimsy, deformed, inferior, pest-infested, diseased and dead cocoons are outrightly rejected, while the only compact, well-formed and tough-shelled cocoons are selected and consigned for preservation. Normally, the selected seed cocoons should weigh above 6.0 g and have fair ratio of male and female pupae. Male cocoons are generally pointed towards posterior end, while female cocoons are oval shaped. Alternately, male and female cocoons may be selected based on the visual examination of pupal sex. A sample of selected seed cocoons is subjected to microscopic examination to ascertain disease freeness.

11.6.2 Preservation of Seed Cocoons

The seed cocoons are stored in well-ventilated ratproof rooms in the form of garlands hanged on the string, iron angle stands, wire-meshed cocoon cages or specially designed cocoon preservation stands. Lower end of each garland is kept at least 2 feet above the ground. Prior to consigning the seed cocoons, the room is thoroughly disinfected by fumigation of formalin and washing with bleaching powder solution.

The garlands are made up of 50–100 cocoons tied in a bunch of 4–5 cocoons at an interval of 2 inches. Normally, the peduncle end of the cocoon is kept upwards for easy moth emergence. These garlands are hanged in such a position that the inter-garland gap is about 3–4 inches and garland-row distance 6 inches to 1 feet (Fig. 11.8).

Pupal diapause is observed in silkworm which necessitates preservation of seed cocoons during adverse climatic conditions. In north-east, the seed cocoons harvested in spring crop are preserved at low altitude in the well-aired rooms or at high

Fig. 11.8 Preservation of oak tasar seed cocoons in the form of garlands

altitude (6000–6500' ASL) from June to February. In north-west, summer or autumn crop is diapausing one, and seed cocoons are preserved at high altitudes (above 8000; ASL) from August to January or March. Optimum temperature and humidity for preservation is 20–25 °C and 60–70% R.H. The following precautions are to be taken care of during preservation of cocoons:

(i) Put thick black cloth curtain on doors and windows to allow diffused light in the room.
(ii) Use exhaust and ceiling fans for proper air circulation.
(iii) Ensure perfect hygienic conditions in the room.
(iv) Rotate cocoon garlands at 180o at least once in a month.
(v) Prevent entry of pests and predators, like rats, lizards, snakes, ants etc. in preservation room.
(vi) Record daily temperature and humidity inside the room.

11.6.3 Grainage Operation

The process of silkworm egg production is referred to as grainage, which involves activities of transportation of seed cocoons from the site of preservation to grainages, photothermal treatment of seed cocoons, moth emergence, moth coupling, oviposition, moth examination and washing and disinfection of eggs.

The seed cocoons produced from diapausing crop, which are preserved for 6–7 months at high altitude, are shifted to grainages situated at low-middle altitudes, as per requirement of silkworm eggs and according to prevailing climatic conditions and leaf sprouting. However, seed cocoons from nondiapausing crop need not to be preserved but instead are processed directly in the grainages after harvest. For uniform and compact emergence, the seed cocoons are subjected to photothermic treatment, i.e. room temperature is gradually increased to 23±2 °C, humidity brought to 75–80% and 16 h light is provided with the help of bulb, tube light or petromax. Moth emergence starts within 10–20 days of treatment. Moth emerges in late hours of evening and continues till midnight, though stray emergence can be seen at any time of day. Maximum emergence is observed between 16–20 h of day (Fig. 11.9). Goel et al. (1993) have found that in case of shortage of either sex moths, they can be preserved at 5–10 °C temperature for 3 days without any adverse effect. Moths normally survive up to about 12 days.

After emergence, male moths fly here and there frantically in search of its female counterparts. Normally, mating takes place within 3–6 h of emergence on cocoon garlands itself. Pandey and Goel (1990) have reported that maximum copulation occurs when male and female moths are kept in the ratio of 4:1 in crowded condition in the baskets. Moths prefer complete darkness with comparative lower temperature and humidity during mating. The mating continues for about 10–12 h. It is reported that there is no correlation between coupling duration and predisposition period as well as hatchability, and mating duration of 4–6 h is quite sufficient to get normal fecundity, optimum fertility and good hatchability. In case of shortage of

Moth emergence   Fully emerged moth   Mating of moths

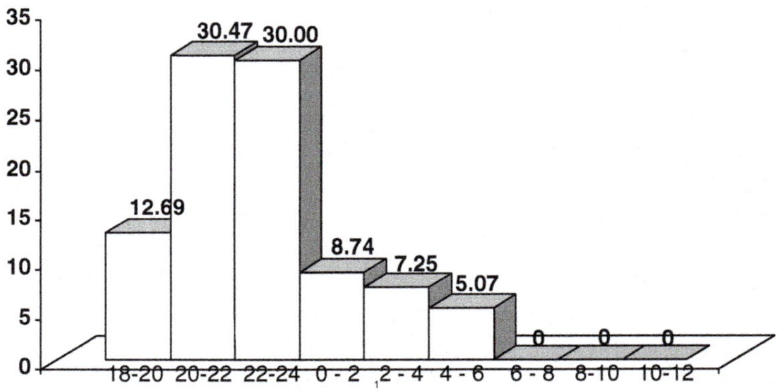

Fig. 11.9 Moth emergence in oak tasar

male moths, they can be used twice or thrice after giving some rest from previous mating. Moths prefer to copulate on the first night of emergence itself, and satisfactory fertilization is obtained among these moths. The greater the time elapsed after male had emerged, the lower is the percent fertilization, i.e. fertility of males decreases with age (Goel et al. 1993).

After 4–6 h mating, male and female moths are separated gently by hand. The abdomen of female moths is gently pressed with the help of fingers for urine discharge, which facilitates easy and smooth egg laying. These females are kept in the nylon net bags, plastic boxes, bamboo baskets, wooden trays or egg laying boxes, after removing the wings from 1 cm from the base, for egg laying. Females lay eggs in the batches of five to ten eggs with gummy substance. Several factors, such as light, temperature, humidity, etc., affect the egg laying capacity of female moths. Complete darkness, 20–22 °C temperature and 70–80% relative humidity in the egg laying room are ideal for optimum egg laying. About 90–95% eggs are laid within 72 h of decoupling, which are considered for incubation purpose (Table 11.6). Egg laying capacity of diapausing pupae is less than that of nondiapausing pupae.

Table 11.6 Relationship between day of egg laying and hatchability

Day of egg laying	No. of eggs laid	Percent eggs laid	Hatching %	Av. weight of egg (g)
1st	68	57	81.0	8.00
2nd	31	26	70.0	7.00
3rd	14	11	57.0	7.00
4th	2	2	18.0	7.00
5th	2	2	13.0	7.00
6th	1	1	10.0	7.00
7th	1	1	0.00	7.00

Generally, mated female moth lives for shorter period than unmated female. Noncopulating female takes longer time to complete egg laying as compared with copulated ones (Goel et al. 1993).

After 72 h of oviposition, females are picked up, cut their posterior portion and smear with the help mortar and pestles. Few drops of sodium hydroxide solution are added in the suspension, which is examined under microscope with x600 magnification. The eggs, which contain pathogen of pebrine, are rejected and destroyed. The disease-free eggs are collected, and dry appendages and other impurities removed and washed thoroughly with mild soap solution (2 %) or 0.4–0.5% solution of NaOH for 40 s followed by rinsing in a series of freshwater for 2 min so as to remove mucanium and other sticky materials. Eggs are surface sterilized in a mixture of 3% formalin and 3% HCl in 1:1 ratio for 30 min followed by washing with freshwater. Eggs are dried under shade by spreading them on a blotting paper in single layer. The dried eggs are collected and packed in muslin cloth bags or perforated plastic boxes. Eggs can be stored at 5±2 °C temperature for about 17 days without any detrimental effect on hatching and larval survivability. Refrigeration is advisable at two stages: (i) one within 48 h of egg deposition, or (ii) just one day before hatching. When removed from refrigerator, the eggs are kept at room temperature for at least 2–3 h and then incubated.

11.6.4 Silk Reeling and Spinning Technology

The post-cocoon sector in oak tasar is in developing stage. Three types of yarns are produced in oak tasar sector – reeled yarn, spun yarn and ghicha yarn.

As the oak tasar is wild variety, the variations in its commercial cocoon characters are quite significant and render production of quality silk of desired standard extremely difficult. After systematic research studies at Central Tasar Research and Training Institute (CTR&TI) Ranchi and Central Silk Technological Research Institute (CSTRI) Bangalore for reeling of oak tasar cocoons, scientific cocoon stifling & softening techniques and improved reeling and spinning devices for production of high quality raw silk have been developed over the years by various workers (Sengupta and Majhi 1982; Tikoo and Goel 1987; Bahl and Pandey 1988; Goel and Tikoo 1990; Manna and Tikoo 1990; Singh et.al. 2003; Goel 2010b).

11.6.5 Oak Tasar Cocoon Cooking Technology

A number of experiments were conducted by various researchers for evolving a suitable cocoon cooking method using various soaking media, boiling and steaming durations and proteolytic enzymes. Cocoon cooking with cocoonase, papain, trypsin, pepsin, Biopril-50, anilozyme-P and bromillin enzymes was tried, and good results with cooking efficiency varying from 91.0 to 95% were obtained. These cooking methods were found suitable for dry system of reeling with *Q. serrata*-fed cocoons produced in north-eastern region. These methods were not found fully suitable with *Q. incana* or *Q. semicarpifolia*-fed cocoons produced in north-western region. However, Central Silk Technological Research Institute, Bangalore, has developed a standard recipe for oak tasar cocoon cooking for wet system of reeling like China, where oak tasar cocoons are reeled like mulberry on wet basins on semi-automatic machines. Initially, oak tasar cocoons are boiled in plain water for about 30–45 min. Cooked cocoons are then placed in lukewarm water (40–45°C) for 30–45 min depending upon shell weight with the recipe (sodium silicate, 15 gpl; soda, 8 gpl; and H_2O_2, 15 cc/l).

After deflossing the upper flossy layer, clear ends are gathered and directly taken into reeling basin for reeling. With this cooking method, cocoons can be reeled in wet reeled conditions.

11.6.6 Oak Tasar Silk Reeling Technology

Oak tasar cocoons can be reeled following dry as well as wet system of reeling, and various reeling appliances have been developed to suit these reeling systems. Initially, the cocoons were used to be reeled on Trivedi reeling machines, but over the time innovations were made to improve the productivity and quality of silk produced, and the machines like CTRS reeling machine and CSTRI reeling cum twisting machines were developed.

Recently, Central Silk Technological Research Institute, Bangalore, has developed mechanical reeling device called "two in one reeling cum twisting machine – model 2" to overcome the disadvantages of earlier versions of reeling machines and to produce the quality silk as compared to Chinese silk (Fig. 11.10).

In this reeling machine, cocoons can be reeled in wet condition. For wet reeling, cold water supply system and a stem inlet provision to the basin to maintain the temperature 40–45 °C has been provided. The reeling machine is provided with Jetteboute attachment for easy casting of filaments. Then the yarn will be passed through the croissure pulleys and finally wound onto the small reels. The production of two in one reeling cum twisting machine – model 2 is around 550–600 g of reeled tasar yarn for 8 h at 80% efficiency.

In addition, CTR&TI Ranchi has developed one machine hand operated wet reeling machine (Fig. 11.11), which can be operated manually, and three persons can produce up to 800 g reeled silk per day. The yarn produced on this machine can directly be used for warp purpose in weaving, and there is no need to impart any twist.

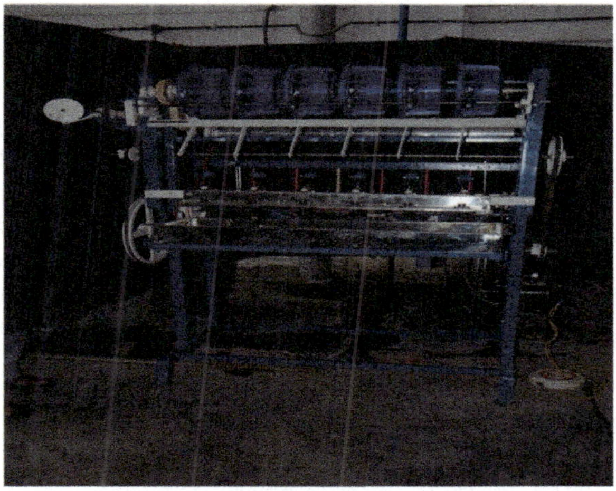

Fig. 11.10 Two in one reeling cum twisting machine – model 2

Fig. 11.11 Hand operated wet reeling machine

11.7 Conclusions

Today, the overall picture of oak tasar culture in India appears to be clearer. Central Silk Board has established different development models for production of oak tasar silk in north-western as well as north-eastern regions through implementation of cluster development projects and micro-projects successfully. Manipur, Mizoram and Nagaland afford great potential for exploitation of oak flora, and Uttarakhand, Jammu and Kashmir and Himachal Pradesh can supplement the oak tasar silk production in the country, if systematically planned scientific approach is adopted for silk production. Various package of practices for oak tasar culture, viz. host plant management, integrated silkworm rearing technology, silk reeling and spinning technology, etc., have been developed, which will not only enhance production and productivity but also improve farmers' income per unit area.

The population residing in the Himalayan belt has weak agriculture system due to topographical disadvantages. Besides, the general socio-economic relevance, which any industry, especially agro-based village and cottage industry, has in this country and need for development of oak tasar lies in the fact that it is grown mostly by the weaker sections of the society living in the remote places and having little or no scope of alternative employment. Thus, oak tasar sector has great promise as an additional income-generating venture because of abundance of natural oak flora, favourable climatic conditions and socio-economic status of the people.

References

Bahl RK, Pandey RK (1988) Wet reeling of Oak Tasar cocoons with improved charkha. Sericologia 28(3):403–410

Bahl RK, Pandey RK (1989) Recruitment of east Himalayan oak (*Quercus serrata)* in Western Himalaya. Indian Forester 115(2):109–112

Bahl RK, Sengupta K, Tikoo BL, Pandey RK, Rao PRT (1985) Altitudinal effect on domestic rearing of oak tasar silkworm *Antheraea proylei* J rearing in Western Himalayas. Sericologia 25(1):55–62

Bahl RK, Sinha RK, Pandey RK, Rao PRT, Sengupta K (1987a) Oak tasar *Antheraea proylei* J rearing in H. P. (India). In: Proceedings of the XV Int. Ser. Cong, Pattaya, Thailand, Section-5, pp 38–41

Bahl RK, Pandey RK, Rao PRT, Sengupta K (1987b) Evaluation of different rearing methods for oak tasar silkworm *A. proylei* J. In: Proceedings of the XV Int. Ser. Cong, Pattaya, Thailand, Section-5, p 120

Bajpai AK, Sinha AK (1993) Research breakthrough in oak tasar. Indian Silk 32(1):6–7

Bhagirath T, Kundu SC, Ibotombi SN (1988) Ultimate karyotype of the hybrid between *Antheraea pernyi* and *A. roylei*. Sericologia 28(1):91–94

Bhat MM (1994) Prospect of oak tasar in the hills of Uttar Pradesh. Indian Silk 32(10):31–35

Bhat MM, Bhattacharya J, Sinha PS, Sengupta K (1987) Effect of different periods of brushing on the rearing performance and cocoon quality of the oak tasar silkworm, *Antheraea proylei* J reared on *Q. semecarpifolia* Smith in U. P. Hills. In: Proceedings of the XV Int. Ser. Cong, Pattaya, Thailand, Section-5, pp 42–46

Chandra JP, Mahendru JL (1977) Propagation of *Quercus incana* by air layering. Indian Forester 103(8):539–541

Das PK, Pandey RK (1991) Scope for the improvement of oak tasar host plant. Indian Silk 29(9):18–21

Dhar SL (1987) Experiments and studies related to oak tasar. Indian Silk 26(1):14–16

Dhar SL, Sinha SS (1996) Stabilization of oak tasar silkworm seed crop in J & K. Indian Silk 34(10):33–35

Dhar SL, Dhar A, Khan MA, Suryanarayan N (2006) Studies on natural sprouting of Oak species and the related raising of Oak Tasar crops in Jammu & Kashmir. In: Proceedings of the Regional Seminar on Prospects & Problems of Sericulture as an Economic Enterprise in North West India: 381–384

Goel RK (1990) Oak tasar culture in Himachal Pradesh. Indian Silk 29(3):39–40

Goel RK (1991) Oak tasar culture in India – its problems and prospects. Sericologia 32(2):243–246

Goel RK (2000a) Grainage practices in Oak Tasar. In: Om AH, Seth MK (eds) Sericulture in India. Bishen Singh Mahendra Pal Singh, Dehradun: 853–865

Goel RK (2000b) Management of quality foliage in Oak Tasar: chemical composition of oak plants. In: Om AH, Seth MK (eds) Sericulture in India. Bishen Singh Mahendra Pal Singh, Dehradun: 867–878

Goel RK (2010a) Oak Tasar silkworm host plants–availability, propagation and management of Oak Flora. In: Proceeding of the workshop on Recent Trends in Oak Tasar Culture, 28–29 October 2010, Palampur: 1–21

Goel RK (2010b) Oak Tasar Cocoon Reeling Technology – an overview. In: Proceedings of the workshop on Recent Trends in Oak Tasar Culture, 28–29 October 2010, Palampur: 109–112

Goel RK, Juyal AC (2009) Technology Descriptor – A Users' Manual. Research Extension Centre, Palampur

Goel RK, Rao JVK (2004) Oak Tasar Culture–Aboriginal of Himalayas. APH Publishing Corporation, New Delhi

Goel RK, Tikoo BL (1990) Efficacy of raw pineapple fruit in softening of Oak Tasar cocoons. Silkworm Inf Bull 3(2):4:6–7

Goel RK, Goel AK, Thangavelu K (1993) Effect of preservation period of seed cocoons and age of males on oviposition and fertilization in oak tasar. J Seric 1(2):39–44

Jolly MS (1972) A new technique of tasar silkworm rearing. Indian Silk 11:5–8

Jolly MS, Narasimhanna MN, Sinha SS, Sen SK (1969) Interspecific hybridization in *Antheraea*. Ind J Heredity 1:45–48

Jolly MS, Sinha BRRP, Sinha SS (1973) Interspecific hybridization in *Antheraea* with special reference to *A. roylei* and *A. pernyi*: Document. In: 12th International Silk Congress, Barcelona: 6

Jolly MS, Chaudhury SN, Sen SK (1975) Non-mulberry sericulture in India. Central Silk Board, Ambica Publishers, Bombay

Joshi MC, Paliwal S, Yadav RS, Kimoti RC, Singhdeo SN (2010) New technique to obtain early foliage for chawki rearing of prepond oak tasar silkworm. In: Proceedings of the workshop on recent trends in Oak Tasar Culture, 28–29 October 2010, Palampur: 32–35

Lal P, Karnataka DC (1993) Effect of orientation of seed sowing and soil mixture on germination behaviour of *Quercus leucotrichophora* (syn *Q. incana* Roxb). Indian Forester 119(2):122–125

Lal P, Singh RK, Rawat GS (1999) Nursery technique of *Q. leucotrichophora:* Standarization of proper methods of seed sowing and optimum irrigation schedule under Dehradun climatic conditions. Indian Forester 125(12):1179–1184

Lethal JS, Joshi V, Singh O (1996) Seed collection and nursery emergence in *Q. semecarpifolia* (Smith). Indian Forester 122(1):85–86

Manna SS, Tikoo BL (1990) Valuation of suitable cooking recipe for Oak Tasar cocoons. Ind J Fibre Tex Res 15:35–36

Mishra PN, Sinha BRRP, Chakravorti S, Khan MA, Ramakant, Kharoo VK (2004) Oak tasar culture and its prospects for future development in north west Himalayas. In: Proceedings of the National Workshop on Potential Strategies for Sustainable Development of Vanya Silks in Himalayan states: 476–482

Negi SS, Naithani HB (1995) Oaks of India, Nepal and Bhutan. M/S International Book Distributors, Dehradun

Pandey RK (1990) Why Oak tasar cocoon is difficult to reel. Indian Silk 29(4):32–34

Pandey RK (1994) Indoor oak tasar silkworm rearing in north-east. Indian Silk 33(10):29–31

Pandey RK, Goel RK (1990) Grainage practices in oak tasar. Indian Silk 29(7):27–28

Pandey RK, Rath PC, Goel RK (1989) Oak tasar silkworm rearing in India. Indian Silk 28(2):13–14

Pandey RK, Naomani MKS, Das PK, Thangavelu K (1991) Advances in oak tasar rearing technology. Indian Silk 30(5):35–36 , 38

Rana B, Prasad B, Nigam MP (1987) Consumption and utilization of food by oak tasar silkworm *Antheraea proylei* Jolly (Lepidoptera: Saturniidae). Sericologia 27(1):11–19

Sengupta K, Majhi SK (1982) Oak tasar *A. proylei* J cocoons processing in India. International Sco. Non-mulberry Silks: 6

Singh TN, Mishra PN (2000) High altitude oak tasar rearing house. Indian Silk 39(8):17–19
Singh NI, Srivastava PK, Singh IN, Singh CS, Singh RY (2001) Recent studies for improvement of oak tasar cultivation in North east India. Bull Ind Acad Seri 5(2):1–8
Singh RL, Devi YR, Devi SK (2003) Enzymological characterization of Pineapple extract for potential application in Oak Tasar (*Antheraea proylei* J) silk cocoon cooking and reeling. Electron J Biotechnol 6(3)
Srivastava PK, Noamani MKR (1995) Cultivation and management of oak tasar food plants with reference to North-eastern India. Workshop on Oak Tasar Culture, Bhimtal
Srivastava PK, Singh NI (1999) Package of practice of oak tasar food plants in North-east India. Bull Ind Acad Seri 3(2):1–6
Srivastava PK, Singh TS, Singh NI, Thangavelu K (2000a) Clonal propagation of *Quercus grifithii* Hook & Thom Ex. Mig. through air layering. Ind J Forestry 25(4):406–410
Srivastava PK, Singh TS, Singh NI (2000b) Clonal propagation of *Q. serrata* Thumb. Syn. *Q. acutissima* Carr. through air layering. Indian Forester 126(8):879–884
Thangavelu, K. (2004). Host Plant wealth of vanya silks in the Himalayan States and strategies for its utilization. In Proc. Nat. Workshop on Potential Strategies for Sustainable Development of Vanya Silks in Himalayan states: 1–8
Tikoo BL (1987) Oak Tasar Culture. Regional Tasar Research Station, Imphal
Tikoo BL, Goel RK (1987) Wet system of reeling in Oak Tasar silk industry. Indian Silk 26(1):23–24
Trelease W (1924) The American Oaks. Mem Natl Acad Sci 20:1–255

Ericulture

12

B.K. Singh and S.A. Ahmed

Contents

12.1 Introduction

The northeastern region of India is blessed with natural hub of sericigenous insects and its host plants. Eri silkworm, *Samia ricini* (Donovan), is considered as the most popular commercially exploited *vanya* silkworm now getting national as well as international importance. Eri silk constitutes 62.88 % of the total non-mulberry raw silk production of 8038 MT in India during 2015–2016. India is one of the biodiversity hotspots among 34 biodiversity hotspots of the world and the northeastern region lies in Indo-Burma hotspot which is the prime issue as far as in situ conservation and commercial exploitation of sericigenous insects are concerned. India produces all types of silk, viz. muga, mulberry, eri, tasar and oak tasar. The common and commercially exploited non-mulberry or vanya silk-producing species are *Antheraea mylitta* Drury, *Antheraea pernyi* Guérin-Meneville, *A. assamensis* Helfer and *Samia ricini* Donovan (Jolly 1985) belonging to the family Saturniidae.

B.K. Singh (✉) • S.A. Ahmed
Central Muga Eri Research & Training Institute, Central Silk Board, Lahdoigarh, Assam, India
e-mail: bijoykumars87@gmail.com

Omkar (ed.), *Industrial Entomology*, DOI 10.1007/978-981-10-3304-9_12

The northeastern region of India located between 21° 57′ to 29° 28′ north latitude and 89° 40′ to 97° 25′ east longitude is one of the internationally recognized biodiversity hotspots known for its rich and unique bio-resources. The region is the abode of various host plants and silkworms producing all types of commercially exploited silks although the major strength of silk industry in the region is by virtue of endemic nature of golden muga and fabulous eri silk. The genus *Samia* contains 19 species from tropical and temperate eastern Asia (Neumann and Peigler 2001; Peigler and Naumann 2003), while Arora and Gupta (1979) reported a single species *S. cynthia* with several subspecies from India including *S. ricini*. Four species of genus *Samia* have been reported from India, i.e. *S. ricini* (domesticated), *S. canningi* (wild progenitor of *S. ricini*), *S. kohlli* (new report from Mizoram) and *S. fulva* (endemic to Andaman and Nicobar Island). Eri silkworm, *S. ricini*, is mainly confined to Northeast India for commercial production. Seven eco-races of eri silkworm have been reported based on the endemic nature and distribution pattern, while six strains of eri silkworm have been isolated from the Borduar and Titabar eco-races based on the larval colour and marking patterns (Debaraj et al. 2001; Singh et al. 2003). Wild counterpart *Samia canningi* (Hutton) is also known to occur in foothills of the region, which provides valuable resource material for cross-breeding (Sharma et al. 2002).

The important food plants of eri silkworm belonging to the family Euphorbiaceae, Araliaceae, Rutaceae, Simaroubaceae and Apocynaceae are exploited by farmers for eri silkworm rearing. Bindroo et al. (2007) reported 24 plant species as host of eri silkworm and designated them as primary, secondary and tertiary based on their extent of utilization and palatability to eri silkworm. Sharma et al. (2002) reported NBR-1 (non-bloomy red) as the most promising castor variety with the highest leaf and cocoon yield, while the promising tapioca varieties include H-97 and H-648. Protection of existing biodiversity, revitalization of the ecosystem, indexing of important biodiversity components and sustainable use of bio-resources are of paramount importance besides conserving the required germplasm as a prerequisite to genetic improvement (Dandin 2005). Dayashankar (1992) reported that castor crop yields about 13,675 kg leaf in Aruna variety supporting rearing of 1368 dfls, producing 438 kg excreta and 4786 kg litter per crop. Singh et al. (2004) also reported yield of 283 kg litters, 200 kg excreta, 40 kg pupa and 6 kg shells per 100 dfls per crop at beneficiary level in Jharkhand under SGSY programme.

Presently, about 2.7 lakh families of the region are engaged in ericulture depending on forest-based food plants. There is an urgent need for sustainable utilization of eri silkworm and its host plants, their products and by-products for ensuring additional income and upliftment of rural economy.

Rearing of eri silkworm (*Samia ricini* Donovan) has been traditionally practiced in Northeast India, and the culture has become inseparable part with the tradition, culture and economy of several communities of the region (Fig. 12.1). Ericulture is predominantly practised in all the northeastern states except Tripura for the production of cocoons and pupae. The protein-rich eri pupa is a favourite delicacy and dietary staple for the Bodo, Rabha, Miri, Kachari, Garo, Khasi, Naga, Adi, Mizo and Synteng tribals of Tibeto-Burman and Indo mongoloid origin of Northeast India

Fig. 12.1 Rearing of muga silkworm at farmer's level in Northeastern region

(Singh and Suryanarayana 2003). The entire gamut of ericulture involves multifarious activities, viz. raising of host plants, production of silkworm seeds, rearing of silk worm, spinning of cocoons into yarn and weaving it to fabrics; and various by-products are generated in these series of activities. In spite of using low-cost and indigenous materials for eri silkworm rearing and spinning of cocoons, there is increasing cost of production in ericulture activities in terms of raising and maintenance of food plants, cost of dfls and rearing cost, etc. In order to cope with increasing cost of production in ericulture activities and also to make the sector economically vibrant, there is an urgent need to augment alternative income sources by effective utilization and value addition of by-products associated with ericulture. Utilization of by-products like silkworm excreta, dried leaves and leaf remnants of host plants and rearing waste, host plant parts, silkworm pupae and moths, etc., are of prime importance as these wastes are the sources of biologically active substances having vast prospects for uses in pharmaceutical, cosmetic, paper and cellulose and also in organic agricultural food industries. Recycling of huge quantity of biomass in ericulture generated from large-scale plantation of eri food plants and utilization in silkworm rearing in an integrated manner would ensure generation of biogas and manure for meeting the rural energy needs and also keep the environment clean besides ensuring additional income to farmers.

12.2 Systematic Position

The eri silkworm *Samia ricini* Donovan belongs to the family Saturniidae under order Lepidoptera and class Insecta. It is a multivoltine breed, having five to six life cycles in a year. The commonly occurring wild eri silkworm – *Samia cynthia* – is either bi- or trivoltine in nature. Both these moths differ slightly in markings and in the amount of white scales in the abdomen. While the cocoon of *S. ricini* is loose and white or brick red, that of *S. cynthia* is compact and light brown in colour. The domesticated eri silkworm *S. ricini* does not occur in the wild. It was originally in a wild state, but due to artificial selection and continuous rearing for decades as is done in Northeast India, it has descended to the present existing form of *S. ricini*. The structure of the genitalia, wing pattern and chromosome number demonstrate that it was derived from its closet wild ally *Samia canningi* (Hutton). The main distinguishing feature of *S. ricini* is the diffusion of the abdominal white tufts, sometimes resulting in individuals with a solid white abdomen.

The systematic position provided to eri silkworm is as follows:

Superfamily:	Bombycoidea
Family:	Saturniidae
Subfamily:	Saturniinae
Genus:	*Samia*
Species:	*ricini*

12.3 Diversity of Eri Silkworm

Eri silkworm apart from domesticated eco-races has a few wild counterparts. Nineteen species have already been reported of genus *Samia* from tropical and temperate region of Southeast Asia (Peigler and Naumann 2003). Out of these species,

Table 12.1 Diversity and distribution pattern of eri silkworm

Species	Distribution
S. cynthia	Peoples Republic of China, Korea and Russia
S. canningi	Southeast Asian mainland
S. pryeri	Japan
S. watsoni	China and North Vietnam
S. vandenberghi	Central Indonesia
S. insularis	Indonesian islands of Java and Sumatra
S. luzonica	Philippines
S. peigleri	Central Indonesia
S. treadawayi	Philippines
S. naumanni	North Central Indonesia
S. abrerai	Java and Bali
S. naessigi	Eastern Indonesia
S. kohlli	Southeast Asia
S. wangi	Southeast China and North Vietnam
S. tetrica	Singapore, Brunei and Indonesia
S. yayukae	Sunda islands of Indonesia
S. fulva	Andaman Islands of India
S. ceramensis	Central Moluccas of eastern Indonesia
S. ricini	Eastern Gangetic Plain, Assam and adjacent sections, Bangladesh

S. ricini, *S. canningi* and *S. fulva* are reported to be found exclusively in India (Peigler and Naumann 2003). Northeast is a natural home of a wide range of flora and fauna including eri silkworm. Wild eri silkworms are polyphagous in nature having a wide range of host food plants. The domesticated *S. ricini* has its wild progenitor, i.e. *S. canningi* (Hutton), which are the permanent inhabitants of Northeast India. *S. fulva*, another wild species, was reported from the evergreen lowland forests of Andaman and Nicobar Islands (Mohanraj et al. 1988; Table 12.1).

Samia ricini is the commercially cultivated multivoltine silkworm. Six homozygous strains were classified from *S. ricini* on the basis of larval colour and body markings, viz. yellow plain (YP), yellow spotted (YS), yellow zebra (YZ), greenish blue plain (GBP), greenish blue spotted (GBS) and greenish blue zebra (GBZ) (Sarmah et al. 2012). These strains produce different cocoons of attractive colours like snow white, cream white, off white, deep brick red and light brick red.

Earlier, 26 eco-races of *S. ricini* have been reported from N.E. region, characterized (Chakravorty et al. 2008) and maintained at the Central Muga Eri Research and Training Institute (CMERTI), Lahdoigarh, Jorhat. Among these, ten accessions (001, 002, 003, 004, 005, 006, 011, 015, 018 and 025) are considered as promising in terms of overall rearing performance (Chakravorty et al. 2008). At present, the accession numbers 001 (Baruduar), 014 (Kokrajhar) 002 (Titabar) and 010 (Diphu) are reared commercially (Table 12.2). The new high yielding improved breed C2 has been developved through hybridization between Borduar and Genung ecoraces (Ahmed et al. 2014).

Table 12.2 Eco-races of eri silkworm and their characteristics features

Accn. No.	Larval body colour	Cocoon colour
SRI-001	Plain and zebra on yellow and blue	White
SRI-002	Plain and zebra on yellow and blue	White
SRI-003	Plain yellow and blue	White
SRI-004	Plain yellow and blue	White
SRI-005	Plain blue	White
SRI-006	Plain yellow and blue	White
SRI-007	Plain yellow	White and brick red
SRI-008	Plain and zebra on yellow and blue	White
SRI-009	Plain and zebra on yellow and blue	White and brick red
SRI-010	Plain and zebra on yellow and blue	White
SRI-011	Plain yellow and blue	White
SRI-012	Plain and spotted on yellow and blue	White
SRI-013	Plain and zebra on yellow and blue	White and brick red
SRI-014	Plain yellow and blue	Brick red
SRI-015	Plain yellow and blue	White
SRI-016	Plain yellow and blue	Brick red
SRI-017	Plain yellow and blue	White and brick red
SRI-018	Plain yellow and blue	White
SRI-019	Spotted on yellow	White
SRI-020	Plain yellow	White
SRI-021	Plain yellow	White
SRI-022	Plain yellow	White
SRI-023	Plain yellow	White
SRI-024	Plain yellow and blue	Brick red
SRI-025	Plain yellow	White
SRI-026	Plain yellow and blue	Brick red

Source: Sarmah et al. (2012)

Two strains of *S. canningi*, i.e. greenish blue plain (GBP) and greenish blue spotted (GBS), are reported. Both the races produced grey to dark brown colour compact cocoons with peduncle. Cocoons are smaller but with stronger moths, compared to those of cultivated species. Hybridization between *S. ricini* and *S. canningi* was conducted in caged condition and the offspring were more vigorous and found to be covered with more powdery substance and mostly preferred kesseru foliages (*Heteropanax fragrans* Seem) (Sarkar et al. 2009). *Samia fulva* is reported to be found in the evergreen low deciduous forests of Andaman Islands.

Characteristic Feature and Rearing Performance The commercially exploited *Samia ricini* is multivoltine and has many eco-races based on place of their collection in different ecological regions, namely, Dhenubhanga, Khanapara and Kokrajhar (Lower Assam), Titabar (Upper Assam), Borduar (Lower Assam), Nongpoh and Mendipathar (Meghalaya) and Sille (Arunachal Pradesh). Further, homozygous

Table 12.3 Characteristic features of ecotypes of eri silkworm

Ecotypes	Availibility origin	Larvae colour	Cocoon colour	Shell weight	Absolute silk yield
Borduar	Borduar (Lower Assam)	YP, YZ, GBP, GPZ	White	Higher	Higher
Titabar	Titabar (Upper Assam)	YP, YS, GBP, GBS	White	Higher	Higher
Dhenubhanga	Dhenubhanga (Lower Assam)	YP, GBP	White	Moderate	Moderate
Kokrajhar	Kokrajhar (Lower Assam)	YP, YZ, GBP, GBZ	Brick Red	Moderate	Moderate
Nongpoh	Nongpoh (Meghalaya)	YP	Pale Cream	Moderate	Moderate
Mendipathar	Mendipathar (Meghalaya)	GBP	White	Moderate	Moderate
Sille	Sille (Arunachal Pradesh)	YP	White	Low	Low

Y yellow, *P* plain, *Z* zebra, *G* green, *B* blue, *S* spotted

Table 12.4 Rearing performance of eri silkworm ecotype

Ecotypes	Egg (numbers)	Hatching (%)	Larvae wf(g)	ERR (%)	Cocoon wf (g)	Shell wf(g)	S R (%)
Borduar	442.0	95.04	6.12	90.06	3.64	0.50	13.74
Titabar	458.91	94.35	6.24	90.52	3.65	0.48	13.15
Dhenubhanga	459.40	92.61	6.01	88.24	3.56	0.47	13.20
Kokrajhar	459.69	95.26	5.94	89.81	3.49	0.51	14.61
Nongpoh	442.74	92.56	5.84	87.31	3.53	0.46	13.03
Mendipathar	455.44	91.77	6.06	89.25	3.61	0.47	13.02
Sille	441.35	92.51	5.88	87.79	3.55	0.46	12.98

strains, viz. greenish blue plain, greenish blue spotted, greenish blue zebra, yellow plain, yellow spotted and yellow zebra based on larval body marking, have been isolated from Borduar and Titabar ecotypes.

The details of characteristic features of ecotypes as well as pure lines of eri silkworm and their rearing performances have been depicted in Tables 12.3, 12.4 and 12.5.

12.4 Life Cycle

Eri silkworms are multivoltine and 5–6 crops can be reared in a year (Table 12.6). The complete life cycle takes about 45 days in summer and 90 days in winter. Best quality cocoons are produced during autumn (October–November) and late spring (May–June). The life cycle of eri silkworm has four stages – egg, larva, pupa encased in cocoon and adult moth. A complete life cycle lasts about 45 days in summer and 90 days in winter as detailed below (Table 12.7).

Table 12.5 Rearing performance of isolated pure lines of eri silkworm

Names of the lines	Egg (numbers)	Hatching (%)	Larvae wf (g)	ERR (%)	Cocoon wf (g)	Shell wf (g)	SR (%)
G.B. plain	451.20	98.31	5.47	89.08	3.42	0.46	13.46
G.B. spotted	452.25	94.0	5.37	90.81	3.41	0.43	13.11
G.B. zebra	442.66	94.37	6.10	90.50	3.59	0.48	13.62
Yellow plain	459.10	94.54	5.49	88.58	3.26	0.47	14.89
Yellow spotted	471.25	94.61	5.43	89.74	3.60	0.47	13.20
Yellow zebra	452.0	93.91	6.15	86.38	3.84	0.51	13.39

Table 12.6 Different rearing crops and corresponding seasons of eri silkworm

#	Crop	Season
1	First	March–April
2	Second	May–June
3	Third	June–July
4	Fourth	August–September
5	Fifth	October–November
6	Sixth	December–March

Table 12.7 Average durations of various stages of eri silkworm during summers and winters

Stages	Summer (days)	Winter (days)
Egg	09	18
Larva	17	44
Pupa	16	24
Adult	03	04
Total	**45**	**90**

The morphology of different stages of the life cycle of eri silkworm has been stated below (Table 12.8).

Egg

The eri silk moth eggs are candid white in colour and oval shaped, with a hard chitinous chorion, measuring 1.5 × 1.0 mm in size. Micropyle is placed in a slight depression at one extremity of the horizontal axis. The pattern of follicular imprints is very distinct. Cells are polygonal and inter-cellular space small with respiratory spines between the cells.

Larva

The newly hatched larva is greenish yellow in colour, elongated and cylindrical in shape about 5.0 × 1.00 mm in size and about 1.5 mg in weight. The body colour changes gradually to pure yellow by the end of the third day. From the third instar onwards, the body colour segregates into yellow, cream, green or blue depending

Table 12.8 Size and colouration of various life stages of eri silkworm

Stages	Colour	Size
Egg	Candid white	1.5 × 1.0 mm
Larva		
1st instar	Yellow	5.0 × 1.0 mm
2nd instar	Yellow	15.0 × 2.0 mm
3rd instar	White	25.0 × 3.4 mm
4th instar	White, green, spotted	32.0 × 4.5 mm
5th instar	Pale cream, white, green	95.0 × 10.0 mm
Pupa	Brown	28.0 × 15.0 mm
Moth	Blackish, brownish, chocolate	25 mm long in male with 130 mm wing span and 30 mm long in female with 150 mm wing span

upon races. The fully mature larva which measures about 90.0 × 15.0 mm is translucent and covered with white powdery substances. Different colour and marking patterns, viz. plain, spotted, greenish blue, zebra and semi-zebra larvae, are found in different strains.

The prothoracic hood of the first instar larva has a black dorsal band, which splits up into a pair of crescent-shaped markings in second and third instars. The marking disappears at the fourth and fifth instars. The tubercles are conspicuous, tubular in shape, bluish at the base and cream colour at the tip

Pupa

The pupae are of obtect adectious type and brown in colour, measuring about 28.0 × 15.0 mm in size and weighing about 2.5 g. The pupa is a prelude to moth stage with all the appendages of future moth such as compound eyes, wings, antennae, legs, genitalia, etc.

Cocoon

Eri cocoons are elongated, soft, flossy peduncle less, open mouthed and unreelable. The cocoon exhibits colour polymorphism being creamy white and brick red. They measure 5.5 × 2.5 cm and weigh about 3 g. The shell weight and shell ratio are about 0.40 g and 13.0 %, respectively.

Moth or Adult

The eri silk moths are stout, brownish or blackish in colour and covered with white scales. The male moth is smaller than the female. The male measures 25 mm long with 130 mm wing span, while the female measures 30 mm with 150 mm wing span. Wings are buff coloured with white-coloured strips in the marginal portion. Wings are covered with scales of different colours and shape. Forewings are longer and narrower than hind wings. The characteristics antimedian line is bright chocolate coloured with a white border on either side and almost runs through the centre. The ocellus in both sexes is crescent shaped and is characteristic of the insect. The hyaline area is almost invisible and located in the most anterior region of the ocellus. The space between the ocellus and post-median line is darker.

12.5 Eri Silkworm Rearing Management

The eri silkworm is multivoltine in nature and can be reared throughout the year in 5–6 crop cycles. Temperature range of 25–28 °C and relative humidity of 85–90 % are optimum. Eggs are generally incubated at 26 °C for uniform hatching. The eggs are oval in shape and white in colour which changes to blue prior to hatching. Hatching generally takes place in the morning between 7.00 and 9.00 A.M. The newly hatched worms are yellowish black in colour, while the mature worms are creamy, white, blue or green in colour. The newly hatched worms are fed on tender leaves and late age worms are fed on mature leave. The worms undergo four moults and have five instars. The larval period lasts for 17–44 days during summer and winter. Different methods of rearing, viz. tray, platform, bunch, etc., are generally followed. After 5–7 days in the fifth instar, the worms stop feeding and start searching for a suitable place for spinning of cocoons. The matured worms are picked up and placed in mountage for spinning of cocoons which are generally made of bamboo (Chandraki) or *jali* made of dry leaves. The worms complete spinning in 2–5 days depending on the season. The cocoons are either white or red in colour measuring 4–5 cm in length and 2–3 cm in width.

Eri silkworm eggs (seed) are produced in the grainage. The seed cocoons are kept in cages or trays and moths emerge in the early morning after 16–22 days of pupation. The moths are blackish brown in colour. Mating takes place after few hours of emergence and the female moths start laying eggs on khorika. About 300–400 eggs are laid by a female moth. The eggs after microscopic examination and confirmation of disease freeness are washed in 2 % formalin solution for 3 min followed by washing in freshwater and drying. With proper incubation of eggs, 80–90 % hatching could be obtained per laying. With an average effective rate of rearing of 80–90 %, cocoon shell yields of 8–10 kg per 100 layings can be obtained with pupal yield of 50–60 kg/100 layings. Each laying with 400 fecundity and 80 % hatching consumes about 10–12 kg of foliage during the larval period that ranges between 17 and 44 days depending on the season. Single cocoon weight and shell weight are about 3.00–3.50 gm and 0.35–0.50 gm, respectively. Some of the important factors necessary for successful rearing are as below:

1. Availability of adequate quantity of quality foliage
2. Adequate rearing and operational space
3. Availability of sufficient rearing appliances
4. Availability of quality seed (dfls)
5. Ideal temperature and humidity
6. Rearing knowledge and managerial skill
7. Maintenance of proper sanitation and hygiene
8. Adoption of effective crop protection and prophylactic measures

Eri silkworms are poikilothermic in nature and hence the prevailing abiotic condition has significant influence on growth, survival, rearing performance and

Table 12.9 Requirements of optimum temperature and relative humidity for eri silkworm rearing

	Summer		Winter	
Stages/activities	Temp (°C)	RH (%)	Temp (°C)	RH (%)
Incubation of eggs	25–26	85–90	25–26	85–90
Larval stage	25–30	75–90	25–28	80–85
Spinning of cocoon	28–30	75–80	26–28	75–80
Storage of seed cocoon	26–28	75–85	26–28	75–80
Pairing of moths	26–28	80–85	26–28	80–85

cocoon productivity. For ensuring successful rearing, the optimum temperature and humidity need to be maintained and stage-wise requirement of optimum temperature and humidity are indicated in Table 12.9.

Rearing House and Appliances

1. Construct low-cost, durable rearing house with locally available materials having size of 25′ × 15′ × 11′ size (L × B × H) with 5′ verandah.
2. Select flat, dry and elevated place for construction of the rearing house.
3. Provide adequate doors, windows and ventilators to ensure proper aeration and light.
4. Procure all equipments and appliances prior to rearing.

Disinfection of Rearing House/Appliances

1. Close windows and ventilators and keep rearing appliances in stack inside the rearing house.
2. Disinfect the rearing house and appliances with 5 % bleaching powder solution.
3. Sprinkle bleaching powder + lime mixture (1:9 ratio) on the floor of the rearing bed.
4. Close all crevices and holes to prevent entry of pest and predators.
5. Maintain strict sanitation and hygiene in and around the rearing house.

Quality Silkworm Seed and Transportation

1. Procure good quality dfls/certified seeds of Borduar/Titabar eco-races and improved breed from government/private agencies.
2. Eggs should be properly packed/transported in specially devised egg transportation boxes for safe transportation.
3. Transportation of eggs during cooler hours of the day.

Incubation of Eggs

1. Place egg boxes/muslin cloth containing eggs in a well-disinfected room.
2. Maintain a room temperature of 25–26 °C and RH of 80–85 % where eggs are kept.
3. Incubate eggs in the incubation chamber/incubator at 25–26 °C and RH of 80–85 %.
4. Conduct black boxing by wrapping eggs with black cloth or paper at pigmentation stage and keep eggs in dark place.

Brushing of Worms

1. Open muslin cloth bag and spread eggs in the tray on the expected day of hatching.
2. Provide tender (glossy), succulent and fresh castor/kesseru leaves when 80 % worms hatched out.
3. Newly hatched larvae crawl on leaves and start feeding.
4. Transfer leaves along with larvae to the rearing tray and provide fresh leaves.
5. Use fine and soft brush/feather for brushing.
6. Select worms hatched during the first 2 days only for rearing.
7. Avoid overcrowding of worms.

Rearing of Young Age Worms

1. Maintain a temperature of 26–28 °C and RH of 80–90 % in the rearing house during chawki stage.
2. Provide tender, succulent and fresh leaves four times/day.
3. Feed tender leaves in whole.
4. Do not expose young worms to extreme heat/cold.
5. Provide no feeding during moult.
6. Handle worms carefully during moult/bed cleaning.
7. Avoid overcrowding of worms in the rearing tray.

Rearing of Late Age Worms

1. Provide medium and mature leaves to fourth and fifth instar worms.
2. Avoid feeding dry, yellow and diseased leaves.
3. Preserve leaf in leaf chamber.
4. Feed late age worms five times/day.
5. Do not provide feeding during moult.
6. Provide first feeding when 80 % of worms come out of moult.

7. Avoid overcrowding of worms on the tray.
8. Keep 300 worms/tray during the final stage.
9. Collect weak, injured, irregular, diseased worms and destroy immediately.

Different Methods of Late Age Rearing

1. Bunch rearing
   (a) Tie 10–12 leaves in bunch and hang vertically on horizontal bamboo/string support.
   (b) Allow worms to feed on a bunch of leaves.
   (c) Replenish foliage by placing fresh bunches.
   (d) Spread bamboo mat below bunches to avoid injury of fallen worms.
   (e) Pick up fallen worms and put on bunches.
2. Platform rearing
   (a) *Rear* late age worms on 3–4 tier bamboo platform of 3′ × 6′ size.
   (b) About 800 worms can be reared per tier.
3. Tray rearing
   (i) *Rear* late age worms on tray of 1.0 m diameter or plastic rearing tray of 2.5′×s3′× size.
   (ii) *Maintain* 300 worms at the 5th stage on the rearing tray of 1.0 m diameter.

Bed Cleaning and Care During Moult

1. Conduct bed cleaning every day during late age rearing.
2. Do not allow piling up of unconsumed foliage in bed.
3. Avoid bed cleaning when worms are in moult.
4. Avoid mechanical injury of worms during bed cleaning.
5. Keep rearing bed dry when worms are in moult.
6. Provide leaves only when 80 % worms cast off their skins and come out of moult.
7. Collect weak, injured, diseased worms regularly and burn/burry to prevent the spread of disease

The stage (instar)-wise duration, ambient temperature, humidity, frequency of feeding and bed cleaning and density of worms per tray of 1.0 m diameter and the amount required for feeding are detailed in Tables 12.10, 12.11 and 12.12.

Collection of Mature Worms

1. Keep mountages, stands and newspapers ready prior to maturation of worms.
2. Collect worm carefully for mounting.
3. Ensure mounting of optimum number of ripe worms.
4. Mount 300 worms/chandrike of 1.0 m diameter.
5. Provide proper ventilation in the mounting hall.

Table 12.10 Instar-wise number of feeding and bed cleaning

Instar	Days	Temp. (°C)	RH (%)	Feeds/day	Frequency of bed cleaning
I	3–4	26–28	80–90	4	1 before moult
II	3–4	26–28	80–90	4	1 before moult
III	3–4	26	80	5	Daily
IV	4–5	25–26	75–80	5	Daily
V	6–7	25–26	70–75	5	Daily

Table 12.11 Instar-wise density of worms/tray (1.0 m diameter)

Instar	Tray size	No. of worms
I	3 ft. × 3 ft. = 9 sq ft	10,000
II	3 ft. × 3 ft. = 9 sq ft	3000
III	3 ft. × 3 ft. = 9 sq ft	1200
IV	3 ft. × 3 ft. = 9 sq ft	600
V	3 ft. × 3 ft. = 9 sq ft	300

Table 12.12 Instar-wise consumption of leaf (fecundity = 400 eggs/dfl)

Instar	Quantity of leaf consumed	Consumption (%)
I	30.90 gm	0.28
II	127.60 gm	1.16
III	492.50 gm	4.48
IV	1330.50 gm	12.12
V	9008.50 gm	81.96
Total	**10990.00 gm**	**100.00**

Harvesting and Assessment of Cocoons

1. Sort out good, double, flimsy, melted, stained cocoons after harvesting.
2. Select good, compact cocoons and keep separately.
3. Reject poor, double, flimsy, melted, stained cocoons.
4. Remove prepupa from cocoons by opening the open end on the third day of spinning.
5. Dry cocoon shells by spreading in sunlight.
6. Preserve cocoons carefully, protecting from lizards, rats and ants.

12.6 Eri Silkworm Diseases and Their Management

Eri silkworm primarily suffers from four major diseases, which are protozoan (pebrine), bacterial (flacherie), viral (grasserie) and fungal (muscardine). The causes, symptoms and control measures of these diseases are mentioned in Table 12.13.

Table 12.13 Causes, symptoms and control measures of four major diseases of the eri silkworm

Disease	Causes	Symptoms	Control measures
Pebrine (protozoan)	1. Through mother moth	1. Larvae loss appetite	1. Effective disinfection of rearing house, appliances and equipments
	2. Improper egg surface sterilization	2. Retarded growth	2. Effective egg surface
	3. Feeding spore-contaminated leaf	3. Unequal in size	3. Sterilization
	4. Accumulation of faecal matter	4. Body slightly shortened	4. Strict microscopic
	5. Contaminated rearing house and appliances	5. Body becomes pale	5. Examination of moths
	6. Inadequate disinfection of rearing house and appliances	6. Presence of spore	6. Proper sanitation and hygiene
			7. Avoid feeding diseased and
			8. Fermented leaf
			9. Adoption of adequate
			10. prophylactic measure
Flacherie (bacterial)	1. High temp. and RH	1. Sluggish movement	1. Effective disinfection of rearing house/ appliances
	2. Poor ventilation	2. Loss of appetite	2. Avoid overcrowding
	3. Poor sanitation	3. Shrinkage of the abdominal segment	3. Maintain optimum temp and rel. humidity
	4. Improper bed cleaning	4. Swollen thorax	4. Avoid feeding diseased and contaminated leaf
	5. Feeding of wet leaves	5. Vomiting of fluid	5. Collect diseased worms and destroy
	6. Accumulation of faeces	6. Loss of clasping power	6. Apply bleaching powder and lime at 1:9 ratio
	7. Feeding of contaminated leaves	7. Body softened and	7. Proper bed cleaning
	8. Overcrowding of worms	8. Rupture	8. Proper sanitation and hygiene
		9. Blackening of thorax	

(continued)

Table 12.13 (continued)

Disease	Causes	Symptoms	Control measures
Grasserie (viral)	1. High temp. and RH	1. Sluggish movement	1. Effective disinfection of rearing house/ appliances
	2. Feeding of polyhedra	2. Worms restless	2. Avoid overcrowding
	3. Contaminated leaves	3. Skin shiny	3. Proper sanitation and hygiene
	4. Poor ventilation and	4. Loss of clasping power	4. Avoid feeding tender leaves to maturing worms
	5. Sanitation	5. Hanging upside down	5. Proper bed cleaning
	6. Improper bed cleaning	6. Intersegmental swelling	6. Avoid accumulation of excreta
	7. Feeding of tender and	7. Skin turns yellowish	7. Destroy diseased worms and sprinkling of bleaching powder + lime at 1:9 ratio
	8. Wet leaves	8. Rectal protrusion	
	9. Accumulation of faeces	9. Soiling of anal region	
	10. Overcrowding of worms		
Muscardine (fungal)	1. Rearing in damp/ shady place	1. Worms sluggish movement	1. Well-aerated room with proper sunlight
	2. Improper bed cleaning	2. Swollen body with fungal spores	2. Avoid feeding wet leaves
	3. Poor ventilation and sanitation	3. Larva mummified at later stage	3. Apply lime and beaching powder mixture at 9:1 ratio

12.7 Eri Silkworm Seed Production Management

Production of quality silkworm seed plays a crucial role in the entire gamut of silk industry, and efficient management of grainage is essential for production of disease-free and quality seed. Hence, there is a need for proper seed organization system in the eri sector for production of quality seed by resorting to seed cocoon selection criteria, quality norms and testing for disease freeness of the lots.

12.7.1 Conventional System of Seed Production

A major portion of the total eri silkworm seed requirement is produced by the farmers themselves in unhygienic manner without following any scientific method of seed production, thereby leading to outbreak of diseases and reduction in quality as well as productivity of eri cocoons. The average yield of green cocoon per 100 dfls was about 35–40 kg with average yield of 5–6 kg shell per 100 dfls revealing less impact on economy from ericulture due to poor quality seed, and hence, ericulture was designated as household subsidiary occupation. Government eri grainages, basic eri seed farms, composite sericulture farms and eri concentration centres (ECCs) have set up eri rearers availing farm facilities as source leaves for eri silkworm rearing due to lack of their own systematic plantation and supplying seed cocoons to grainages for commercial seed production, thereby fulfilling certain portion of required eri seed.

12.7.2 Organized System of Seed Production

Of late, ericulture has been considered as one of the most promising kinds of sericulture in the northeastern states including some nontraditional states because of the initiation of the Central Silk Board and Directorate of Sericulture of the concerned states towards creation of infrastructural support under the Catalytic Development Programme (CDP) and adequate production and supply of quality basic eri seed by the Eri Silkworm Seed Organization (ESSO) by following the three-tier seed multiplication system. Under the existing seed organization system, the Central Muga Eri Research and Training Institute (CMER&TI) and Lahdoigarh/Regional Eri Research Station (RERS), Mendipathar, Meghalaya, produce P-4/P-3 basic seeds and are further multiplied by P-2 basic seed stations, viz. Hosur (Tamil Nadu) and Azara (Assam) along with Adopted Seed Rearers (ASRs) following basic grainage norms to produce and supply basic P-1 seeds rearing of which are undertaken by ASRs and subsequently multiplied for production and supply of commercial seeds by the concerned DOS/NGOs with support of ASRs and private graineurs. In order to ensure production of disease-free and high-quality seeds, efficient management of grainage is essential with advanced planning, effective selection and utilization of seed cocoons, timely picking of emerged moths, strict microscopic examination of mother moths, maintenance of accounts of grainage expenses and adoption of adequate prophylactic measures.

12.7.3 Role of Grainage in Production of Quality Seeds

1. Production of disease-free and quality seeds through efficient management.
2. Proper selection of seed cocoons, timely picking of emerged moths, strict microscopic examination and supply of dfls.
3. Maintain hygiene and adoption of adequate prophylactic measures.
4. Cost-effective production of disease-free layings (dfls).
5. Maintenance of grainage record and accounts.

12.7.4 Present Constraints Associated with Eri Seed Production

1. Major portion of required seed produced by farmers without adopting prescribed technological norms
2. Poor quality/productivity of seed cocoons
3. Lack of quality consciousness and awareness of quality seeds
4. Inadequate rearing management and complacency with small-scale rearing
5. Crop loss due to diseases during adverse seasons
6. Underutilization of potential seed production infrastructure in the state sector

12.7.5 Location of Grainage Hall

1. Elevated, well-aerated place free from waterlogging.
2. Well connected, accessible and located in area of intense sericultural activities
3. Proper orientation of the grainage house to maintain optimum temperature and RH
4. Availability of electricity, telephone, water services and required equipments

12.7.6 Size of Grainage Hall

1. A grainage size of 40′ × 25′ is ideal for commercial grainage of processing 50,000 seed cocoons (10,000 dfls).
2. A grainage hall of 20′ × 18′ with 6 ft verandah is ideal for production of 5000 dfls per operation by the private graineurs at manageable level.
3. A model eri grainage with production capacity of 2.00 lakh dfls/year should be of 2000 sq ft with sufficient equipments and manpower.

12.7.7 Disinfection and Maintenance of Grainage Room and Appliances

1. Drench the grainage hall/appliances thoroughly with 5 % bleaching powder solution.
2. Spray 2 % formaldehyde solution and 0.5 % slaked lime mixture at 1 litre per 2.5 m^2 and seal the hall for 48 hours.
3. Always use freshly prepared disinfectant/solution.
4. Complete the disinfection at least 3 days before consigning seed cocoons.
5. Maintain hygienic conditions in and around the grainage building.

12.7.8 Proper Selection and Transportation of Seed Cocoons

1. Harvest cocoons on 5th or 6th day of spinning in summer and 8–9 days in winter.
2. Procure and transport cocoons after complete pupation/turn dark brown in colour.
3. Transport seed cocoons on the 7th–9th day of spinning in summer and 12th–14th day in winter.
4. Avoid cocoon transportation at prepupal stage.
5. Seed cocoons should be selected/purchased only after confirming complete disease freeness of the lots through sample pupal testing.
6. Live pupae should be above 80 % in the lot.
7. No. of seed cocoons/kg should be 300–350.
8. Average cocoon yield should be 40–50 kg per 100 dfls.
9. Transport seed cocoons during cooler hours of the day.
10. Transport seed cocoons in suitable bamboo basket and perforated carton boxes/plastic crates
11. Avoid exposure to direct sunlight, heat, rain and jerks during transportation.

12.7.9 Proper Selection and Transportation of Seed Cocoons

1. Harvest cocoons on the 5th or 6th day of spinning in summer and 8–9 days in winter.
2. Procure and transport cocoons after complete pupation/turn dark brown in colour.
3. Transport seed cocoons on the 7th–9th day of spinning in summer and 12th–14th day in winter.
4. Avoid cocoon transportation at prepupal stage.
5. Transport seed cocoons during cooler hours of the day in suitable bamboo basket and perforated carton boxes/plastic crates.
6. Avoid exposure to direct sunlight, heat, rain and jerks during transportation.

12.7.10 Selection of Seed Cocoons

1. Seed cocoons should be selected/purchased only after confirming complete disease freeness of the lots through sample pupal testing.
2. Enforce selection of seed cocoons based on the fixed norms. Norms for selection of seed cocoons for P-2 and P-1 levels of multiplications are given in Table 12.14.

Table 12.14 Quality norms for selection of seed cocoons for P-2 and P-1 levels of multiplications

Multiplication level	Parameters	Norms
P-2	Fecundity (no./dfl)	350–400
	Hatching (%)	85–90
	Cocoon yield/100 dfls (kg)	50–60
	Cocoon yield/100 dfls (no.)	18,000–20,000
	Average cocoon wt.(gm)	2.90–3.5
	No. of seed cocoon/kg	280–350
	Pupation rate (%)	85–90
	Average shell wt. (gm)	0.45–0.55
	Shell ratio (%)	14.0–16.0
P-1	Fecundity (no./dfl)	300–350
	Hatching (%)	80–85
	Cocoon yield/100 dfls (kg)	45–50
	Cocoon yield/100 dfls (no.)	15,000–18,000
	Average cocoon wt. (gm)	2.50–3.00
	Pupation rate (%)	80–85
	No. of seed cocoon/kg	350–400
	Average shell wt. (gm)	0.40–0.50
	Shell ratio (%)	12.50–14.00

12.7.11 Preservation of Seed Cocoons

1. Preserve seed cocoons in single layer in plastic trays of 2′ × 3′ size @ 2 kg seed cocoon per tray.
2. Preserve seed cocoon at temperature of 25 ± 3 °C and RH of 80 ± 5 % with L:D of 12 h:12 h.
3. Avoid overcrowding of cocoons while preserving in trays/cage.
4. Protect seed cocoons from natural enemies like ants and rats.

12.7.12 Monitoring Moth Emergence

1. Pupal period is about 17–18 days in summer and 25–35 days in winter. The moth emergence starts in the early morning hours and may continue up to 9:00 A.M.
2. Place the emerged moths in mating cages at 50:50 (male/female) ratio.
3. Put the cages in a dark and well-aerated place.
4. Keep the excess moths under nylon net or moth cage in a well-aerated place.

12.7.13 Ensure Effective Coupling

1. Collect the newly emerged healthy male and female moths in the morning and keep at a ratio of 50:50 (male/female) in the coupling cages.
2. Allow the moths to mate naturally at the room temperature conditions.
3. Ensure 8–10 h of coupling.
4. Ensure semi-dark condition of grainage room during coupling and oviposition.
5. Place the female moths in nylon bag/khorika after decoupling to lay eggs.

12.7.14 Care During Oviposition (Egg Laying)

1. After decoupling allow the gravid female moths to freely lay eggs on the khorika or fill them in the nylon pouch measuring 16 × 9 × 3 cm for egg laying.
2. Hang the khorikas 15–20 cm apart in cool, dark and shady places, whereas stack of nylon pouches filled with female moths in trays should be kept in stand.
3. Ensure dark and optimum conditions with temperature of 26–28 °C and 80–85 % relative humidity in the oviposition room.
4. Allow oviposition for 3 days.

12.7.15 Mother Moth Examination

1. Conduct microscopic examination of the mother moth on 4th day of oviposition.
2. Resort to examination of the individual mother moth for seed crop.
3. Examine a minimum of five microscopic fields of each sample.
4. Conduct mother moth examination following Fujiwara technique for both basic and commercial seeds.

Fujiwara's Method of Mother Moth Examination

A. *Infrastructural facilities required*:
   (i) Separate processing and testing unit
   (ii) Running water supply
   (iii) Uninterrupted power supply
   (iv) Facility for disposing testing materials, moths, debris, etc.
B. *Preparations required*:
   (i) Prepare 0.6 % potassium carbonate solution (prepare by dissolving 6 gm of K_2CO_3 in 1 litre water). To test 100 samples, about 8.5 litres of 0.6 % K_2CO_3 is required.
   (ii) 2 % bleaching powder solution for disinfection of testing materials.
   (iii) *Detergent* solutions for mopping the testing room floor after the operations.

C. *Process of mother moth examination*:

Step 1: Collection of Abdomen Samples for Examination

(i) Cut the abdomen portion of the female moth after 76 h of egg laying.
(ii) Collect the abdomen of the single/individual mother moth for basic seed and sample of 20 pooled abdomens for mass moth examination.
(iii) Label the samples with appropriate details like lot number, laid on, etc.

Step 2: Homogenization

(i) Add 80 ml of 0.6 % K2CO3 solution in crushing jar containing 20 abdomen samples.
(ii) Crush the sample in a mixture grinder at 10,000 rpm for 2 min (to liberate spores from tissues).
(iii) Pour the contents in cleaned 250 ml beaker and allow to stand for 2–3 min to facilitate filtration.

Step 3: Filtration

(i) Place a thin layer of absorbent cotton on a funnel.
(ii) Slowly filter the liquid portion of homogenate to 100 ml capacity centrifuge tube for 1–2 min.
(iii) Label the centrifuge tube with appropriate lot numbers.

Step 4: Centrifugation and Dissolving Sediments

(i) Level the volume of filtrate in centrifuge tube by adding 0.6 % K2CO3 solution.
(ii) Centrifuge the filtrate at 3000–3500 rpm for about 3–4 min.
(iii) Slowly discard the supernatant solution without disturbing the sediment ring at the bottom of centrifuge tube.
(iv) Add 2–3 drops of 0.6 % K2CO3 solution to the sediments and dissolve it well using glass rod.

Step 5: Smear Preparation

(i) Use glass rod or cleaned stick for placing thin smear on micro-slide and cover it with coverslips.
(ii) Put two samples on each micro-slide.
(iii) Smear should not be too thick nor too thin to avoid overflowing and quick drying.

Step 6: Microscopic Examination

(i) Examine the smear under 600 x magnification of microscope.
(ii) Examine five microscopic fields per smear.
(iii) Pebrine spores appear as oval, shining body.
(iv) The spores exhibit Brownian movement.
(v) Record the intensity of infection of pebrine spores.
(vi) Discard the lots detected with pebrine spores.

Egg Surface Sterilization and Drying

1. Egg surface sterilization should be done by washing in 0.4 % bleaching powder solution for 2–3 min followed by washing with soap and running water.
2. After washing, dry the eggs in shades by placing on blotting paper in single layer.
3. After proper drying, wrap the eggs in muslin cloth bag/perforated egg boxes.

Transportation of Eggs

1. Avoid stuffing the egg packets inside handbags or polythene bags.
2. Transport eggs in specially designed wooden egg boxes of 21 × 11 × 3 cm sizes.
3. Ensure adequate ventilation/air circulation during transportation.
4. Put label on boxes indicating source, race and quantity with expected date of hatching.

Maintain Hygiene in Grainage

1. Disinfect moth testing area and appliances with 2 % formalin or 5 % bleaching powder solution.
2. Collect wings, cut/crushed moth along with debris/refuses and treat with 2 % formalin or 5 % bleaching powder and dump in a soak pit away from the grainage building.
3. Dispose the pierced cocoons immediately after grainage operation.
4. Use hand gloves, masks, apron and footwear during grainage operation.
5. Disinfect the foot by stepping on a foot mat soaked with 2 % formaldehyde solution.
6. Sprinkle bleaching powder and slaked lime mixture at the entrance and around the grainage hall.
7. Proper mother moth examination following Fujiwara technique for both basic and commercial seeds.
8. Resort to examination of the individual mother moth for seed crop.

   Examine a minimum of five microscopic fields of each sample.

12.8 Challenges and Future Strategies

The eri silk sector has got immense scope for expansion and commercialization to capture not only domestic but also in international market through brand management and promoting itself as natural ahimsa silk variety. However, the sector has few challenges to make the sector more vibrant and full-time livelihood avenue.

Specific Problems and Issues

1. Small-scale rearing by farmers, lack of systematic plantation, separate rearing house/appliances and seed production in conventional system
2. Need for systematic multiplication of seed following critical quality norms and moth testing for confirming disease freeness
3. Need for adequate crop protection measures during adverse seasons
4. Need to identify a few basic seed farms for maintaining the races with rigid selection process
5. Lack of strict disease monitoring mechanism
6. Use of jali prepared from dry leaves as mountage with cocoon quality

Suggestion for Improvement

1. Popularization of high yielding eri C2 breed among farmers to improve quality and productivity.
2. Popularization of NBR and GAUCH-1 varieties of castor among farmers to improve quality and productivity.
3. Raising of kesseru and *Ailanthus* sp.-based plantation involving forest and agriculture departments as eri silkworm food plants.
4. Popularization of low-cost bamboo platform rearing device among farmers.
5. Popularization of collapsible plastic mountage against traditional *jali*.
6. Expansion of organized seed multiplication system involving the Central Silk Board and Department of Sericulture.
7. Introduction of quality management systems like ISO 9001:2008 certification for production good quality eri seeds.
8. Norms for seed cocoons, seed, etc., to be standardized and seed certification should be followed as per provision under the Seed Act.
9. Involvement of more adopted seed rearers and private graineurs for quality seed production.
10. Effective introduction of mobile testing facility for disease monitoring.
11. Effective dissemination of technologies to the farmers through extension programmes, viz. field days, technology demonstrations, krishimelas, exhibitions, awareness programmes, etc.
12. Capacity building and training of farmers, state department staff, trainers training programme and skill development training programmes need to be conducted on a regular basis.
13. Implementation of extension oriented programmes like the institute village linkage programme (IVLP), vanya cluster programmes, etc., to create awareness among the farmers.

12.9 Conclusions

The eri silk is found to be a lucrative livelihood venture in the northeastern region. In recent years, ericulture has been introduced in other non-traditional states of the country also by virtue of its multifarious properties, assured crops, softness, thermal properties, blending abilities with other fibres and food value of its protein-rich pupae. Major thrust and efforts are required for making the value chain and marketing linkages as farmers friendly and economically viable venture through product diversification and utilization of by-products. The need of the hour is to put concerted efforts by different stakeholders for effective implementation of developmental schemes and dissemination of proven and adoptable technologies among the farmers besides marketing and credit support in rural areas to develop microenterprises. Strengthening of existing infrastructure, encouragement of public-private participation for quality seed production supported by adequate R&D coupled with effective management of seed production and rearing activities including post-cocoon technologies supported by diversification of products will ensure in creating marketing avenues both at national and international levels. This will not only boost eri raw silk production in the country but also act as sustainable source of income generation and livelihood earning for the grass root level stakeholders.

References

Ahmed SA, Sarmah MC, Sarkar BN, Giridhar K, Singha BB (2014) Package of practices for eri silkworm C2 breed. Central Silk Board, Bangalore

Arora GS, Gupta IJ (1979) Taxonomic studies of some of the Indian non-mulberry silk moths (Lepidoptera: Saturniidae). Mem Zool Surv India 16(1):1–64

Bindroo BB, Singh Tiken N, Sahu AK, Chakravorty R (2007) Eri silkworm host plants. Indian Silk 2007:13–16

Chakravorty R, Chaoba Singh K, Sarkar BN, Mech D, Neog K, Sarmah MC (ed and Comp) (2008) A monograph on catalogue on Eri Silkworm (Samia ricini) germplasm

Chaoba Singh K, Suryanarayana N (2003) Eri pupae: a popular cuisine too. Indian Silk 41(12):57–58

Dandin SB (2005) National strategy for conservation and sustainable use of seri-biodiversity: a felt need. Indian Silk 2005:5–10

Dayashankar KN (1992) performance of Eri silkworm, *Samia cynthia ricini* Boisduval on different host plants and economics of rearing on castor. M.Sc. (Agri.) thesis, UAS, Bangalore, pp 86

Debaraj Y, Sarmah MC, Datta RN, Singh LS, Das PK, Benchamin KV (2001) Field trial of elite crosses of eri silkworm, *Philosamia ricini* Hutt. Indian Silk 40(2):15–16

Jolly MS (1985) Species differentiation in the genus *Antheraea*. Sericologia 25(1):84–94

Mohanraj P, Veenakumari K. Nauman S (1988) *Samia fulva Jordan,* 1911 from the Andaman Islands, India (Indian Ocean)-preimaginal instars, host plants and taxonomic notes (Lepidoptera:Saturniidae)-NF 19(1):51–63

Neumann S, Peigler RS (2001) Four new species of the silk moth genus Samia (Lepidoptera:Saturniidae). Nachr Entomol Ver Apollo N E 22(2):75–83

Peigler RS, Naumann, S. (2003). A revision of the Silkmoth Genus *Samia,* 288 pp 148

Sarkar BN, Chutia BC, Chakraborty R (2009) Hybridization and rearing performance of samia ricini donovan and Samia canningie Hutton. In: National conference on vanya silk, 28–30 January 2009, NASSI, Organized by CMER&TI, Lahdoigarh, Central Silk Board, Jorhat, Assam

Sarmah MC, Ahmed SA, Sarkar BN (2012) Research & technology development, byproduct management and prospects in Eri culture – a review. Munis Entomol Zool 7(2):1006–1016

Sharma MC, Debaraj Y, Singh BK, Das PK, Suryanarayana N (2002) Status paper on eri host plant Germplasm. In: Proceedings: National workshop on Sericultural Germplam Management and Utilization, Hosur, Tamil Nadu, 6–7 February 2002, pp 59–61

Singh BK, Debaraj Y, Sarmah MC, Das PK, Suryanarayana N (2003) Eco-races of eri silkworm. Indian Silk 5:7–10

Singh BK, Sahu J, Krishna Rao JV, Das PK (2004). Utilization of by-products of eri silkworm *Samia ricini* (Donovan). In: Proceedings: National Workshop on potential and strategies for sustainable development of vanya silk in the Himalayan States, Dehradun, pp 462–466

Muga Sericulture

13

Babulal, A.A. Siddiqui, and S. Chouhan

Contents

13.1 Introduction

China and India are the only countries which produce mulberry as well as nonmulberry silks, viz. mulberry, tropical tasar, temperate tasar, Eri and Muga silk. Muga silk is the pride of India being the endemic to Assam and adjoining areas. Nonmulberry silks are referred to as Vanya silk. The global silk production was recorded 178,039 metric tonnes (MT) during the year 2014 against the production of 159,718 MT in 2013 (source: ISC website (http://inserco.org). The raw silk production in India increased from 26,480 MT in 2013–2014 to all-time high of 28,708 MT in 2014–2015. The mulberry raw silk production increased by 9.8 %

Babulal (✉) • S. Chouhan
Central Sericultural Research and Training Institute, Central Silk Board, National Highway 1-A, Gallandar, Pampore 192121, Jammu and Kashmir, India
e-mail: babulal.csb@gmail.com; sahadevjodhpur@gmail.com

A.A. Siddiqui
Former Scientist, Central Silk Board, E – 7, Abrar Nagar, Kalyanpur, Lucknow 226022, Uttar Pradesh, India
e-mail: aasiddiqui.csb@gmail.com

Omkar (ed.), *Industrial Entomology*, DOI 10.1007/978-981-10-3304-9_13

from 19,476 MT in 2013–2014 to 21,390 MT in 2014–2015. The production of Vanya silk during the year 2014–2015 was 7318 MT and 7004 MT in 2013–2014, which showed 4.5 % increase in production. Exclusively bivoltine mulberry, Eri and Muga silks achieved recorded production of 3870 MT, 4726 MT and 158 MT, respectively, during 2014–2015. However, tasar silk production of 2434 MT during 2014–2015 as compared to 2619 MT in 2013–2014 and 7.1 % decrease were recorded. Silk production is an excellent example of healthy biological interaction between primary producer (host plant) and consumer (silkworm) that is an integral part of ecosystem (Peigler 1996).

The five silkworm species commercially exploited for silk production in India are as under:

Silk variety	Silkworm	Category
Mulberry	*Bombyx mori* L.	Mulberry
Tropical tasar	*Antheraea mylitta* D.	Nonmulberry
Temperate tasar	*Antheraea proylei* J.	Nonmulberry
Eri	*Samia cynthia* Donovan	Nonmulberry
Muga	*Antheraea assama* Ww.	Nonmulberry

Sericultural activities are being taken up in 27 states in India. Muga silk production activities are confined only to northeastern states, viz. Assam, Meghalaya, Arunachal Pradesh, Manipur and Mizoram (Giridhar et al. 2007). The muga culture activities have also started in West Bengal, Andhra Pradesh and Uttarakhand at low scale.

The contribution of muga silk in national silk production is only (0.06 %). The muga silk on commercial scale is produced in Assam (93.04 %), Meghalaya (5.04 %), Arunachal Pradesh (1.22 %) and little amounts (0.20 %) in each Manipur, Mizoram, Nagaland and West Bengal (Giridhar et al. 2007).

Different species of silkworms produce different types of silk of different colours and qualities. Silk is a durable fabric and fine and subtle to touch having the quality of being warm in winter and cool in summer. The muga silk glitters in its shimmering natural golden colour. The distinctiveness and smooth feel of muga appeal to fashion and home designers all over the world. The silk is expensive and used by Assamese people to furnish homes. Vibrant Sualkuchi sarees, mekhala and chaddars are traditional items made from muga silk in northeast region. Besides, the use of muga silk as a substitute of Zari is finding favour with reputed weavers.

Muga culture originated first in Brahamputra Valley of Assam (Chaudhary 1981). The moist tropical and subtropical deciduous forests are suitable for luxuriant plant growth. The plantation of *Persea bombycina* Kost. abundantly in the areas of upper Assam, like North Lakhimpur, Sibsagar, Dibrugarh and Jorhat, and at low scale in some parts of lower Assam, Nagaland and Meghalaya, are raised for commercial utilization for muga silk production (Plate 13.1). Seed crops are raised mainly in Kamrup and Goalpara districts of Assam and Garo Hills of Meghalaya on *Litsea monopetala* (Roxb.) Pers. Reeling and weaving activities are confined to only district Kamrup, particularly Sualkuchi.

The muga silk industry is facing certain problems and production trends are not much encouraging. The reasons for low production and slow pace of growth of

Plate 13.1 Rearing of muga silkworm at farmers level in Assam

muga silk industry are attributed to unfavourable climatic conditions, inadequate supply of disease-free silkworm seed, devastating flood and deforestation of naturally grown muga food plants (Thangavelu and Chakraborty 1986). The silk production is not increasing to commensurate with its demand and thereby gap in silk demand and supply is existing. The yield gap in muga cocoon production in Assam in seed and commercial crops has been reported to be 20 % and 50 %, respectively (Barah et al. 2003). The yield gap can be bridged by increasing muga silk production through effective transfer of technologies, augmentation of muga food plants in private sector in systematic manner, strengthening of extension network for rearing and advancement in post-cocoon technology and also taking steps to spread the muga culture in other states of the country to utilize the available vast flora. It was previously reported that the quality of leaves directly influences the health, growth and survival of silkworms.

13.2 Host Plants

The muga silkworms feed on a variety of plant species, which are widely distributed in tropical and subtropical areas in nature. The food plants on priority of utilization for muga silk cultivation are grouped into three categories, like primary, secondary and tertiary food plants.

Persea bombycina Kost. and *Litsea monopetala* (Roxb.) Pers. are primary food plants. Secondary food plants are *Litsea salicifolia* Roxb., *Cinnamomum glaucescens*, *Cinnamomum obtusifolium* Nees, *Actinodaphne angustifolia* Bl., *Actinodaphne obovata* Bl., *Celastrus monosperma* Roxb., *Gmelina arborea* Roxb., *Litsea citrata* Blume., *Litsea nitida* Roxb., *Magnolia sphenocarpa* Roxb. and *Zanthoxylum rhetsa* DC.

Table 13.1 Food plants of muga silkworm (*Antheraea assama* Ww.)

S. No.	Botanical names	Common names (Assamese language)	Family
	Primary		
1.	*Litsea monopetala* (Roxb.) Pers. (Syn. *Litsea polyantha* Juss.)	Soalu	Lauraceae
2.	*Persea bombycina* Kost.	Som	Lauraceae
	Secondary		
1.	*Actinodaphne angustifolia* BI.	Bangla	Lauraceae
2.	*Actinodaphne obovata* BI.	Patihonda	Lauraceae
3.	*Celastrus monosperma* Roxb.	Bhumloti	Celastraceae
4.	*Cinnamomum glaucescens* Nees Ex Wall.	Gansarai	Lauraceae
5.	*Cinnamomum obtusifolium* Nees	Chhamejam	Lauraceae
6.	*Gmelina arborea* Roxb.	Gamari	Verbenaceae
7.	*Litsea citrata* Blume.	Mejankari	Lauraceae
8.	*Litsea nitida* Roxb.	Kathalua	Lauraceae
9.	*Litsea salicifolia* Roxb.	Dighloti	Lauraceae
10.	*Magnolia sphenocarpa* Roxb.	Chapa/Panchapa	Magnoliaceae
11.	*Zanthoxylum rhetsa* DC.	Bajramoni	Rutaceae
12.	*Ziziphus jujuba* Lamk.	Bogori/Ber	Rhamnaceae

Thangavelu et al. (1988)

Ziziphus jujuba Lamk. is a tertiary food plant (Thangavelu et al. 1988); *Cinnamomum glanduliferum* was reported as an additional host plant for muga silkworm (Raja and Samson 1991). The details of food plants (Table 13.1) are given below.

The majority of food plants belong to the family Lauraceae, which has 6 genera and 2000 species, and most of the members are trees or well-formed shrubs. The family has economic significance for utilization of its plant species for muga silk cultivation. The distribution of muga host plant in Myanmar, India, Indonesia and Malaysia has been reported by Jolly et al. (1979). Brandis (1972) and Isa and Thangavelu (1988) reported the occurrence of *Litsea monopetala* (Roxb.) Pers. in Doon Valley and its adjoining areas. The other food plants, like *Persea odoratissima*, *Persea gamblei* and *Litsea glutinosa*, are also available in Uttarakhand.

Litsea monopetala (Roxb.) Pers. is distributed in the region extending northwest from Punjab up to salt range along the foot hills of Himalaya ascending to 3000 m above mean sea level (AMSL) eastward to Northeast India. The plant is known with different names (Table 13.2) in different parts of the country and confirms its availability and awareness for utility in different purposes.

Litsea monopetala (Roxb.) Pers. is a small- to medium-sized plant and attains a height of up to 21 m with girth of 1.8 m (Plate 13.2). The plant is very common in shrub forests in Dehradun and Saharanpur districts. The systematic position of *Litsea monopetala* (Roxb.) Pers. and *Machilus bombycina* King is given below (Table 13.3).

The leaves of *Litsea monopetala* (Roxb.) Pers. are oblong, ovate or obovate, glabrous above, rusty tomentose and strongly reticulate beneath, with its tip acute or

Table 13.2 Local names of host plants

Sl. No.	Language	Local name
1.	Hindi	Meda, Katmara, Patoia, Kakuri, Karankawa
2.	Bengali	Bara Kukurchita
3.	Marathi	Ranamba, rapamba
4.	Telugu	Naramamidi
5.	Tamil	Maidalagadil, Muchaippeyetti, Picinabattaw
6.	Punjabi	Rain, Gwa Haerin, Maida Lakdi
7.	Lepcha	Sunyok-kung, Sapot-kung
8.	Assamese	Muga, Hoanlu
9.	Nepali	Ratmariti, Kadmeya
10.	Oriya	Baghoari, Kulya, Bastura

Source: Anon (1962), Wealth of India

Plate 13.2 Full-grown soalu plant

Table 13.3 Systematic position of *Litsea monopetala* and *Machilus bombycina* plants

Kingdom	Plantae	Kingdom	Plantae
Division	Phanerogams	Division	Phanerogams
Subdivision	Angiosperm	Subdivision	Angiosperm
Class	Dicotyledonae	Class	Dicotyledonae
Subclass	Polypetalae	Subclass	Polypetalae
Order	Laurales	Order	Laurales
Family	Lauraceae	Family	Lauraceae
Genus	*Litsea*	Genus	*Machilus*
Species	*monopetala* (Roxb.) Pers.	Species	*bombycina*
	Syn. *polyantha* Juss.		Syn. *Persea* King (Kost.)

rounded. The lateral nerves have 5–10 pairs, and the petiole measures 1.27–2.54 cm long, with a stout umbel inflorescence. It has 5–6 pedicelled flowers. The plant is dioecious with five membranous bracts, 9–13 stamens and hairy filaments. Its fruit is ovoid, 6–7 mm long on a small perianth base.

Machilus bombycina King is an evergreen middle-sized tree with spreading dark grey, rather rough branches and simple, exstipulate, petiolate, alternate, entire, obtuse leaves, with upper surface glabrous and lower pubescent; length-breadth ratio of leaf is 2.5 to 5.0 × 0.8 to 2.0 in., elliptic lanceolate to ovate lanceolate. The lateral nerves is 6–8 on either side, with the inflorescence being a panicle. The yellow small flowers are bisexual and hypogynous, with the perianth having 2 persistent, imbricate whorls and with 12 stamens in 4 whorls (the stamens of the third whorl bear a pair of lateral glandular outgrowths, and the inner most fourth whorl is transformed into staminode). Filaments are hairy at the base. Anthers are adnate, opening by a valve from below upwards; there is one carpel and the one-celled ovary has a single ovule. The style is terminal, stigma has three lobes, and the placentation is marginal. The fleshy berry, globose fruit is 6–7.5 mm, with the plant flowering from December to March and bearing fruits from March to May.

The bark of *Litsea monopetala* (soalu plant) is used in Indian medicine. It also has diverse utility in industries, like candle, agarbatti, soap, fuel, cosmetics and plywood (Anon 1962). It is distributed exclusively in foothills and low hill regions, particularly in the areas of Doon Valley, Ramnagar, Nainital, Askot, Lansdowne, Mussoorie and also Haldwani, Bhimtal and Pithoragarh in Kumaon (Sengupta et al. 1995). The plant covers 0.1 % of total vegetation in Uttarakhand (Isa and Thangavelu 1988). Troops (1921) gave information on various uses of this plant species for house building, oars, furniture, agriculture implements, boxes for transportation of tea and firewood due to its high calorific value and in internal construction in the form of plywood. The bark of plant contains irritant substance and tannin, which is used for the treatment of diarrhoea due to its stringent property. The powder of bark and roots is used in external applications for pains, bruises and contusions and also in fractures in animals. Besides, leaves are used for feeding the cattle. The seed contains 21 % fat used for candle manufacture and preparation of ointments for rheumatism (Troops 1921). The knowledge of complex characters, like leaf weight and its components, which show less variability to environmental conditions, is useful for improvement of leaf yield/leaf weight.

Siddiqui et al. (1993) have reported that suitable plants play an important role in increasing the productivity of silk because silk quality, silk production, fecundity and survival rate of silk worms are generally determined by leaf quality. Many morphovariants of both food plants are available in nature. The leaves of 50 morphovariants are palatable to silkworm and some are least palatable. The detail studies pertaining to taxonomical and biochemical bioassay and post-cocoon parameters are depicted in Tables 13.4 and 13.5.

Nutritional Status of Som and Soalu

The nutritional status of the leaves is directly related to the silkworm nutrition and depends upon the level of moisture, protein, total carbohydrates and minerals.

Table 13.4 Key characteristics of different morphovariants of som

Name of morphotypes	Taxonomical	Biochemical	Bioassay	Post-cocoon parameters
S_1	Leaf length= 14.3 cm Leaf width=1.2 cm Single leaf weight 1.1 g No. Of stomata=24 Length of stomata =125.8 µcm Width of stomata = 100.5 µcm	Protein= 10.11 % Carbohydrate= 79.29 % Lipid=6.60 % Sugar=5.80 mg/100 mg Vit. A=85.00 mg/100 mg Vit. C= 15.00 mg/100 mg Chlorophyll=4.43 mg/100 mg Lignin=11.21 %	Moderate palatable and effective rate of rearing (62 %) Prolong larval period: Larval weight=13.33 g Cocoon weight= 5.65 g Shell weight= 0.436 g S.R. = 7.62 %Absolute silk yield= 75.40 g/ 300 larvae	Filament length= 318 mtrs. Silk weight= 0.20 g Waste weight=0.16 g Recovery = 55.46% Denier= 5.60 No. of breaks= 1.70
S_2	Leaf length=14.0 cm Leaf width = 4.4 cm Single leaf weight = 1.13 g No. of stomata =12 Length of stomata = 125 µcm Width of stomata =100 µcm	Protein= 10.93 % Carbohydrate= 77.95 % Lipid= 6.66 % Sugar= 6.00 mg/100 mg Vit. A= 65.00 mg/100 mg Vit. C= 20.00 mg/100 mg Lignin= 14.13 %	Less palatable, moderate ERR (60.67 %) Prolong larval period: Larvae weight = 12.75 g Cocoon wt.=5.55 g Shell wt.= 0.44 g S.R. =7.72 % Absolute silk yield= 75.24 g/300 larvae	Filament length=316 mtrs. Silk wt.=0.19 g Waste wt.=0.22 g Recovery = 46.11% Denier=5.40 No. of breaks= 2.00

(continued)

Table 13.4 (continued)

Name of morphotypes	Taxonomical	Biochemical	Bioassay	Post-cocoon parameters
S_3	Leaf length= 7.3 cm Leaf width= 1.6 cm Single leaf weight= 0.63 g No. of stomata = 13 Length of stomata = 137.5 μμm Width of stomata = 87.5 μcm	Protein =11.33 % Carbohydrate= 78.57 % Lipid=5.42 % Sugar= 8.90 mg/1.mg Vit. A= 38.00/100 mg Vit. C= 40.00 mg/100 mg Chlorophyll= 3.44 mg/100 mg Lignin= 8.92 %	Most palatable, high effective rate of rearing (71.67 %) Normal larval period: Larval wt.= 11.80 g Cocoon wt.= 5.52 g S. R = 7.89 % Absolute silk yield= 86.92 g /300 larvae	Filament length =313 mtrs. Silk wt.= 0.19 g Waste wt.= 0.21 g Recovery = 47.40% Denier= 5.35 No. of breaks=2.10
S_4	Leaf length= 8.5 cm Leaf width= 2.9 cm Single leaf weight= 0.37 g No. of stomata = 15 Length of stomata =150 μcm Width of stomata =112.5 μcm	Protein= 10.19 % Carbohydrate= 79.19 % Lipid= 5.72 % Sugar= 6.20 mg/100 mg Vit. A = 36.00 mg/ 100 mg Vit. C= 15.00 mg/100 mg Chlorophyll= 1.87 mg/100 mg Lignin= 10.74 %	Palatable, high effective role of rearing (68.33 %) Normal larval period: Larval wt.= 12.20 g Cocoon wt.= 5.41 g Shell wt.= 0.42 g S.R. = 7.91% Absolute silk yield=81.229/300 larvae	Filament length= 342 mtrs. Silk wt.= 0.21 g Waste wt.= 0.19 g Recovery = 52.18% Denier=5.25 No. of breaks =1.60
S_5	Leaf length= 6.9 cm Leaf width= 2.4 cm Single leaf weight= 0.48 g No. of stomata= 14 Length of stomata= 87.5μcm Width of stomata= 62.5 μcm	Protein= 10.08 % Carbohydrate= 78.98 % Lipid= 6.97 % Sugar= 6.90 mg/100 mg Vit. A = 73.00 mg/ 100 mg Vit. C= 16.00 mg/100 mg Chlorophyll= 3.03 mg/100 mg Lignin= 13.00 %	Palatable, high effective rate of rearing (70.0 %) Normal larval period: Larval wt.= 12.25 g Cocoon wt.= 5.42 g Shell wt.= 0.42 g S.R. = 7.79 % Absolute silk yield=82.969/300 larvae	Filament length= 326 mtrs. Silk wt.= 0.20 g Waste wt.= 0.18 g Recovery = 52.18% Denier=5.35 No. of breaks =2.00

S_6	Leaf length= 8 cm Leaf width= 3.5 cm Single leaf weight= 0.64 g No. of stomata= 13 Length of stomata= 150µcm Width of stomata= 100 µcm	Protein= 8.52 % Carbohydrate= 79.95 % Lipid= 5.51 % Sugar= 5.90 mg/100 mg Vit. A = 58.00 mg/ 100 mg Vit. C= 13.00 mg/100 mg Chlorophyll= 3.36 mg/100 mg Lignin= 10.46 %	Palatable, high effective rate of rearing (70.0 %) Normal larval period: Larval wt.= 12.17 g Cocoon wt.= 5.50 g Shell wt.= 0.46 g S.R. = 8.60% Absolute silk yield=97.559/300 larvae	Filament length= 340 mtrs. Silk wt.= 0.21 g Waste wt.= 0.18 g Recovery = 53.43% Denier=5.20 No. of breaks =1.60
S_7	Leaf length= 8 cm Leaf width= 4 cm Single leaf weight= 0.71 g No. of stomata= 16 Length of stomata= 150µcm Width of stomata= 100 µcm	Protein= 10.14 % Carbohydrate= 77.90 % Lipid= 7.46 % Sugar= 7.70 mg/100 mg Vit. A = 29.00 mg/ 100 mg Vit. C= 10.00 mg/100 mg Chlorophyll= 4.26 mg/100 mg Lignin= 10.32 %	Palatable, moderate effective rate of rearing (63.37 %) Normal larval period: Larval wt.= 13.29 g Cocoon wt= 5.49 g Shell weight= 0.43 g S.R. =7.89% Absolute silk yield=75.439/300 larvae	Filament length= 333 mtrs. Silk wt.= 0.21 g Waste wt.= 0.18 g Recovery = 50.88% Denier=5.30 No. of breaks =1.60
S_8	Leaf length= 7 cm Leaf width= 3.3 cm Single leaf weight= 0.40 g No. of stomata= 15 Length of stomata= 137µcm Width of stomata= 100 µcm	Protein= 11.88 % Carbohydrate= 80.10 % Lipid= 5.72 % Sugar= 5.80 mg/100 mg Vit. A = 42.00 mg/ 100 mg Vit. C= 9.00 mg/100 mg Chlorophyll= 4.63 mg/100 mg Lignin= 7.94 %	Palatable, moderate effective rate of rearing (61.0 %) Normal larval period: Larval wt.= 12.17 g Cocoon wt.= 5.68 g Shell weight= 0.45 g S.R. =7.90% Absolute silk yield=76.549/300 larvae	Filament length= 302 mtrs. Silk wt.= 0.18 g Waste wt.= 0.27 g Recovery = 45.90% Denier=5.30 No. of breaks =1.60

Table 13.5 Biochemical compositional variation of soalu

Morphotype	Moisture (%)	Protein (%)	Carbohydrate (mg/100 g)	Nitrate (%)
M1	72.04	20.0	165.0	3.55
M2	69.32	21.0	150.0	3.16
M3	65.80	24.0	115.0	1.18
M4	63.11	19.0	158.0	1.23
F1	74.41	25.0	240.0	1.84
F2	62.63	19.0	123.0	0.91
F3	62.74	20.0	142.0	0.62
F4	61.72	23.0	204.0	1.25
F5	66.86	25.8	186.0	1.80
F6	64.24	22.0	280.0	2.22

N.B. M1 to M4 stand for male morphovariants and F1 to F6 stand for female morphovariants

Significant differences were noticed in above mentioned and other constituents among the different morphovariants of som and soalu plants of muga silkworm host plants (Tables 13.4 and 13.5). The findings of Siddiqui et al. (2000) and Siddiqui's (2012) are in the agreement with the statement. Parpieve (1968) also reported the favourable effects of moisture content in leaves on the palatability and assimilability of nutrients and considered that the moisture content may serve as one of the criteria in estimating the leaf quality. Paul et al. (1992) established higher degree of correlation between larvae and moisture content.

13.3 Muga Silkworm

Muga silk is produced from the cocoons of muga silkworm, *Antheraea assama* Ww., synonym *Antheraea assamensis* (Plate 13.3) Helfer, which belongs to the family Saturniidae under the insect order Lepidoptera. The systematic position of *Antheraea assama* Ww. is given below:

Phylum: Arthropoda
Class: Insecta
Order: Lepidoptera
Superfamily: Bombycoidea
Family: Saturniidae
Genus: *Antheraea*
Species: *assama*

This silkworm is multivoltine and endemic sericigenous species to Northeast India. It is reared in semi-domestic conditions, viz. rearing in outdoor condition and cocooning in indoor conditions to extract silk in northeast region of India. It has six generations in a year, viz. Kotia (October–November), Jarua (December–January), Chotua (March–April), Jethua (May–June), Aherua (June–July) and Bhodia

(August–September), as per Assamese calendar. Kotia (October–November) and Jethua (May–June) are two main commercial crops, and the rest are seed crops.

(a) Effective rate of rearing and economic traits in different muga silkworm crops are given as under:

Parameters	Jarua	Jethua	Aherua	Bhodia	Kotia
ERR (%)	20–30	48–55	25–35	30–40	55–60
Cocoon wt. (g)	4.10	5.20	4.50	4.50	5.80
Shell wt. (g)	0.28	0.45	0.35	0.33	0.57
Silk filament length (mt.)	204	400	225	300	500
Shell ratio (%)	5.49	8.27	6.36	6.50	8.39

Fifth stage

Plate 13.3 Different stages of larvae of *Antheraea assama*

Muga silkworm is a holometabolous insect. The entire life cycle of this insect, from egg, to larval and pupal stages, to adult, lasts for about 50 days in summer and 120 days in winter in the northeast (Thangavelu et al. 1988). Female moth lays 150–230 brownish-grey eggs. The larvae undergo four moults and pass through five instars.

13.4 Life Cycle of Muga Silkworm

Egg
The egg is oval, dorsoventrally flattened and bilaterally symmetrical. It is about 2.4–2.7 mm in length, 2.5 mm in diameter and 0.00073 mg in weight. The eggs are brownish grey in colour at the time of oviposition and they appear light green or creamy. The duration of egg stage is 7 days in summer and 16 days in winter.

Larva
The larvae hatch in the morning hours by nibbling through the egg shell. Soon after hatching they eat a portion of the egg and disperse in search of food. The larvae undergo four moults and pass through five instars. A newly hatched larva is 0.7–1.2 cm in length and approximately 0.0069–0.007 g in weight. The colour of the body is black with distinct yellow lines at intersegmental region. The body tubercles are black in colour and are provided with setae. The larva prefers to feed on tender leaves of the host plant. After feeding for about 3–4 days in summer and 6–8 days in winter, the larva enters the first moult and transforms itself into the second instar larva.

The second instar larva is light yellow in colour and about 1.4–1.8 cm in length and approximately 0.083– 0.0912 g in weight. The tubercle colour is blue. This stage lasts from 3–5 days in summer and 7–10 days in winter; thereafter the second instar larva enters into the third stage.

The third instar larva is green in colour and about 1.8–2.5 cm in length just after moulting. The larvae are approximately 0.442–0.632 g in weight. This stage lasts for 5–7 days in summer and 10–13 days in winter. The tubercle colour is violet.

The fourth instar larva is dorsally dark green and ventrally light green in colour, about 2.5–3.5 cm in length and approximately 2.00–3.5 g in weight. The head is triangular in shape with distinct epicranial sutures and copper colour. Small black eyespots on the lateral side of the head are prominent. The lateral line is distinct and yellow in colour. The clasper is triangular in shape with a copper black margin. The tubercles are red in colour. This stage lasts 7–10 days in summer and 12–15 days in winter.

The fifth instar larva is approximately 4–5.5 cm in length and approximately 4.121–5.213 g in weight. The dorsal surface of the body is light green, and ventral surface is deep green in colour. Eyespots are prominent. The body tubercles are

Table 13.6 The colour of tubercles in different instars

Instar	Colour of tubercles
First	Black
Second	Blue
Third	Violet
Fourth	Red
Fifth	Brick red

Thangavelu et al. (1988)

brick red in colour and with setae. The lower portion of the lateral line is yellow and the upper portion is chocolate in colour. Three pairs of thoracic and four pairs of abdominal legs are well developed. Spiracles are black in colour. This stage lasts for 10–12 days in summer and 16–19 days in winter. The mature larva is 10–15 g in weight. The female larvae are larger and heavier than the male larvae.

The different stages of muga silkworm can be identified by the colour change of tubercles in every instar (Table 13.6).

There is variation in cocoon, shell and pupal weight and silk content in different seasons. The male moths are brown to dark brown, while female moths are yellowish light brown. The wild population in natural habitat exhibits colour polymorphism, and three different colour morphs in the larval stage, viz. yellow blue and orange colour morph, are observed among the normal green muga population. Blue colour larvae are bigger in size than normal green larvae. The orange colour larvae have a coating of white powdery scales on the body surface. Unlike tubercles in green larvae, the tubercles of yellow larvae are white in colour. The wild larvae have lower survival rate in the plains. It is known that the allelic expression of colour in insect is largely due to alteration of extrinsic and/or intrinsic factors influencing pigment metabolism. The silkworms have two salivary glands in their body, and during formation of cocoon, a viscous gummy fluid comes out of their mouth, which hardens on exposure to air in fine threads called silk. It is obtained commercially from the cocoons, in which silkworms spin around themselves before passing into the resting stage of their life cycle, the pupa.

Pupa

On maturity, the larva spins the cocoon with silk filaments around its body, after selecting a suitable site for pupation. The pupa is brown to dark brown in colour with harder skin. Sex markings are prominent in pupa, which makes it easier to determine the sex in the pupal stage than in the larval stage. The female pupa has a fine longitudinal line on the eighth abdominal segment, whereas such marking is absent in the male pupa. The pupal stage lasts for 14 days in summer and 40 days in winter.

Moth

The moths emerge from the cocoons at dusk and continue till dawn. The male and female moths exhibit distinct sexual dimorphism (Plates 13.4 and 13.5). In male moth, the tips of the forewings have sharp curve, while in female it is not there. The

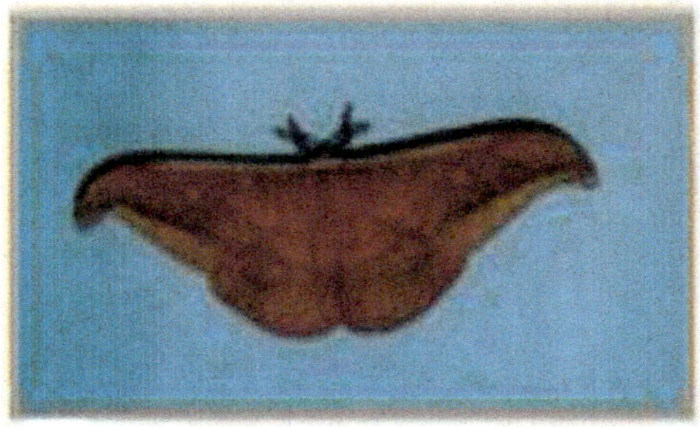

Plate 13.4 Male moth of *Antheraea assama*

Plate 13.5 Female moth of *Antheraea assama*

Plate 13.6 Pairing of *Antheraea assama*

antennae of the male moths are broader than those of females. The abdomen of male moths is narrow and small, whereas in females it is broad, larger and swollen. The colouration of the wings and the body of male moths is brown to dark brown, while in the female it is yellowish light brown. The male moth flies actively and copulates with the stationary female (Plate 13.6). Coupling lasts overnight. After decoupling, the female moth lays eggs on dry twigs called 'Khorika'. The moth is a nonfeeding stage and dies within 7–12 days after its emergence.

The duration in different developmental stages of muga silkworm

	Duration (in days)				
Developmental stages	Jarua	Jethua	Aherua	Bhodia	Kotia
Egg	14–16	9–10	7–8	7–8	8–9
Larva	52–60	27–38	22–30	22–32	28–35
Pupa	32–40	20–28	18–26	17–25	21–28
Moth	10–15	8–10	6–9	6–8	8–10

13.5 Exploitation of Muga Silkworm Through Breeding

India has the monopoly in the world to produce lustrous golden muga silk, as muga silkworm is endemic to northeast India rather than in Assam. Unlike mulberry, tasar and Eri silkworms, muga silkworm has no ecoraces due to continuous breeding in the single population observed which further indicates genetic uniformity in the population, viz. low egg laying, low cocoon formation and low absolute silk yield.

Genetic uniformity poses a threat to the survival of a population/species as a result of natural selection according to Darwin's theory. It is heterogeneity on which selection operates in times of environmental as well as ecological changes. On the basis of these studies, immediate attention is needed to create variation in muga silkworm for further breeding programme and evolution of improved varieties and lines for higher silk yield (Siddiqui 2012).

13.6 Practices Followed in Muga Silk Industry

13.6.1 Host Plant Cultivation

(a) *Propagation through seeds*: In nature, som and soalu plants are propagated through seeds. Som seeds are collected during March and April, while soalu seeds are collected during May and June. The pulp of the seed is washed off by kneading two or three times in water and dried under shade for a few hours and sowed the seeds soon after collection. The seed remains viable for a very short period and requires to be sown early, and it can be stored in moist seedbed under low temperature to prolong the viability.

(b) *Preparation of nursery bed and raising seedlings* : Generally seeds are sown directly for raising plantation. But the present trend is to grow seedlings in the nursery and transplant them in the field to reduce the period of establishment and save wastage of precious seed material. Seedbeds of convenient size are prepared in soil, ploughed twice or thrice followed by proper levelling and raised up to a height of 20 cm to avoid waterlogging. A heavy dose of well-decomposed farmyard manure (FYM) (30 tons per hectare) is applied. Muga host plants grow well in sandy loam soil, acidic with pH between 4.5 and 5.0. High land and sloppy land with good drainage system are suitable for muga plantation. Seeds are sown in lines in the prepared beds at a spacing of 15 cm in the row and 15 cm between rows at a depth of about 2 cm.

(c) *Germination of seed* **:** Germination starts after 4 weeks of sowing. Normally one seed gives rise to one seedling only, but polyembryonate seedlings are also known to occur.

(d) *Raising tube seedlings* **:** The seedlings are raised in polythene tubes. The seeds after collection are kept on moist bed under tree shade covered with gunny bags followed by sprinkling of water to maintain the moisture. The germination starts after 4 weeks. The germinated seeds are sorted out and sown individually in polythene tubes of 9 × 6 inch size of 150–220 gauze, filled with rooting media and kept under tree shade. A mixture of sand, soil and farmyard manure (FYM) in the ratio of 1:1:1 is used as the rooting media for raising seedlings in the polythene tubes.

(e) *Transplantation*: The seedlings raised in nursery or polythene tubes are maintained for about 9–12 months till they attain a height of 1 m. The raised plants are planted in pits 1 × 1 × 1 ft. filled with 0.5 ft. of farmyard manure with 10 g lindane. The season from July to September is ideal for transplantation.

(f) *Spacing for som and soalu plantation* **:** The spacing of 4 × 4 m for som and 5 × 5 m for soalu in square system of plantation is ideal, which accommodates 625 and 400 plants/ha., respectively.

Vegetative Propagation Due to free breeding in som and soalu morphovariants, available seeds are heterozygous and plants raised through these seeds are not preferred by the farmers. With the increasing demand of suitable plant varieties of som and soalu for muga silkworm rearing and also for optimizing biomass production, the importance of vegetative propagation was felt, and varied techniques like air layering (Isa et al. 1989), stem cutting (Chaudhary 1965), single leaf bud cutting and juvenile shoot cutting (Siddiqui et al. 1996) were tried. But these techniques are not viable at commercial level. In the meantime, clonal seed orchards of preferred morphovariants were suggested in isolated place to obtain homozygous seeds, since micropropagation is yet to take momentum in the industry (Siddiqui 2012).

The pest and diseases affecting the muga host plants are discussed below.

Pest and Disease of Primary Muga Food Plants, viz. Som and Soalu
The important pests and diseases of som and soalu are described below:

A. Insect pests
   (a) *Sucking pests*: Pests like thrips, aphids and jassids cause damage to leaves of som and soalu by sucking the plant sap and cause wilting of leaves. The symptoms are leaf margin rolling followed by curling and wilting of leaves. The sucking pests are effectively controlled by the application of 5 % solution of any systemic insecticide like Rogor.

(b) *Stem borer*: The carpenter worm, *Zeuzera multistrigata* Moore (Cossidae: Lepidoptera), is a major pest and causes extensive damage to som and soalu plantations. A single adult female moth lays more than 1000 eggs scattered on the tree trunk. The larvae emerging from the eggs bore through the bark, make a tunnel and start intensive feeding of middle portion of tree trunk. The plants infested by stem borer with extensive tunnelling break off at the time of heavy wind/cyclone. Fumigants, like chloroform, benzene, nuvan, etc., are used to control the stem borer by plugging the larval entry hole with cotton ball soaked in fumigants followed by plastering with mud.

(c) *Leaf miner*: Caterpillars, viz. semiloopers, cause extensive damage to the leaves of som and soalu. They feed voraciously and leave large holes on the leaf surface. Spray of 10 % Thiodan is very effective for the control of semiloopers.

(d) *Leaf galls*: *Paropsylla bessoni* (Psyllidae: Homoptera) induces galls in soalu leaves. Gall is a malignant tumour-like growth which makes leaves unsuitable for muga silkworm rearing. Field sanitation and cultural practices, like removal of gall-infested leaves and destroying them in fire, are very effective to reduce the incidence of galls.

(e) *Leaf roller*: *Pleuroptya scinisalis* (Lepidoptera: Pyralidae), commonly known as leaf roller, is a serious pest of soalu. The female moth lays clusters of eggs on the ventral surface of the leaves. The larvae develop inside the leaf rolls where the faecal materials also accumulate. Mechanical control is very effective in reducing the incidence of leaf roller. Besides, spraying of 10 % Thiodan reduces the incidence of leaf roller.

(f) *Cricula trifenestrata* (Lepidoptera: Saturniidae) is locally known as 'Amphutukani' in Assam. The larvae of the pest voraciously feed on leaves of som plant causing drastic loss of foliage. This pest is controlled mechanically by collection and killing of caterpillar and eggs. Cocoons of the pest are collected and burnt to stop further multiplication that is also very effective and accurate method for controlling the pest. Besides, 0.05 % phosphamidon is also used effectively to control the pest.

B. Diseases

(a) *Grey blight*: Grey blight of leaf in som is caused by a fungus *Pestalotiopsis disseminata*. The infected leaves become dry with dark greyish colour on the soft tender leaves, and leaves become unsuitable for rearing of muga silkworm.

Control measures: Spray of 1 % Bordeaux mixture is effective for control of the fungus. Besides, picking and burning of diseased leaves is also more effective.

(b) *Leaf spot*: The leaf spot is caused by *Phyllosticta perseae* Ell. & Mart. It infects the leaves and a number of circular or irregular dark brown spots with pale yellow margin are developed which make the leaves unsuitable for silkworm rearing.

Control measures: The disease is controlled effectively by spraying 0.2 % Difolatan.

(c) *Leaf curling*: The mosaic symptoms, occasional curling of leaves and stunted growth of the plants, are caused by virus. The leaf curling makes the leaves unsuitable for silkworm rearing.

Control measures: The affected leaves are collected and burnt. On high infection, the plants are removed and burnt for control and further spread of the disease.

(d) *Red rust*: The disease is caused by algae, *Cephaleurus* sp. (class, Chlorophyceae), which is an intercellular parasite. On older leaves, orange yellow patches on the upper surface of the leaf are developed which makes the leaves unsuitable for muga silkworm rearing.

Control measures: Removal of the affected twigs and their burning is very effective for control of disease. Besides, application of Bordeaux mixture after pruning is very effective for control of the disease.

13.6.2 Muga Silkworm Grainage and Rearing Technology

The grainage and rearing technology is described below:

A. Grainage

The grainage technology of Thangavelu et al. (Thangavelu and Sahu 1983) is followed for seed production. The technology is described as under:

(a) *Garlanding of seed cocoons*

The seed cocoons are selected and garlands are made of 50 cocoons per garland. These garlands are hung keeping distance of 18–24 cm between the garlands.

(b) *Emergence*

The moths emerged on the 15th–17th day after spinning in different seasons from dusk till late night hours. The male and female moths are picked up on the basis of distinct morphological features and colour separately and are placed in equal ratio for pairing in basket.

(c) *Pairing*

The pairings are tied on Khorika made up of straw. The male moths are de-paired after 6 hours of pairing and female moths are allowed for egg laying. The female moths lay eggs in the evening hours and continue till morning. The eggs laid up to 3 days are considered for laying.

(d) *Mother moth examination*

After egg laying, female moths are examined under a microscope for trans-ovarian disease, viz. pebrine, for which abdominal portion of the mother moth is crushed in 2 % potassium hydroxide (KOH) solution and smear is made with the help of mortar and pestle and examined under 600 x magnifications with the help of a compound microscope.

(e) *Harvesting of eggs*

The disease-free layings are selected and eggs are detached from Khorika and surface sterilization is done with 2 % formaline solution for 2 min and dried under shade.

Plate 13.7 Commercial setup of grainage of *Antheraea assama*

(f) *Seed supply*
The disease-free layings are supplied to the farmers/government agencies in small bags prepared from markin cloth in cooler hours. The different steps of grainage are presented in Plate 13.7.

B. Incubation of eggs
The silkworm eggs are incubated under laboratory conditions at 25 ± 2 °C temperature and 75 ± 5 % relative humidity in semi-dark room in paper plates kept in basin.

C. Preparation for rearing
The base around the plants selected for rearing is cleaned before brushing the worms. The bleaching powder is sprinkled around the plants. The dead twigs/branches and yellow leaves are removed from the plant. The wasp nests are also destroyed. The rearing appliances are disinfected with 2 % formaldehyde solution. The disease-free eggs are put in paper plate along with a leaf.

D. Brushing of worms
The hatching of worms occurs early in the morning from 5 am to 8 am, and plates containing hatched worms are tied on host plants with the help of thread. Tender leaves are put in the paper plate for crawling worms over leaves and spread over the leaves of the plant.

E. Protection during rearing
The worms are protected by nylon net to prevent the loss on account of pests and predators attack. The muga silkworms after feeding have the tendency to come down from the tree, so a barrier of thatch grass around the trunk of host plants is

tied to avoid crawling of the worms down the barrier. The dead worms are collected and placed in a container containing 2 % formaldehyde solution and buried in a small pit away from the rearing site. The worms are transferred from one plant to another after complete eating of foliage. The constant vigil is made during the entire period. The worms pass last urine and come down from the plant when it attains maturity during dusk hours. The mature worms are collected and placed for cocooning in cocoonage, called Jali, made from semi dried twigs of mango and jackfruit and placed in aerated room.

F. Transfer of worms

The worms are transferred from one plant to another plant after consumption of foliage with the help of chaloni. The constant watch is kept at the rearing site to drive away the predators, like wasps, birds, bats, owl and monkeys. Gummy traps are used for wasps and *Uzi fly* for avoiding the loss.

G. Harvesting

The cocoons are harvested from the Jalies after 7–8 days and placed in single layer in trays. Parameters, viz. hatching (%), larval period and weight of larvae in different instars, cocoon yield per disease-free laying (dfl), cocoon weight, shell weight, pupae weight, effective rate of rearing (%), shell ratio percentage and absolute silk yield/300 larvae, are recorded during different seasons. The different steps of rearing are presented from Plates 13.8, 13.9 and 13.10.

H. Indoor rearing

The muga silkworm rearing is done in outdoor conditions. To check the initial loss of worms, indoor rearing was practised by inserting twigs of food plant in bottles filled with water to maintain freshness of the leaves. The indoor rearing up to third stage practice could not yield good results. The brushing in outdoor conditions is currently in practice.

Plate 13.8 Rearing of *Antheraea assama* in progress

Plate 13.9 Cocooning of *Antheraea assama*

Plate 13.10 Cocoons of *Antheraea assama*

13.7 Natural Enemies and Diseases of Muga Silkworm

The natural enemies and diseases of muga silkworm are given below:

(a) *Natural enemies*: The major parasites of muga silkworm are *Uzi fly* (*Exorista sorbillans*) and a braconid fly (*Apanteles glomeratus*), which cause considerable damage in cocoon crop. The predators of muga silkworm consist of several invertebrate and vertebrate species. The invertebrate are hornet (*Vespa orientalis*), yellow wasp (*Polistes hebraeus*), fire ant (*Solenopsis* sp.), carpenter ant (*Camponotus* sp.), red ant (*Oecophyllia smaragdina*), Pentatomidae (*Eocanthecona furcellata*), Reduviid (*Reduvius cincticrus*), Mantis (*Hierodula westwoodi*) and spiders.

The vertebrate predators are garden lizard (*Calotes versicolor*), wall lizard (*Hemidactylus flaviviridis*), red whiskered bulbul (*Pycnonotus jocosus*), house sparrow (*Passer domesticus*), rat (*Rattus rattus*), mouse (*Mus musculus*), snakes (*Bungarus fasciatus* and *Naja naja*), fox (*Vulpes bengalensis*), jackal (*Canis aureus*), flying fox (*Pteropus giganteus*), bat, monkey (*Macaca mulatta*, *M. radiate*), white cheek bulbul (*Pycnonotus leucogenys*), house myna (*Acridotheres tristis*), passeriform (*Amandava amandava*), white wagtail (*Motacilla alba*), drongo (*Dicrurus macrocercus*), golden oriole (*Oriolus xanthornus*), house crow (*Corvus splendens*), jungle crow (*Corvus macrorhynchos*), common owl (*Glaucidium radiatum*), screech owl (*Tyto alba*), kite (*Milvus migrans*) and tee pie (*Dendrocitta vagabunda*).

(b) *Uzi fly menace in muga culture and management practices*: *Uzi fly* (*Exorista sorbillans*), a tachinid, looks like a common housefly but is lightly bigger in size. The increase in the *Uzi fly* population in the Northeast India has adversely affected the indigenous muga silk industry where over 50–60 % of muga silkworm rearings are affected. *Uzi fly* maggots often fail to come out from the body of the silkworm larva before it starts spinning. This happens when the ovipositor by the *Uzi fly* occurs on the late fifth instar silkworm and the *Uzi fly* pupates within the silkworm cocoon; subsequently the adult fly fails to emerge from the silken cocoon leading to the death of *Uzi fly*, which is limited to 5 % destruction of the population.

Control measures: The muga silkworm rearing is outdoor where very effective measures cannot be adopted for control of the *Uzi fly*. The chemicals used as repellant of *Uzi fly* are also not very effective. The use of nylon nets during silkworm rearing can reduce the infestation of the *Uzi fly*.

(c) *Diseases of muga silkworm*: Muga silkworms are susceptible to various diseases caused by protozoan, bacteria, virus and fungus which are described as under:

(i) *Pebrine disease*: Pebrine is caused by a protozoan, *Nosema* sp. It is transmitted from the infected mother moth to the offspring by transovarian transmission. Besides, it is also transmitted through secondary infection. When the infection is primary, most of the worms die in the second and third instars, and in secondary infection, the worms spin the cocoon of inferior quality.

Control measures: The disease is controlled by resorting to strict mother moth examination and rearing of only disease-free layings. Besides, proper monitoring of the disease in each instar is also very effective for check of the disease.

(ii) *Flacherie*: Flacherie is caused by a virus followed by a secondary infection of bacteria. The incidence of the disease is high in summer months. Sudden fluctuations in temperature, bad weather, unsuitable leaf conditions and high water content in the leaf are the main causes for the virulence to the disease. The infected larvae become lethargic and motionless. The haemolymph of the heavily infected larva becomes black in colour. The larvae spin flimsy cocoon and die at the chrysalis stage when the infections are at the last larval stage. High atmospheric temperature and high humidity are favourable for the spread of this bacterial infection. During summer 20 to 30 % larval mortality occurs due to flacherie.

Control measures: Proper care of larvae at the early stage, protection of larvae from sun and rain, the use of disease-free laying and procurement of seed cocoons from the healthy zones are effective measures to minimize the incidence of the disease.

(iii) *Grasserie*: Grasserie is caused by virus. The haemolymph of the infected larva turns milky, and when examined under the microscope, numerous hexagonal crystals appear in the suspension. The incidence of the disease is high in summer when the atmosphere becomes highly humid.

The disease during Aheua, Bhodia (summer) and Jarua (winter) crops is more. The crop loss due to this disease in these seasons varies from 20 to 25 %. During summer and winter, the climatic conditions are quite unfavourable, and wide fluctuation in temperature within the short period of time affects the growth and development of silkworm. Besides, the quality of leaves also plays a vital role for the spread of grasserie disease. In the fifth stage silkworm feeds on tender, succulent and diseased leaves; the incidence of the grasserie is observed more. It will be better to avoid rearing on tall and old trees and ensure removal of tender and succulent leaves to reduce incidence of grasserie disease.

(iv) *Muscardine*: Incidence of muscardine is very less in muga silkworm rearing. The disease is caused by fungus. The proper surface disinfection will reduce the chances of the disease, as fungal spores will be destroyed.

(v) *Rectal protrusion*: The rectal part of the alimentary canal of larva comes out, and claspers loss the griping power and fall down and die. No effective control measures have been reported for the disease.

13.8 Post-Cocoon Analysis

The local advisory committee fixes the prices of different grades of the cocoons on the basis of size, compactness, silk content and reelability. Four thousand cocoons yielding 1 kg silk yarn are considered 'A' grade, 5000 cocoons yielding 1 kg silk yarn as B grade and other cocoons as C grade. The muga silk reeling involved cocoon stifling, degumming and reeling. The muga cocoons which are compact and leathery in structure and continuous silk filament vary from 350 to 450 m with 4–5 breaks. The reelability is low only 40–45 %. The performance of existing reeling machines and quality assessment of silk yarn yield on these machines is given below.

Plate 13.11 Golden Muga silk

Type of reeling machine	Quality parameters					
	Silk yarn production per 8 hours (gm)	Denier	Twist per inch	Tenacity (gm/denier)	Percent elongation at break	Breaking strength (gm)
Bhir	137.4	45.0	2.0 'Z'	4.66	23.00	209.80
Choudhury	160	50.3	5.0 'S'	4.63	30.90	233.20
Trivedi	118.6	71.6	5.0 'S'	4.00	N.A.	286.50
Bharali	76.1	96.7	4.0 'Z'	4.30	33.10	416.70
CMERS	114.8	63.2	3.6 'S'	3.70	32.20	231.80
Golden muga	105.3	64.3	2.5 'S'	4.59	34.40	295.70

The muga yarn is presented in Plate 13.11.

13.9 Conclusions

There are high scopes of muga culture in India, and there is a need to extend the culture in other parts of the country by augmentation of food plants and rearing for production of commercial cocoons. There is an urgent need to develop the muga silk industry for more benefits to the farmers, reelers and weavers associated with their livelihood with the culture, which will strengthen the rural economy by uplifting socio-economic condition of the farmers.

Acknowledgements Authors are grateful for the support provided by Smt. Nivedita Khanduri, T.A., ZSSO, Majra, Dehradun, Uttarakhand.

References

Anon (1962). Wealth of India. Vol. VI L-MP, pp 154–155

Barah A, Mech D, Suryanarayana N (2003) Yield gap and constraint analysis in Muga Culture. Abstracts. In: National conference on tropical sericulture for global competitiveness held at CSR&TI, Mysore on 5–7 November 2003, pp 107–108

Brandis D (1972) The forest flora of North West and Central India, pp 381

Chaudhary SN (1965) Studies on the propagation of Som (*Persea bombycina*) by stem cutting with the aid of growth regulators. Indian Agric 19(1):82–83

Chaudhary SN (1981) Muga silk industry. Directorate of Sericulture and Weaving, Govt. of Assam, Guwahati, India, pp 1–178

Giridhar K, Mahanta JC, Deole AK (2007) Raw silk production. Indian Silk 46(6):43–44

Isa MD, Thangavelu K (1988) North West Sub-Himalayan Belt-A potential zone for muga culture. Indian Silk 27:7–11

Isa M, Yadav G, Thangavelu K (1989) Seasonal effect on the rooting of air layering in *Machilus bombycina* King(som). Indian J Seric 28(2):139–144

Jolly MS, Sen SK, Sonwalkar TN, Prasad GK (1979) Non mulberry Silk – *FAO manual*, Agriculture Service Bull. Central Silk Board, Bangalore, pp. 124–125

Parpieve BA (1968) Shalk 39:15–17

Peigler RS (1996) Catalogue of parasitoid of Saturniidae of the world. J Res Lepidoptera 33:121

Paul DC, Subha Rao G, Deb DC (1992) J Insect Physiology 38(3):229–235

Raja R, Samson MV (1991) *Cinnamomum glanduliferum* Meissn as an primary host plant for *Antheraea assama* Ww. Indian J Seric 34(1):54–56

Sengupta AK, Srivastava AK, Raja R, Negi BK (1995) Muga –new strides. Indian Silk 34:28–32

Siddiqui AA (2012) Book on biodiversity of muga silkworm host plant and their utilization C.S.B. Bangalore

Siddiqui AA, Raja R, Basumatary BK, Sengupta AK (1993) Role of biotechnology in Muga food plants improvement. Indian Silk 32(1):33–36

Siddiqui AA, Basumatary BK, Babulal, PK D (1996) Juvenile shoot cutting and large scale plantation of Som. Indian Silk 35(6&7):15–17

Siddiqui AA, Babulal, Bhattacharya A, PK D (2000) Nutritional status in morphovariants of som. Indian Silk 37(8):27–29

Thangavelu K, Chakraborty AK (1986) Present status and future prospects of Muga Silk Industry. Indian Silk 24(9):10, 11, 23–26

Thangavelu K, Sahu AK (1983) Domestication of muga silkworm a major breakthrough in muga culture. Sericologia 23(2–3):153–158

Thangavelu K, Chakraborty AK, Bhagowati AK, Isa MD (1988) Handbook of muga culture. Central Silk Board, Bangalore, pp 1–101

Troops RS (1921) Silviculture of Indian trees: Lauraceae to Coniferales, III:154–155

Insect Crop Pollinators

14

M.S. Khan and Manish Kumar Yogi

Contents

14.1 Introduction

Pollination is a prerequisite for fruit and seed set, and in majority of the crop species, this vital service is performed by animals, the insects being the most important among them (Shivanna 2015). Data collected from 200 countries revealed that nearly 75 % of the 115 principal food crops of the world rely on animals for the potential produce, whereas about 25 % are wind- and self-pollinated (Klein et al. 2007). According to Crane and Walker (1983), a significant share of human food in tropical countries comes from insect-dependent crops. Insect pollination services in crop ecosystems increase the yield (Losey and Vaughan 2006; Gallai et al. 2009) and underpin $361 billion of crop production throughout the world (Lautenbach et al. 2012).

For maintaining pollination services, diverse fauna of bees (order Hymenoptera) is of greater importance (Kremen et al. 2002), the most important among them being the members of family Apidae (Klein et al. 2007) especially the honey bees. The honey bees have been well recognized by the people as productive insects since ancient times for their common valuable products, honey and wax. The honey bees also provide royal jelly, bee venom, propolis and pollen which along with honey and wax made them as perfect industrial insects. The honey bees are also an integral

M.S. Khan (✉) • M.K. Yogi
Department of Entomology, College of Agriculture, G.B. Pant University of Agriculture & Technology, Pantnagar 263145, Uttarakhand, India
e-mail: sarfrazms65@yahoo.co.in

Omkar (ed.), *Industrial Entomology*, DOI 10.1007/978-981-10-3304-9_14

component of the natural pollination systems of diverse flora including various crop species, and the benefits obtained through their pollination services are several times more than what we get together from honey and wax. In areas or in commercial crops where existing natural pollinating insect fauna is scarce, desired pollination levels are achieved through managed honey bee pollination systems to get maximum potential produce with good quality. The managed pollination systems using honey bees, stingless bees and some other solitary bees have now become a trade in many parts of the world.

The most commonly used bee species in managed pollination services are the European honey bee, *Apis mellifera* L., and the Asian honey bee, *Apis cerana* Fabricius. In the last few decades, there has been an increasing demand for pollination services in agriculture which in turn has led to a significant increase in population of managed honey bee hives the world over (Aizen et al. 2008). The honey bees are necessary for the production of many crops including apples, almonds, sweet cherries, plums, prunes, cucumbers, squash, pumpkins, melons, etc., and even one third of our daily diet depends on honey bees (Mussen 2007). However, the honey bees are not always the most efficient pollinators (Kearns and Inouye 1997). Another group of the bees from the family Apidae are the stingless bees that are considered important pollinators of many crops in the tropical and subtropical areas of the world (McGregor 1976) besides pollinating many wild plant species and can be taken up as an alternate to the honey bee pollination in future (Slaa et al. 2006).

In situations where honey bees are less in number or are not efficient pollinators, other pollinators like the bumble bees and many solitary wild bees from genera *Megachile*, *Osmia*, *Nomia* and *Xylocopa* have found more skilled pollinators than the honey bees, for example, in tomato, alfalfa, watermelon and coffee cropping systems in countries like Australia, the USA, Costa Rica and Indonesia (Heard 1999; Kremen et al. 2002, 2007; Klein et al. 2003, 2007; Ricketts 2004; Greenleaf and Kremen 2007). Next to the bees, the flies especially the syrphid flies (order Diptera) are considered as the most important flower visitors performing pollination services in many plants. Other insect groups, namely, butterflies (order Lepidoptera), wasps (order Hymenoptera) and the beetles (order Coleoptera), have shown their importance as potential pollinators in many plant species. Potential of insects for crop pollination is reviewed in this chapter with a view to signify their role to enhance crop production.

14.2 Crop Pollination by Different Groups of Insects

Bees are well-known insect pollinators providing pollination services. Among bees, social bees (honey bee, stingless bee and bumble bee) and solitary bees (leafcutter bee, mason bee, carpenter bee, alkali bee, digger bee and blue-banded bee) are important and effective pollinators of many crops; besides bees, syrphid fly, butterfly, wasp and some beetles are potential pollinators in agriculture and horticulture ecosystem (Table 14.1; Fig. 14.1). The major insect species having global importance as pollinators for crops grown for direct human consumption have been listed by Klein et al. (2007) and shown in Table 14.2.

Table 14.1 Major insect pollinators enhancing crop production in field/greenhouse

Sr. No.	Pollinator group	Scientific name	Pollinated crop	References
1	Honey bee	*Apis mellifera*	Coffee	Roubik (2002)
			Pumpkins	Walters and Taylor (2006)
		Apis cerana	Radish	Partap and Verma (1994)
			Cabbage and cauliflower	Verma and Partap (1994)
			Buckwheat	Rahman and Rahman (2000)
		Apis florea	Onion	Abrol (2010)
2	Stingless bee	*Trigona carbonaria*	Sweet pepper	Occhiuzzi (2000)
		Scaptotrigona aff. *depilis*	Cucumber	Santos et al. (2008)
3	Bumble bee	*Bombus impatiens*	Tomato	Morandin et al. (2001)
		Bombus terrestris	Hot pepper	Kwon and Saeed (2003)
4	Wild bee	*Amegilla holmesi*	Tomato	Bell et al. (2006)
		Amegilla chlorocyanea	Tomato	Hogendoorn et al. (2006)
		Megachile rotundata	Alfalfa	Goettel et al. (1991)
			Canola	Pitts-Singer and Cane (2011)
			Carrot	Tepedino (1997)
			Low-bush blueberry	MacKenzie (2009)
		Osmia cornuta	Almond and apple	Vicens and Bosch (2000)
			Pear	Maccagnani et al. (2003)
		Osmia lignaria	Cherry	Bosch et al. (2006)
			Almond and apple	Bosch and Kemp (2001)
		Osmia cornifrons	Tart cherry	Biddinger et al. (2013)
		Osmia aglaia	Raspberry and blackberry	Cane (2008b)
		Xylocopa virginica	Blueberry	Sampson et al. (2004)
		Xylocopa pubescens	Honeydew melon	Sadeh et al. (2007)
		Nomia melanderi	Alfalfa	Cane (2008a)
5	Syrphid fly	*Eristalis tenax*	Sweet pepper	Jarlan et al. (1996)
6	Butterfly	*Meneris tulbaghia*	Fynbos sp.	Johnson and Bond (1994)
7	Wasp	*Blastophaga psenes*	Fig	Nefdt and Compton (1996)
8	Beetle	*Cyclocephala* sp.	*Cyclanthus bipartitus*	Beach (1982)

(Source: Khan and Srivastava, 2013)

Fig. 14.1 Major insect pollinators

Table 14.2 List of insect pollinators for global crops that are grown for direct human consumption

Sr. No.	Pollinator group	Species
1	Honey bee	*Apis cerana*, *Apis dorsata*, *Apis florea* and *Apis mellifera*
2	Stingless bee	*Melipona quadrifasciata*, *Melipona favosa*, *Melipona subnitida*, *Nanotrigona perilampoides*, *Nanotrigona testaceicornis*, *Trigona cupira*, *Trigona iridipennis*, *Trigona terminata*, *Trigona minangkabau*, *Trigona toracica* and *Scaptotrigona depilis*
3	Bumble bee	*Bombus affinis*, *Bombus californicus*, *Bombus hortorum*, *Bombus hypnorum*, *Bombus impatiens*, *Bombus lapidarius*, *Bombus pascuorum*, *Bombus sonorus*, *Bombus terrestris* and *Bombus vosnesenskii*
4	Wild bee	*Amegilla holmesi*, *Amegilla chlorocyanea*, *Megachile rotundata*, *Megachile addenda*, *Osmia cornuta*, *Osmia lignaria*, *Osmia cornifrons*, *Osmia aglaia*, *Xylocopa virginica*, *Xylocopa pubescens*, *Centris tarsata*, *Nomia melanderi*, *Halictus tripartitus*, *Andrena ilerda*, *Anthophora pilipes*, *Creightonella frontalis* and *Habropoda laboriosa*
5	Syrphid fly	*Eristalis tenax* and *Eristalis cerealis*
6	Wasp	*Blastophaga psenes*
7	Beetle	*Carpophilus hemipterus* and *Carpophilus mutilatus*

Source: Klein et al. (2007)

14.2.1 Honey Bees

The honey bees (genus *Apis*) being social in nature require ample food (pollen and nectar) to nourish the brood and other nest mates. For this, they need to make frequent visits to the flowering plants. As such, the honey bees have a wide foraging range, and by doing this they perform pollination in diverse plant species including cross-pollinated crop plants in agriculture, horticulture and other ecosystems. The morphological, ecological and biological adaptations in the honey bees often make them reliable, economical and, in most instances, more efficient pollinators in a variety of plant species contributing in increased yield along with improved quality of the produce. Seed production of many vegetables (asparagus, carrots, celery, onions, radishes and turnips) and forage crops (alfalfa, clovers, trefoil and vetch) is coupled with honey bee pollination. Honey bee pollination is also associated with the fruit-bearing ability, size of fruits and proper maturity of fruits/berries. A number of ornamental plants also require honey bee pollination to produce the fruits for birds and other beneficial animals (Mussen 2007).

In absence of honey bees, the yields of some fruit, seed and nut crops are decreased up to 90 % (Southwick and Southwick 1992). The managed honey bee hives increased the fruit size, seed number and fruit shape (Free 1993) and ensured the better crop pollination (Klein et al. 2007). Honey bees, mainly *A. mellifera*, are the most economical and valuable pollinators of the crops (Watanabe 1994; Roubik 2002), and the worldwide beekeeping industry makes honey bees as the most important commercial pollinators (Richards 2001). Supplementing natural pollination with *A. mellifera* colonies ensured the maximum fruit size and fruit yield of pumpkins (Walters and Taylor 2006).

Introduction of *A. cerana* colony in caged condition gave the greater pod set, number of seeds per pod and seed weight for radish (*Raphanus sativus* L.) plants, which were 45 %, 42 % and greater than 45 %, respectively, than for open-pollinated plants (Partap and Verma 1994). Fruit sets on cauliflower and cabbage plants pollinated by *A. cerana* were also documented 20 % and 27 % higher productivity, respectively, than on open-pollinated plants (Verma and Partap 1994). Rahman and Rahman (2000) investigated the effect of honey bee (*A. cerana indica*) on seed set and yield of buckwheat (*Fagopyrum esculentum* Moench) and demonstrated that yield was highest (1.42 t/ha) in bee-pollinated followed by in open-pollinated (0.93 t/ha) and self-pollinated plants (0.64 t/ha).

Besides the domesticated honey bees (*A. mellifera* and *A. cerana*), the commonly occurring wild honey bee species, *Apis dorsata* Fabricius and *Apis florea* Fabricius, also play significant role in pollination of many plant species including some important field crops. Crane (1991) reported that in orchard and field crops, the dwarf honey bee, *A. florea*, is an efficient pollinator. Higher seed production was obtained in onion (*Allium cepa* L.) plants visited by *A. florea* than those excluded from their visits (Abrol 2010). Likewise in coffee, the wild honey bee *A. dorsata* dominated (58 % of total) the flower visitors and is the major pollinator of this crop in South India (Krishnan et al. 2012). In Lyallpur, Pakistan, more fruit set and increased number of seeds were observed in Kinnow mandarin (*Citrus reticulata* Blanco) branches when they were accessible to *A. dorsata* and *A. florea* (Manzoorul-Haq et al. 1978). In India, *A. mellifera* and *A. cerana* are being commercially used for pollination services in fruits (apple, litchi, pear, peach, plum, etc.) and vegetables (radish, cabbage, cauliflower, etc.).

14.2.2 Stingless Bees

The stingless bees have several advantages over the honey bees (Inoue et al. 1984; Delfinado-Baker et al. 1989; Kakutani et al. 1993) that enable them to work as more efficient pollinators in the honey bee crisis situations (Heard 1999). Pollination efficiency (fruit set after a single flower visit) of *Trigona* (*Lepidotrigona*) *terminata* Smith in coffee was reported significantly higher among the 15 native bee species of Indonesia, and the species is considered as the most efficient pollinator with 80 % fruit set (Klein et al. 2003). Can-Alonso et al. (2005) suggested that the native *Trigona nigra* Schwarz is the potentially efficient pollinator of avocado crop. Due to short foraging range, the stingless bees are considered as efficient pollinators for protected cultivation. Occhiuzzi (2000) reported that *Trigona carbonaria* Smith efficiently pollinated the sweet pepper giving 11 % higher fruit weight and 34 % higher seeds/fruit under glass greenhouse conditions in Australia.

In tomato (var. Rodas), *Melipona quadrifasciata* (Lepeletier) was as good pollinator as the yield achieved through hand pollination (Sarto et al. 2005), and the use of stingless bees has been advocated for commercial greenhouse tomatoes and cucumber in Brazil (Santos et al. 2008, 2009). In India, the stingless bee, *Tetragonula iridipennis* (Smith), is used by the farmers for pollination in coconut plantations.

14.2.3 Bumble Bees

Depending upon their tongue length and suitability of the flowers, the bumble bees are able to pollinate a wide range of temperate crops since they can forage in very cold conditions (Willmer et al. 1994). The short-tongued bumble bee, *Bombus terrestris* (L.), pollinates oilseed rape, while species with medium or long tongues (common carder bee, *Bombus pascuorum* (Scop.), or garden bumble bee, *Bombus hortorum* (L.)) are considered necessary to pollinate field beans and red clover (Bohart 1972; Fussell and Corbet 1991). The bumble bees were found better pollinators than the honey bees, and hence they are widely used for pollination in tomatoes (*Lycopersicon esculentum* Mill.) grown in greenhouses (Banda and Paxton 1991; Dafni 1998).

Morandin et al. (2001) recorded 50–70 % more fruit set in tomato flowers visited by the bumble bee, *Bombus impatiens* Cresson. Using the *B. impatiens* for pollination in pepper also gave significantly higher fruit mass (Meisels and Chiasson 1997). Kwon and Saeed (2003) recorded *B. terrestris* as efficient pollinator of hot pepper, *Capsicum annuum* L., cultivated in greenhouse as it produced 47.8 % more number of seeds and 27.2 % greater fruit weight as compared to control. Earlier

Abak et al. (1997) observed 19 % higher weight of fruit and 52 % more seeds in pepper pollinated by the bumble bee, *B. terrestris*.

In Europe, *B. terrestris* colonies in tomato greenhouses subsequently replaced the manual pollination since 1987 (Van Heemert et al. 1990). Several species of *Bombus* are commercially available in the USA and have a dramatic effect on the greenhouse tomato (Kueneman 1995). *B. impatiens* is a highly successful commercially available pollinator for greenhouse crops, viz., tomatoes, *L. esculentum* (Kevan et al. 1990); muskmelons, *Cucumis melo* L. (Fisher and Pomeroy 1989); and sweet peppers, *Capsicum annuum* L. (Meisels and Chiasson 1997). In Southern Ontario, bumble bee (*B. impatiens*) pollination substituted the hand pollination of commercial tomato crops and created a new multimillion-dollar supply business (Morandin et al. 2001).

In self-fertile tomatoes, bumble bees sonicated (buzz pollination, a process in which the bee grasps the flower tightly and rapidly vibrates the anthers by its flight muscles) the flowers, and the resulted cross-pollination increased the fruit set by 45 % and fruit weight by 200 % (Greenleaf and Kremen 2006). The bumble bee hives in addition to the honey bee colonies in pear orchards improved the percentage of fruit set and fruit size (Zisovich et al. 2012).

14.2.4 Solitary/Wild Bees

Native wild bees provided noticeably adequate crop pollination in the shortages of honey bees (Kearns et al. 1998). As the honey bees are threatened by colony collapse disorder, it is envisaged that native bees will act as alternate pollinating agents in agriculture. The experiments by Winfree et al. (2007) at 23 farms in New Jersey and Pennsylvania resulted that native wild bees alone provide sufficient pollination at more than 90 % of the farms for watermelon crops. Gemmill-Herren and Ochieng (2008) found *Xylocopa caffra* (L.) and *Macronomia rufipes* (Smith) as most frequent pollinators in eggplant and used them in pollination effectiveness tests. The solitary bees currently managed for pollination of various crops are the leafcutter bee (*Megachile rotundata* (F.)), alkali bee (*Nomia melanderi* Cockerell) and several mason bees (*Osmia* spp.; Maccagnani et al. 2007). Different types of wild bees with their pollinated crops are described here in brief.

14.2.4.1 Blue-Banded Bees

In Australia, the blue-banded bee, *Amegilla holmesi* Rayment, is an efficient substitute to mechanical pollination for greenhouse tomatoes (Bell et al. 2006). *Amegilla chlorocyanea* (Cockerell) pollination provided considerably heavier and seedier tomatoes compared with wand pollination with a total increase of 20–24 % yield (Hogendoorn et al. 2006).

14.2.4.2 Leafcutter Bees

The most extensively and commercially used leafcutter bee is the *M. rotundata* that has revolutionized alfalfa seed production in the USA and Western Canada (Goettel et al. 1991; Bosch and Kemp 2005). *M. rotundata* is also used extensively for hybrid seed production of canola (*Brassica napus* L.) in Western Canada (Pitts-Singer and Cane 2011). Caged *M. rotundata* effectively pollinated carrot (Tepedino 1997) and canola (Soroka et al. 2001) for hybrid seed production. In Northern Alberta, the use of *M. rotundata* in red clover, *Trifolium pratense* L., gave higher seed yield as compared to control where this pollinator was not used (Fairey et al. 1989). *M. rotundata* improved wild low-bush blueberry (*Vaccinium angustifolium* Aiton) fruit set by 30 % over open pollination. Fruit set was significantly higher in plots visited by migratory *M. rotundata* (forage provided before blueberry bloom) than in plots visited by non-migratory *M. rotundata* (no forage provided before bloom) (Stubbs and Drummond 1997). *M. rotundata* is being used successfully to pollinate *V. angustifolium* particularly in New Brunswick and Quebec (MacKenzie 2009).

14.2.4.3 Mason Bees

European orchard bee, *Osmia cornuta* (Latreille), is considered as an efficient pollinator of almond and apple (Vicens and Bosch 2000). In northern and southern Honshu, desired pollination in apple has been achieved successfully through managed pollination using horn-faced bee, *Osmia cornifrons* (Radoszkowski) (Bohart 1972). Maccagnani et al. (2003) reported that pear (*Pyrus communis* L.) fruit set and seed set were highest in the caged trees with *O. cornuta*, followed by open-pollinated trees and caged trees without bees. Over a 5-year period, the use of blue orchard bee, *Osmia lignaria* Say, in sweet cherry (*Prunus avium* L.) in Utah increased the yield twofold (Bosch et al. 2006), whereas Biddinger et al. (2013) reported *O. cornifrons*-pollinated tart cherry (*Prunus cerasus* L.) trees gave higher cherries than the control orchard. Berry bee, *Osmia aglaia* Sandhouse, shows promise as a manageable and effective pollinator for commercial raspberries and blackberries (*Rubus*) (Cane 2008b).

14.2.4.4 Large Carpenter Bees

The native *Xylocopa* bees of Australia are effective pollinators of tomatoes (Hogendoorn et al. 2000). Eastern carpenter bee, *Xylocopa virginica* L. pollination enhanced fruit set and seed production of ocotillo, *Fouquieria splendens* Engelm and Virginia bluebell, *Mertensia virginica* (L.) Pers. (Enz 2001). Sampson et al. (2004) study shows that transfer of rabbiteye blueberry (*Vaccinium ashei* Reade) pollen by male carpenter bees (*X. virginica*) during floral robbery resulted in greater fruit set. Sadeh et al. (2007) reported that *Xylocopa pubescens* Spinola pollination increased greenhouse-grown honeydew melon (*Cucumis melo* L.) fruit set threefold as compared to honey bee (*A. mellifera*) pollination. The large carpenter bees also effectively pollinate cucurbitaceous and bean crops.

14.2.4.5 Small Carpenter Bees

The small carpenter bee, *Ceratina* sp., was reported to be the major pollinator of *Acacia* hybrid (Sornsathapornkul and Owens 1998) and teak (Tangmitcharoen and Owens 1997) at a plantation in Central Thailand. Both studies indicated that these pollinators contributed to a high amount of selfing because this insect mostly visited flowers within the same inflorescence or within the same tree (Tangmitcharoen et al. 2009). Hussain et al. (2016) reported *Ceratina smaragdula* (Fabricius) as a potentially important pollinator of leguminous and cucurbit crops in north-western Pakistan. Small carpenter bees are efficient pollinators of many crops such as beans, cowpeas, apples and coffee, and are likely to contribute to increased agricultural productivity (Eardley et al. 2009).

14.2.4.6 Alkali Bees and Mining Bees

The only managed ground-nesting alkali bee, *N. melanderi*, is used for effective pollination of alfalfa (*Medicago sativa* L.) crop grown for seed in the Western USA and New Zealand (Maeta 1990; Cane 2002, 2008a). Gardner and Ascher (2006) surveyed apple orchards in New York and found that *Andrena* spp. (mining bees) are the most numerous among available honey bees and native bee species.

14.2.5 Syrphid Flies

Syrphid flies are frequent flower visitors of wild plants and agricultural crops and are often considered as the second most important group of pollinators after the bees. It is considered that bees are able to carry a greater volume of pollen on their bodies, but flies are able to compensate for this by making a greater number of flower visits (Larson et al. 2001). Jarlan et al. (1996) tested European hover fly, *Eristalis tenax* (L.), as a pollinator of sweet pepper under glasshouse conditions in southern Quebec and found that seed setting and fruit weight were higher in *E. tenax*-visited flowers compared to unvisited flowers. Furthermore, the duration of hover flies visits significantly increased both seed set and fruit weight. Jarlan et al. (1997) reported that pepper flowers visited by syrphid flies produced larger seed sets than those non-visited flowers.

In alpine zones of New Zealand, syrphid flies are the most common visitors to flowers (Campbell et al. 2010), whereas Morris (2000) observed that the syrphid flies, *Melanostoma fasciatum* (Macquart) and *Melangyna novaezelandiae* (Macquart), frequently visit the flowers in agri-ecosystems in New Zealand. The orchid, *Govenia utriculata* (Sw.) Lindl., is entirely visited and pollinated by syrphid flies of the genus *Sapingogaster* (Pansarin 2008).

14.2.6 Butterflies

Butterflies are nectar/pollen feeders of both wild and cultivated plant species and sometimes are considered important pollinators (Munyuli 2012). *Thymelicus flavus* Brunnich (small skipper butterfly) visit many flower species showing floral constancy to greater extent (Goulson et al. 1997). Butterflies are the primary pollinators of peacock flowers, *Caesalpinia pulcherrima* (L.) Sw., and the pollen are carried primarily on their wings. Of the numerous species that visit the flowers, members of Papilionidae are the most important pollinators (Cruden and Hermann-Parker 1979).

The butterfly, *Meneris tulbaghia* (L.), is a potential candidate for conservation as it is the exclusive pollinator of many red-flowered *Fynbos* species. Such kind of specific interaction for pollination is hardly found in plants (Johnson and Bond 1994). Pollination of *Gladiolus* (Iridaceae, Crocoideae) by lepidopteran insects comprises two entirely different sets of pollinators, night-flying moths (Noctuidae or Sphingidae) and butterflies (evidently only one species of Satyridae) (Goldblatt and Manning 2002).

14.2.7 Wasps

Agaonine fig wasps are the specialist pollinators of fig trees. The fig trees' ovaries are the feeding sites (galled seeds) of fig wasps; these wasps oviposit into the ovaries via their styles and also transfer pollens between trees. The production of fig seeds is dependent on the number of accessible ovaries and the average number of female wasps entering each fig (Nefdt and Compton 1996). Such an interaction of insects and plant seems most specialized case of mutualism (Herre et al. 1996) as pollination solely depends on female fig wasps (Molbo et al. 2003).

The morphological adaptations of different species of wasp to fig trees are associated with active and passive pollination. In active pollination, the wasps possess specialized structures for carrying pollen in the external part of the thorax and the front legs (Ramirez 1969) and show distinctive behaviour for collecting and depositing pollen (Frank 1984). The male flowers in actively pollinated figs are relatively small and less numerous (Galil and Meiri 1981). In passive pollination, figs have relatively higher ratios of anthers to female flowers and produce much more pollen, and their mature anthers tend to dehisce naturally, thereby facilitating the passive collection of pollen (pollen adheres to various parts of the body surface) by their pollinators (Ramirez and Malavasi 1997).

14.2.8 Beetles

Beetles are potential pollinators and visitors of a variety of flowers since earliest times and are also observed in advanced angiosperms (Gottsberger 1977). *Annona* spp. with characteristic odours of flowers and wide floral chambers are exclusively pollinated by dynastid scarab beetles, *Cyclocephala* spp. (Gottsberger 1989). Likewise, scarabaeid beetle, *Amphicoma* sp., is the primary pollinator of *Anemone coronaria* L., *Papaver rhoeas* L., *Ranunculus asiaticus* L. and *Tulipa agenensis* Redo, and the halictid bee *Lasioglossum marginatum* Br. and anthophorid bee *Synhalonia plumigera* Kohl are the secondary pollinators in these plant species (Dafni et al. 1990).

Floral odours play an important role as an attractant to pollinating beetles (*Cyclocephala* sp. and *Erioscelis* sp.) in *Dieffenbachia longispatha* Engler and Krause, a tropical flowering plant. These beetles fly to the inflorescence in darkness and become covered with pollen as they crawl up the spadix, and then the beetles fly to the nearest female inflorescence and pollinate the flower (Young 1986). The flowers of *Orchidantha inouei* Nagam and Sakai (Lowiaceae, Zingiberales), lacking floral nectar and having dung-like odour, are pollinated by the scarabaeid dung beetles, *Onthophagus* spp. (Sakai and Inoue 1999). Beach (1982) experiments on *Cyclanthus bipartitus* Poit. (stemless perennial plant) with bagged inflorescences showed that no seeds are produced in the absence of *Cyclocephala* (Scarabaeidae) beetles, which suggests that *Cyclanthus* is obligately allogamous.

14.3 Conclusion

In this chapter, important insect pollinators providing pollination services in different agricultural and horticultural ecosystems are reviewed. The pollination potential of various insect pollinators and the increase in crop yields have been demonstrated with suitable examples. In addition, the industrial use of insects in terms of pollination services in various crops has been highlighted. For optimum utilization of available indigenous insect fauna for sustaining plant biodiversity and crop production enhancement, systematic studies on individual crop and their pollinator relationships are required to obtain more precision in pollination management of major crops.

References

Abak K, Dasgan H, Ikiz O, Uygun N (1997) Pollen production and quality of pepper grown in unheated greenhouses during winter and the effect of bumble bees (*Bombus terrestris*) pollination on fruit yield and quality. Acta Hortic 437:303–307

Abrol DP (2010) Foraging behaviour of *Apis florea* F., an important pollinator of *Allium cepa* L. J Apic Res 49(4):318–325

Aizen MA, Garibaldi LA, Cunningham SA, Klein AM (2008) Long-term global trends in crop yield and production reveal no current pollination shortage but increasing pollinator dependency. Curr Biol 18:1572–1575

Banda HJ, Paxton RJ (1991) Pollination of greenhouse tomatoes by bees. Acta Hortic 288:194–198

Beach JH (1982) Beetle pollination of *Cyclanthus bipartitus* (Cyclanthaceae). Am J Bot 69(7):1074–1081

Bell MC, Spooner-Hart RN, Haigh AM (2006) Pollination of greenhouse tomatoes by the Australian bluebanded bee *Amegilla* (*Zonamegilla*) *holmesi* (Hymenoptera: Apidae). J Econ Entomol 99(2):437–442

Biddinger DJ, Joshi NK, Rajotte EG, Halbrendt NO, Pulig C, Naithani KJ, Vaughan M (2013) An immunomarking method to determine the foraging patterns of *Osmia cornifrons* and resulting fruit set in a cherry orchard. Apidologie 44(6):738–749

Bohart GE (1972) Management of wild bees for the pollination of crops. Annu Rev Entomol 17:287–312

Bosch J, Kemp WP (2001) How to manage the blue orchard bee as an orchard pollinator, Handbook No. 5. Sustainable Agriculture Network, Beltsville

Bosch J, Kemp WP (2005) Alfalfa leafcutting bee population dynamics, flower availability, and pollination rates in two Oregon alfalfa fields. J Econ Entomol 98(4):1077–1086

Bosch J, Kemp WP, Trostle GE (2006) Bee population returns and cherry yields in an orchard pollinated with *Osmia lignaria* (Hymenoptera: Megachilidae). J Econ Entomol 99(2):408–413

Campbell D, Bischoff M, Lord JM, Robertson AW (2010) Flower color influences insect visitation in alpine New Zealand. Ecology 91(9):2638–2649

Can-Alonso C, Quezada-Euan JJG, Xiu-Ancona P, Moo-Valle H, Valdovinos-Nunez GR, Medina-Peralta S (2005) Pollination of 'criollo' avocados (*Persea americana*) and the behaviour of associated bees in subtropical Mexico. J Apic Res 44:3–8

Cane JH (2002) Pollinating bees (Hymenoptera: Apiformes) of US alfalfa compared for rates of pod and seed set. J Econ Entomol 95(1):22–27

Cane JH (2008a) A native ground-nesting bee (*Nomia melanderi*) sustainably managed to pollinate alfalfa across an intensively agricultural landscape. Apidologie 39(3):315–323

Cane JH (2008b) An effective, manageable bee for pollination of *Rubus* bramble fruits, *Osmia aglaia*. Acta Hortic 777:459–464

Crane E (1991) *Apis* species of tropical Asia as pollinators, and some rearing methods for them. Acta Hortic 288:29–48

Crane E, Walker P (1983) The impact of pest management on bees and pollination. Tropical Development and Research Institute, College House, Wrights Lane, London

Cruden RW, Hermann-Parker SM (1979) Butterfly pollination of *Caesalpinia pulcherrima*, with observations on a psychophilous syndrome. J Ecol 67(1):155–168

Dafni A (1998) The threat of *Bombus terrestris* spread. Bee World 79:113–140

Dafni A, Bernhardt P, Shmida A, Ivri Y, Greenbaum S, O'Toole C, Losito L (1990) Red bowl-shaped flowers: convergence for beetle pollination in the Mediterranean region. Israel J Bot 39(1-2):81–92

Delfinado-Baker M, Baker EW, Phoon ACG (1989) Mites (Acari) associated with bees (Apidae) in Asia, with description of a new species. Am Bee J 129:609–613

Eardley CD, Gikungu MW, Schwarz MP (2009) Bee conservation in Sub-Saharan Africa and Madagascar: diversity, status and threats. Apidologie 40:355–366

Enz J (2001) Nectar robbers and non-nectar robbers: their visitation rates and effects on seed set in Virginia bluebells, *Mertensia virginica* (L.). In: Pers. Botany 2001 Conference, Plants and People, August 12–16, 2001, Albuquerque, NM

Fairey DT, Lefkovitch LP, Lieverse JAC (1989) The leafcutting bee, *Megachile rotundata* (F.): a potential pollinator for red clover. J Appl Entomol 107(1–5):52–57

Fisher RM, Pomeroy N (1989) Pollination of greenhouse muskmelons by bumble bees (Hymenoptera: Apidae). J Econ Entomol 82(4):1061–1066

Frank SA (1984) The behavior and morphology of the fig wasps *Pegoscapus assuetus* and *P. jimenezi*: descriptions and suggested behavioral characters for phylogenetic studies. Psyche 91:298–307

Free JB (1993) Insect pollination of crops, 2nd edn. Academic Press, London

Fussell M, Corbet SA (1991) Forage for bumble bees and honey bees in farmland: a case study. J Apic Res 30:87–97

Galil J, Meiri L (1981) Number and structure of anthers in fig syconia in relation to behaviour of the pollen vector. New Phytol 88:83–87

Gallai N, Salles JM, Settele J, Vaissiere BE (2009) Economic valuation of the vulnerability of world agriculture confronted with world pollinator decline. Ecol Econ 68:810–821

Gardner KE, Ascher JS (2006) Notes on the native bee pollinators in New York apple orchards. J N Y Entomol Soc 114(1):86–91

Gemmill-Herren B, Ochieng AO (2008) Role of native bees and natural habitats in eggplant (*Solanum melongena*) pollination in Kenya. Agric Ecosyst Environ 127(1):31–36

Goettel MS, Richards KW, Schaalje GB (1991) Bioassay of selected fungicides for control of chalkbrood in alfalfa leafcutter bees, *Megachile rotundata*. Apidologie 22(5):509–522

Goldblatt P, Manning JC (2002) Evidence for moth and butterfly pollination in *Gladiolus* (Iridaceae-Crocoideae). Ann Mo Bot Gard 89(1):110–124

Gottsberger G (1977) Some aspects of beetle pollination in the evolution of flowering plants. In: Plant systematics and evolution, vol 1. Springer, Vienna, pp 211–226

Gottsberger G (1989) Beetle pollination and flowering rhythm of *Annona* spp. (*Annonaceae*) in Brazil. Plant Syst Evol 167(3):165–187

Goulson D, Ollerton J, Sluman C (1997) Foraging strategies in the small skipper butterfly, *Thymelicus flavus*: when to switch? Anim Behav 53:1009–1016
Greenleaf SS, Kremen C (2006) Wild bee species increase tomato production and respond differently to surrounding land use in Northern California. Biol Conserv 133(1):81–87
Greenleaf SS, Kremen C (2007) Wild bees enhance honey bees' pollination of hybrid sunflower. Proc Natl Acad Sci 103(37):13890–13895
Heard TA (1999) The role of stingless bees in crop pollination. Annu Rev Entomol 44(1):183–206
Herre EA, Machado CA, Bermingham E, Nason JD, Windsor DM, McCafferty SS, VanHouten W, Bachmann K (1996) Molecular phylogenies of figs and their pollinator wasps. J Biogeogr 23:521–530
Hogendoorn K, Steen Z, Schwarz MP (2000) Native Australian carpenter bees as a potential alternative to introducing bumble bees for tomato pollination in greenhouses. J Apic Res 39:67–74
Hogendoorn K, Gross CL, Sedgley M, Keller MA (2006) Increased tomato yield through pollination by native Australian *Amegilla chlorocyanea* (Hymenoptera: Anthophoridae). J Econ Entomol 99(3):828–833
Hussain A, Abdulaziz SA, Mohamed S, Michael SE (2016) Notes on the nesting biology of the small carpenter bee *Ceratina smaragdula* (Hymenoptera: Apidae) in northwestern Pakistan. Fla Entomol 99(1):89–93
Inoue T, Sakagami SF, Salmah S, Nukmal N (1984) Discovery of successful absconding in the stingless bee *Trigona* (*Tetragonula*) *laeviceps*. J Apic Res 23:136–142
Jarlan A, De Oliveira D, Gingras J (1996) Pollination of sweet pepper (*Capsicum annuum* L.) in green-house by the syrphid fly (*Eristalis tenax* L.). Acta Hortic 437:335–339
Jarlan A, De Oliveira D, Gingras J (1997) Pollination by *Eristalis tenax* (Diptera: Syrphidae) and seed set of greenhouse sweet pepper. J Econ Entomol 90(6):1646–1649
Johnson SD, Bond WJ (1994) Red flowers and butterfly pollination in the fynbos of South Africa. In: Plant-animal interactions in Mediterranean-type ecosystems, vol 31. Springer, Dordrecht, pp 137–148
Kakutani T, Inoue T, Tezuka T, Maeta Y (1993) Pollination of strawberry by the stingless bee, *Trigona minangkabau*, and the honey bee, *Apis mellifera*: an experimental study of fertilization efficiency. Res Popul Ecol 35:95–111
Kearns CA, Inouye DW (1997) Pollinators, flowering plants, and conservation biology. Bioscience 47:297–307
Kearns C, Inouye D, Waser N (1998) Endangered mutualisms: the conservation biology of plant-pollinator interactions. Annu Rev Ecol Syst 29:83–112
Kevan PG, Clark E, Thomas VG (1990) Insect pollinators and sustainable agriculture. Am J Altern Agric 5(1):13–22
Khan MS, Srivastava P (2013) Insect pollinators and crop pollination, Research Bulletin No. 189. Directorate of Research, G.B. Pant University of Agriculture & Technology, Pantnagar, pp 70
Klein AM, Steffan-Dewenter I, Tscharntke T (2003) Bee pollination and fruit set of *Coffea arabica* and *C. canephora* (Rubiaceae). Am J Bot 90:153–157
Klein AM, Vaissiere BE, Cane JH, Steffan-Dewenter I, Cunningham SA, Kremen C, Tscharntke T (2007) Importance of pollinators in changing landscapes for world crops. Proc R Soc Lond B 274:303–313
Kremen C, Williams NM, Thorp RW (2002) Crop pollination from native bees at risk from agricultural intensification. Proc Natl Acad Sci 99(26):16812–16816
Kremen C, Williams NM, Aizen MA, Gemmill-Herren B, LeBuhn G, Minckley R, Packer L, Potts SG, Roulston T, Steffan-Dewenter I, Vazquez DP, Winfree R, Adams L, Crone EE, Greenleaf SS, Keitt TH, Klein A-M, Regetz J, Ricketts TH (2007) Pollination and other ecosystem services produced by mobile organisms: a conceptual framework for the effects of land-use change. Ecol Lett 10:299–314

Krishnan S, Kushalappa CG, Shaanker RU, Ghazoul J (2012) Status of pollinators and their efficiency in coffee fruit set in a fragmented landscape mosaic in South India. Basic Appl Ecol 13(3):277–285

Kueneman T (1995) The pollinator. Greenhouse Production News, (Sept), pp 8–10

Kwon YJ, Saeed S (2003) Effect of temperature on the foraging activity of *Bombus terrestris* L. (Hymenoptera: Apidae) on greenhouse hot pepper (*Capsicum annuum* L.). Appl Entomol Zool 38(3):275–280

Larson BMH, Kevan PG, Inouye DW (2001) Flies and flowers: taxonomic diversity of anthophiles and pollinators. Can Entomol 133:439–465

Lautenbach S, Sepplett R, Leibscher J, Dormann C (2012) Spatial and temporal trends of global pollination benefit. PLoS One 7(4):e35954

Losey JE, Vaughan M (2006) The economic value of ecological services provided by insects. Bioscience 56(4):311–323

Maccagnani B, Ladurner E, Santi F, Burgio G (2003) *Osmia cornuta* (Hymenoptera, Megachilidae) as a pollinator of pear (*Pyrus communis*): fruit-and seed-set. Apidologie 34(3):207–216

Maccagnani B, Burgio G, Stanisavljevic LZ, Maini S (2007) *Osmia cornuta* management in pear orchards. Bull Insectolo 60(1):77–82

MacKenzie K (2009) Pollination practices and the use of bees in *Vaccinium* crops. Acta Hortic 810:527–538

Maeta Y (1990) Utilization of wild bees. Farming Japan 24(6):13–22

Manzoorul-Haq, Rafie-ul-Din M, Ghaffar A (1978) Effect of insect pollination on fruit bearing in Kinnow mandarin (*Citrus reticulata*), and physical and chemical properties of the fruit. J Apic Res 17(1):47–49

McGregor SE (1976) Insect pollination of cultivated crop plants, Agriculture Handbook No. 496. US Department of Agriculture, Washington, DC

Meisels S, Chiasson H (1997) Effectiveness of *Bombus impatiens* Cr. as pollinators of greenhouse sweet peppers (*Capsicum annuum* L.). Acta Hortic 437:425–429

Molbo D, Machado CA, Sevenster JG, Keller L, Herre EA (2003) Cryptic species of fig-pollinating wasps: implications for the evolution of the fig–wasp mutualism, sex allocation, and precision of adaptation. Proc Natl Acad Sci 100(10):5867–5872

Morandin LA, Laverty TM, Kevan PG (2001) Bumble bee (Hymenoptera: Apidae) activity and pollination levels in commercial tomato greenhouses. J Econ Entomol 94(2):462–467

Morris MC (2000) Coriander (*Coriandrum sativum*) "companion plants" can attract hover flies, and may reduce infestation in cabbages. N Z J Crop Hortic Sci 28:213–217

Munyuli T (2012) Assessment of indicator species of butterfly assemblages in coffee–banana farming system in central Uganda. Afr J Ecol 50:77–89

Mussen EC (2007) Don't underestimate the value of honey bees! Notes of University of California, Davis, pp 1–2

Nefdt RJ, Compton SG (1996) Regulation of seed and pollinator production in the fig-fig wasp mutualism. J Anim Ecol 65:170–182

Occhiuzzi P (2000) Stingless bees pollinate green-house *Capsicum*. Aussie Bee 13:15

Pansarin ER (2008) Reproductive biology and pollination of *Govenia utriculata*: a syrphid fly orchid pollinated through a pollen-deceptive mechanism. Plant Species Biol 23(2):90–96

Partap U, Verma LR (1994) Pollination of radish by *Apis cerana*. J Apic Res 33(4):237–241

Pitts-Singer TL, Cane JH (2011) The alfalfa leafcutting bee, *Megachile rotundata*: the world's most intensively managed solitary bee. Annu Rev Entomol 56:221–237

Rahman A, Rahman S (2000) Effect of honey bee (*Apis cerana indica*) pollination on seedset and yield of buckwheat (*Fagopyrum esculentum*). Indian J Agr Sci 70(3):168–169

Ramirez W (1969) Fig wasps: mechanisms of pollen transfer. Science 163:580–581

Ramirez W, Malavasi J (1997) Fig wasps: mechanisms of pollen transfer in Malvanthera and Pharmacosycea figs. Rev Biol Trop 46:1635–1640

Richards AJ (2001) Does low biodiversity resulting from modern agricultural practice affect crop pollination and yield? Ann Bot 88(2):165–172

Ricketts TH (2004) Tropical forest fragments enhance pollinator activity in nearby coffee crops. Conserv Biol 18:1262–1271

Roubik DW (2002) The value of bees to the coffee harvest. Nature 417:708

Sadeh A, Shmida A, Keasar T (2007) The carpenter bee *Xylocopa pubescens* as an agricultural pollinator in greenhouses. Apidologie 38(6):508–517

Sakai S, Inoue T (1999) A new pollination system: dung-beetle pollination discovered in *Orchidantha inouei* (Lowiaceae, Zingiberales) in Sarawak, Malaysia. Am J Bot 86(1):56–61

Sampson BJ, Danka RG, Stringer SJ (2004) Nectar robbery by bees *Xylocopa virginica* and *Apis mellifera* contributes to the pollination of rabbiteye blueberry. J Econ Entomol 97(3):735–740

Santos SA, Roselino AC, Bego LR (2008) Pollination of cucumber, *Cucumis sativus* L. (Cucurbitales: Cucurbitaceae), by the stingless bees *Scaptotrigona* aff. *depilis* moure and *Nannotrigona testaceicornis* Lepeletier (Hymenoptera: Meliponini) in greenhouses. Neotrop Entomol 37(5):506–512

Santos SB, Roselino AC, Hrncir M, Bego LR (2009) Pollination of tomatoes by the stingless bee *Melipona quadrifasciata* and the honey bee *Apis mellifera* (Hymenoptera, Apidae). Genet Mol Res 8(2):751–757

Sarto MCL, Peruquetti RC, Campos LAO (2005) Evaluation of the Neotropical stingless bee *Melipona quadrifasciata* (Hymenoptera: Apidae) as pollinator of greenhouse tomatoes. J Econ Entomol 98:260–266

Shivanna KR (2015) Management of pollination services to enhance crop productivity. In: Plant biology and biotechnology. Springer, Karnataka, pp 697–711

Slaa EJ, Chaves LAS, Malagodi-Braga KS, Hofstede FE (2006) Stingless bees in applied pollination: practice and perspectives. Apidologie 37(2):293–315

Sornsathapornkul P, Owens JN (1998) Pollination biology in a Tropical *Acacia* Hybrid (*A. mangium* Wild. × *A. auriculiformis* A. Cunn. ex Benth). Ann Bot 81:631–645

Soroka JJ, Goerzen DW, Falk KC, Bett KE (2001) Alfalfa leafcutting bee (Hymenoptera: Megachilidae) pollination of oilseed rape (*Brassica napus* L.) under isolation tents for hybrid seed production. Can J Plant Sci 81:199–204

Southwick EE, Southwick L (1992) Estimating the economic value of honey bees (Hymenoptera: Apidae) as agricultural pollinators in the United States. J Econ Entomol 85:621–633

Stubbs CS, Drummond FA (1997) Management of the alfalfa leafcutting bee, *Megachile rotundata* (Hymenoptera: Megachilidae), for pollination of wild lowbush blueberry. J Kansas Entomol Soc 70:81–93

Tangmitcharoen S, Owens JN (1997) Floral biology, pollination, pistil receptivity, and pollen-tube growth of teak (*Tectona grandis* Linn. f.). Ann Bot 79:227–241

Tangmitcharoen S, Tasen W, Owens JN, Bhodthipuks J (2009) Fruit set as affected by pollinators of teak (*Tectona grandis* L. f.) at two tree spacings in a seed orchard. Sonklanakarin J Sci Technol 31(3):255–259

Tepedino VJ (1997) A comparison of the alfalfa leafcutting bee (*Megachile rotundata*) and the honey bee (*Apis mellifera*) as pollinators for hybrid carrot seed in field cages. Acta Hortic 437:457–461

Van Heemert C, De Ruijter A, Van den Eijnde J, Van der Steen J (1990) Year round production of bumble bee colonies for crop pollination. Bee World 71(2):54–56

Verma LR, Partap U (1994) Foraging behaviour of *Apis cerana* on cauliflower and cabbage and its impact on seed production. J Apic Res 33(4):231–236

Vicens N, Bosch J (2000) Pollinating efficacy of *Osmia cornuta* and *Apis mellifera* (Hymenoptera: Megachilidae, Apidae) on 'red Delicious' apple. Environ Entomol 29(2):235–240

Walters SA, Taylor BH (2006) Effects of honey bee pollination on pumpkin fruit and seed yield. Hortscience 41(2):370–373

Watanabe ME (1994) Pollination worries rise as honey bees decline. Science 265:1170

Willmer PG, Bataw AM, Highes JP (1994) The superiority of bumblebees to honey bees as pollinators: insect visits to raspberry flowers. Ecol Entomol 19:271–284

Winfree R, Williams NM, Dushoff J, Kremen C (2007) Native bees provide insurance against ongoing honey bee losses. Ecol Lett 10(11):1105–11113

Young HJ (1986) Beetle pollination of *Dieffenbachia longispatha* (Araceae). Am J Bot 73(6):931–944

Zisovich AH, Goldway M, Schneider D, Steinberg S, Stern E, Stern RA (2012) Adding bumblebees (*Bombus terrestris* L., Hymenoptera: Apidae) to pear orchards increases seed number per fruit, fruit set, fruit size and yield. J Hortic Sci Biotechnol 87(4):353–359

Insects as Food

15

Geetanjali Mishra and Omkar

Contents

15.1 Introduction

Conservative estimates indicate that world population is likely to reach nine billion by the year 2050. This increase in population would further place an untenable strain on the land owing to the need for increased agricultural production as well as housing and other basic amenities. As the increase in population continues unabated, the biggest question facing the world is how to feed these growing numbers adequately in the years to come. The possibility of increasing agricultural production is also seriously curtailed due to the limitation of available land suitable for it. The other option of venturing to the increased consumption of non-vegetarian food is fraught with even greater damage to the ecosystem. Under the current global food system, 30% of ice-free land is used for rearing large livestock, and the increasing population would require us to at least double it. This would in turn lead to soil degradation due to overgrazing, erosion and compaction, water pollution due to run off and augmentation of greenhouse gas emissions through production of methane gases and ammonia wastes. What other options for increasing food production to feed the many hungry mouths in the near future then lie with us?

Insects have been an important component of human diet across time and land zones. Holt (1885) mentions that instances and examples of insect consumption can be found from each and every part of the human-inhabited globe across different eras of human civilization. Historically, human food records show that like their

G. Mishra (✉) • Omkar
Department of Zoology, University of Lucknow, Lucknow 226007, Uttar Pradesh, India
e-mail: geetanjalimishra@hotmail.com; omkaar55@hotmail.com

Omkar (ed.), *Industrial Entomology*, DOI 10.1007/978-981-10-3304-9_15

closest primate relatives, human diets have been largely omnivorous ranging from leafy matter, roots, fruits and nuts to animals including insects. In view of the gatherer lifestyle of early man, it seems highly likely that gathering of not only fruits, nuts, roots, etc. but also insects could have been a component of the early life as well as diet of human populations.

What is however interesting is that a large number of historical records as well as text book accounts, while mentioning that ancient man hunted, gathered and scavenged a large variety of food items, have consistently failed to mention insects as one of them (Diamond 1991; Jones et al. 1992; Ingold 1994). Bodenheimer (1951) mentioned that the ignorance by scholars of such a major diet source for early man as well as in many cultures is baffling, despite the mention of such instances in the accounts of a number of travellers, naturalists and anthropologists. It is worth wondering, whether this complete ignorance of well-established facts has to do with the largely Western world squeamishness with insect consumption. Being exceptions to the trend, Bodenheimer (1951) and Defoliart (2012) have contributed extensively in discussing the importance of edible insects historically and also in detailing the many insect species that are consumed as food by the native populations in Australia, Africa, Asia and Americas, particularly the tropics.

While the ubiquity of insects as food is well recognized and accepted in many parts of the world, yet, a large part of the populace is affected by the Western affliction of disgust for consumption of insects or entomophagy. Western countries per se look down upon the consumption of insects leading to an outlook that has led to their reduced and rather infrequent consumption as well as restricted discussions in the programmes of international organizations and donor agencies that deal with food security issues (http://www.fao.org/forestry/edibleinsects/65424/en/). This is also a reason for lack of promotion of entomophagy (DeFoliart 1999; Yen 2009).

Bequaert (1921) and Bodenheimer (1951) can be credited with providing an overview of entomophagic practices the world over, with the latter making extensive listings of the various insects consumed as food. However, while these workers played a major role in recognizing entomophagy as an acceptable food practice the world over, they did not encourage its adoption in the non-entomophagy world. It was in particular that DeFoliart, who first as an editor of the *Food Insects Newsletter* (DeFoliart et al. 2009) and then via an online biography (DeFoliart 2012), focussed the attention of the masses on the massive potential, nutritional content and ecosustainability of insects. Reviews on entomophagy practices in different regions (Cherry 1991; Ramos-Elorduy et al. 2002; van Huis 2003; Yhoung-Aree and Viwatpanich 2005) as well as on the nutritional aspects of entomophagy (Bukkens 1997) have been published. Around 1900 insect species are consumed across the globe and the number continues to increase (http://www.ent.wur.nl/UK/Edible+insects/Worldwide+species+list/). The FAO branch of the UN has taken active steps since the beginning of this decade for assessing the role as well as potential of insects in providing food security to both humans and their livestock (http://www.fao.org/forestry/edibleinsects/74848/en/).

This review explores the entomophagic practices across the globe as well as whether insects should be actively considered as a sustainable and viable food resource for the massive human as well as animal population of the near future.

15.1.1 Entomophagic Practices Across the Globe

A huge number as well as variety of insects is consumed across the globe. Below is an overview of regions from where maximum incidence of entomophagy has been investigated and published.

15.1.1.1 Africa

Amongst a huge variety of insects consumed, caterpillars and termites (winged sexuals) are the most extensively consumed insects in Africa.

Out of a minimum of 22 families, more than 65 species of insects are consumed as food in Congo (Gomez et al. 1961). Of the total animal protein produced in Congo, insects account for a total of 10%. Similarly, they account for 30% of game, 47% for fishing, only 1% for fish culture, 10% for grazing animals and 2% for poultry (Gomez et al. 1961). In some districts of Congo, insect diets make up 37% of the total animal protein diet. Of the 30+ insect species consumed in Kwango and Kwilu districts of Congo, *Cirina forda* along with two more saturniid larvae are a major export component. In Southern Congo, at least 35 species of caterpillars are consumed (Malaisse and Parent 1980). Dried caterpillars of 23 species (including 17 Saturniidae) on analysis revealed that crude protein content accounted for 63.5% per 100 g, with iron levels being 335% of the daily intake.

In Angola, the commonly consumed termite, *Macrotermes subhyalinus*, and palm weevil larva, *Rhynchophorus phoenicis*, were found to have a high calorific value, 613 and 561 kcal/100 g, respectively, and the latter along with saturniid caterpillar, *Usta terpsichore*, had high levels of zinc, thiamine and riboflavin (Oliveira et al. 1976).

Most Nigerians have consumed or heard of consumption of insects, with usually the rural populations being more ready to admit the occurrence of such entomophagic practices versus the "educated elite". It was first recommended by Fasoranti and Ajiboye (1993) that malnutrition in Nigeria could be resolved through entomophagy. They also stated the necessity of the mass multiplication of insects versus exploitation of the natural sources for a continued and sustained supply. Mass rearing is also essential since the fast-occurring losses of natural habitats could constrain the availability of a food source. In case of *Anaphe venata*, the reduction of its host plant, *Triplochiton scleroxylon*, owing to excessive deforestation led to a drastic reduction in its availability in Nigeria (Ashiru 1988). These larvae are an extremely good source of fat (Ashiru 1988), having 611 kcal/100 g, but are also responsible for seasonal ataxia syndrome (Adamolekun 1993).

Entomophagy has been found not only to reduce the incidence of malnutrition but also to preserve biodiversity. The opening of national parks and sanctuaries to the local populations for harvesting of insects for food consumption has led to the decrease in poaching incidences in Malawi (Munthali and Mughogho 1992). This programme began in 1990, and it has been found that using gross margin analysis, insect collection and beekeeping (within parks) produced twice or more profits than many agricultural crops (Munthali and Mughogho 1992). What was also interesting was that these enterprises did not compete for labour with the other agricultural activities, leading to added advantage of higher earnings in the absence of labour conflicts. This was also the underlying reason behind low poaching incidences and could be a worthwhile step to consider and emulate for wildlife conservation.

During November–February, when food sources run low in Zambia, insects form a major source of nutrients, in comparison to the other available fruits and fresh mushrooms, containing only 2 g and 1 g protein, respectively. Silow's (1976) fieldwork recognized and acknowledged the importance of caterpillars as food in Zambia. Caterpillars are consumed as roasted snacks or as meat in porridge as main meals. The Mbunda tribe consumes 31 insects and markets 7. Most Zambian tribes find the meat of termites (*Macrotermes*) more delectable than meat of animals, birds and fishes.

More than 60 species of insects belonging to 15 families of 6 orders are consumed in Zambia. Honeybees from multiple genera produce multiple nutritional as well as commercial products (Mbata 1995). Honey not only is a food but also a liquor source for the people of Zambia. Wax is used in candle making and for conditioning the stretched skins on traditional drums. The immature stages (larvae and pupae) are consumed in multiple forms, viz. raw, boiled, roasted and fried. White (1961) notes that it is sacrilege and highly offending to interfere with a persons' marked hives in the Luvale tribe of Zambia. Not only honeybees but also caterpillars from Saturniidae family, locally known as *mumpa* (Bemba term), are highly treasured by people in Zambia and are a major animal protein source in areas of abundance (Holden 1991). Rich pickings (of about 20 l/day) for a week can garner a price that is equal to or more than a month's salary for a general worker in Zambia.

At least 40 species of insects belonging to 25 genera, 14 families and 7 orders are used as foods in Zimbabwe. Insects gathered from the wild are used as relish on the daily stiff cereal porridge consumed in rural diets. Insects are one of the cheapest sources of protein found here and play a major role in averting kwashiorkor (Chavunduka 1975). Fresh caterpillars are used as relishes and else their dried version as consumed. The marked decline in insect availability is not strongly correlated with the decline in their food trees, indicating over exploitation as a cause. *Gonimbrasia belina* an important food item has price similar to fresh beef. Winged termites are also consumed as raw, grilled or fried post removal of wings. *Brachytrupes membranaceus* (Fig. 15.1) is the most commonly consumed cricket and has witnessed an increase in its populations because of its adaptability to new agroecosystems (Wilson 1989) and has turned into a significant pest. McGregor (1991) also remarked on the abundant sale of these crickets in urban markets. Up to a 100 crickets can usually be collected in a day with women and children contributing to it (Gelfland 1971).

15.1.1.2 Asia and Oceania

In India, in northeastern India, pupa of wild silkworms, especially the Eri silkworm, *Samia ricini*, is considered a food connoisseur item (Chowdhury 1982). Ericulture being a cottage industry also results in cocoons as byproducts. In Manipur, insect consumption forms the cheapest source of protein (Gope and Prasad 1983) and should thus be encouraged.

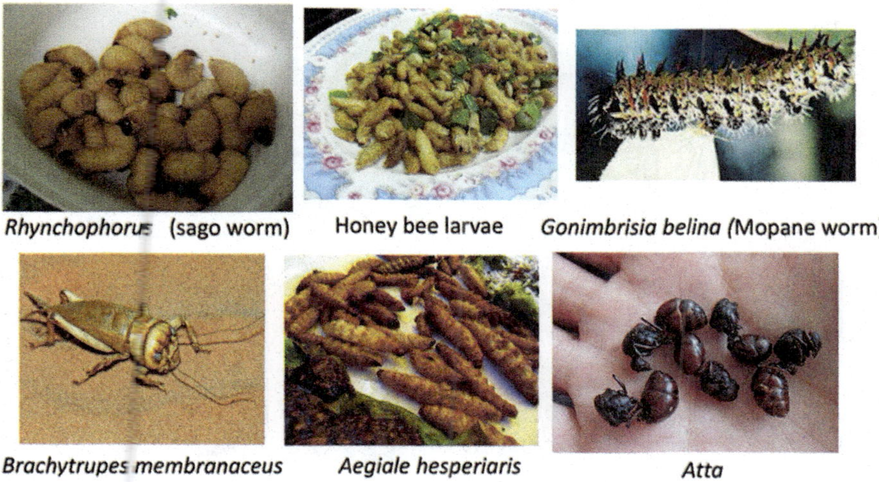

Fig. 15.1 Some much loved edible insects

Sungpuang and Puwastien (1983) found that in Ubon, Thailand, 20–60 g of insects are consumed per day. The Ministry of Public Health of Thailand has, in 1987, published a booklet highlighting six insect species that should be consumed to battle malnutrition in rural areas. At least 80 species of insects from 35 families are consumed in Thailand with details available from identification, collection to preparation and marketing (Vara-asavapati et al. 1975). Many insects, such as wasps, bamboo caterpillars, crickets and locusts, are considered as gourmet delicacies and now grace the finest restaurants (Yhoung-Aree et al. 1997). The Thai government has been hugely responsible for promulgating insect consumption, particularly during times of insect abundance such as locust plagues. Thai farmers began collecting grasshoppers/locusts as food in 1983 (Expat World 1992), and the price increased from 12 cents per kg in 1983 to $2.80 per kg in 1992, with earnings from locusts being double that of agricultural produce.

In China, usage of insects as food has been highlighted in different time frames (Luo 1997). Mass production of two insects, viz. ant *Polyrhachis vicina* (Zhang et al. 2008) and larvae of *Musca domestica vicina*, for edible consumption is being attempted. Sale of ant and related food items sustain an almost $100 million economy (Kantha 1994).

Ricefield grasshoppers (*Oxya yezoensis* and *Oxya japonica*), stir fried and seasoned, form one of the most preferred foods of Japanese society, both in earlier and modern times. These grasshoppers, which had once nearly disappeared due to excess pesticide use, have now reappeared as luxury items (Mitsuhashi 1997). "Hachinoko", bee or wasp larvae, is consumed raw, boiled in soy sauce or overboiled rice, is another delicacy in Japan and subject to the modernity of present day times is even available canned. Canned wasps weighing 65 g were selling for 1000 yen (= $8.00) in 1988. The late Emperor Hirohito was especially fond of the wasp-rice dish, which he is purported to have polished off even in the event of lack of

Fig. 15.2 Traditional and modern gastronomic delights made from insects

appetite following surgery (Mitsuhashi 1997). Larval Trichoptera, which are aquatic insects inhabiting gravel river beds and also known as "zazamushi", are also consumed in large quantities in Japan. Mitsuhashi (1997) has provided updated information on annual prices and processed quantities of edible species in Japan.

Papua New Guinea is also home to a number of edible insect species, chief amongst which is the sago grub, *Rhynchophorus ferrugineus papuanus* (Fig. 15.1). It is not only widely available and highly relished but it is also usually the centre of attraction of grub festivals. Connoisseurs advertise its texture as tender and taste as sweet with nutty overtones. Even the Europeans, who are reluctant consumers of insects, are fond of these grubs (Mercer 1993). They form a good 30 % of the protein intake and are a good nutrient source of fat, iron and zinc. These insects are found in the rotting pith of sago palms Mercer (1994). The production of these insects is environmentally efficient and a major source of rural income.

A similarly efficient system is the harvesting of cerambycid grub, *Hoplocerambyx severus*, from *Anisoptera polyandra* logs in Papua New Guinea (Mercer 1994). The population is so dense that almost 100 grubs can be gathered from a single log in less than a quarter of an hour.

Aborigines of Australia are voracious consumers of a number of insects, such as witchetty grubs (Cossidae), bogong moth (Noctuidae) and bardee larva (Cerambycidae), honeypot ants and honey and brood of the stingless bees. These food items have recently caught the fancy of high-end restaurants and are making their appearance on "couture" dishes (*Christian Science Monitor*, 1991; Fig. 15.2). Witchetty grubs are believed to be a gourmet's delight when lightly roasted in hot ashes (Tindale 1966).

15.1.1.3 Latin America

Leafcutter ants (*Atta* spp.), palm weevil larvae (*Rhynchophorus* spp.) and bee and wasp brood (Apidae and Vespidae) are highly recommended for their taste and flavour not only amongst the indigenous tribe of Latin America but also the outsiders. Palm weevil, *R. palmarum*, is believed to have immense mass production and marketing potential (DeFoliart 1993) and has been "semi-cultivated" by native people in many countries of the world. The fungus-feeding *Atta* ants are widely preferred edible items in South America (DeFoliart 1997). They also have ecological impact in the rain forest, since its consumption of a cultured fungus has let them tap into a limitless supply of cellulose (Hodgson 1955).

In South America, edible insects are commonplace in rural as well as urban markets. They sometimes are so much in demand that sometimes the price of certain insects, such as immature stages of the ants, *Liometopum apiculatum* and *Liometopum occidentale* var. *luctuosum* (escamoles), has gone up to 1000 pesos per kilogram (in 1981). Current rates for these insect larvae and also for white maguey worms (Fig. 15.1) are $25 per plate in many restaurants. Digging out these underground ant nests, which are regarded as private property, is hard work. Post harvesting of nests, the nest is covered with dried grass, or fresh weeds, in order to maintain an environment suitable for survival and regrowth of the colony (Ramos-Elorduy et al. 1989). Escamole hunting can lead to ant season income being higher than annual incomes in rural areas (de Conconi 1982).

Stingless bees (*Melipona, Scaptotrigona, Trigona*) are cultivated for honey as well as brood in small clay jars kept near the walls of houses and in small hollowed trunks (de Conconi 1982). Wasp brood enclosed in combs is usually sold (de Conconi 1982). These nests are first collected from nature when they have just begun (foundation combs) and hung in roofs of farm homes and barns till they reach the desired size, which may in case of *Polybia occidentalis bohemani* reach up to 1 m.

"Ahuahutle" or Mexican caviar-producing aquatic hemipterans (five species of Corixidae, one species of Notonectidae) are reared in alkaline lakes and have been for centuries farmed (de Conconi 1982). Oviposition traplines made of shore grass are used to trap bugs as well as their eggs every 3 weeks. The harvest of this delicacy has been affected by the pollution in lakes, but still it is traditionally harvested. Stuffed in tortillas and tamales, it forms a local delicacy served by restaurants. It is also exported to Germany and Great Britain as fish and bird feed (Ramos-Elorduy and Pino 1990).

Larva of the giant skipper butterfly, *Aegiale hesperiaris* (white agave or maguey worm), as well of *Comadia redtenbacheri* (pink worm of the maguey, the red agave worm) found on agave cactus, is much in demand in Mexico, the United States, Canada, France and Japan in fresh as well as canned form (Ramos-Elorduy and Pino 1990). The latter is consumed fried, as fillings in tortilla, in rice or tomato soup or in roasted and ground form (Ramos-Elorduy and Pino 1990).

In Mexico, more than 20 species of grasshoppers and locusts are sold and consumed widely (deConconi 1982). Their preparation involves cooking with onion, garlic and chilli powder and take the taste of the ingredients that they are cooked with.

Alate leafcutter ants (*Atta* spp.; Fig. 15.2), known as hormigas culonas or big-bottomed ants, are a national delicacy in Colombia and at par in gastronomic value with Russian caviar or French truffles. It has been stated that toasted ants are the crème de la crème of Colombian cookery. There is a huge literature available on the *Atta* ants in Brazil (Kevan and Bye 1991).

Insects of 22 genera belonging to seven orders are used as a food source by tribes along the Colombia-Venezuela border region. Though abundance was not a criterion for entomophagy, the decrease in game and aversion to consuming recently domesticated animals had led to an increase in entomophagy (Ruddle 1973). On the other hand, Dufour (1987) found that the most important insects in the diet of Tukanoans are the ones whose aggregations are large and predictable in nature.

15.1.2 Why Eat Insects?

15.1.2.1 Alternative Protein Sources

Increased wealth and better financial status has led to an increase in the global food demand (Msangi and Rosegrant 2011), particularly that of meat (Tillman et al. 2011). An increase in wealth is expected to lead to a rise in per capita consumption by 9% in countries belonging to the higher-income bracket and by 50% in China by the year 2000. This would lead to an increase in the demand for livestock feed by 48 and 158%, in higher-income countries and China, respectively (Msangi and Rosegrant 2011). It has been reported that each kilogram increase in production of animal protein requires a corresponding sixfold increase in plant protein (Pimentel and Pimentel 2003; Trostle 2008).

This along with the rapidly increasing human population will lead to an untenable strain on the agricultural sector, and the earth will not be able to sustain. This would also cause a global escalation in prices of agricultural crops, which will peter down to that of animal protein, with a predicted 30% rise in 50 years (2000–2050; Nelson et al. 2009). The same study also indicates that climate change would further aggravate the situation, causing an additional 18–21% rise in prices. The need for more biofuels in conjunction with decrease in agricultural productivity would further fuel up the situation. As it is, the rate of increase in productivity of land is slowly reaching a plateau (Alston et al. 2009). Thus, the need for alternative protein sources, such as cultured meat (Fayaz Bhat and Fayaz 2011), seaweed (Fleurence 1999), and mini-livestock (Paoletti 2005), would build up rapidly.

15.1.2.2 Benefits of Mini-livestock

15.1.2.2.1 Greenhouse Gas and Ammonia Emissions

Livestock production and transportation are believed to contribute a substantial share (18%) of human activity resulting greenhouse gas (GHG) emissions (Steinfeld

Fig. 15.3 Greenhouse gas emissions by macro- and mini-livestock (www, http://entomo.farm/insect-farming-systems/industrial-systems-insect-farming/)

et al. 2006). Enteric fermentation and manure leads to a substantial 31 and 6% contribution, respectively, to global methane emissions. Similarly, 65% of the N_2O released into the atmosphere owing to human activity is due to prolific used of fertilizers and manure. A kilogram each of beef, pork and chicken contributes 14.8, 3.8 and 1.1 kg of CO_2, respectively (Fiala 2008).

Insects such as cockroaches (Blaberidae and Blattinae), termites (Isoptera) and scarab beetles (Scarabaeidae) also produce GHGs (Hackstein and Stumm 1994). However, commonly reared edible insect species such as the yellow mealworm (*Tenebrio molitor*), the house cricket (*Acheta domesticus*) and the migratory locust (*Locusta migratoria*) emit much less GHG compared to the current major sources of animal protein (Oonincx et al. 2010; Fig. 15.3).

15.1.2.2.2 Feed Conversion Ratio

The amount of feed that is converted into meat or feed conversion ratios (FCRs) is an important factor to consider as increased demand for meat is inextricably linked to a similar increase in that of grains, water, etc.

FCRs for chicken (2.5), pork (5) and beef (10) were calculated in the United States following long-term studies (Smil 2002a, b). On the other hand, FCR values are relatively economical for insects such as *Acheta domesticus* (0.9–1.1; Nakagaki and deFoliart 1991; 1.7 for fresh weight; Collavo et al. 2005). Also, the proportion of edible portions (edible weight) of the livestock differs vastly amongst macro-livestock and insects. However, in chicken, pork and beef, 55, 55 and 40%, respectively, of live weight can be consumed (Flachowsky 2002; Smil 2002a, b). Insects can usually be eaten whole, with the exception of its indigestible chitin exoskeleton leading to 80% edible weight (Nakagaki and deFoliart 1991). Thus crickets are 2, 4 and 12 times more efficient than chickens, pigs and cattle, respectively (Fig. 15.4). The protein content per kg body weight for poultry, pork, beef, cricket nymph and

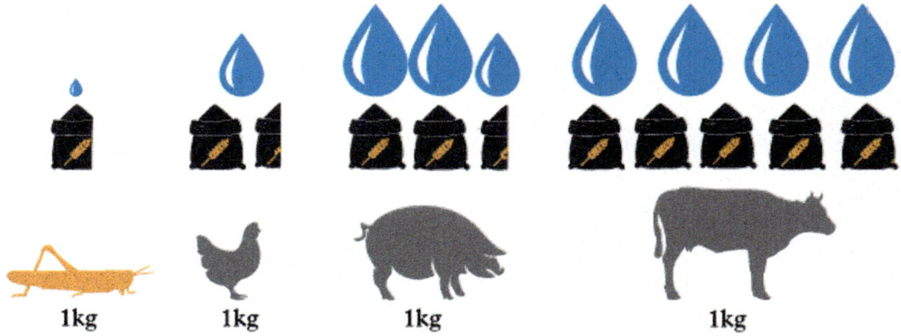

Fig. 15.4 Food conversion ratios by macro- and mini-livestock (www, http://entomo.farm/insect-farming-systems/industrial-systems-insect-farming/)

adults is 200, 150, 190, 154 and 205 g, protein per kilogram edible weight, respectively (Flachowsky 2002; Finke 2002).

15.1.2.2.3 Zoonoses

When animals are reared under high-density conditions, it increases multifold the risks of diseases gaining epidemic proportion as well as the emergence of antibiotic-resistant pathogens. Avian influenza (H5N1), foot-and-mouth disease, bovine spongiform encephalopathy (BSE) and classical swine fever are amongst some such infectious diseases that have led to severe global economic losses (King et al. 2006). These diseases in some cases, as the mad cow disease, have been known to be transmitted to humans (King et al. 2006) and also being a cause of cardiovascular disease and cancer (Pan et al. 2012). The taxonomical distance of humans from insects makes rearing and consuming of insects a very low-risk area as far as disease transference is concerned.

15.1.2.2.4 Water Use

Water is an increasingly rare commodity internationally and of its total virtual flow, nearly half of the volume is involved in raising feed crops (Chapagain and Hoekstra 2003). Beef in particular requires 22,000–43,000 l/kg of meat produced (Chapagain and Hoekstra 2003; Pimentel et al. 2004). The increased animal protein diet has already resulted in water scarcity issues in China from 1961 to 2003 (Liu et al. 2008a, b). On the other hand, virtual water flows for insect that are to be reared as food sources are going to be much lower. One reason for this is that many of these insects are drought resistant, such as yellow and lesser mealworm (Ramos-Elorduy et al. 2002), and have efficient FCRs.

15.1.2.3 Insects as Feed Ingredients

Insects are also potential replacements of fish meal and oil or at least major components of animal diets. Meal and oil from fish and soybean are major components of feed for aqua-organisms as well as livestock. In 2008, 19% of the global fish production, usually from small pelagic forage fish (Tacon and Metian 2008), was used in the production of fish meal and oil (FAO 2010). Aquaculture, which accounted a minor portion (4%) of the fish supplies across the globe, accounted for 38% in 2008 (FAO 2010), with future growth rates being estimated at 8% annually. Marine over-exploitation (FAO 2010) also adds to the cost of producing fish oil and meal (Deutsch et al. 2007; Tacon and Metian 2008). An increment in global demand for soy products has also led to a subsequent rise in cost of soybean (Trostle 2008). As these traditional protein sources become exorbitantly priced and their attainment prohibitive, the search for alternative sources has gained momentum (FAO 2012). Insects form a not so alien, cheap and massively available protein source. Insects which show much promise for mass production are black soldier fly, the common house fly, the yellow mealworm, the lesser mealworm, silkworm (*Bombyx mori*) and several grasshopper species (Anand et al. 2008).

Black soldier fly larvae develop by converting manure into body mass. They are rich in protein (42%) and fat (35%), making them suitable as livestock (Newton et al. 2005) and fish feed (Bondarik and Sheppard 1984). House fly maggots that can be reared on poultry manure can also be used as poultry feed in (Zuidhof et al. 2003; Awoniyi et al. 2004; Hwangbo et al. 2009; Teguia and Beynen 2005). Their pupae have high levels of proteins (61%; El Boushy 1991). The rearing techniques need to be fine-tuned and preferably automated to cater to the large demand from aquatic, poultry and livestock feed requirements.

15.1.2.4 Farming Insects

Other than being harvested from the wild, which is one of the most common means of procuring edible insects, they can also be farmed. They can, like silkworms and honeybees, be reared for purposes other than edible and also consumed as a byproduct (Bodenheimer 1951). Cochineal dye is also obtained from a farmed insect (*Dactylopius coccus*). They are usually found on prickly pear plants and, other than being collected from the wild, can be grown on prickly pear fences around houses. In Mexico, they are grown in plastic environment-controlled microtunnels, on prickly pear (Aldama-Aguilera et al. 2005).

Other than being farmed, insects can also be semi-cultivated by performing environmental manipulations. Examples of this are (a) harvesting eggs of aquatic hemipterans from artificial oviposition sites in lakes, (b) cutting palm trees to trigger egg laying by palm weevils followed by harvesting of larvae and (c) manipulating host tree distribution, preservation and abundance, along with artificial inoculation (Van Itterbeeck and van Huis 2012).

House crickets, palm weevil, giant water bug (*Lethocerus indicus*) and water beetles (Jäch 2003) are farmed for consumption in many parts of the world. While these commercial methods may suffice for human consumption in the current date, they will not be enough for meeting the current let alone the future demands from the feed industry. The primary challenge is developing an automated large-scale rearing method, which would provide quantity as well as quality hygienically and under sterile conditions (Bolckmans 2010). Sterility is of prime importance as pathogens have been known to seriously interfere with attempts to do commercial rearing (Szelei et al. 2011).

While it is not so difficult to use large-scale industrial methods for mass production of edible insects (Kok et al. 1988), its economic feasibility depends on the labour costs incurred. Automation is definitely going to enhance the quality and quantity of the product leading to more production in less space as well as reduced chances of microbial contamination by humans (Parker 2005). Silkworm, termite and drugstore beetle can be used for the dual purpose of recycling waste material as well as production of food (Katayama et al. 2008).

15.1.2.5 Conservation

Insects are largely a non-domesticated resource, with only few species being reared and most being gathered. Gathering from the wild carries the advantage of them being relatively free from pesticides. However, human greed has led to the overexploitation and subsequent local disappearance of many species (Nel and Illgner 2001; Madibela et al. 2009), an example being the mopane caterpillars (*Imbrasia belina;* Fig. 15.1). One method that has been found to deal with this problem is the placing of local embargos for certain periods (Mbata et al. 2002), but it is found to afflict sustainable harvesting (Akpalu et al. 2009).

A major edible insect of the Central African Republic, *Imbrasia oyemensis*, has been threatened due to commercial exploitation of its host plants (Vantomme et al. 2004). The solution to this has been the protection of at least one seed tree for 10 hectares of logged forests, but it is not the satisfactory solution to the scarcity of the insects.

Deforestation, water pollution and bush burning are a major cause of reduction in availability of edible insects in Nigeria (Agbidye et al. 2009). Fourteen edible insects have been documented as threatened owing to overexploitation or ecosystem degradation (Ramos-Elorduy 2006). Higher demand as well as incompetent, insincere and illegal harvesters are also responsible for overexploitation of edible insects (Cesard 2004). Multiple causes of ecosystem degradation such as pollution and pesticide use are recognized as reasons for insect insufficiency (Ramos-Elorduy 2006). In France, development of river banks has led to the drastic decline in mayfly (*Ephoron virgo*) populations, which were abundant till the mid-1980s (Cesard 2010). To effectively conserve relevant insect populations, detailed investigations into their effect on livelihoods, ecosystem, legislation, etc. are required.

Other than this, aiding and sustaining insect reproduction and survival by creating suitable habitats, sustainable harvesting and rearing can help conserve and sustain insect populations and even post their utilization as food. Semi-captive or sheltered rearing of insects is one such tool, which reduces attack by pathogens, parasitoids and predators (Silow 1976; Latham 1999; Ngoka et al. 2008).

15.1.2.6 Controlling Insects by Using Them as Feed and Food

Not only non-pest insects but also pests can be exploited as both food and feed (Cerritos and Cano-Santana 2008; DeFoliart 1997; Latham 1999). Locusts are one such example, the consumption of which was encouraged by the government in Thailand during their infestation (Yhoung-Aree et al. 1997). Sometimes, unintentional control can also take place, as is in the case of edible grasshoppers, which sell

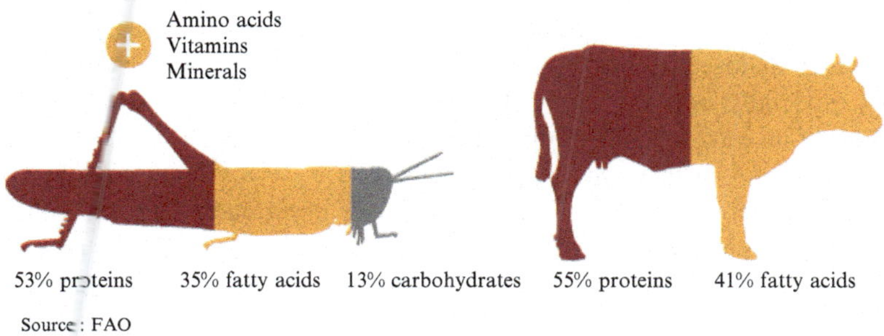

Fig. 15.5 Nutritional composition of macro- and mini-livestock (www, http://entomo.farm/insect-farming-systems/industrial-systems-insect-farming/)

for more than the crop value at times in Niger (van Huis 2003). In Laos, electric-light water traps are placed facing ricefields, with the primary aim of capturing edible insects and not controlling pests (van Huis 2013). In certain instances, control by pesticides results in lesser profits than physical control via harvesting; grasshopper *Sphenarium purpurascens* is a pest of corn, bean, alfalfa, squash and broad bean in Mexico, Guatemala and some Caribbean islands (Cerritos and Cano-Santana 2008). Also, in the Philippines, harvesting of several edible insect species, e.g. a migratory locust (*Locusta migratoria manilensis*), a mole beetle (*Gryllotalpa* sp.), a June beetle (*Leucopholis irrorata*) and the Korean bug (*Palembus dermestoides*), is reported to serve as a control strategy (Adalla and Cervancia 2010).

During the winter of 1988/1989, locusts invading Kuwait were sprayed with insecticides, even though the local population consumed them (DeFoliart 1997). Samples had high levels of phosphorus- and chlorine-containing pesticide residues (Saeed et al. 1993).

An example of biological control in conjunction with entomophagy is that of weaver ants, *Oecophylla*, which are effective predators of many pest species in orchards (Van Mele 2008). Their queen brood is highly popular in many Eastern countries, and making available as cat food and sucrose for the ants in mango orchards increases both biological control of many pests as well as doubles the brood size and number (Offenberg and Wiwatwitaya 2009).

15.1.2.7 Nutrition

The nutritional value of many of the edible insects has been of interest to many researchers, with that of almost 1900 such insects having been recorded (Bukkens 1997; DeFoliart 1992; Fig. 15.5). However, these values are not constant and are known to vary with life stage and diet (Oonincx and Dierenfeld 2011).

Most edible insects are complete diets, with many of the amino acids missing in cereal and legume diets, such as lysine, being provided by them (Bukkens 1997). *Imbrasia belina*, *Rhynchophorus phoenicis*, *Oryctes rhinoceros* and *Macrotermes bellicosus* contain all essential amino acids (Ekpo 2011). The use of such nutritious insects can be a major step in reducing malnutrition (Allotey and Mpuchane 2003; Ohiokpehai 2003). The amount of fat and the ratio of saturated/unsaturated fatty acids are better than even fish (DeFoliart 1991; Bukkens 1997).

Not only substantial amount of protein but also micronutrients and trace metal are provided by insects (Michaelsen et al. 2009). In countries facing deficiencies of metals, especially iron and zinc (Müller and Krawinkel 2005), encouraging consumption of insects, such as termites and crickets (Christensen et al. 2006), can be extremely helpful, as they are loaded with these.

Chitin, chitosan and chitooligosaccharides are purported to enhance immunity (Lee et al. 2008; Muzzarelli 2010; Reese et al. 2007; Xia et al. 2011), promote the growth of friendly microflora and prevent pathogenic activity (Khempaka et al. 2011; Liu et al. 2010; Muzzarelli 2010). The reduction in allergies in developing countries has been attributed to the exposure to more chitin-containing intestinal parasites due to lower hygiene levels (Brinchmann et al. 2011). This begs the point, whether increased consumption of chitin through promotion of insects as food in early childhood may protect against allergy later.

15.1.2.8 Food Safety

While insects form an extremely nutritious, low-polluting, cost-effective mode of meeting the global food scarcity situation, which is sure to follow in the coming years, it is essential to keep in mind the potential consumer and animal health issues. The Hazard Analysis and Critical Control Point System is a preventive system adopted by the Codex Alimentarius Commission (FAO/WHO 2008). This is essential, as consumption of insects has not been without incidence. A case in point is the ataxic syndrome that occurs after the consumption of the African silkworm *Anaphe venata* (Adamolekun 1993). It is only after thorough detoxification via heat treatment can this insect be safely consumed (Nishimune et al. 2000). In most insects that are toxic in their raw form, heating, boiling or sundrying makes them less toxic (Akinnawo et al. 2002).

Quick, even drying of mopane caterpillar being post degutted, boiled, dried, and stored can save it from fungal attack (Mpuchane et al. 2000; Simpanya et al. 2000). A heating step is sufficient for inactivation of Enterobacteriaceae; however, spore-forming bacteria, most probably introduced through soil, survive this treatment. Attempts to find noninstrumental approach for insect preservation seem more practical and promising (Klunder et al. 2012).

15.1.3 Preservation and Processing

Preservation of edible insects, especially those that are not mass reared and only seasonally available, increases storage and access for the rest of the year. Wasp larvae, weaver ant brood, silkworm pupae, giant water bugs, crickets and grasshoppers are quite commonly canned.

Preservation is also important as many of the freshly caught insects have a limited shelf life, post which they get spoilt. Early preservation increases their reach and also provides a large number of tastes. Weaver ant larvae and pupae and stink bugs are placed on ice in markets.

In tropical countries, fresh caterpillars undergo prior processing before they hit the market. In Zambia, caterpillar processing for long-term storage in the household or for sale involves the following steps: (*a*) eviscerating (degutting) live caterpillars soon after they are collected from the foliage of host plants; (*b*) roasting the eviscerated caterpillars over hot coals, from bonfires set up in the woodlands, until the setae and spine body adornments are burned off and the caterpillars become hardened; (*c*) sundrying the roasted caterpillars until they are crispy; and (*d*) packaging the sun-dried caterpillars in sacks or other materials (Mbata et al. 2002).

Producing conventional consumer foods such as crackers, muffins, sausages and meat loaf from termites and lake flies enhances their consumption (Ayieko et al. 2010). Caterpillars on being mixed with sorghum and Bambara nuts provided a protein-rich diet for children of more than 10 years of age (Allotey and Mpuchane 2003). The nutritional content of the insects differs with the processing method; degutting improves crude protein and digestibility, whereas cooking lowers it and roasting on hot coal enhances mineral content (Madibela et al. 2007). Smoke drying is known to reduce cholesterol levels in many caterpillars (Edijala et al. 2009).

Not only does the nutritional content change for the better with processing but is also likely to influence the sensibilities of the consumer, increasing its acceptance (Damodaran 1997).

15.1.4 Commercialization

In many tropical countries, edible insects are preferred over conventional meat despite their higher costs, owing to the status of the former as a delicacy. For example, the presence of the mopane caterpillar and grasshoppers in the market reduces the sale of beef, which is much lower in price (Quin 1959; Agea et al. 2008). Lowering of the costs by doing away with middlemen and high market dues may further promote the sale of insects. Another factor that affects their sale is the low shelf life (Cesard 2004); though processing can make the shelf life longer, it reduces the price.

The positive aspect of insect retail is that most of the collection is done by non-skilled personnel providing them a means of major earning during insect season. This along with the collection timings (dawn or dusk) allows them to pair this source of income with other sources (Mbetid-Bessane 2005; Meyer-Rochow et al. 2008).

15.1.5 Consumer Acceptance

Prior to commercialization of insect production, it is essential to know and understand the prejudices, the preferences and the barriers in society for their use as food and feed. While in some instances such as aquaculture, provision of insects as feed would probably not change societal perception (since it is the natural food), doing the same in poultry and cattle industry may be objected too.

While culturally insects are acceptable in large parts of the world, and the current global trends are also promoting gourmet insect food, there still are many taboos, which need to be addressed (Lawal and Banjo 2007). Food acceptance is a result of the satiation of the five senses as well as cultural and societal conditioning. Furthermore the innate reluctance to open up to new experiences, especially food such as insects which seem alien, leads to complication in the promulgation of entomophagy. While the nutrient value and environmental friendliness can act as convincing reasons (Fischer and Frewer 2009; Frewer et al. 2011) to try insects as food, the battle takes place against the innate conditioning.

Proper promotion, along with better processing and presentation of insects can increase the acceptance of this food of the future (Martins and Pliner 2005). A case in point of overcoming feelings of aversion is the promotion of Sushi in the Western world. Providing conditions leading to informed choice would also enable conversion (Houghton et al. 2006).

For persons concerned about animal welfare, usual aspects such as animal density per unit surface area are not of concern when it comes to insects. However, aspects such as pain perception (Kang et al. 2010) and cognition abilities (Elwood 2011) are recognized in insects and so good care of them is also required.

15.2 Conclusions

In a world where the ways of the Western world are copied and accepted all over, entomophagy has been shifted to the back burner despite it being an innate part of many cultures. But as meat-centric diets become increasingly less sustainable both economically and environmentally, there is need to opt for other strategies to fulfil our nutrient requirements and also stimulate our taste buds, at the same time being ecofriendly and sustainable. Insects form the ideal food solution for the future.

Opting for entomophagy would not only lead to reduction in land use and greenhouse gas problems but also do away with nutrition-related challenges, simply owing to both the quantity and quality of insects produced much more easily than other livestock. High nutrient value, high edible weight, easy rearing, low GHGs, possible degradation of garbage, etc. all make insects the food and feed of the future. "Chirpy food" is the only way forward.

References

Adalla CB, Cervancia CR (2010) Philippine edible insects: a new opportunity to bridge the protein gap of resource-poor families and to manage pests. In: Durst PB, Johnson DV, Leslie RN, Shono K (eds) Edible forest insects: humans bite back. Food Agriculture Organisation Regional Office. Asia and the Pacific, Bangkok, pp 151–160

Adamolekun B (1993) Anaphe venata entomophagy and seasonal ataxic syndrome in southwest Nigeria. Lancet 341:629

Agbidye FS, Ofuya TI, Akindele SO (2009) Some edible insect species consumed by the people of Benue State Nigeria. Pak J Nutr 8:946–950

Agea JG, Biryomumaisho D, Buyinza M, Nabanoga GN (2008) Commercialization of Ruspolia nitidula (Nsenene grasshoppers) in Central Uganda. Afr J Food Agric Dev 8:319–332
Akinnawo OO, Abatan MO, Ketiku AO (2002) Toxicological study on the edible larva of Cirina forda (Westwood). Afr J Biomed Res 5(1, 2):43–46
Akpalu W, Muchapondwa E, Zikhali P (2009) Can the restrictive harvest period policy conserve mopane worms in Southern Africa? A bio-economic modelling approach. Environ Dev Econ 14:587–600
Aldama-Aguilera C, Lianderal-Cazares C, Soto-Hernandez M, Castillo-Marquez LE (2005) Cochineal (Dactylopius coccus Costa) production in prickly pear plants in the open and in microtunnel greenhouses. Agrociencia 39:161–171
Allotey J, Mpuchane S (2003) Utilization of useful insects as food source. Afr J Food Agric Nutr Dev 3:1–6
Alston JM, Beddow JM, Pardey PG (2009) Agricultural research, productivity, and food prices in the long run. Science 325:1209–1210
Anand H, Ganguly A, Haldar P (2008) Potential value of acridids as high protein supplement for poultry feed. Int J Poult Sci 7:722–725
Ashiru MO (1988) The frequency distribution of eggs and larvae of Anaphe venata Butler (Lepidoptera: Notodontidae) on Triplochiton scleroxylon K Schum. Int J Trop Insect Sci 9(05):587–592
Awoniyi TAM, Adetuyi FC, Akinyosoye FA (2004) Microbiological investigation of maggot meal, stored for use as livestock feed component. J Food Agric Environ 2:104–106
Ayieko MA, Oriamo V, Nyambuga IA (2010) Processed products of termites and lake flies: improving entomophagy for food security within the Lake Victoria region. Afr J Food Agric Nutr Dev 10:2085–2098
Bequaert J (1921) Insects as food. How they have augmented the food supply of mankind in early and recent times. J Am Mus Nat Hist 21:191–200
Bodenheimer FS (1951) Insects as human food: a chapter of the ecology of man. Junk, The Hague, 352 pp
Bolckmans KJF (2010) New, novel, innovative and emerging applications of insect rearing. Symposium No. 5. 12th workshop of the arthropod mass rearing and quality control working group of the IOBC, Vienna, Austria, 19–22 October
Brinchmann BC, Bayat M, Brøgger T, Muttuvelu DV, Tjønneland A, Sigsgaard T (2011) A possible role of chitin in the pathogenesis of asthma and allergy. Ann Agric Environ Med 18:7–12
Bukkens SGF (1997) The nutritional value of edible insects. Ecol Food Nutr 36:287–319
Cerritos R, Cano-Santana Z (2008) Harvesting grasshoppers Sphenarium purpurascens in Mexico for human consumption: a comparison with insecticidal control for managing pest outbreaks. Crop Prot 27:473–480
Cesard N (2004) Harvesting and commercialisation of kroto (*Oecophylla smaragdina*) in the Malingping area, West Java, Indonesia. In: Kusters K, Belcher B (eds) Forest products, livelihoods and conservation: case studies of non-timber forest product systems, vol 1. Central International Forest Research, Jakarta
Cesard N (2010) Vie et mort de la manne blanche des riverains de la Saône. Etudes Rurales 185:83–98
Chapagain AK, Hoekstra AY (2003) Virtual water flows between nations in relation to trade in livestock and livestock products. Value Water Research Report Series No. 13, U.N. Education, Science, Cultural Organ, Institute. Water Education, Delft, the Netherland. 60 pp
Chavunduka DM (1975) Insects as a source of protein to the African. Rhodesia Sci News 9:217–220
Cherry R (1991) Use of insects by Australian aborigines. Am Entomol 32:8–13
Chowdhury SN (1982) Eri silk industry (ed) Gov. of Assam
Christensen DL, Orech FO, Mungai MN, Larsen T, Friis H, Aagaard-Hansen J (2006) Entomophagy among the Luos of Kenya: a potential mineral source? Int J Food Sci Nutr 57:198–203
Collavo A, Glew RH, Huang Y-S, Chuang L-T, Bosse R, Paoletti MG (2005) House cricket small-scale farming. See Ref 115:519–544

Damodaran S (1997) Food proteins: an overview. In: Damodaran S, Paraf A (eds) Food proteins and their applications. Marcel Dekker, New York, pp 1–21

DeFoliart GR (1991) Insect fatty acids: similar to those of poultry and fish in their degree of unsaturation, but higher in the polyunsaturates. Food Insects Newsl 4:1–4

DeFoliart G (1992) Insect as human food Gene DeFoliart discusses some nutritional and economic aspects. Crop Prot 11:395–399

DeFoliart GR (1993) Hypothesizing about palm weevil and palm rhinoceros beetle larvae as traditional cuisine, tropical waste recycling, and pest and disease control on coconut and other palms-can they be integrated? Principes 37(1):42–47

DeFoliart G (1997) An overview of the role of edible insects in preserving biodiversity. Ecol Food Nutr 36:109–132

DeFoliart GR (1999) Insects as food: why the Western attitude is important. Annu Rev Entomol 44:21–50

DeFoliart G (2012) The human use of insects as a food resource: a bibliographic account in progress. http://www.food-insects.com

DeFoliart G, Dunkel FV, Gracer D (2009) The food insects newsletter: chronicle of a changing culture. Aardvark, Salt Lake City, 414 pp

Deutsch L, Gräslund S, Folke C, Troell M, Huitric M, et al (2007) Feeding aquaculture growth through globalization: exploitation of marine ecosystems for fishmeal. Glob Environ Chang 17:238–249

Diamond J (1991) The rise and fall of the third chimpanzee. Random House, London

Dufour DL (1987) Insects as food: a case study from the northwest Amazon. Am Anthropol 89(2):383–397

Edijala JK, Egbogbo O, Anigboro AA (2009) Proximate composition and cholesterol concentrations of Rhynchophorus phoenicis and Oryctes monoceros larvae subjected to different heat treatments. Afr J Biotechnol 8:2346–2348

Ekpo KE (2011) Effect of processing on the protein quality of four popular insects consumed in Southern Nigeria. Arch Appl Sci Res 3:307–326

El Boushy AR (1991) House-fly pupae as poultry manure converters for animal feed: a review. Bioresour Technol 38:45–49

Elwood RW (2011) Pain and suffering in invertebrates? Inst Lab Anim Res J 52:175–184

Expat World (1992) The newsletter of international living (vol 4, No. 1, January 1992, P.O. Box 1341, Raffles City, Singapore 9117

Fasoranti JO, Ajiboye DO (1993) Some edible insects of Kwara state, Nigeria. Am Entomol 39(2):113–116

Food and Agriculture Organization (FAO) (2010) The state of world fisheries and aquaculture 2010. FAO, Fish Aquaculture Department, Rome http://www.fao.org/docrep/013/i1820e/i1820e00.htm

Food and Agriculture Organization (FAO) (2012) Assessing the potential of insects as food and feed in ssuring food security. Presented at Technical Consultation Meet, FAO, Rome, Italy, January 23–25

Food and Agriculture Organization/World Health Organization (FAO/WHO) (2008) Guideline for the validation of food safety control measures. FAO/WHO, Rome

Fayaz Bhat Z, Fayaz H (2011) Prospectus of cultured meat—advancing meat alternatives. J Food Sci Technol 48:125–140

Fiala N (2008) Meeting the demand: an estimation of potential future greenhouse gas emissions from meat production. Ecol Econ 67:412–419

Finke MD (2002) Complete nutrient composition of commercially raised invertebrates used as food for insectivores. Zoo Biol 21:269–285

Fischer ARH, Frewer LJ (2009) Consumer familiarity with foods and the perception of risks and benefits. Food Qual Prefer 20:576–585

Flachowsky G (2002) Efficiency of energy and nutrient use in the production of edible protein of animal origin. J Appl Anim Res 22:1–24

Fleurence J (1999) Seaweed proteins: biochemical, nutritional aspects and potential uses. Trends Food Sci Technol 10:25–28

Frewer LJ, Bergmann K, Brennan M, Lion R, Meertens R et al (2011) Consumer response to novel agri-food technologies: implications for predicting consumer acceptance of emerging food technologies. Trends Food Sci Technol 22:442–456

Gomez P, Halut R, Collin A (1961) Production of animal proteins in the Congo. Bull Agric Congo Belge 52:689–815

Gope B, Prasad B (1983) Preliminary observation on the nutritional value of some edible insects of Manipur. J Adv Zool 4(1):55–61

Hackstein JH, Stumm CK (1994) Methane production in terrestrial arthropods. Proc Natl Acad Sci USA 91:5441–5445

Hodgson EF (1955) An ecological study of the leaf-cutting ant, Atta cephalotes. Ecology 36:298–304

Holden S (1991) Edible caterpillars – a potential agroforestry resource? Food Insects Newsl 4(2):3–4

Holt VM (1885) Why not eat insects?. Field & Tuer

Houghton JR, van Kleef E, Rowe G, Frewer LJ (2006) Consumer perceptions of the effectiveness of food risk management practices: a cross-cultural study. Health Risk Soc 8:165–183

Hwangbo J, Hong EC, Jang A, Kang HK, Oh JS et al (2009) Utilization of house fly–maggots, a feed supplement in the production of broiler chickens. J Environ Biol 30:609–614

Ingold T (1994) Compendium encyclopedia of anthropology. Routledge, London

Jäch MA (2003) Fried water beetles: Cantonese style. Am Entomol 49:34–37

Jones S, Martin RD, Pilbeam DR, Bunney S (eds) (1992) The Cambridge encyclopedia of human evolution. Cambridge University Press, Cambridge

Kang K, Pulver SR, Panzano VC, Chang EC, Griffith LC et al (2010) Analysis of Drosophila TRPA1 reveals an ancient origin for human chemical nociception. Nature 464:597–600

Katayama N, Ishikawa Y, Takaoki M, Yamashita M, Nakayama S et al (2008) Entomophagy: a key to space agriculture. Adv Space Res 41:701–705

Kevan PG, Bye RA (1991) Natural history, sociobiology, and ethnobiology of Eucheira socialis Westwood (Lepidoptera: Pieridae), a unique and little-known butterfly from Mexico. Entomologist

Khempaka S, Chitsatchapong C, Molee W (2011) Effect of chitin and protein constituents in shrimp head meal on growth performance, nutrient digestibility, intestinal microbial populations, volatile fatty acids, and ammonia production in broilers. J Appl Poult Res 20:1–11

King DA, Peckham C, Waage JK, Brownlie J, Woolhouse MEJ (2006) Epidemiology infectious diseases: preparing for the future. Science 313:1392–1393

Klunder HC, Wolkers-Rooijackers J, Korpela JM, Nout MJR (2012) Microbiological aspects of processing and storage of edible insects. Food Control 26:628–631

Kok R, Lomaliza K, Shivhare US (1988) The design and performance of an insect farm/chemical reactor for human food production. Can Agric Eng 30:307–317

Latham P (1999) Edible caterpillars of the BasCongo Region of the Democratic Region of the Democratic Republic of Congo. Antenna 23:135–139

Lawal OA, Banjo AD (2007) Survey for the usage of arthropods in traditional medicine in Southwestern Nigeria. J Entomol 4:104–112

Lee CG, Silva CAD, Lee J-Y, Hartl D, Elias JA (2008) Chitin regulation of immune responses: an old molecule with new roles. Curr Opin Immunol 20:684–689

Liu Q, Tomberlin JK, Brady JA, Sanford MR, Yu Z (2008a) Black soldier fly (Diptera: Stratiomyidae) larvae reduce *Escherichia coli* in dairy manure. Environ Entomol 37:1525–1530

Liu J, Yang H, Savenije HHG (2008b) China's move to higher-meat diet hits water security. Nature 454:397

Liu P, Piao XS, Thacker PA, Zeng ZK, Li PF et al (2010) Chito-oligosaccharide reduces diarrhea incidence and attenuates the immune response of weaned pigs challenged with *Escherichia coli* K881. J Anim Sci 88:3871–3879

Luo ZY (1997) Insects as food in China. Ecol Food Nutr 36(2–4):201–207

Madibela OR, Seitiso TK, Thema TF, Letso M (2007) Effect of traditional processing methods on chemical composition and in vitro true dry matter digestibility of the mophane worm (Imbrasia belina). J Arid Environ 68:492–500

Madibela OR, Mokwena KK, Nsoso SJ, Thema TF (2009) Chemical composition of mopane worm sampled at three different sites in Botswana and subjected to different processing. Trop Anim Health Prod 41:935–942

Malaisse F, Parent G (1980) Les chenilles comestibles du Shaba méridional (Zaïre). Naturalistes Belges 61(1):2–24

Martins Y, Pliner P (2005) Human food choices: an examination of the factors underlying acceptance/rejection of novel and familiar animal and nonanimal foods. Appetite 45:214–224

Mbata KJ (1995) Traditional use of arthropods in Zambia. Food Insects Newsl 8(1):5–7

Mbata KJ, Chidumayo EN, Lwatula CM (2002) Traditional regulation of edible caterpillar exploitation in the Kopa area of Mpika District in northern Zambia. J Insect Conserv 6:115–130

Mbetid-Bessane E (2005) Commercialization of edible caterpillars in Central African Republic. Tropicultura 23:3–5

McGregor J (1991) Woodland resources: ecology, policy and ideology. An historical case study of woodland use in Shurugwi Communal Area, Zimbabwe. PhD Dissertation, Loughborough University of Technology. (Introduction and several orders and families)

Mercer CWL (1993) Insects as food in Papua New Guinea. In: Hardouin J, Stievenart C (eds) Invertebrates (Minilivestock) Farming., pp 157–162

Mercer CWL (1994) Sago grub production in Labu swamp near Lae – Papua New Guinea. Klin (J Forestry Papua New Guinea) 5(2):30–34

Meyer-Rochow VB, Nonaka K, Boulidam S (2008) More feared than revered: insects and their impact on human societies (with some specific data on the importance of entomophagy in a Laotian setting). Entomol Heute 20:3–25

Michaelsen KF, Hoppe C, Roos N, Kaestel P, Stougaard M et al (2009) Choice of foods and ingredients for moderately malnourished children 6 months to 5 years of age. Food Nutr Bull 30:343–404

Mitsuhashi J (1997) Insects as traditional foods in Japan. Ecol Food Nutr 36(2–4):187–199

Mpuchane S, Gashe BA, Allotey J, Siame B, Teferra G, Dithlogo M (2000) Quality deterioration of phane, the edible caterpillar of an emperor moth Imbrasia belina. Food Control 11:453

Msangi S, Rosegrant MW (2011) Feeding the future's changing diets: implications for agriculture markets, nutrition, and policy. In: 2020 conference: leveraging agriculture for improving nutrition and health. International Food Policy Research Institute, Washington, DC

Müller O, Krawinkel M (2005) Malnutrition and health in developing countries. CMAJ 173:279–286

Munthali SM, Mughogho DE (1992) Economic incentives for conservation: beekeeping and Saturniidae caterpillar utilization by rural communities. Biodiv Conserv 1(3):143–154

Muzzarelli RAA (2010) Chitins and chitosans as immunoadjuvants and non-allergenic drug carriers. Mar Drug 8:292–312

Nakagaki BJ, deFoliart GR (1991) Comparison of diets for mass-rearing *Acheta domesticus* (Orthoptera: Gryllidae) as a novelty food, and comparison of food conversion efficiency with values reported for livestock. J Econ Entomol 84:891–896

Nel E, Illgner P (2001) Tapping Lesotho 's' white gold': inter-basin water transfer in Southern Africa. Geography 86(2):163–167

Nelson GC, Rosegrant M, Koo J, Robertson R, Sulser T et al (2009) Climate change: impact on agriculture and costs of adaptation. Food policy report. International Food Policy Research Institute, Washington, DC

Newton L, Sheppard C, Watson DW, Burtle G, Dove R (2005) Using the black soldier fly, *Hermetia illucens*, as a value-added tool for the management of swine manure. Report for Mike Williams, Dir. Animal Poultion Waste Management Center, North Carolina State University, Raleigh
Ngoka BM, Kioko EN, Raina SK, Mueke JM, Kimbu DM (2008) Semi-captive rearing of the African wild silkmoth Gonometa postica (Lepidoptera: Lasiocampidae) on an indigenous and a non-indigenous host plant in Kenya. Int J Trop Insect Sci 27:183–190
Nishimune T, Watanabe Y, Okazaki H, Akai H (2000) Thiamin is decomposed due to Anaphe spp. entomophagy in seasonal ataxia patients in Nigeria. J Nutr 130:1625–1628
Offenberg J, Wiwatwitaya D (2009) Sustainable weaver ant (*Oecophylla smaragdina*) farming: harvest yields and effects on worker ant density. Asia Myrmecol 3:55–62
Ohiokpehai O (2003) Nutritional aspects of street foods in Botswana. Pak J Nutr 2:76–81
Oliveira JS, de Carvalho JP, De Sousa RB, Simao MM (1976) The nutritional value of four species of insects consumed in Angola. Ecol Food Nutr 5(2):91–97
Oonincx DGAB, Dierenfeld ES (2011) An investigation into the chemical composition of alternative invertebrate prey. Zoo Biol. 29:1–15
Oonincx DGAB, van Itterbeeck J, Heetkamp MJW, van den Brand H, van Loon JJA, van Huis A (2010) An exploration on greenhouse gas and ammonia production by insect species suitable for animal or human consumption. PLoS One 5:e14445
Pan A, Sun Q, Bernstein AM, Schulze MB, Manson JE et al (2012) Redmeat consumption and-mortality: results from two prospective cohort studies. Arch Intern Med 172:1134845
Paoletti MG (2005) Ecological implications of minilivestock: potential of insects, rodents, frogs and snails. Science, Enfield, 648 pp
Parker AG (2005) Mass-rearing for sterile insect release. In: VA D, Heindrichs J, AS R (eds) Sterile insect technique. principles and practice in area-wide integrated pest management. Springer, Dordrecht, pp 209–232
Pimentel D, Pimentel M (2003) Sustainability of meat-based and plant-based diets and the environment. Am J Clin Nutr 78(Suppl. 3):660S–663S
Pimentel D, Berger B, Filiberto D, Newton M, Wolfe B et al (2004) Water resources: agricultural and environmental issues. Bioscience 54:909–918
Ramos-Elorduy J (2006) Threatened edible insects in Hidalgo, Mexico and some measures to preserve them. J Ethnobiol Ethnomed 2:1–10
Ramos-Elorduy J, Pino JM (1990) Contenido calórico de algunos insectos comestibles de México. Soc Quim Mex 34(2):56–68
Ramos-Elorduy de Conconi J, Bourges Rodríguez H, Pino Moreno JM (1982) Valor nutritivo y calidad de la proteína de algunos insectos comestibles de México. Folia Entomológica Mexicana, Mexico
Ramos-Elorduy J, Moreno P, Ramos-Elorduy JMJ, Moreno JMP (1989) Los insectos comestibles en el México antiguo: Estudio etnoentomológico (No. 641.696 R3I55 F1219. 3. F7)
Ramos-Elorduy J, González EA, Hernández AR, Pino JM (2002) Use of Tenebrio molitor (Coleoptera: Tenebrionidae) to recycle organic wastes and as feed for broiler chickens. J Econ Entomol 95(1):214–220
Reese TA, Liang H-E, Tager AM, Luster AD, Rooijen NV et al (2007) Chitin induces accumulation in tissue of innate immune cells associated with allergy. Nature 447:92–96
Ruddle K (1973) The human use of insects: examples from the Yukpa. Biotropica 5:94–101
Saeed T, Dagga FA, Saraf M (1993) Analysis of residual pesticides present in edible locusts captured in Kuwait. Arab Gulf J Sci Res 11:1–5
Silow CA (1976) Edible and other insects of Mid-Western Zambia, Studies in Ethno-Entomology II. Almqvist/Wiksell, Stockholm, 223 pp
Simpanya MF, Allotey J, Mpuchane SF (2000) Amycological investigation of phanen and edible grasshopper of an emperor moth. Imbrasia belina J Food Prot 63:137–140
Smil V (2002a) Eating meat: evolution, patterns, and consequences. Popul Dev Rev 28:599–639
Smil V (2002b) Worldwide transformation of diets, burdens of meat production and opportunities for novel food proteins. Enzym Microb Technol 30:305–311

Steinfeld H, Gerber P, Wassenaar T, Castel V, Rosales M (eds) (2006) Livestock's long shadow: environmental issues and options. Food Agric. Organ, Rome, 319 pp

Sungpuag P, Puwastien P (1983) Nutritive values of protein food sources of rural people: insects. J Nutr Assoc Thailand 17:5–12

Szelei J, Woodring J, Goettel MS, Duke G, Jousset FX et al (2011) Susceptibility of North-American and European crickets to *Acheta domesticus* densovirus (AdDNV) and associated epizootics. J Invertebr Pathol 106:394–399

Tacon AGJ, Metian M (2008) Global overview on the use of fish meal and fish oil in industrially compounded aquafeeds: trends and future prospects. Aquaculture 285:146–158

Teguia A, Beynen AC (2005) Alternative feedstuffs for broilers in Cameroon. Livest Res Rural Dev 17:34 http://www.lrrd.org/lrrd17/3/tegu17034.htm

Tindale NB (1966) Insects as food for the Australian aborigines

Trostle R (2008) Global agricultural supply and demand: factors contributing to the recent increase in food commodity prices. Economic research service report WRS-0801, US Department of Agriculture, Washington, DC, pp. 1–30, July 2008 rev

van Huis A (2003) Insects as food in sub-Saharan Africa. Insect Sci Appl 23:163–185

Van Huis A (2013) Potential of insects as food and feed in assuring food security. Annu Rev Entomol 58:563–583

Van Itterbeeck J, van Huis A (2012) Environmental manipulation for edible insect procurement: a historical perspective. J Ethnobiol Ethnomed 8:1–19

Van Mele P (2008) A historical review of research on the weaver ant Oecophylla in biological control. Agric For Entomol 10:13–22

Vantomme P, Göhler D, N'Deckere-Ziangba F (2004) Contribution of forest insects to food security and forest conservation: the example of caterpillars in Central Africa. Odi Wildl Policy Brief 3:1–4

Vara-asavapati V, Visuttipart J, Maneetorn C (1975) Edible insects in north-east Thailand. Res

White CMN (1961) Elements in Luvale beliefs and rituals (No. 32). Rhodes-Livingstone Institute by the Manchester University Press and in the USA by Humanities Press, New York

Wilson KB (1989) The ecology of wild resource use for food by rural southern Africans: why it remains so important. Paper presented to the conference at the destruction of the environment and the future of life in the Middle East and Africa, 14–17 July 1989. (Introduction)

Xia W, Liu P, Zhang J, Chen J (2011) Biological activities of chitosan and chitooligosaccharides. Food Hydrocoll 25:170–179

Yen AL (2009) Entomophagy and insect conservation: some thoughts for digestion. J Insect Conserv 13:667–670

Yhoung-Aree J, Puwastien P, Attig GA (1997) Edible insects in Thailand: an unconventional protein source? Ecol. Food Nutr. 36:133–149

Yhoung-Aree J, Viwatpanich K (2005) Edible insects in the Laos PDR, Myanmar, Thailand, and Vietnam. See Ref 115:415–440

Zhang CX, Tang XD, Cheng JA (2008) The utilization and industrialization of insect resources in China. Entomol Res 38(s1):S38–S47

Zuidhof MJ, Molnar CL, Morley FM, Wray TL, Robinson FE et al (2003) Nutritive value of house fly (*Musca domestica*) larvae as a feed supplement for turkey poults. Anim Feed Sci Technol 105:225–230

Entomoceuticals

16

Geetanjali Mishra and Omkar

Contents

16.1 Introduction

Humans have drawn inspiration from both animals and plants to fuel and sustain their growth as they lay increasing claim over Earth and its resources. Of all life forms, insects have been one of the most easily accessible resources and inspirations as they form one of the most ubiquitous groups on Earth. The associations so formed amongst insects and humans have ranged from harmful to beneficial.

The harmful associations in forms of pests of crops, vectors of disease, nuisance value and sources of fears and phobias have been much highlighted in research as well as in the minds of the populace in general. However, the beneficial associations of insects and humans have been underplayed with the exception of their silk-, honey- and lac-producing abilities. Their role as biological control agents for pest management and as sources of cheap yet high-quality nutrition source has only come into the limelight in recent years. Other than these benefits, the potential role of insects in medical sciences as therapies and cures has been massively ignored.

Most of us do not realize that over 70% drugs in the market are derived from natural sources/compounds. This is primarily because natural compounds have evolved to optimally perform functions such as binding to specific target proteins or interacting with membranes, amongst others. Most of these compounds are however

G. Mishra (✉) • Omkar
Department of Zoology, University of Lucknow, Lucknow 226007, Uttar Pradesh, India
e-mail: geetanjalimishra@hotmail.com; omkaar55@hotmail.com

Omkar (ed.), *Industrial Entomology*, DOI 10.1007/978-981-10-3304-9_16

of plant origins. Trowell (2003) points out that there are at least 16 times as many insect species as there are plant species, yet plant chemistry has been studied 7000 times as much as insect chemistry when comparing the amount of research per species.

Of all compounds produced in nature, toxic substances or venoms in particular are attractive lead compounds for drugs owing to their cyto- and neurotoxicity as well as their microbicidal properties. Insects in particular are potential prime leaders in such research because of the many such compounds that they produce. So most of what makes insects scary is also what makes them medically relevant!

Unfortunately, as previously stated, such beneficial medical applications of insects, which should be studied under the field of medical entomology, are most often neglected, while the attention remains firmly focused on disease-causing insects. Owing to this, a new field has been created in entomology to deal with the potential uses of insects as medicines and has been christened as "entomotherapy" and/or "pharmaceutical entomology". The former is a broader term including the use of live and dead insects as well as crude and refined extracts and concoctions from insects for the treatment of human diseases/ailments. The latter on the other hand more specifically deals with identification, purification and utilization of lead compounds from insects as medicines.

While most agree on the immense potential for identifying and finding medically relevant drugs from insects, owing to the massive insect chemical biodiversity, there are relatively less publications on the topic. Only a few workers such as Costa-Neto, (2005) and Ramos-Elorduy et al. (1988) have suggested that insect-derived compounds are the new frontier for prototype drugs. Laurent et al. (2005) have mentioned in their review a few defensive insect chemicals that have been pursued for practical medicinal applications. Dossey (2010) has recently published a detailed review on the chemical weaponry of insects and its medicinal potential.

Large-scale survey on arthropods and their medicinal potential and uses, especially in cancer research, has been led by a few groups (Petit 1996; Pettit et al. 1968, 2010; Trowell 2003). The Trowell group which began systematic surveys into arthropod-based drug discovery at Australia's Commonwealth Scientific and Industrial Research Organisation (CSIRO) was so encouraged by the potential that it launched a company Entocosm (Trowell 2003). Pharmaceutical giant Merck is also not far behind in the discovery of a scorpion toxin which works as an immunosuppressant (Costa-Neto 2005). In 1991, the National Biodiversity Institute of Costa Rica (INBio) entered into a US$1 million agreement with Merck aimed at helping Costa Rica develop resources to protect its biodiversity effectively and to test insect extracts for their efficacy against infections, AIDS, cancer and inflammatory conditions (Raynor 1993).

While the potential is known and recognized by workers in the field, large-scale awareness of the potential and efficacy of insect-based therapies and medicines amongst the common populace and even academic or government institutions is severely lacking. The present chapter thus aims to bring to the attention of the entomologists and the people from drug industry the potential of using insects as therapists and doctors.

16.2 Traditional Use of Insects as Cures

The use of insects as medicines dates far back in time, with records of insect- and spider-derived medicines existing back to the sixteenth century BC, in Ebers Papyrus of Egypt (Weiss 1947). Chinese traditional medicine has been using larvae of *Bombyx mori L.* for a greater part of the last 3000 years (Zimian et al. 1997). The father of surgery in India, Susruta (ca. 600 BCE), is reported to have used ant mandibles as biodegradable sutures (Rastogi 2011; Fig. 16.1).

The wound healing properties of larvae of some dipteran flies have been known for centuries (Sherman et al. 2000). In *Naturalis historiae*, Pliny the Elder mentions some entomotherapeuticals (insect-derived medicines) employed in the first century AD (Carrera 1993). Dioscorides has also mentioned some insect remedies (Morge 1973). Namba et al. (1988) have discussed 54 kinds of crude drugs derived from insects listed in the "herbal" *Jingshi Zhenglei Daguang Bencao* (AD 1108), edited during the Chinese Song dynasty (AD 960–1280). The list was extended to 73 species in Li Shizhen's *Bencao Gangmu* (*Compendium of Materia Medica*) published in 1578, and 11 additional species were added in the *Supplement to Compendium Materia Medica* by Chao Xueming in 1756 (Chen 1994). Today, 143 medicinal insects are used in China (Zimian et al. 1997). Marie (1955) made an outstanding contribution to the history of entomotherapy by describing the use of 33 kinds of medicinal insects.

Pliny the Elder explained that cricket (possibly *Acheta domesticus*) was good for catarrhs and inflammations of the amygdales, for earache (if applied jointly with soil in which it lived) and for weak sight and it also had diuretic properties (Weiss 1947). Dioscorides in *Materia Medica* states that "… grasshoppers in a suffumigation relieve under a dysury especially such as is incident to the female sex" and that *Locusta africana* is a good antidote for scorpion poison (Dioscorides 1547).

Across countries, the use of insects in traditional or alternative medicine, as we call it, is well reported. The native population of Native America is known to extensively use insects as medicines (Costa-Neto 2002, 2005; Ferreira et al. 2009; Ratcliffe et al. 2011). In Brazil, up to 42 insect species from 9 orders in the various states of Brazil (Costa-Neto 2002) are used in medicines. Ramos-Elorduy et al. (1988) enlisted 43 species of insects (from 16 families and 6 orders) being used in Aztec traditional medicine as well as modern Mexican culture. Several examples of

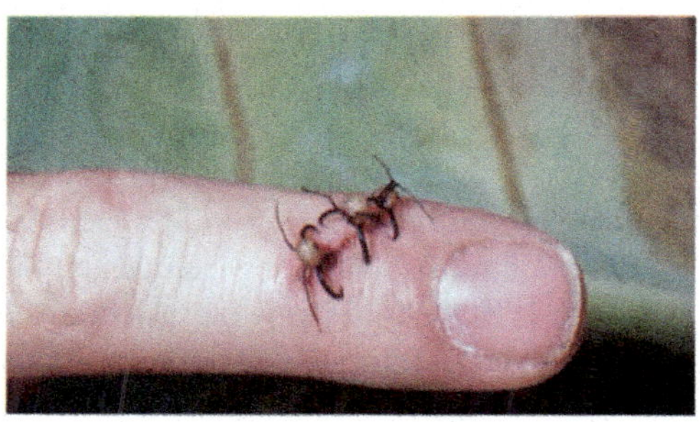

Fig. 16.1 Driver ant (*Dorylus*) heads being used as sutures

insect use in traditional Asian medicine exist in the literature with East Asian traditional healing systems utilizing a huge number of insects (Read 1935; Pemberton 1999; Oudhia 2005). Insects are also utilized in China for a number of industrial processes including mass production of drugs and pesticides (Zhang et al. 2008). In one of the largest world market of traditional medicines in South Korea, two of the three most used medicines were obtained from *Scolopendra* and silkworm infected with *Beauveria bassiana* (Pemberton 1999). Oudhia (2005) in his massive compendium has enlisted a huge number of arthropods being used in traditional medicine systems in India, and the diversity can be gauged by the very fact that in a single state of Chhattisgarh, over 500 species of insects and arachnids are used by over 3500 healers. Europe remains the poorest in the use of insects as medicines when compared to the rest of the world.

The orders most used are Coleoptera, Hymenoptera, Orthoptera and Homoptera. Of 411 medicinal insect species recorded worldwide, 92.6% are utilized to relieve internal diseases and 353 disorders: for skin (57), digestive (58), respiratory (34), kidney (22), reproductive (31), circulatory (31), nervous (28), neuromuscular (13), eye (21), bone (19), immunological (6), hearing (4), endocrinological (4) and other diseases (43) (Motte-Florac et al. 2002). We attempt below to list a few examples of the major orders of insects, which are traditionally used as cures.

16.2.1 Orthoptera

Pulverized cockroaches are a treatment for epilepsy (Ratcliffe 1990). In many parts of Brazil, the hind legs of the common cricket, *Acheta domesticus* (L.), are variously used to treat dandruff, asthma, eczema, earache, vomits in children, intestinal parasites, fever, rheumatism, children that urinate in bed, children slow in talking, urine retention, oliguresis, pterygium, etc. (Costa-Neto and Ramos-Elorduy 2006). The gut contents of mole crickets are smeared on feet to ward off foot infections (Fasoranti 1997). Rice green leafhopper, *Nephotettix nigropictus* (Stal.), is used to treat diseases such as gonorrhoea by applying a freshly crushed paste of green leafhoppers to the affected area. In the Indian state of Manipur, nymphs and adults of *Gryllotalpa orientalis* Burmeister are used as well as roasted for asthma, sprain and external infections, Also, *Locusta migratoria* (L.) is used as nutritional supplement and for blood fortification and relief of chronic cough (Singh 2014).

16.2.2 Homoptera

Oil from several bugs is used in Mexico to treat skin diseases caused by *Mycobacterium tuberculosis*, while mealybugs are boiled until a sticky mass and used for treatments as varying from lesions and burns to cleaning teeth. In India, the abdomen is rubbed on the white patches of the skin (leucoderma) 2–3 times daily with positive results visible in a weeks' time (Singh 2014). Even the bedbug is used for reducing pain and inflammation in the extremities (Singh 2014).

16.2.3 Hymenoptera

The large stinging ant with the most painful sting of any ant in the world, *Paraponera* sp., is used by people in Brazil to treat rheumatism and backaches. The venom from these ants contains a potent peptide neurotoxin called poneratoxin, which may be amenable to pharmacological adaptation as a painkiller.

Ammonia-like fumes produced by rubbing weaver ant, *Oecophylla smaragdina* Fabricius, are inhaled by Tamil labourers in India to relieve symptoms of common cold (Veeresh 1999). Tribal people in Koraput District of Orissa eat the brood of *O. smaragdina*, to keep the body and mind cool in hot summer, and utilize the workers as food so as to improve their eyesight (Anon 2009). The Paniyan tribes of Kerala in India use the mud taken from the interior of unidentified ant nests to treat rabies (Wilsanand et al. 2007). *Oecophylla smaragdina* is also documented to be useful in the treatment of cold and cough in Australia (Crozier et al. 2010 and references therein). Chinese medicinal ant, *Polyrhachis lamellidens* Smith, found throughout southern China and Taiwan, has been extensively used as folk medicine for the treatment of rheumatoid arthritis, chronic hepatitis and ageing (Cheng et al. 2001, and references therein). Most ant species used in entomotherapy also comprise part of the normal diet of local people.

Ancient European literature mentions the use of ant extracts for treatment of sore eyes, weak vision and cataracts (Lockhart 2000 and references therein).

16.2.4 Diptera

The common housefly is used to treat skin infections in the forms of boils caused by *Staphylococcus aureus*, a strain of which has become antibiotic resistant leading to massive MRSA outbreaks and urging the need for rapid discovery of new antibiotics. In such a scenario, insects could provide strong new leads. The houseflies are crushed and applied directly to the boils. Tsetse fly paste is applied to the bites of the tsetse fly itself after making an incision, so as to protect the persons from sleeping sickness, perhaps a crude precursor to vaccine production (Antonio 1994).

16.2.5 Coleoptera

During the seventeenth century, people in Europe believed that many kinds of insects had some healing power (Wigglesworth 1976). Examples include the belief that the oil obtained from the larvae of the May beetle, *Melolontha vulgaris* (L.), can be used topically on scratches and other wounds and as a cure for rheumatism and that the adult beetles soaked in wine are helpful in treating anaemia (Ratcliffe 1990). Blister beetles have been used by the Chinese for removing warts and cancer treatment (Moed et al. 2001; Kunert et al. 2008) and by the Greeks for enhancing sexual libido (Moed et al. 2001); the latter treatment could not be confirmed in modern investigations. Larvae (or grubs) of the flower beetle *Protaetia brevitarsis*

Lewis (Scarabaeidae) have been used to treat hepatic cancer, liver cirrhosis, hepatitis, breast cancer and inflammatory disease in Asia (Lee et al. 2014 and references therein). The darkling beetle, *Palembus dermestoides* (Fairmaire) (Coleoptera: Tenibrionidae), is used in Brazil to treat sexual impotence as well as asthma, tuberculosis and arthritis.

There are numerous such documented as well as undocumented cultural uses of insects across the globe through all centuries. However, the advance of medical science and the onslaught of "modern knowledge" have led to the suppression of folk knowledge regarding insects as medicines (Holt 1992). It is now time to revisit it and realize the massive potential that both entomotherapy and pharmaceutical entomology hold. For this, we need to go back to the roots and use traditional knowledge as a source. Ethnoentomology studies conducted by the modern-day anthropologists as well as going through ancient texts will help. Such practices have actually begun to help in modern-day medicine with many treatments being slowly and steadily researched, accepted and approved globally. The next section deals with the current as well as probable use of insects in modern-day practices.

16.3 Modern-Day Entomotherapy Practices

16.3.1 Live Insects

Here in this section, we will discuss the many ways in which live insects have been used as cures or in therapies for multiple ailments. Leading the way in insect therapy is the much studied use of maggots for wound healing followed by the use of honey bees for relief from muscular as well as joint pain.

16.3.1.1 Maggots

Incidences of use of maggots of flies (usually blow flies, in particular *Lucilia sericata* (Meigen)) for healing wounds can be traced back to the Middle Ages as well as to many parts of the world (Asia, South America and Australia; Whitaker et al. 2007). The Maya, for example, are said to have used maggots for therapeutic purposes a thousand years ago (Zimmer 1993). Their use in cleaning of deep wounds is primarily because of their predilection for necrotic tissue. Fly larvae are able to abet the healing of wounds via (1) larval secretions that aided breakage of fibronectin and collagen into smaller fragments which promotes fibroblast aggregation and tissue repair (Horobin et al. 2003), (2) consumption of necrotic tissue, (3) release of antibacterial substances and (4) destruction of ingested bacteria (Whitaker et al. 2007).

The maggots that are currently employed in wound healing are grown under sterile conditions. Multiple studies indicate that while maggot therapy does not affect the rate or timing of healing vis-a-vis conventional hydrogel dressing, it did significantly increase the debridement process, though the pain score was high for such patients (Courtenay et al. 2000; Dumville et al. 2009). Maggot therapy has been found effective in curing chronic wounds (Sherman et al. 2000; Steenvoorde et al. 2007; Turkmen et al. 2010; Fig. 16.2).

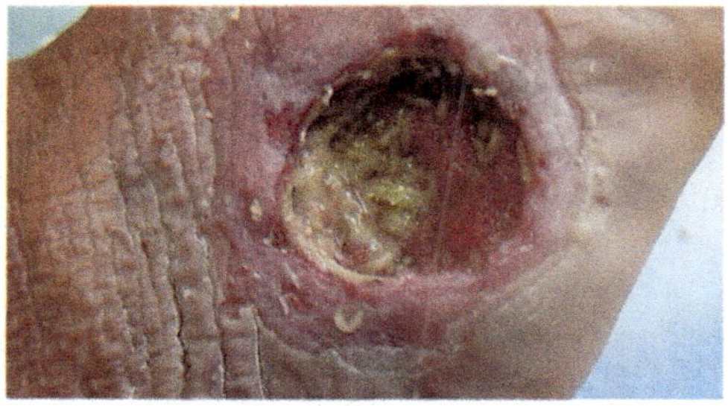

Fig. 16.2 Maggot therapy in a diabetic foot

16.3.1.2 Apitherapy

Honey bees are like the "Kamadhenu cows" of the insect world. They not only keep our agricultural and horticultural produce high via pollination but also provide amazing nutritional substances in the form of honey, royal jelly and propolis as well as beeswax for myriad purposes. Even the venom of the honey bees has medicinal properties.

Bee venom is a compound containing immunoreactive and neuroactive peptides, enzymes, glucose, fructose and water. Owing to its anti-inflammatory nature, bee venom is being used as a cure for arthritis (Park et al. 2004; Son et al. 2007), multiple sclerosis (Castro et al. 2005) as well as pain (Luo et al. 1998). The application of bee venom ranges from the more refined (extraction followed by injection) to the crude (direct stings or apipuncture; Son et al. 2007). Apipuncture has an antinociceptive effect on chemically induced pain in rats (Kwon et al. 2005). Several clinical trials have been conducted on the effect of apipuncture on arthritis in Korea but with differing results (Lazner et al. 1999; Lee et al. 2005; Fig. 16.3).

16.3.2 Insect Products and Extracts

16.3.2.1 Insect Products

Insects, through their biological and behavioural activities, are known to produce a large number of products, such as silk, honey, lac, propolis, beeswax, etc. Of these some have medicinal properties and have primarily been a part and parcel of the traditional grandmother cures.

Honey

Honey, also known as liquid gold, and a product with high nutritional content, has historically been used in many cultures as a cure for ailments ranging from wounds, infections and bowel disorders. Though honey composition varies with the bee species as well as floral characteristics, yet it always possesses antioxidant and antimicrobial properties. p-Hydroxybenzoic acid, naringenin, pinocembrin and chrysin are four major antimicrobial and antioxidant compounds of honey, with even the carbohydrates being antimicrobial in nature (Estevinho et al. 2008). Honey is largely considered as anti-mutagenic (Wang et al. 2002).

Fig. 16.3 Apitherapy in progress

Most studies on the medicinal efficacy of honey have been conducted on its wound healing properties, which are attributed to its osmotic properties (that moisturize the wound bed and reduce the risk of maceration) and anti-inflammatory property (inhibits fibrin which slows down wound repair; Fig. 16.4). The efficacy of honey as a wound healer has varied with studies, but all studies have found it to be an equal or a better healing agent than conventional therapies (Al-Waili and Saloom 1999; Moore et al. 2001; Marshall et al. 2005; McIntosh and Thomson 2006). Honey is also regarded as an excellent burn salve than conventional dressings (Fig. 16.4). Other than these, honey has also been used as a cure for many infectious diseases (Kilicoglu et al. 2006; Nilforoushzadeh et al. 2007), skin conditions such as dermatitis and dandruff (Al-Waili 2001), gastrointestinal disorders (Prakash et al. 2008) and allergic rhinitis.

Royal Jelly

Royal jelly is secreted from the mandibular and hypopharyngeal glands of the worker bees and is extensively used in Oriental medicine. It is known to stimulate production of collagen and bone (Miyata 2007), restore estrogenic properties (Mishima et al. 2005), promote healing in the tympanum (Calli et al. 2008) and liver (Kamakura et al. 2001) and suppress diseases by enhancing immune function and stimulating antibody production (Mannoor et al. 2009).

Propolis

Propolis is a polyphenol-rich plant-based hive sealant formed by the bees. It is being used in Egypt and Greece for medicinal purposes since antiquity as anti-inflammatory, antimicrobial, antiviral, estrogenic and antineoplastic (Song et al. 2002; Khalil 2006; Farnesi et al. 2009) agents. It has been studied for its potential as a filler for dental carries (Libério et al. 2009).

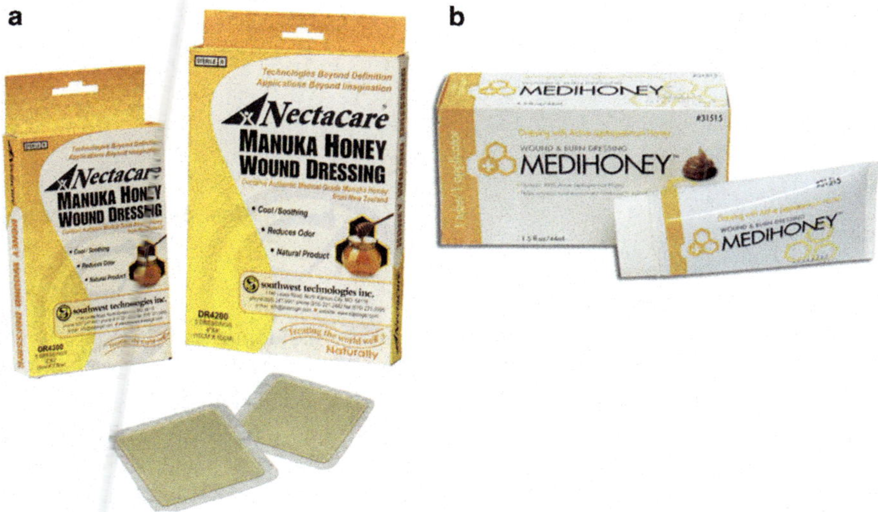

Fig. 16.4 Honey (**a**) wound dressings and (**b**) burn salves

Insect Venom

Insect venom or defence chemicals hold great medicinal potential. Investigations on a whole range of defence chemicals have thrown up rather interesting examples and potential chemicals for use as medicines.

Bee venom is one such potential medicine, which is anti-inflammatory with modulation of pain perception. Injections of bee venom are becoming increasingly commonplace in alternative medicines. Another venom that has great potential is the ant venom, which is anti-complement (Altman et al. 1984) and used in the treatment of arthritis. Isolated venom components such as melittin, apamin and adolapin were found to be anti-inflammatory and are also being individually used as medicines. Masroporan-1, a component of hornet venom, is anti-inflammatory as well as antimicrobial.

One of the most exciting recent discoveries in stick insect defensive chemistry is that of a novel compound parectadial (Butenandt et al. 1961) from the species *Parectatosoma mocquerysi* of Madagascar (Burks 2007; Dossey et al. 2006; Fig. 16.5). The defensive secretion from *P. mocquerysi* is reported to cause reddening and peeling of the skin, without any pain or irritation (Dossey et al. 2006). These effects suggest that it may possess some pharmacologically useful properties. Perillyl alcohol and perillaldehyde are components which have both been explored for use against cancer in a number of studies (Elegbede et al. 2003; Fernandes et al. 2005; Stratton et al. 2010). Several studies have shown perillyl alcohol to have anticancer properties, such as chemopreventive activity against skin carcinogenesis and related skin damage (Stratton et al. 2010), as well as activity against lung cancer (Yeruva et al. 2007) and breast cancer cells (Yeruva et al. 2010). It has also been the subject of several clinical trials (Ripple et al. 2000; da Fonseca et al. 2008).

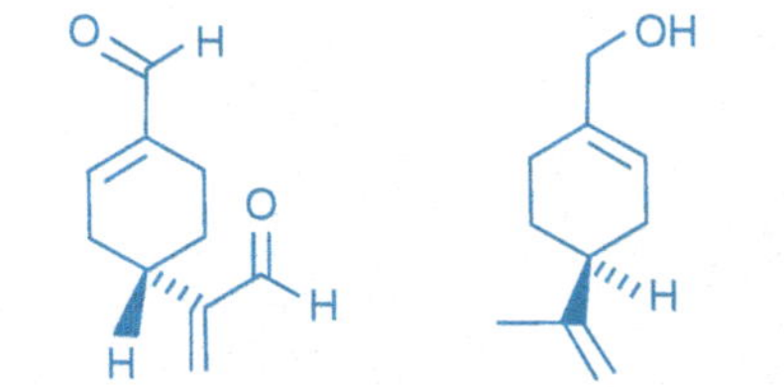

(4S)-Parectadial *S*-Perillyl alcohol

Fig. 16.5 Important components of insect venom (Dossey et al. 2006)

16.3.2.2 Insect Extracts

Recent research demonstrates that social insects including ants possess well-developed immune systems and disease resistance ability (Rosengaus et al. 1999; Fefferman et al. 2007; Stow and Beattie 2008; Wilson-Rich et al. 2009). Antibiotics produced by the paired metapleural glands of ants are secreted externally (Beattie et al. 1986) and provide protection against pathogenic fungi and bacteria (Fernández-Marín et al. 2006; Poulsen et al. 2006). It is suggested that high microbial parasites and pathogen pressure have led to the evolution of immune proteins in social insects (Viljakainen et al. 2009) including ants (Schluns and Crozier 2009). Extract of *P. dives* is used as a sexual stimulant, to enhance immunity and to treat rheumatism in China (Chen and Alue 1994).

A variety of antimicrobial (bactericidal/fungicidal) compounds which provide protection against environmental pathogens have been isolated and characterized from social insects, particularly ants (see review by Schluns and Crozier 2009). Two antibacterial peptides synthesized in the ant *Myrmecia gulosa* (Fabr.) in response to bacterial infection have been characterized (Macintosh et al. 1998). Extracts of the Eurasian ant, *Formica aquilonia* Yarrow, are found to exhibit antioxidant and anti-inflammatory properties (Piao et al. 2009). Also, 15 novel peptides exhibiting anti-bacterial and insecticidal properties have been isolated from the venom of the predatory ant, *Pachycondyla goeldii* (Forel) (Orivel et al. 2001).

Fungus-growing ants are found to ward off fungal parasites by means of mutualistic, antibiotic-producing bacteria which they harbour on special structures located on their cuticle (Currie et al. 2006). Recent studies demonstrate that ant-associated actinomycetes are highly diverse and are rarely specific in their inhibition of other microbes (Sen et al. 2009). Investigations of the proteome of the obligate intracellular endosymbiotic bacteria *Blochmannia floridanus*, found in the gut of carpenter ants, may lead to the production of new antibiotics that can target Gram-negative bacteria (Zientz et al. 2006; Ruiz et al. 2008).

Pharmacological investigations and clinical trials indicate that certain components of the ant *P. lamellidens* are active in the treatment of diseases including arthritis, rheumatism, liver ailments and asthma (Cheng et al. 2001 and references therein). Extracts of the *P. dives* demonstrate inhibition in ferric nitrilotriacetate-induced nephrotoxicity in laboratory rats (Ma et al. 1997). Further, the analgesic and anti-inflammatory activities of extracts of *P. lamellidens* have been demonstrated by Kou et al. (2005), while from the same species Jiang et al. (2008) have identified two polyketides, suggested to have potential in the treatment of rheumatoid arthritis.

16.4 Conclusions

Insects thus seem to harbour massive medicinal potential that can be harnessed to revolutionize medicine in the years to come. The success of the insects in all realms indicates their adaptability to multiple situations, a situation which reflects in their physiology and biochemistry. Thus, the compounds derived from insects are likely to be more potent. Antibiotics derived from insects have shown their immense efficacy against MRSA, which is currently the bane of doctors the world over. In a time when most existing medicines are reaching their sell by date, a huge influx of new potent chemicals can be obtained from insects. The age of antibiotics can now be replaced by the age of "entoceuticals".

References

Altman RD, Schultz DR, Collins-Yudiskas B, Aldrich J, Arnold PI, Brown HE (1984) The effects of a partially purified fraction of an ant venom in rheumatoid arthritis. Arthritis Rheum 27(3) 277–284

Al-Waili NS (2001) Therapeutic and prophylactic effects of crude honey on chronic seborrheic dermatitis and dandruff. Eur J Med Res 6(7):306–308

Al-Waili NS, Saloom KY (1999) Effects of topical honey on post-operative wound infections due to gram positive and gram negative bacteria following caesarean sections and hysterectomies. Eur J Med Res 4(3):126–130

Anon (2009) Strange food habits of tribals of Boipariguda. The Hindu Online Edition of India's National Newspaper

Antonio TM (1994) Insects as remedies for illnesses in Zaire. The Food Insects Newsl 7(3):4–5

Beattie AJ, Turnbull CL, Hough T, Knox RB (1986) Antibiotic production-a possible function for the metapleural glands of ants (Hymenoptera, Formicidae). Ann Entomol Soc Am 79:448–450

Burks R (2007) Insect's Venom eyed for cancer defense. Chem Eng News 85

Butenandt A, Beckmann R, Hecker E (1961) Über den sexuallockstoff des seidenspinners, i. der biologische test und die isolierung des reinen sexuallockstoffes bombykol. Hoppe-Seyler's Zeitschrift für physiologische Chemie 324(1):71–83

Calli C, Tugyan K, Oncel S, Pınar E, Demirtaşoglu F, Calli AO et al (2008) Effectiveness of royal jelly on tympanic membrane perforations: an experimental study. J Otolaryngol Head Neck Surg 37(2)

Carrera M (1993) A entomologia na Histo'ria natural de Plı'nio. Revista Brasileira de Entomologia 37(2):387–396

Castro HJ, Mendez-Inocencio, JI, Omidvar B, Omidvar J, Santilli J, Nielsen HS, Pavot AP, Richert JR, Bellanti JA (2005) A phase I study of the safety of honeybee venom extract as a possible treatment for patients with progressive forms of multiple sclerosis. Allergy Asthma Proc 26(6):470–476. OceanSide Publications, Inc

Chen Y (1994) Ants used as food and medicine in China. The Food Insects Newsl 7(2):1, 8–10

Chen Y, Alue RD (1994) Ants used as food and medicine in China. The Food Insects Newsl 7:1–10

Cheng Y, Tseng C, Wang H, Liao C (2001) Component analysis of black ant (*Polyrhachis lamellidens*) extracts from supercritical fluid extraction. J Food Drug Anal 9:72–78

Costa-Neto EM (2002) The use of insects in folk medicine in the state of Bahia, northeastern Brazil, with notes on insects reported elsewhere in Brazilian folk medicine. Hum Ecol 30:245–263

Costa-Neto EM (2005) Entomotherapy, or the medicinal use of insects. J Ethnobiol 25(1):93–114

Costa-Neto E, Ramos-Elorduy J (2006) Los insectos comestibles de Brasil: Etnicidad, diversidad e importancia en la alimentación. Boletín de la Sociedad Entomológica Aragonesa 38:423–442

Courtenay M, Church JC, Ryan TJ (2000) Larva therapy in wound management. J R Soc Med 93(2):72–74

Crozier RH, Newey PS, Schlüns EA, Robson SK (2010) A masterpiece of evolution–Oecophylla weaver ants (Hymenoptera: Formicidae). Myrmecological News 13(5)

Currie CR, Poulsen M, Mendenhall J, Boomsma JJ, Billen J (2006) Coevolved crypts and exocrine glands support mutualistic bacteria in fungus-growing ants. Science 311(5757):81–83

da Fonseca CO, Schwartsmann G, Fischer J, Nagel J, Futuro D, Quirico-Santos T, Gattass CR (2008) Preliminary results from a phase I/II study of perillyl alcohol intranasal administration in adults with recurrent malignant gliomas. Surg Neurol 70(3):259–266

Dioscorides Anazarbeus (1547) Materia Medcinale. Florence

Dossey AT (2010) Insects and their chemical weaponry: new potential for drug discovery. Nat Prod Rep 27(12):1737–1757

Dossey AT, Walse SS, Rocca JR, Edison AS (2006) Single-insect NMR: a new tool to probe chemical biodiversity. ACS Chem Biol 1(8):511–514

Dumville JC, Gill W, Martin Bland J, Cullum N, Dowson C, Iglesias C, Mitchell JL, Andrea Nelson E, Soares MO, Torgerson DJ (2009) Larval therapy for leg ulcers (VenUS II): randomised controlled trial. BMJ 338:b773

Elegbede JA, Flores R, Wang RC (2003) Perillyl alcohol and perillaldehyde induced cell cycle arrest and cell death in BroTo and A549 cells cultured in vitro. Life Sci 73(22):2831–2840

Estevinho L, Pereira AP, Moreira L, Dias LG, Pereira E (2008) Antioxidant and antimicrobial effects of phenolic compounds extracts of Northeast Portugal honey. Food Chem Toxicol 46(12):3774–3779

Farnesi AP, Aquino-Ferreira R, JONG DD, Bastos JK, Soares AEE (2009) Effects of stingless bee and honey bee propolis on four species of bacteria. Genet Mol Res 8(2):635–640

Fasoranti JO (1997) The place of insects in the traditional medicine of southwestern Nigeria. The Food Insects Newsl 10(2):1–5

Fefferman NH, Traniello JF, Rosengaus RB, Calleri DV II (2007) Disease prevention and resistance in social insects: modeling the survival consequences of immunity, hygienic behavior, and colony organization. Behav Ecol Sociobiol 61(4):565–577

Fernandes J, Fonseca CD, Teixeira A, Gatrass CR (2005) Perillyl alcohol induces apoptosis in human glioblastoma multiforme cells. Oncol Rep 13(5):943–948

Fernández-Marín H, Zimmerman JK, Rehner SA, Wcislo WT (2006) Active use of the metapleural glands by ants in controlling fungal infection. Proc R Soc Lond B Biol Sci 273(1594):1689–1695

Ferreira FS, Brito SV, Ribeiro SC, Saraiva AA, Almeida WO, Alves RR (2009) Animal-based folk remedies sold in public markets in Crato and Juazeiro do Norte, Ceará, Brazil. BMC Complement Altern Med 9(1):1

Holt VM (1992 [orig. 1885]) Why not eat insects? Pryor Publications, Kent

Horobin AJ, Shakesheff KM, Woodrow S, Robinson C, Pritchard DI (2003) Maggots and wound healing: an investigation of the effects of secretions from Lucilia sericata larvae upon interactions between human dermal fibroblasts and extracellular matrix components. Br J Dermatol 148(5):923–933

Jiang ZH, Yang QX, Tanaka T, Kouno I (2008) Bicyclic polyketide lactones from Chinese medicinal ants, Polyrhacis lamellidens. J Nat Prod 71(4):724–727

Kamakura M, Mitani N, Fukuda T, Fukushima M (2001) Antifatigue effect of fresh royal jelly in mice. J Nutr Sci Vitaminol 47(6):394–401

Khalil ML (2006) Biological activity of bee propolis in health and disease. Asian Pac J Cancer Prev 7:22–31

Kilicoglu B, Kismet K, Koru O, Tanyuksel M, Oruc MT, Sorkun K, Akkus MA (2006) The scolicidal effects of honey. Adv Ther 23(6):1077–1083

Kou J, Ni Y, Li N, Wang J, Liu L, Jiang ZH (2005) Analgesic and anti-inflammatory activities of total extract and individual fractions of Chinese medicinal ants *Polyrhachis lamellidens*. Biol Pharm Bull 28(1):176–180

Kunert M, Søe A, Bartram S, Discher S, Tolzin-Banasch K, Nie L et al (2008) De novo biosynthesis versus sequestration: a network of transport systems supports in iridoid producing leaf beetle larvae both modes of defense. Insect Biochem Mol Biol 38(10):895–904

Kwon YB, Ham TW, Kim HW, Roh DH, Yoon SY, Han HJ et al (2005) Water soluble fraction (< 10 kda) from bee venom reduces visceral pain behavior through spinal α 2-adrenergic activity in mice. Pharmacol Biochem Behav 80(1):181–187

Laurent P, Braekman JC, Daloze D (2005) Insect chemical defense. In: The chemistry of pheromones and other semiochemicals II. Springer, Berlin Heidelberg. pp 167–229

Lazner F, Gowen M, Kola I (1999) An animal model for pycnodysostosis: the role of cathepsin K in bone remodelling. Mol Med Today 5(9):413–414

Lee JD, Park HJ, Chae Y, Lim S (2005) An overview of bee venom acupuncture in the treatment of arthritis. Evid Based Complement Alternat Med 2(1):79–84

Lee J-E, Jo D-E, Lee A-J, Park H-K, Youn K, Yun E-Y, Hwang J-S, Jun M, Kang BH (2014) Hepatoprotective and antineoplastic properties of Protaetia brevitarsis larvae. Entomol Res 44(6):244–253

Libério SA, Pereira ALA, Araújo MJA, Dutra RP, Nascimento FR, Monteiro-Neto V et al (2009) The potential use of propolis as a cariostatic agent and its actions on mutans group streptococci. J Ethnopharmacol 125(1):1–9

Lockhart GJ (2000) Medicinal ants. In: Ants, and other great medicines – an unpublished book (edited and PDF of part placed online by Jacobson AL in 2007), 3–15. Downloaded from www.arthurleej.com/Ants.pdf on 12 August 2010

Luo C, Chen J, Li HL, Li JS (1998) Spatial and temporal expression of c-Fos protein in the spinal cord of anesthetized rat induced by subcutaneous bee venom injection. Brain Res 806(2):175–185

Ma Y, Wang X, Zhao Y, Kawabata T, Okada S (1997) Inhibitory effects of Chinese ant extract (CAE) on nephrotoxicity induced by ferric-nitrilotriacetate (Fe-NTA) in Wistar rats. Res Commun Mol Pathol Pharmacol 96(2):169–178

Mackintosh JA, Veal DA, Beattie AJ, Gooley AA (1998) Isolation from an ant *Myrmecia gulosa* of two inducible O-glycosylated proline-rich antibacterial peptides. J Biol Chem 273:6139–6143

Mannoor MK, Shimabukuro I, Tsukamotoa M, Watanabe H, Yamaguchi K, Sato Y (2009) Honeybee royal jelly inhibits autoimmunity in SLE-prone NZB× NZW F1 mice. Lupus 18(1):44–52

Marie R (1955) Contribution a l'histoire des insectes en the'rapeutique. Ph.D. Dissertation (Pharmacy), Strasbourg University, Strasbourg

Marshall C, Queen J, Manjooran J (2005) Honey vs povidone iodine following toenail surgery. WOUNDS UK 1(1):10

McIntosh CD, Thomson CE (2006) Honey dressing versus paraffin tulle gras following toenail surgery. J Wound Care 15(3):133

Mishima S, Suzuki KM, Isohama Y, Kuratsu N, Araki Y, Inoue M, Miyata T (2005) Royal jelly has estrogenic effects in vitro and in vivo. J Ethnopharmacol 101(1):215–220

Miyata T (2007) Pharmacological basis of traditional medicines and health supplements as curatives. J Pharmacol Sci 103(2):127–131

Moed L, Shwayder TA, Chang MW (2001) Cantharidin revisited: a blistering defense of an ancient medicine. Arch Dermatol 137(10):1357–1360

Moore OA, Smith LA, Campbell F, Seers K, McQuay HJ, Moore RA (2001) Systematic review of the use of honey as a wound dressing. BMC Complement Altern Med 1(1):1

Morge G (1973) Entomology in the western world in antiquity and in medieval times. In: Smith RF (ed) History of entomology, Annual Reviews, Palo Alto, pp 37–80

Motte-Florac E, Ramos-Elorduy J, Stepp JR, Wyndham FS, Zarger RK (2002) Is the traditional knowledge of insects important?. In: Ethnobiology and biocultural diversity: proceedings of the 7th International Congress of Ethnobiology, Athens, Georgia, USA, October 2000. International Society of Ethnobiology, c/o University of Georgia Press, pp 207–224

Namba T, Ma Y-H, Inagaki K (1988) Insect-derived drugs in the Chinese song dynasty. J Ethnopharmacol 24:247–285

Nilforoushzadeh MA, Jaffary F, Moradi S, Derakhshan R, Haftbaradaran E (2007) Effect of topical honey application along with intralesional injection of glucantime in the treatment of cutaneous leishmaniasis. BMC Complement Altern Med 7(1):1

Orivel J, Redeker V, Le Caer J-P, Krier F, Revol-Junelles A-M, Longeon A, Chaffotte A, Dejean A, Rossier J (2001) Ponericins, new antibacterial and insecticidal peptides from the venom of the ant *Pachycondyla goeldii*. J Biol Chem 276(21):17823–17829

Oudhia P (2005) Medicinal herbs of Chhattisgarh. India having less known traditional uses, VII. Thura

Park HJ, Lee SH, Son DJ, Ki Wan O, Kim KH, Song HS, Kim GJ, Goo Taeg O, Yoon DY, Hong JT (2004) Antiarthritic effect of bee venom: inhibition of inflammation mediator generation by suppression of NF-κB through interaction with the p50 subunit. Arthritis Rheum 50(11):3504–3515

Pemberton RW (1999) Insects and other arthropods used as drugs in Korean traditional medicine. J Ethnopharmacol 65(3):207–216

Pettit GR (1996) Progress in the discovery of biosynthetic anticancer drugs 1. J Nat Prod 59(8):812–821

Pettit GR, Hartwell JL, Wood HB (1968) Arthropod neoplastic agent. Cancer Res 28(10):2168–2169

Pettit GR, Ye Q, Herald DL, Hogan F, Pettit RK (2010) Antineoplastic agents. 573. Isolation and structure of Papilistatin from the papilionid butterfly *Byasa polyeuctes termessa*. J Nat Prod 73(2):164–166

Piao C, Tachikawa E, Hu Y, Sato Y, Yang P, Yamamoto K, Shinada T, Suzuki K (2009) In vitro pharmacological activities of the extracts from red ant *Formica aquilonia* as potential therapeutic agents. J Transl Med 26(2):61–67

Poulsen M, Hughes WOH, Boomsma JJ (2006) Differential resistance and the importance of antibiotic production in Acromyrmex echinatior leaf-cutting ant castes towards the entomopathogenic fungus Aspergillus nomius. Insect Soc 53(3):349–355

Prakash A, Medhi B, Avti PK, Saikia UN, Pandhi P, Khanduja KL (2008) Effect of different doses of Manuka honey in experimentally induced inflammatory bowel disease in rats. Phytother Res 22(11):1511–1519

Ramos-Elorduy J, Pino JM, Romero LA (1988) Determinación del valor nutritivo de algunas especies de insectos comestibles del Estado de Puebla. Ann Inst Biol UNAM 58(1):355–372

Rastogi N (2011) Provisioning services from ants: food and pharmaceuticals. Asian Myrmecology 4:103–120

Ratcliffe BC (1990) The significance of scarab beetles in the ethnoentomology of non-industrial indigenous peoples. In the Proceedings of the 1st International Congress of Ethnobiology 2, pp 159–185. Museu Paraense Emı'lio Go¨eldi, Bele'm

Ratcliffe NA, Mello CB, Garcia ES, Butt TM, Azambuja P (2011) Insect natural products and processes: new treatments for human disease. Insect Biochem Mol Biol 41(10):747–769

Raynor W (1993) Tropical forests and pharmaceuticals. Pan Exch 3(3)

Read BE (1935) Insect drugs. Peking Nat Hist Bull 94:8–85

Ripple GH, Gould MN, Arzoomanian RZ, Alberti D, Feierabend C, Simon K, Binger K, Tutsch KD, Pomplun M, Wahamaki A, Marnocha R, Wilding G, Bailey HH (2000) Phase I clinical and pharmacokinetic study of perillyl alcohol administered four times a day. Clin Cancer Res 6:390–396

Rosengaus RB, Traniello JF, Chen T, Brown JJ, Karp RD (1999) Immunity in a social insect. Naturwissenschaften 86(12):588–591

Ruiz N, Gronenberg LS, Kahne D, Silhavy TJ (2008) Identification of two inner-membrane proteins required for the transport of lipopolysaccharide to the outer membrane of *Escherichia coli*. Proc Natl Acad Sci 105(14):5537–5542

Schluns H, Crozier RH (2009) Molecular and chemical immune defenses in ants (Hymenoptera: Formicidae). Myrmecological News 12:237–249
Sen R, Ishak HD, Estrada D, Dowd SE, Hong E, Mueller UG (2009) Generalized antifungal activity and 454-screening of Pseudonocardia and Amycolatopsis bacteria in nests of fungus-growing ants. Proc Natl Acad Sci 106(42):17805–17810
Sherman RA, Hall MJR, Thomas S (2000) Medicinal maggots: an ancient remedy for some contemporary afflictions. Annu Rev Entomol 45(1):55–81
Singh OL (2014) Medicinal insects of Manipur. Int J Res Sci Technol 1(1):1–5
Son DJ, Lee JW, Lee YH, Song HS, Lee CK, Hong JT (2007) Therapeutic application of anti-arthritis, pain-releasing, and anti-cancer effects of bee venom and its constituent compounds. Pharmacol Ther 115(2):246–270
Song YS, Park EH, Hur GM, Ryu YS, Lee YS, Lee JY, Kim YM, Jin C (2002) Caffeic acid phenethyl ester inhibits nitric oxide synthase gene expression and enzyme activity. Cancer Lett 175:53–61
Steenvoorde P, Jacobi CE, Van Doorn L, Oskam J (2007) Maggot debridement therapy of infected ulcers: patient and wound factors influencing outcome–a study on 101 patients with 117 wounds. Ann R Coll Surg Engl 89(6):596–602
Stow A, Beattie A (2008) Chemical and genetic defenses against disease in insect societies. Brain Behav Immun 22(7):1009–1013
Stratton SP, Alberts DS, Einspahr JG, Sagerman PM, Warneke JA, Curiel-Lewandrowski C, Myrdal PB, Karlage KL, Nickoloff BJ, Brooks C, Saboda K (2010) A phase 2a study of topical perillyl alcohol cream for chemoprevention of skin cancer. Cancer Prev Res 3(2):160–169
Trowell S (2003) Drugs from bugs: the promise of pharmaceutical entomology. Futurist 37(1):17–19
Turkmen A, Graham K, McGrouther DA (2010) Therapeutic applications of the larvae for wound debridement. J Plast Reconstr Aesthet Surg 63(1):184–188
Veeresh GK (1999) Pest ants of India. In: Vander Meer RK, Jaffe K, Cedena A, (eds) Applied myrmecology – a world perspective. Westview studies of insect biology. Westview Press, Boulder, pp 17–24
Viljakainen L, Evans JD, Hasselmann M, Rueppell O, Tingek S, Pamilo P (2009) Rapid evolution of immune proteins in social insects. Mol Biol Evol 26(8):1791–1801
Wang XH, Andrae L, Engeseth NJ (2002) Antimutagenic effect of various honeys and sugars against Trp-p-1. J Agric Food Chem 50(23):6923–6928
Weiss HB (1947) Entomological medicaments of the past. J N Y Entomol Soc 55(2):155–168
Whitaker IS, Twine C, Whitaker MJ, Welck M, Brown CS, Shandall A (2007) Larval therapy from antiquity to the present day: mechanisms of action, clinical applications and future potential. Postgrad Med J 83(980):409–413
Wigglesworth VB (1976) Insects and the life of man. Halsted Press, New York
Wilsanand V, Preema V, Rajitha P (2007) Therapeutics of insects and insect products in South Indian traditional medicine. Indian J Pharm 6:563–568
Wilson-Rich N, Spivak M, Fefferman NH, Starks PT (2009) Genetic, individual, and group facilitation of disease resistance in insect societies. Annu Rev Entomol 54:405–423
Yeruva L, Pierre KJ, Elegbede A, Wang RC, Carper SW (2007) Perillyl alcohol and perillic acid induced cell cycle arrest and apoptosis in non small cell lung cancer cells. Cancer Lett 257(2):216–226
Yeruva L, Hall C, Elegbede JA, Carper SW (2010) Perillyl alcohol and methyl jasmonate sensitize cancer cells to cisplatin. Anti-Cancer Drugs 21(1):1
Zhang CX, Tang XD, Cheng JA (2008) The utilization and industrialization of insect resources in China. Entomol Res 38(s1):S38–S47
Zientz E, Beyaert I, Gross R, Feldhaar H (2006) Relevance of the endosymbiosis of Blochmannia floridanus and carpenter ants at different stages of the life cycle of the host. Appl Environ Microbiol 72(9):6027–6033
Zimian D, Yonghua Z, Gao X (1997) Medicinal insects in China. Ecol Food Nutr 36:209–220
Zimmer C (1993) The healing power of maggots. Discover 17

Mass Production of Biocontrol Agents of Insect Pests

17

Pradyumn Kumar, Jaswinder Kaur, J.C. Sekhar, Soujanya P. Lakshmi, and S.B. Suby

Contents

17.1 Introduction

With the increased need and awareness of integrated pest management concept among the farmers, there is increasing emphasis on the utilization of biocontrol agents for the management of pests. Though their demand is increasing, yet their availability is far from sufficient. The biocontrol agents, particularly parasitoids and predators, have very short life span and they cannot be stored for long. Their transportation also requires certain specific conditions which are difficult and expensive. These are some of the important reasons that private sectors are not much interested in them. Of late, there have been quite a few inventions for the automation in the mass production of biocontrol agents which have removed bottlenecks for their economic and efficient production. Development of *Corcyra* rearing cage (Kumar and Kumar 2001) that keeps the most common parasitoid, *Bracon* spp. at bay, insect handling device (Kumar and Jalali 1993) for capturing the reared insects in the

P. Kumar (✉) • J. Kaur • J.C. Sekhar • S.P. Lakshmi • S.B. Suby
Indian Institute of Maize Research, Indian Agricultural Research Institute Campus, New Delhi 110012, India
e-mail: pradyumn.kumar@gmail.com; jasspau@yahoo.com; jcswnc@rediffmail.com; soujanyak.scientist@gmail.com; subysb@gmail.com

Omkar (ed.), *Industrial Entomology*, DOI 10.1007/978-981-10-3304-9_17

laboratory and transferring them in desired cages, egg cleaning device (Kumar et al. 2008) for *Corcyra cephalonica* eggs which prevents the workers from inhalation of hazardous insect scales and improves the efficiency of collection and cleaning, UV-C sterilization chamber for the sterilization of *Corcyra* eggs before parasitization, oviposition cage for *Helicoverpa armigera* (Kumar and Ballal 1990) which provides the desired climatic conditions for the mating during summer and winter and suitable substrate for oviposition, *Spodoptera litura* rearing cage (Kumar and Ballal 1991), versatile insect rearing cage (Kumar et al. 2011) and *Chrysoperla* oviposition cage are some of the technologies to name a few, which have facilitated the mass production of biocontrol agents.

Now, the biocontrol agents can be mass produced at small-scale/cottage industry level just on the lines of sericulture or apiculture. Simple, low-cost and down-to-earth technology can be used for their mass production. They can be marketed in the region where they are produced. The simple methods suitable for their transportation have been developed. The production of biological control in cottage industries is also likely to increase their acceptability in the rural areas. The learning of production technology and planning of production so as to synchronize with the time of its marketing and use in the field are essential aspects to venture into this trade.

17.2 Mass Production of Host Insects

17.2.1 *Corcyra cephalonica*

Rice meal moth, *Corcyra cephalonica* Stainton (Lepidoptera: Galleriidae), is a stored grain pest on a variety of food grains and processed foods. The eggs and larvae of this insect have been found suitable fictitious host for many kinds of parasitoids, predators and some entomopathogenic nematodes. Mass rearing of this insect is, therefore, prerequisite for the commercial production of several biocontrol agents. Eggs are rendered sterilized by irradiation and then used for mass multiplication of parasitoids, like *Trichogramma* spp., and predators like *Chrysoperla carnea*, *Mallada boninensis*, *Orius* spp. and *Cyrtorhinus lividipennis*. *Corcyra* eggs are also successfully used for the rearing of *Chelonus blackburni*, an egg-larval parasitoid of cotton bollworms. *Goniozus nephantidis*, *Eriborus* spp. *Stenobracon deesae*, *Apanteles* spp., *Bracon* spp., etc. and Entomopathogenic nematode, such as *Steinernema feltiae* (*Neoaplectana carpocapsae*), are also multiplied on *Corcyra* larvae. *Corcyra* can be easily reared on any coarse grain, but mass rearing requires selection of suitable rearing medium, congenial conditions, protection from natural enemies and efficient and safe methods of handling moths, scales and eggs.

17.2.1.1 Food and Food Consumption

Sorghum, millet, maize, rice, wheat, groundnut, rice bran and wheat bran are generally used as rearing medium. They are used in various combinations or otherwise. Yeast is used as an additive to improve the quality of the moths. The development was reported faster in ground grains rather than in whole grains. Eleven food media were studied for commercial production of *C. cephalonica*, and food efficiency index (FEI) was calculated by dividing the product of percent emergence and egg weight by average development period. FEI was highest in case of sorghum followed by pearl millet, maize, pearl millet + rice husk + wheat bran, sorghum + rice husk, wheat bran and rice, while in other food media, it was extremely low (Kumar and Kumar 2002).

There have been several reports on quantitative requirement of food for the larval development. The ratio varies from 1000 eggs per 1.5 kg of sorghum to 5000 eggs per kg rice. The most acceptable being 2000 eggs per kg food medium. In less crowded condition, higher percentage of eggs develops into moth. Development period was found to increase with the increase in crowd.

17.2.1.2 Environmental Conditions

The optimum condition for incubation was observed to be 25–30 °C and 75–90 % RH. The incubation period was found to be reduced by increasing temperature but 40 °C proved lethal. The larval development is also greatly influenced by temperature and humidity. At low relative humidity, the number of instars increases, hence the development period. Larval development was indifferent to photoperiod; however, moths preferred to stay in dark.

17.2.1.3 Mating and Fecundity

Both the sexes are known to secrete pheromones. Mating occurs during night, usually once; however, multiple mating is recorded both in males and females. Males are reported to mate more often than females. Oviposition occurs 2 h after mating. The fecundity is directly correlated with the weight of females. Fecundity is also influenced by relative humidity. An average fecundity of 350 eggs per female is obtained under ambient conditions.

17.2.1.4 Rearing Methodology

Tribolium spp. is commonly present in the rearing medium. Larval food is, therefore, freed from it and other stored grain pests. Heat sterilization in oven is a common practice in the insectaries and reduces the moisture content of the food medium, thus making it less suitable for any stored grain pests. Alternatively, freshly ground grains should be used for charging the boxes for infestation with *Corcyra*. The stored grain pests get destroyed by sheering force and heat generated between the grinding stones during the grinding. Four kilogram of this medium is uniformly mixed with 0.4 cc of *Corcyra* eggs and kept in wooden boxes (45 × 25 × 20 cm.). The lids of the boxes have two windows (10 × 10 cm.). These windows are covered by double layer of fine copper wire mesh (400 squares/inch^2). This arrangement prevents the contamination of *Corcyra* culture from *Bracon* spp. (Kumar and Kumar 2001). *Corcyra* larvae are known to emit some kairomone which attracts *Bracon* spp. The system of using trays covered with cloth for *Corcyra* rearing, therefore, renders the culture vulnerable for *Bracon* infestation, which is a serious scourge to the culture. The boxes are periodically prepared as per the requirement of culture and stacked in the racks. Temperature of 28±1 °C is maintained by using air

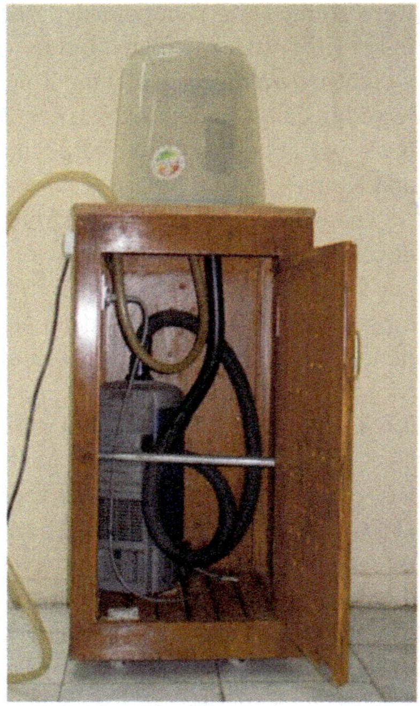

Fig. 17.1 Moth handling device

conditioners or heat convectors, as the need may be. Both the cooling and heating appliances drastically reduce the humidity. Room cooler intercepted by humidistat is used very successfully to maintain 75±5% RH. Room cooler in close chamber does not alter the room temperature significantly. Under such conditions, the moth emergence starts after 30 days and continues for over 45 days. Some of the eggs are laid by the emerging moths inside the cages, before they are collected. These eggs develop later on residual food and the moth emergence continues even after 45 days. It is advisable to cease moth collection after 45 days and discard this weak population. It is also advisable to collect the moth twice a day to reduce the chance of egg laying inside the rearing cages (Fig. 17.1).

17.2.1.5 Moth Handling

Moth collection is most time-consuming and health hazardous activity in the mass production of *Corcyra*. In an effort to reduce the manual labour, Parshad (1977) suggested covering stack of trays containing *Corcyra* culture with alkathene sheets. The moths after emergence slide down into the converging funnel-like sheet from where they are collected into the oviposition cage. The system has got some serious limitations. A large number of moths preferred to stay in the dark trays rather than sliding down; hence there is a very poor recovery of moths. Secondly, the moths that remain in trays mate and lay the eggs and thus cause overcrowding of larvae in the trays. Further, in this closed system excessive moisture gets accumulated which then promotes the growth of mould. Jalali and Singh (1989) used suction pump for the collection of moth, but the suction pump gets clogged with the scales hence of little use. Kumar and Jalali (1993) developed a huge aspirator, wherein the scales of moths were collected in the dust bag of vacuum cleaner. Yet more improved moth handling device was later developed by Kumar and Sekhar (2012) by which its efficiency was tremendously increased and at the same time the workers do not inhale hazardous scales. The suction pressure of the vacuum cleaner is adjusted using a pressure regulator at the top of outer container. With the help of this device, it is possible to collect the moths at the rate of 1000 moths per hour directly into the oviposition cage housed inside the outer container. The oviposition cage, when full, is replaced by another cage and the cage with moths is kept in a tray for egg laying.

17.2.1.6 Collection and Cleaning of Eggs

The eggs laid in the oviposition cages percolate down through the wire mesh in the collection tray. The eggs are mixed with lots of scales shed by moths. The eggs are separated from scales and collected in a container by another device called 'egg collection and cleaning device' for immaculate clean eggs developed by Kumar et al. (2008).

The entire content from egg collection tray is sucked into the system where the eggs, scales and broken wings and legs of moths are fluidized. The scales, being lighter in weight, are sucked up in the column and get collected in the dust bag of the vacuum cleaner. The eggs being heavier remain in the lower part of the column which get dropped into the container placed below the system, when the vacuum cleaner is switched off. The suction pressure in the column is adjusted by pressure regulator provided in the column. The scales are very light in weight and thus contaminate the air and cause respiratory problems. The egg collection and cleaning device offers easy collection of clean eggs and safe disposal of scales (Fig. 17.2).

17.2.1.7 Prevention from Natural Enemies

The *Corcyra* production is considerably affected by variety of natural enemies and competitors in the culture. All the stages are affected by them. More often than not, rearing medium is infested with *Tribolium* and other stored grain pests. These stored grain pests compete with *Corcyra*. The larvae of *Tribolium* also prey upon *Corcyra* larvae. It is, therefore, advisable to grind sorghum a day before using it. The stored grain pests in sorghum get destroyed during grinding. *Bracon* is yet another serious pest of *Corcyra* culture. Apart from double-layered wire mesh in the windows of the cage, *Bracon* needs to be regularly monitored against the light. Further, the double door system in the culture room is also recommended to reduce the chances of invasion of parasitoids of *Corcyra*. A small glass window in the culture room smeared with poisoned sticky material helps monitoring as well as killing of the invading *Bracon* adults. Nearly ten species of spiders are reported feeding on *Corcyra* moths. A general cleanliness in the culture room obviates this problem. Rat is another potential pest, both for the culture and the food medium. It is, therefore, important to have the doors, windows, rat proof, etc. Where such an arrangement is not possible, regular use of rat traps scares them away. House lizards feed upon both eggs and the moths of *Corcyra*. Oviposition cages are to be specially kept in the area free

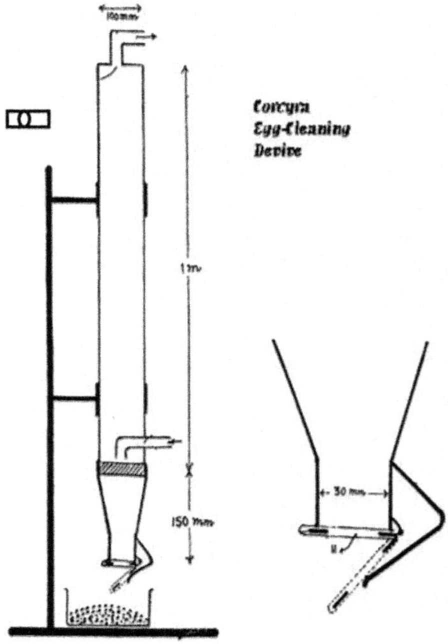

Fig. 17.2 Egg collection and cleaning device

from the approach of lizards. Dead moths, if left for long, provide suitable substrate for the development of mites in the culture. Mite problem is commonly encountered when the oviposition cages are not properly cleaned or the moths are not collected from the cages for several days. If the *Corcyra* culture gets infested with mites, it can be cured by treating it with sulphur powder. General cleaning of laboratory keeps most of the natural enemies like mites, ants, etc. at bay.

17.2.2 *Helicoverpa armigera*

For the mass production of *Helicoverpa* larvae, culturing of the insect in utmost clean condition is essential. A little negligence may lead to viral epidemic in the culture.

17.2.2.1 Initiation of Culture

Helicoverpa larvae can be collected from any of the variety crops infested by this pest in the fields. Depending upon the crop season, they can be collected from cotton, gram, pigeon pea, tomato, marigold, sunflower, etc. Since the phenomenon of cannibalism occurs in them, they are to be kept in isolation immediately after collection. For large-scale collection of larvae, 20 cm x 1 cm plastic tube pieces are used. A small cotton plug is pushed in the centre of the tube, and larva along with a leaf bit is inserted and plugged. Three larvae can be inserted on each side of the tube in this fashion. All of them are separated by cotton plug. Thus a tube can hold six larvae. In the laboratory, the larvae are transferred individually in glass vials having semisynthetic diet.

The following constituents are required for preparing semisynthetic diet for *Helicoverpa* larvae.

Bengal gram powder	105.00 g
Ascorbic acid	3.25 g
Methyl paraben	2.00 g
Sorbic acid	1.00 g
Streptomycin sulphate	0.50 g
Yeast	10.00 g
Multivitaplex	2 Caps.
Viteoline (vitamin A)	2 caps.
Formalin (10 %)	2 mL
Agar-agar	12.75 g
Distilled water	780 mL

All the ingredients, except vitamins and agar-agar, are mixed in blender using half the quantity of water. Agar is separately mixed in a container in the remaining water and boiled. The boiled agar is poured in the blender and mixed thoroughly. Now the vitamins are added and mixed. The diet is poured in the containers.

Since the field collected larvae are of assorted size/age, they take different times for pupation. They are observed every alternate day for the pupation. Some of the larvae might be parasitized or infected by some pathogens. The diseased and abnormal pupae are discarded. The normal looking pupae from healthy larvae are collected and washed in 0.05 % sodium hypochlorite solution.

17.2.2.2 Sexing of Pupae

The pupae can be separated based on their sex by observing the distance between the gonopore and anal pore on their ventral side. The distance in female is relatively more than that of male. Female pupae are slightly bigger in size.

17.2.2.3 Oviposition Cage

The oviposition cage described by Kumar et al. (2011) is a simple collapsible black cotton cloth cage as shown in the figure below (Fig. 17.3). The cage is kept in a tray having little water sufficient to moisten the base of the cage. The cloth is kept wet by water which rises up to the height of few inches by capillary action. This arrangement keeps the inner of the cage cool and humid, the two conditions ideal for pupae and adults.

Fifty pairs of moths are kept in each cage. Moths are provided with 10% honey solution in small Petri dish. Every day, the moths are transferred to other cage. The eggs are laid on inner wall of the cage. During winter, the cage described earlier is kept in the dry circular tray. The tray along with the cage is kept in a container having 200 mL water in the base. The container is closed with the lid and kept warm using table lamp or kept in a warm room (27 °C). The water increases the humidity inside the container.

Fig. 17.3 *Helicoverpa* oviposition cage

17.2.2.4 Extraction of Eggs

After removing the moths, the egg-borne cloth cage is reversed and dipped in a 0.05% sodium hypochlorite solution in the bathtub of a washing machine. After dipping for 5 minutes, the cloth is churned for 2 min. The outer chorion layer of eggs gets dissolved in sodium hypochlorite solution, thereby facilitating the detachment of eggs from the cloth. The eggs also get surface sterilized by the solution.

A sieve is fixed to the drainpipe and the water is drained out. The residual eggs in the cloth are collected by adding more water and agitating second time. The eggs along with scales and other derbies are collected in the sieve. The entire content of the sieve is transferred in a bucket full of water. The eggs being heavier sink in the base, and the scales and other debris float. The floating material is decanted. The eggs are again collected by sieving the remaining water with eggs. They are washed with tap water and transferred on a small clean piece of cloth. The eggs are kept for hatching at 27 °C and 70% RH.

17.2.2.5 Group Rearing of Early-Stage Larvae

For the first 5 days, the larvae can be group reared as the cannibalism is observed after second instar. The neonate larvae can be reared on any of the following way.

17.2.2.6 Semisynthetic Diet

About 50 ml hot diet is poured in 15 cm Petri dish and allowed to cool. The neonate larvae are reared in Petri dish inverted over a glass plate. A sterilized cotton cloth is spread between the Petri dish and glass plate. The cloth provides requisite ventilation and offer suitable ambience to the larvae. One Petri dish can hold 80–100 larvae for 4 days. The larvae feed and grow under the Petri dish. The excreta fall on

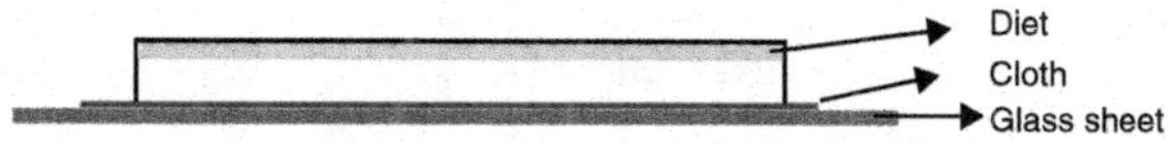

Fig. 17.4 Rearing of early-stage larvae

the tissue paper and do not mix with the diet. After 5 days about 10% of the larvae are used for maintaining healthy culture and the rest are sent for NPV unit (Fig. 17.4).

17.2.2.7 Rearing on Gram

Gram are soaked in 0.01 % formalin solution. They are then pounded in mortar and pestle. Neonate larvae are put in them in a bowl (approx. 500 ml.) covered by a thick piece of sterilized cloth. Trapping of excess of moisture should be avoided otherwise fungus develops on the grains. Larvae can be grown for 5 days this way.

17.2.2.8 Rearing on Ladyfinger

Okra fruits are washed thoroughly and dried. They are cut transversely into 3–4 pieces and kept in a bowl. About 100 neonate larvae or 2-day-old eggs are sprinkled over the cut fruits and covered by a thick cloth. After 5 days, the larvae are collected by sterilized brush. Ten per cent of the five-day-old larvae are used for maintaining the healthy culture. The remaining 90% are sent to NPV unit.

17.2.2.9 Larval Rearing for the Culture

Wide mouth, 35–40 ml glass containers are used. The height of the container should not be more than 5–6 cm. Freshly prepared diet (7–8 ml) is poured in the container and allowed to cool. One 5-day-old larva is kept in each container. The containers are inverted in tray having 1 cm layer of sterilized sand. The larvae feed on the diet stick to the ceiling of the container. The excreta fall on the sand. The water content of the excreta gets absorbed by the sand, thus preventing the mould formation on it. On completion of the development of larval stage, the prepupa falls down and pupates in the soil. The containers are lifted and pupae are collected and washed. Natural conditions are simulated in the rearing method. The diet does not get mixed with the excreta.

17.2.2.10 Handling of Pupae

Pupae are collected and thoroughly washed with water. They are then treated with 0.05% sodium hypochlorite solution for 5 minutes and rinsed with tap water, dried and stored for 5 days in 75% RH and 27 °C. The pupae are segregated based on their sex and kept in the oviposition cage for emergence.

17.2.3 *Spodoptera litura*

Spodoptera litura is a polyphagous lepidopterous pest of great economic significance. *Nuclear polyhedrosis virus* of *S. litura* (SlNPV) is an effective ecofriendly

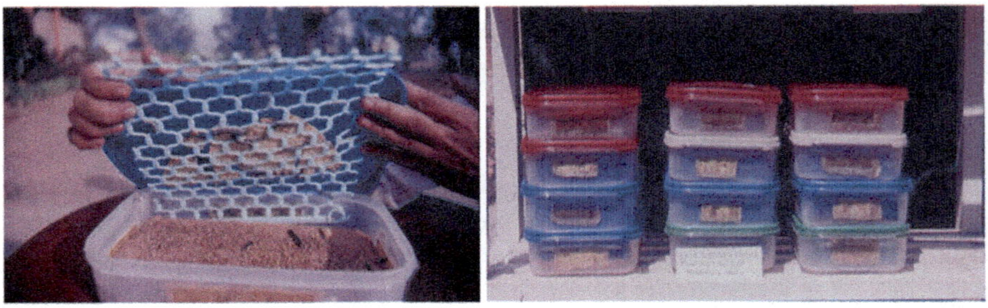

Fig. 17.5 *Spodoptera litura* rearing cage

control measure. SlNPV can be produced only in vivo; hence mass production of host insect is imperative for the mass production of this viral pesticide.

17.2.3.1 Rearing of Larvae

The larvae for the nucleus culture can be obtained from field crop infested with *S. litura*. Unlike *H. armigera*, there is no cannibalism in this insect; hence it can be group reared. The semisynthetic diet used for the rearing of *H. armigera* larvae can be used for the rearing of *S. litura* larvae too. Early instar larvae of *S. litura* are used in the same manner as *H. armigera* larvae are reared. About 90% of 4-day-old larvae are transferred to SlNPV production unit. The remaining 10% are used for *S. litura* culture.

After 4 days the larvae are reared in a rectangular cage designed by Kumar and Ballal (1991). The cage is made of bread box provided with windows in all four sides, sealed with fine wire mesh for ventilation. About 1 centimetre thick layer of sterilized sand is kept in the floor of the box. A plastic mesh, slightly bigger than the width of the cage, is placed in the box in such a way that it makes an upward curve (Fig. 17.5).

The semisynthetic diet poured in tray and cooled. A big piece of diet is cut and placed on the upward curve of the plastic mesh as shown in the figure. Approximately fifty 4-day-old larvae are placed over the sand. The larvae migrate to the diet and start feeding. The excrement falls to the sand. The moisture is absorbed in the sand, thus preventing mould formation. The diet quantity can be observed from the window. When consumed completely, fresh diet is placed on the plastic mesh. When the larvae are about to pupate, they move deep in the soil and pupate. When all the larvae are pupated, they are collected and washed in 0.05% solution of sodium hypochlorite. The conditions in cage simulate the natural conditions for the larvae. In nature, the larvae feed from underneath of leaves and pupate in the soil. The diet does not mix with the excreta.

17.2.3.2 Oviposition

2.5 L glass jar is lined with butter paper. Sterilized pupae are kept in it. A cotton swab soaked in 50% honey solution is provided in a small Petri plate as an adult food. The eggs are laid in masses on the butter paper which are cut and used for culture.

17.3 Mass Production of Biological Control Agents

17.3.1 *Trichogramma* spp.

The 'trichocards' are generally postcard size cards, bearing host eggs parasitized by *Trichogramma* sp. Name of the species, date of exposure to the parasitoids and likely date of emergence of parasitoids are printed on one of the sides of the card. The eggs are glued to the other side. The cards are punched so that it can be easily cut into small pieces at the time of release of the *Trichogramma* sp. like a stamp ticket.

17.3.1.1 Glueing of Eggs

A stencil of hard plastic sheet is made by cutting slots in the manner so that each slot fit onto the punched strip. The card is positioned under the stencil, and the glue is pasted on the stencil with the help of paint brush. The care has to be taken for the thickness of the glue smear. The glue thickness should be such that only a very small portion of the egg comes in contact with the glue and most of the surface area of egg is available to the parasitoid for parasitization and its emergence. Soon after the glue is smeared, the card is transferred to a tray. Clean *Corcyra* eggs are taken in a strainer and sprinkled over the card. Eggs get glued on the strips. Each card can hold 1 cm^3 eggs (approximately 20, 000).

17.3.1.2 Sterilization of Eggs

To prevent the embryonic development in *Corcyra* eggs, the embryo is killed by exposing the cards to UV rays. The cards are exposed to UV tube light of 30 W at a distance of 35 cm for a period of 10 minutes. During this period the glue also gets dried up. The process is done in UV-C sterilization chamber. This chamber has semi-circular drawers of 35 cm radius, provided with flanges which prevent cards from sliding over other cards as shown in Fig. 17.6. Each chamber is provided with a 30 W UV-C tube light in the centre of its ceiling. A timer is also provided separately for each tube light. This arrangement equally exposes all the eggs.

17.3.1.3 Parasitization

The date of exposure is marked on the card. Two cards are kept in a polythene bag (30 × 15 cm) so that their egg side is outside. Two strips of parasitized trichocards from which wasps are likely to emerge and are inserted in the bag. A small strip of polythene sheet smeared with honey, folded and stapled in such a way so that smeared surface forms the inner surface of the loop is also kept inside which provides food to the emerging parasitoids. The bags are slightly inflated by making space with the help of the hand and tied with rubber band. The wasps emerge from the trichocards, feed on honey, mate and parasitize the eggs glued on cards. Such bags are kept in a controlled conditioned room thoroughly illuminated with fluorescent light. After 4–5 days, the eggs turn steal grey. They can be stored for 12–15 days at 10 °C. During this period they must be despatched for field releases. Further storage period will reduce the efficacy of the parasitoid.

Fig. 17.6 UV-C sterilization chamber

17.3.1.4 Packaging and Transportation

The cards must be handled very gently. The couplets of the cards facing egg side are stored by sandwiching a soft tissue paper in between them. This arrangement prevents damage of eggs during transit. The stack of ten couplets is kept in one bag (30 × 15 cm). Such bags can be kept with ice pack in an insulated box for transport.

Different species of *Trichogramma* are recommended for different pests. The rearing technology for all species is by and large same. *Trichogramma chilonis* is effective against *Helicoverpa* in cotton, tomato, maize and for sugarcane borers. *T. japonicum* is used against rice stem borer, rice leaf folder, sugarcane top and shoot borer. *T. brasiliensis* is used for cotton boll worms. *T. embryophagum* is specific to the codling moth of apple.

17.3.1.5 Planning of Production

Planning is first and foremost activity for any project. Keeping magnitude of the production and time of requirement in mind, the host culture has to be planned so as to pace the production and meet the target. Since the shelf life of the parasitoid is very short, the production should be synchronized with the time and quantity of requirement.

17.3.1.6 Method of Release

It is advisable to synchronize the release programme after monitoring the onset of moths by the use of pheromone traps, light traps, visual observation, etc. The tricho-card strips are cut into pieces and stapled in the underside of the leaf. The release of parasitoids should be done preferably in evening hours. During night, the parasitoid will emerge and search the host eggs. The parasitization may be completed before the next noon. The care should be taken that the pesticide spray is not done immediately before or after the release.

17.3.2 *Chrysoperla carnea*

Chrysoperla carnea is neuropteran, very voracious predator of eggs of variety of insects and soft-bodied insects like aphids, whiteflies, mealy bugs, scales, leafhoppers, spider mites, neonate larvae, etc. The larva devours about 11,000 spider mites or 300–400 aphids to complete its development. This general predator is a very potential biocontrol agent in regulating insect populations in the field. Adults are soft-bodied green insect of about a centimetre long. They are attracted to light.

17.3.2.1 Oviposition Cage

Adults are kept in the oviposition cage. It is rectangular wooden box and its interior is lined with net cloth. The top of the cage is provided with a sliding lid lined with black cloth. The lid has free sliding movement on both sides breadthwise. The females prefer to lay eggs on the black cloth; net cloth deters egg laying on it. Such an arrangement induces egg laying on the ceiling of black cloth. One of the edges of breadth of cage is provided with a comb with tooth upside, positioned in such a way that when the lid is slid over the comb, the eggs laid on the cloth pass in between the teeth of the comb and do not get injured, while the comb teeth prevent escape of adults. Every day the lid laden with eggs is replaced by pushing it with another lid, thereby replacing an egg laden lid with a new lid. Eggs are scrapped from the cloth using razor.

17.3.2.2 Rearing of Larvae

The eggs are kept in Petri dishes at 27 °C and 75% relative humidity. The eggs of *Corcyra* are sterilized by exposing them to ultraviolet rays. After 3 days, one egg of *Chrysoperla* is kept with about 2000 eggs of *Corcyra* in a small plastic container (approx. 10 ml). Since there is cannibalism in this insect, group rearing is not possible. Small injection vials can also be used for the rearing of *Chrysoperla* In 4 days, the eggs are consumed by the larva hatched from *Chrysoperla* egg. Fresh *Corcyra* eggs are provided after every 14-day interval till the larva pupates into a small off-white ball. A small piece of cardboard is kept in the vial just before the pupation. The larva prefers to pupate on cardboard bit. It requires 28–30 days for the newly hatched larva to pupate. The pharate *Chrysoperla* emerge out from the neat hole cut by it which is then moult into a winged adult.

17.3.2.3 Transportation and Release

The predator is transported and released in the form of eggs or neonate larvae. Eggs are mixed with rice husk to keep them separate from each other. Neonates are kept with *Corcyra* eggs.

17.3.3 Nuclear Polyhedrosis Virus of *Helicoverpa armigera*

17.3.3.1 NPV Production Unit

Semisynthetic diet of *Helicoverpa armigera* used for healthy culture is used for infected larvae too. Infected larvae can be reared in 30–35 mL plastic containers or even 20 mL injection vials. The former are unbreakable while the later are available at much lower cost. Each vial is provided with 7–8 mL of diet. NPV solution is prepared having ten million PIBs per litre. This solution is filled in a drip bottle having a discharge regulating device. Two drops of this solution are added in each of the containers arranged in tray. The surface of the diet is smeared with the dilute solution of NPV so as to have about 600 polyhedral inclusion bodies (PIBs). Five-day-old larva collected from healthy culture unit is kept in each vial. For 4–5 days the larva feed on diet mixed with NPV and grow to full size. If higher concentration of PIBs is used, the early death comes to the larvae. This results in the formation of less quantity of virus. This also occurs if smaller size larvae are used. It is, therefore, very essential to use suitable size of larvae and the desired concentration of virus so as to get the optimum yield of virus. The larva then ceases to feed and become still. It eventually die but the PIBs continue to increase in numbers. The skin turns soft and smooth. They are then collected after 6–7 days. If the larvae overstay on diet, they rupture and the PIBs get mixed with diet. Each full bloated larva yields approximately six billion PIBs.

The infected larval cadavers are collected in an earthen pot and allowed to putrefy for 5–7 days. Water is sprinkled over the rotten mass to avoid desiccation and faster putrefaction. The rotten mass which awfully smells is then grinded in a grinding stone operated by electrical motor. This will release polyhedra from the putrefied tissue. The content is then diluted in tap water and sieved in the muslin cloth. All PIBs are filtered out in the water. PIBs are completely extracted by adding more water in the cloth. The filtrate is centrifuged at 6000 rpm for 30 minutes. All the PIBs are settled in the base. They can be stored in the concentrated form. The concentration of PIBs is measured using haemocytometer.

17.4 Conclusions

As pesticide application accelerates in an attempt to increase agricultural production, there is a parallel increase in direct and indirect hazardous side effects of the same. In response to these, the advocacy of low pesticide usage along with integrated pest management is reaching its peak and thus making biological control a hugely sought-after option for pest management. Thus the created increased demand of biocontrol agents has to be met successfully and promptly. We thus require good technology in place for their low-cost mass production. This would not only aid farmers but also be a side business or a small-scale industry for educated youth, further increasing its relevance in developing nations.

References

Jalali SK, Singh SP (1989) A new method of *Corcyra cephalonica* Stainton moth collection. Entomon 14(3&4):281–282

Kumar P, Ballal CR (1990) An improved oviposition cage and egg collection technique for the mass rearing of *Heliothis armigera* (Hb.). Indian Journal of Plant Protection 18:255–259

Kumar P, Ballal CR (1991) Improved technique for mass rearing of *Spodoptera litura* (Lepidoptera:Noctuidae). Entomon 16:59–62

Kumar P, Jalali SK (1993) A hygienic and efficient method of Corcyra moth and egg collection. Int J Pest Manage 39(1):103–105

Kumar P, Kumar S (2001) Foolproof cage for rearing *Corcyra cephalonica*. Indian J Entomol 63(3):322–324

Kumar P, Kumar S (2002) Evaluation of different rearing media for Corcyra cephalonica (Stainton). J Biol Control 16(2):117–120

Kumar P, Sekhar JC (2012) An insect handling device, Indian Patent no. 252363/2012. http://www.allindianpatents.com/patents/252363-an-insect-handling-device

Kumar P, Kumar S, Puri SN (2008) An efficient device for collecting and immaculate cleaning of insect eggs mixed with loose scales, Indian Patent No. 213744/2008. http://www.allindianpatents.com/patents/213744-egg-cleaning-device

Kumar P, Suby SB, Sekhar JC, Sai Kumar R (2011) A collapsible insect rearing cage, Indian Patent Application no. 0923/DEL/2011

Parshad B (1977) A new method for the maintenance of culture of Corcyra cephalonica Stainton with least manual labour. Indian J Entomol 37:303–306

Printed by Printforce, the Netherlands